D1422381

High Power Microwaves
Second Edition

Series in Plasma Physics
Series Editor:

Steve Cowley, Imperial College, UK and UCLA, USA

Other books in the series:
Controlled Fusion and Plasma Physics
K Miyamoto

Plasma Electronics: Applications in Microelectronic Device Fabrication
T Makabe and Z Petrović

An Introduction to Inertial Confinement Fusion
S Pfalzner

Aspects of Anomalous Transport in Plasmas
R Balescu

Non-Equilibrium Air Plasmas at Atmospheric Pressure
K H Becker, R J Barker and K H Schoenbach (Eds)

Magnetohydrodynamic Waves in Geospace: The Theory of ULF Waves and their Interaction with Energetic Particles in the Solar-Terrestrial Environment
A D M Walker

Plasma Physics via Computer Simulation (paperback edition)
C K Birdsall and A B Langdon

Plasma Waves, Second Edition
D G Swanson

Microscopic Dynamics of Plasmas and Chaos
Y Elskens and D Escande

Plasma and Fluid Turbulence: Theory and Modelling
A Yoshizawa, S-I Itoh and K Itoh

The Interaction of High-Power Lasers with Plasmas
S Eliezer

Introduction to Dusty Plasma Physics
P K Shukla and A A Mamun

The Theory of Photon Acceleration
J T Mendonça

Laser Aided Diagnostics of Plasmas and Gases
K Muraoka and M Maeda

Reaction-Diffusion Problems in the Physics of Hot Plasmas
H Wilhelmsson and E Lazzaro

Series in Plasma Physics

High Power Microwaves
Second Edition

James Benford
Microwave Systems Inc.
Lafayette, California, USA

John A. Swegle
J-Two
Walnut Creek, California, USA

Edl Schamiloglu
University of New Mexico
Albuquerque, New Mexico, USA

Taylor & Francis
Taylor & Francis Group
New York   London

CRC Press
Taylor & Francis Group
6000 Broken Sound Parkway NW, Suite 300
Boca Raton, FL 33487-2742

© 2007 by Taylor & Francis Group, LLC
CRC Press is an imprint of Taylor & Francis Group, an Informa business

QM LIBRARY
(MILE END)

International Standard Book Number-10: 0-7503-0706-4 (Hardcover)
International Standard Book Number-13: 978-0-7503-0706-2 (Hardcover)

Library of Congress Cataloging-in-Publication Data

Benford, James.
   High power microwaves / James Benford, John A. Swegle, and Edl Schamiloglu. -- 2nd ed.
        p. cm.
   Includes bibliographical references and index.
   ISBN-13: 978-0-7503-0706-2 (alk. paper)
   ISBN-10: 0-7503-0706-4 (alk. paper)
   1. Microwave devices. 2. Microwaves. I. Swegle, John Allan, 1952- II. Schamiloglu, Edl. III. Title.

TK7876.B44 2007
621.381'3--dc22   2006026489

Visit the Taylor & Francis Web site at
http://www.taylorandfrancis.com

and the CRC Press Web site at
http://www.crcpress.com

Dedication

For

Hilary and Eloise Benford

Al, Gerry, and Jennifer Swegle

Elmira Schamiloglu

Preface

This second edition is changed from the 1992 first edition in significant ways:

- It is a textbook, intended for classroom use or self-study, with problem sets. Teachers can obtain solutions from the authors.
- Chapter 2, on HPM Systems, is a new chapter on this complex activity, with a detailed example that we call SuperSystem.
- "Ultrawideband Systems" (Chapter 6) is another new chapter, a survey of a class of high power radiators, with very different technologies and applications, that has fully emerged since the first edition.
- Our new HPM formulary contains a handy compilation of frequently used rules of thumb and formulas.
- Every chapter is rewritten, not merely updated.
- Despite the new material, we have kept the length of this new edition about the same as the first. See the Microwave Sciences Web site for new problems, updates, and errata: http://home.earthlink.net/~jbenford/index.html.

This book is meant to communicate a wide-angle, integrated view of the field of high power microwaves (HPM). Our treatment of HPM is actually rather brief; by limiting the book's length to a manageable number of pages, we hope to encourage the reader to explore the field, rather than skipping to select sections of narrower interest, as happens with lengthier tomes. Our presentation is broad and introductory with the flavor of a survey; however, it is not elementary. For the reader seeking greater detail, we have provided an extensive set of references and guidance to the literature with each chapter.

We anticipate that our readers will include researchers in this field wishing to widen their understanding of HPM; present or potential users of microwaves who are interested in taking advantage of the dramatically higher power levels being made available; newcomers entering the field to pursue research; and decision makers in related fields, such as radar, communications, and high-energy physics, who must educate themselves in order to determine how developments in HPM will affect them.

This book has its origin in many short courses we have taught in the U.S. and Western Europe. Over time, as we have continued to update and widen the scope of our classes, our outlook in the field of HPM has expanded considerably. Although brought to the field by our initial investigations with particular HPM sources, we have found ourselves asking questions with increasingly more global implications, which this book is intended to address:

- How does HPM relate historically and technically to the conventional microwave field?
- What applications are possible for HPM, and what key criteria will HPM devices have to meet in order to be applied?
- How do high power sources work? Are there really as many different sources as the nomenclature seems to indicate (No!)? What are their capabilities, and what limits their performance?
- Across the wide variety of source types, what are the broad fundamental issues?

In addressing these questions, we feel it has profited us, and ultimately the reader, that our perspectives are largely complementary: our primary research interests lie with different source types. One of us (James Benford) works largely as an experimentalist, while the other (John Swegle) is a theorist. Edl Schamiloglu does both.

Every reader is encouraged to read the introduction (Chapter 1), where we outline the historical trends that have led to the development of HPM and compare the capabilities of HPM to those of conventional microwaves. A thrust of this book is that the field can be divided into two sectors, sometimes overlapping: applications-driven and technology-driven. Chapter 2 is entirely new and deals with both perspectives. Chapter 3 directly treats the applications of HPM. The other chapters are focused on technologies. Chapter 4, on microwave fundamentals, is a guide to the major concepts to be used in later chapters. Chapter 5, on enabling technologies, describes the equipment and facilities surrounding the sources in which microwaves are generated. These are the elements that make the system work by supplying electrical power and electron beams to the source, radiating microwave power into space, and measuring microwave properties. Chapter 6 deals with the ultrawideband technologies. Chapters 7 to 10 detail the major source groups.

We express our thanks and appreciation for the help of our colleagues in the preparation of this book:

Gregory Benford	Steven Gold
Dominic Benford	Bill Prather
David Price	Edward Goldman
Keith Kato	David Giri
Carl Baum	William Radasky
Jerry Levine	Douglas Clunie
Kevin Parkin	Charles Reuben
Bob Forward	Richard Dickinson
George Caryotakis	Bob Gardner
Mike Haworth	Daryl Sprehn

Yuval Carmel   Sid Putnam
Mike Cuneo   Ron Gilgenbach
Larry Altgilbers

Special thanks go to Hilary Benford, who tirelessly compiled the manuscript and figures.

James Benford
John Swegle
Edl Schamiloglu
July 2006

Authors

James Benford is president of Microwave Sciences, Inc., in Lafayette, California. He does contracting and consulting in high power microwaves (HPM). His interests include HPM systems from conceptual designs to hardware, microwave source physics, electromagnetic power beaming for space propulsion, experimental intense particle beams, and plasma physics. He earned a Ph.D. in physics at the University of California–San Diego. He is an IEEE fellow and an EMP fellow. He has taught 25 courses in HPM in nine countries.

John A. Swegle, a staff member at Lawrence Livermore National Laboratory and formerly a staff member at Sandia National Laboratories, also works as an independent consultant in the field of HPM. He has taught short courses on HPM in the U.S., Europe, and China, and his previous research work included HPM sources, magnetic insulation in pulsed power systems, and particle beam transport. His current work in the field concentrates on the analysis of HPM systems. He earned a Ph.D. in applied physics at Cornell University.

Edl Schamiloglu is the Gardner-Zemke Professor of Electrical and Computer Engineering at the University of New Mexico. His interests include the research and development of intense beam-driven HPM sources, in addition to their effects on networked infrastructure. He is an IEEE fellow and an EMP fellow.

Contents

1

Introduction

This new edition of *High Power Microwaves* is a substantial departure from the 1992 first edition.[1] That work was a technical monograph that reflected the activities and trends up to that time. High power microwaves (HPM) has moved from being a promising technology to implementation in several applications. HPM systems are being built, studied, and applied not only in the U.S., Russia, and Western European countries such as the U.K., France, Germany, and Sweden, but also in China and developing nations such as India, Taiwan, and South Korea.

In this new volume, we have adopted a *systems point of view*. A significant change in the HPM community has been increasing emphasis on optimizing the entire system. It is widely realized in the HPM community that only by viewing HPM systems as integrated devices may the output of sources exceed present levels of power and pulse energy. Further, the community of potential HPM users demands it. One can no longer separate the system into discreet constituent components and optimize them separately. In adopting a systems point of view, one begins with a basic understanding of the constraints imposed by the application. Then one identifies subsystem component classes and how they interact and properly takes account of the requirements of ancillary equipment.

To get the flavor of this new approach, we advise the reader to begin by reading two chapters before reading the specific technical chapters that follow. Chapter 2, on HPM systems, describes how to conceptualize an HPM system by making choices of components based on a standard methodology. Chapter 3, on HPM applications, sets out the requirements for such systems.

Another major difference between the two editions is that this edition is a textbook intended to be used by students in HPM courses or for self-study by the technically trained. Therefore, we have introduced problems for most of the chapters. (Readers who wish to acquire solutions for these problems, or additional problems, should contact the authors at http://home.earthlink.net/~jbenford/index.html.)

There are several other innovations:

- An HPM formulary giving rules of thumb that the authors have found useful
- A new chapter on ultrawideband systems

1.1 Origins of High Power Microwaves

HPM has emerged in recent years as a new technology allowing new applications and offering innovative approaches to existing applications. A mix of sources that either push conventional microwave device physics in new directions or employ altogether new interaction mechanisms has driven the quantum leap in microwave power levels. HPM generation taps the enormous power and energy reservoirs of modern intense relativistic electron beam technology. Therefore, it runs counter to the trend in conventional microwave electronics toward miniaturization, with solid-state devices intrinsically limited in their peak power capability.

Our definition of HPM is:

- Devices that exceed 100 MW in peak power
- Devices that span the centimeter- and millimeter-wave range of frequencies between 1 and 300 GHz

This definition is arbitrary and does not cleanly divide HPM and conventional microwave devices, which have, in the case of klystrons, exceeded 100 MW. The HPM devices we consider here have reached powers as high as 15 GW.

HPM is the result of the confluence of several historical trends, as shown in Figure 1.1. Microwaves were first generated artificially by Hertz in the 1880s. Radio came into use at lower frequencies in the early 20th century with the advent of gridded tubes. In the 1930s, several investigators realized that higher frequencies could be obtained by using resonant cavities connected to electrical circuits, and the first cavity device, the klystron, was produced in 1937. This was followed by a burst of activity during the Second World War that included the extrapolation of the magnetron and the invention of the traveling wave tube (TWT) and the backward wave oscillator (BWO). Modulators that frequently used gridded power tubes powered all these sources. In the 1960s, the cross-field amplifier was developed. Thereafter, the 1970s saw the strong emergence of lower-power, but extremely compact, solid-state–based microwave sources. By this time, microwave tube technology was oriented toward volume production, and the research effort was curtailed.

In the 1950s, efforts to control thermonuclear fusion for energy production led to a detailed understanding of the interaction between particles and waves, and ultimately to the requirement for new tube developments using gyrotrons for higher average power at frequencies that have climbed to over 100 GHz. In the 1960s, electrical technology was extended with the introduction of *pulsed power,* leading to the production of charged particle beams with currents in excess of 10 kA at voltages of 1 MV and more. These intense beams were applied to the simulation of nuclear weapons' effects,

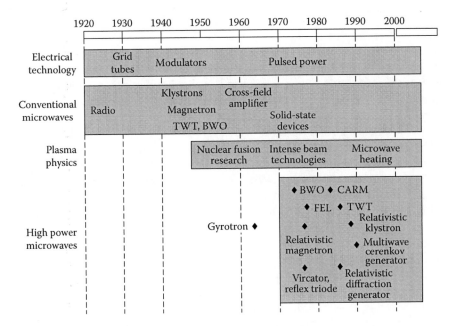

FIGURE 1.1
Historical origins and emergence of high power microwaves.

inertial confinement fusion, and other studies of high-energy density physics. The availability of intense relativistic electron beams was the last piece in the puzzle, allowing the knowledge of wave–particle interaction gained in the study of plasma physics to be put to use in the generation of microwaves. As a consequence of this genealogy, the culture of the HPM community lies closer to those of the plasma physics and pulsed power communities than to that of the conventional microwave tubes. One result of this is that the HPM technology was initially slow to adopt materials and surface and vacuum techniques that are well established in the microwave tube community and essential to resolving the pulse-shortening challenge that HPM faces.

1.2 High Power Microwave Operating Regimes

The first HPM sources were descendants of conventional microwave sources such as the magnetron, the backward wave oscillator, and the traveling wave tube. In these devices, increased power comes from using higher operating currents and stronger beam–field couplings within the interaction region.

Using high voltages to produce so-called relativistic electron beams, meaning electron energies comparable to or greater than the 510-keV rest energy of an electron, has had several profound consequences for HPM. The most important is the introduction of new devices, such as the *virtual cathode oscillator* or *vircator* and the *relativistic klystron*, which depend fundamentally on the very high beam currents that accompany high voltages. Another is the further exploitation to devices based explicitly on relativistic effects, most prominently the *gyrotron*. Finally, there is the stronger energy-, as opposed to velocity-, dependent tuning of the output frequencies in devices such as the free-electron laser (FEL) and cyclotron autoresonant maser (CARM).

The domain of HPM also includes *impulse sources*, which are very short duration, high power generators of broad bandwidth radiation. Usually referred to as *ultrawideband* (UWB) devices, they typically generate ≈ 1 GW or so with a duration of about 1 nsec, and only a few cycles of microwaves launched. Bandwidth is therefore of the same order as the frequency, typically 1 GHz.[2] Such pulses are generated by direct excitation of an antenna by a fast electrical circuit, instead of the earlier method of extracting energy from an electron beam. Although the peak output power levels of UWB sources can be comparable to narrowband sources, the energy content is considerably smaller because of very short pulse lengths.

The heroic age of HPM, until the 1990s, had the character of a "power derby," during which enormous strides were made in the production of high power levels at increasingly higher frequencies. Researchers in the former Soviet Union/Russia achieved their world-record successes using a series of new devices — the multiwave Cerenkov and diffraction generators and the relativistic diffraction generator (multiwave Cerenkov generator (MWCG), multiwave diffraction generator (MWDG), and relativistic diffraction generator (RDG) — based on large interaction regions many microwave wavelengths in diameter. In the U.S., relativistic magnetrons and klystrons at lower frequencies and FELs at higher frequencies generated the high powers. A measure of the success of these efforts is the product of the peak microwave power and the square of the frequency, Pf^2. Figure 1.2 shows the general history of the development of microwave sources in terms of this factor. Conventional microwave devices (tubes) made three orders of magnitude progress between 1940 and about 1970, but thereafter only minor improvements were made. Conventional devices, the klystron in particular, continue to make advances, albeit slowly. HPM devices began at $Pf^2 \sim 1$ and have progressed upward an additional three orders of magnitude in the ensuing 20 years. The device with the highest figure of merit produced to date is the FEL, with a radiated output power of 2 GW at 140 GHz.

Figure 1.3 shows the peak power generated by a representative sample of high power sources as a function of the frequency. It appears that peak power of many sources falls off roughly as f^2 at high frequencies; however, no clear trend emerges for the lower frequencies below 10 GHz. Why Pf^2? The roughly Pf^2 scaling for a given source type reflects the fact that power extracted from resonant cavity devices is proportional to their cross section, which scales with wavelength squared. In addition, the power in a waveguide, which has a cross section proportional to λ^2, is limited by a

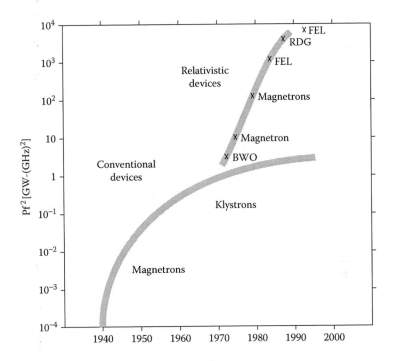

FIGURE 1.2
Growth of microwave devices in terms of quality factor Pf^2.

breakdown electric field. Another meaning is that power density on the target of a microwave beam transmitted is proportional to Pf^2 for a fixed antenna aperture (because antenna gain is $\sim f^2$; see Chapter 5), so it is a good parameter to rank sources for directed energy and power beaming over a distance.

The early heroic age of HPM ended in the 1990s with a sobering realization that devices were fundamentally limited above a peak power of ~10 GW and pulse energy of 1 kJ. These parameters were achieved through considerable effort and the added benefit of sophisticated three-dimensional computational modeling tools that were not available earlier. The term *pulse shortening* came into vogue, encompassing the various causes for the decrease in pulse length as peak power increases, with radiated energy staying roughly constant as power climbed. This has led to a reassessment of the way the HPM community has been developing its sources, as well as a concentration on improving them. Since then, much of the focus has been on improving existing sources: better surface cleanliness and vacuum environment, cathodes that can deliver the requisite electron current densities at low electric fields yet yield minimal plasma, and better designs to ensure that the intense electron beam impacts as little of the electrodynamic structure as possible as it gives up its kinetic energy and passes into the collector. There has been a significant decline in research on new source configurations.

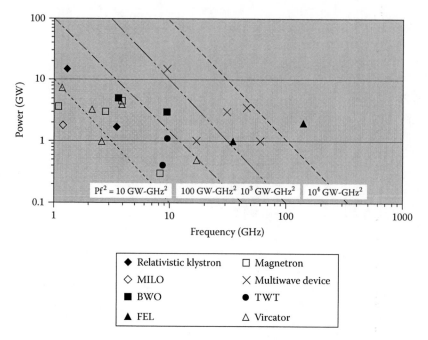

FIGURE 1.3
Peak powers of types of pulsed HPM sources. Values of quality factor Pf^2 vary over several orders of magnitude.

The ultimate limits on HPM source peak power are not well known. They are set by trade-offs between the usual factors that limit conventional tubes, breakdown, and mode competition, and factors unique to HPM, such as intense beam–field interactions and evolution of plasmas from surfaces and diodes. Electrical pulses with power up to about 10 TW are available from a single pulse generator, and one can buy 1-TW generators (a laboratory-based device, not suitable for mobile applications) for a few million dollars, commercially. At a moderate extraction efficiency of 10%, one could therefore expect a peak power of 100 GW. We expect that such powers can be achieved, but only if there is a need for the power and a willingness to accept the accompanying size and cost. To do so requires substantial improvement in understanding the limitations of specific devices and overcoming pulse shortening and source-specific problems such as spurious mode generation.

Another set of issues limits HPM average power achieved in repetitive operation. The convention in HPM has been to quote device efficiency in terms of the *peak power efficiency* of the source, defined as the ratio of the peak microwave power to the electron beam power at that moment. This is to be compared with the *energy efficiency*, which would be the ratio of the microwave pulse energy to the electron beam pulse energy. Aggressive research programs aimed at gaining a detailed understanding of large-signal behavior have been rewarded in certain cases with instantaneous power efficiencies of 40 to 50%; under most circumstances, however, power effi-

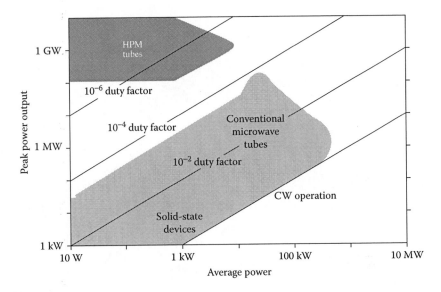

FIGURE 1.4
Peak vs. average power for microwave sources with duty factors.

ciency is in the vicinity of 10%. Energy efficiencies are usually much lower, due to the device falling out of resonance and pulse shortening at high power levels. Therefore, energy efficiencies are seldom quoted in the literature.

A comparison of the peak and average power characteristics for conventional and HPM sources is given in Figure 1.4. The peak power of a device is plotted against its average power. While conventional devices span an enormous range in this parameter space, the HPM devices have not been developed to produce substantial average power levels. This is because conventional sources have been developed to be the workhorses for specific applications, such as radar and particle acceleration, while the HPM sources have not, as yet. The Stanford Linear Accelerator Center (SLAC) klystrons are the highest power devices of conventional origin (~100 MW peak, 10 kW average). The relativistic magnetron is the highest average power HPM device (1 GW peak, 6 kW average). The *duty factor* for HPM sources — the product of the pulse length and the pulse repetition rate — is of the order of 10^{-6} (at most 10^{-5}), whereas for conventional devices it varies from 1 (continuous operation, called continuous wave, CW) to about 10^{-4}. We expect that the impetus provided by HPM applications will spur advances in repetitive operation, and average powers will approach 100 kW. There are as yet no applications for *both* high peak and high average power.

Another comparison of conventional and HPM sources is shown in Figure 1.5. Considering only pulsed devices, i.e., leaving out continuous operation, the conventional tubes typically produce ~1 MW for ~1 μsec, giving a joule per pulse. Klystrons for accelerators produce energetic pulses at 1 to 10 MW. The SLAC klystron has developed to 67 MW, 3.5 μsec, yielding 235 J. This borders on our definition of HPM (>100 MW). HPM sources produced tens

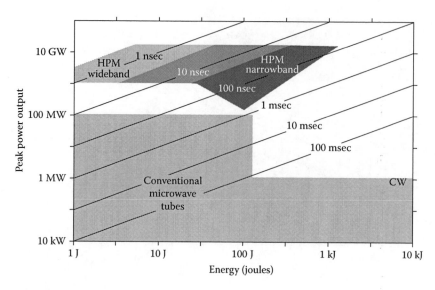

FIGURE 1.5
Peak power and energy from continuous and pulsed (durations shown) microwave sources, narrowband and wideband.

of joules in the 1970s, hundreds of joules in the 1980s, and surpassed a kilojoule in the 1990s (MWCG and relativistic klystron amplifier [RKA]). These developments were achieved by increasing peak power at ~100-nsec pulse duration. Further increasing pulse duration, the obvious route to higher energies, faces pulse shortening.

Most of the devices shown in Figure 1.5 are narrow bandwidth. The wideband impulse sources, with very short pulse durations, fall in the region of 100 MW, <1 J.

The increasing use of HPM sources will also raise the issues of reliability and lifetime. With HPM sources predominantly functioning in a research environment to date, important reliability and lifetime issues have received very little attention. For example, the >10^{-6} Torr vacuum levels of most HPM experiments contrast with the <10^{-7} Torr levels of conventional tubes, where higher pressures are known to cause lifetime problems. Still farther in the future lie the sophisticated issues of bandwidth, gain, linearity, phase and amplitude stability, and noise levels. These can be quite important in the context of a specific application.

The parameter spaces of the current applications of HPM are shown in Figure 1.6. The power levels for directed energy and short-pulse radar vary widely. The highest power requirements are for earth-to-space power beaming (earth-to-space, space-to-space, and space-to-earth, primarily average power driven) and directed energy weapons. It is notable that first-generation HPM weapons approaching deployment are high in average power, operating continuously, not as a series of pulses in a burst (see Section 3.2.3). These are confined mostly to the frequency regime below 10 GHz, in part

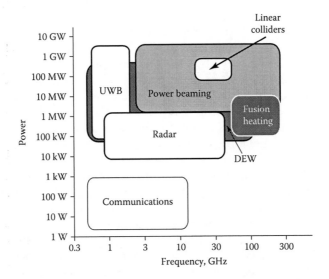

FIGURE 1.6
Domains of HPM applications. "Power" is typically peak power, but in some cases, such as fusion heating and some directed energy weapons (DEWs), it is average power. Regions are schematic; specific schemes vary widely.

by atmospheric absorption. Linear colliders (particle accelerators) require higher frequencies (~30 GHz) at about 100 MW. Electron cyclotron resonance heating of plasmas is a much higher frequency application (over 100 GHz), with a continuous or average power requirement of up to 10 MW. Some of these applications are approaching maturity, as described in Chapter 3.

The most stressful applications are military, because military platforms have severe requirements for onboard volumes and masses. Historically, the most comprehensive HPM weapons research programs were in the U.S. and Russia, especially during the Cold War era. There are now significant activities in Europe, with the U.K., France, Sweden, and Germany the most prominent. In Asia, China and India have weapons programs. In Japan, interest is centered on gyrotrons for plasma heating, klystrons for accelerators, and FELs for a broad range of applications that span the frequency spectrum. The Chinese are also investigating gyrotrons, including some innovative configurations, and FELs at laboratories in, for example, Chengdu, Beijing, and Shanghai. To accelerate its program, China has purchased HPM technology from Russia.

1.3 Future Directions in High Power Microwaves

The field of HPM is maturing. Decades of investigation have led to a sharpening of the research focus, created a commercial supplier base, and generated international interest in the technology. While earlier experiments have demonstrated very high power levels over a broad range of frequencies, users and the research community have fixed on a few systems with very versatile features and relatively compact dimensions.

In our first edition we made seven predictions about HPM's future directions. By our accounting, two of those seven predictions have borne out. Revisiting them tells us something about the changes that have occurred in the last decade and perhaps will occur in the next decades.

We learn more from failed predictions than from successful ones. First, the wrong predictions: We predicted that the peak power would exceed 100 GW, because 10 GW had been achieved by the RKA and MWCG. But this was not to be. The most important change that has occurred in the last decade is that *the power derby is over*. Peak power itself is not the driving factor in HPM, although it is always important. In the 1990s it became clear that peak power is limited by plasma formation and breakdown inside the sources. Substantial efforts have occurred in the last decade to attack the pulse-shortening problem, and they have been somewhat successful.

Pulse energy is more important for some applications, and at the highest powers energy is limited by pulse shortening. We predicted that most sources would achieve a microsecond pulse — some sources 10 μsec — and that sources not capable of microsecond operation would not be further developed. In retrospect, this did not happen because we did not understand that pulse shortening is such a serious limitation. The reality today is that sources are limited to the range of 100 J to 1 kJ at most, meaning 1 GW for a microsecond or 10 GW for 100 nsec. The limits are set by a plasma production in the sources and by breakdowns at high internal electric field. These limitations are discussed in Chapters 3, 7, and 8.

We predicted 100-kW average powers would be common, approaching 1 MW, because ~10 kW had been achieved in the early 1990s and higher repetition rates had been demonstrated for a few sources. But it did not happen because neither higher repetition rates nor higher peak powers were pursued. At present there appear to be no applications that require repetition rates above about 100 Hz.

In the 1990s phase locking of both amplifiers and oscillators had been demonstrated at high power (2 GW) operation in a module of seven sources, together with good phase stability and phase control (1 to 10°). We predicted phase locking would be extended to gangs of 100 sources, by which we meant that modules of about 10 sources would be built and that a group of ~10 such modules could reach ~100 GW. These would couple to antenna arrays to produce extremely high radiated powers with very narrow beams. This did not occur because no application required it. Technology exists but the need does not.

Now for a minor success. A decade ago most sources had power efficiencies of 10 to 20% with energy efficiencies less than 10% due to pulse shortening. We predicted the sources would develop to 50% energy efficiency. That has occurred, but only for continuous gyrotrons, not for pulsed sources.

The problem is that although theoretical efficiencies are high, in practice few devices have been truly optimized. Efficiency is low because it is hard to maintain resonance throughout the electrical pulse; so efficiency is low at the beginning and trailing ends of the pulse.

Another prediction we can be proud of is the merging of HPM and conventional microwave technologies and their research communities. They both follow the demands of applications. This has driven them to work together, as can be seen from two volumes uniting the two communities, sponsored by the U.S. Department of Defense.[3,4] We predict that this community will in the future produce better sources and HPM full-scale systems.

We anticipate that the two fields of HPM and conventional microwave devices, which have constituted different communities, technologies, and traditions, will converge further in the future. HPM can benefit from many aspects of conventional technology. Perhaps the most important of these is clean environments, most especially improved surface and vacuum techniques. Another area of importance is the control of breakdown.

The techniques of production technology, in which quality control is preeminent in building reliable and maintainable devices for a long lifetime, will be transferred from the conventional tube houses into the HPM community. This may lead to performance improvements in commercial HPM sources, usually for HPM testing facilities, in the next decade. Most types (magnetrons, vircators, BWOs, klystrons, reltrons, and gyrotrons) are available commercially, and we expect that many other source types with improved parameters will become available as applications take on more reality and the market for these sources grows. There is a substantial international market in HPM equipment, with Russia and the U.S. as the major suppliers. For example, Russia's military export company is offering buyers the opportunity to participate in the co-development of an HPM weapon, RANETS-E, that would operate in the X-band at 1-GW peak power, 5-nsec pulse duration, and 100-Hz repetition frequency (see Chapter 3). For a future extrapolation, see SuperSystem, Section 2.5.

HPM systems will follow the demands of the applications. We expect that flatness of the driving voltage pulse will become very important in the future because it is necessary for some applications, such as linear colliders, and it improves the overall energy efficiency of the device. Some devices will need broad bandwidth, especially if HPM is coupled with electronic warfare. Fundamental properties of sources, such as phase stability, will have to be improved if electronic warfare systems are to be improved by using high power sources. The use of modulations on the microwave power envelope will necessitate the development of amplifiers rather than oscillators. All applications will require reliability, a virtue little developed in HPM to date. For military applications, compact, low-mass, and transportable systems will be essential for deployments other than a fixed site. Mass and volume are driven by either system elements ancillary to the source — such as magnets, beam collection, and cooling in high average power applications — or prime and pulsed power on one end or antennas for high power radiation into air on the other.

Ultimately, the ability of HPM to achieve maturity will depend upon its ability to serve applications with a true market. These markets are only beginning to emerge and could include many defense applications, particle acceleration, industrial processes, or power beaming.

Following the extraordinary technical successes achieved by HPM in the last 30 years, HPM is finding applications driven by real needs. We remain confident that the international effort to expand the HPM technology and its emerging applications will help this new domain of electromagnetics continue to flourish in the years ahead.

Further Reading

HPM is a broad enough field that finding the appropriate references is not simple. There are several books in English[2-6] and an excellent review article by Gold.[7] The basic book on conventional tubes is by Gilmour.[8] The best reference in the technical journals is *IEEE Transactions on Plasma Science*, especially the special biannual issues dealing with HPM sources, and there are special issues on pulsed power as well. Otherwise, papers on HPM appear primarily in the *Journal of Applied Physics, Physics of Fluids B, Physical Review A*, and the *International Journal of Electronics*. The proceedings of the International Conferences on High Power Particle Beams, known popularly as, for example, BEAMS '06, contain state-of-the-art reports on HPM sources driven by pulsed power electron beam generators. UWB work is chronicled in IEEE special issues and in *Ultra-Wideband, Short-Pulse Electromagnetics*, a continuing series from Kluwer Academic/Plenum Publishers.

References

1. Benford, J. and Swegle, J., *High Power Microwaves*, Artech House, Norwood, MA, 1992.
2. Giri, D.V., *High-Power Electromagnetic Radiators*, Harvard University Press, Cambridge, MA, 2004.
3. Barker, R.J. and Schamiloglu, E., Eds., *High-Power Microwave Sources and Technologies*, IEEE Press, New York, 2001.
4. Barker, R.J. et al., *Modern Microwave and Millimeter-Wave Power Electronics*, IEEE Press, New York, 2005.
5. Granatstein, V.L. and Alexeff, I., Eds., *High Power Microwave Sources*, Artech House, Norwood, MA, 1987.
6. Gaponov-Grekhov, A.V. and Granatstein, V.L., Eds., *Applications of High Power Microwaves*, Artech House, Norwood, MA, 1994.
7. Gold, S., Review of high-power microwave source research, *Rev. Sci. Instrum.*, 68, 3945, 1997.
8. Gilmour, A.S., *Microwave Tubes*, Artech House, Norwood, MA, 1986.

2

Designing High Power Microwave Systems

2.1 The Systems Approach to High Power Microwaves

One of our primary goals in writing this book is to develop in the reader a systems perspective toward high power microwaves (HPM). This chapter describes that perspective and its application to the design and construction of HPM systems. To illustrate some of the points in this chapter, we draw upon our experience in assembling systems for various applications.

Our operational definition is that a system is an *orderly working totality*. More concretely, an HPM system can be represented by the simple block diagram shown in Figure 2.1. From left to right, we have:

1. A prime power subsystem that generates relatively low power electrical input in a long-pulse or continuous mode

2. A pulsed power subsystem that takes the low power/long-pulse electrical power, stores it, and then switches it out in high power electrical pulses of much shorter duration

3. A microwave source in which the short-duration, high power electrical pulses are transformed into electromagnetic waves

4. Perhaps a mode converter to tailor the spatial distribution of electromagnetic energy to optimize transmission and coupling to an antenna

5. An antenna, which directs the microwave electromagnetic output, essentially compressing that output spatially into a tighter, higher-intensity beam

As presented, the arrows in the figure depict the flow of power and energy within the system, from the prime power generator on the input end to the radiation of microwaves from the antenna on the output end. In assembling such a system, *the selection of individual components is subordinate to the overall goal of optimizing the system as a whole for a chosen purpose*. Note that this is the outline of a system; later diagrams will show much greater complexity.

FIGURE 2.1
Components of an HPM system.

For example, *power conditioning*, which indeed is required for burst-mode operation, is missing.

The need for adopting a *system approach* stems from the nature of the process of designing and building a *complex* system that must meet the *requirements* posed by a user's application within a set of sometimes competing *constraints*. In addressing complexity, while striving to meet requirements within a set of constraints, one must learn to step back and view the system from end to end, considering the effect that each subsystem has on the other subsystems, both upstream and downstream in the sense of the power flow depicted in Figure 2.1. Before we characterize our systems approach, let us consider these three motivating elements: complexity, requirements, and constraints. The logical line of development is to follow a systems approach consisting of the steps below:

- Determine the goals in terms of a clearly stated application, typically performance (as distinct from cost and schedule).
- Define the objective criteria against which you will measure the achievement of your goals.
- Analyze various candidate systems.
- Select a system configuration and the subsystem configurations and component choices that realize that system.
- Construct the subsystems and integrate into the final system.

Unfortunately, these logical steps can rarely be performed in this order because, in practice, the system designer performs some or all of these functions simultaneously throughout the process. The reason for this is rather circular: problems cannot be adequately formulated until they are well understood, but they cannot be well understood until approximately solved. Thus, the two processes of *understanding the problem* and *devising solutions* are intertwined. Design of a system is extremely iterative. We can therefore see that *systems thinking involves considerable art as well as science; it makes heavy use of intuition based on experience.*

To get an idea of the complicated nature of system design, consider a case in which the application provides a rather straightforward set of overall requirements that specify the following: (1) the peak energy per pulse to be delivered to a target at a given range, (2) the pulse repetition rate within a burst of pulses, (3) the length of the burst of pulses, (4) the burst repetition rate, (5) the system volume, (6) the system mass, and (7) the maximum angular slew rate for redirecting the beam. To design a system to meet those seven requirements, we have a great deal of freedom.

Return to the system block diagram of Figure 2.1, which has subsystems for prime power, pulsed power, a microwave source, mode converter, and antenna. We have multiple candidate technologies for each of those six block elements, and the realization of some of those choices may involve making component choices within the subsystem and individual part choices within the components as well. We will return to this hierarchy of system–subsystem–component–part later in the chapter. Each block element accepts certain input parameters from the subsystem upstream of it in the energy flow depiction of the figure, and provides certain output parameters to the subsystem downstream. For example, a microwave source accepts electrical input characterized by the voltage, current, pulse length, and repetition rate from the pulsed power, and it provides microwave output in the form of a certain power, pulse length, and repetition rate. It also provides additional parameters — mass and volume contributions, platform-relevant parameters such as form factor and center of mass, and operational restrictions such as operating temperature, humidity, and vibration level — that in sum characterize the system as a whole. Individual components and parts have their own input and output parameters, and they provide their individual contributions to system mass, volume, and linear dimensions. Further, there are constraints on voltage, current, frequency, and efficiency. Some of those constraints come from individual parts of each component, e.g., you cannot buy an off-the-shelf gas thyratron switch for your pulsed power subsystem that handles over 40 kV.

Viewing the whole system branching tree structure, under each subsystem there is a set of component choices and under each component its individual parts, some of which have subbranches for different part choices, all contributing input and output and system parameters to the design calculations. This compounded complexity is called *combinatoric explosion*. The optimization process is made even more difficult by two factors. First, the system designer will not be an expert in the design of every possible system element and component, so that, in effect, he runs the risk of ignoring certain design options out of ignorance. Therefore, there is a design team. Second, not every component exists as a well-characterized off-the-shelf product. Some have to be designed as custom items to the specifications set by the overall system designer. Thus, it is difficult to characterize in advance, because one is not entirely certain what items are going to be available.

Considering the daunting complexity, one comes to see the value of the systems approach to design, which we will describe in this chapter. It is easier to begin by scoping the system with reduced scaling models, so that before going to the trouble of getting components designed, you can make choices about whether they are even in the ballpark. From an initial systems analysis you can do a better job of setting requirements.

Systems design involves optimization of the overall system, as distinct from piecemeal suboptimization of the elements of the system. *Suboptimiza-*

tion is the common failure mode in systems, and it is a problem that occurs with high frequency. Common ways to suboptimize are:

- *Win the battle/lose the war:* You optimize a subsystem to the detriment of overall system performance.
- *Hammer looking for a nail:* Your solution was decided in advance, so that all solutions employ your chosen subsystem or component.
- *A little knowledge is a dangerous thing:* Operating with limited knowledge, you miss or mischaracterize a better solution.

To avoid suboptimization, the system designer has to think one level up and one level down from his particular viewpoint at any step in the process. This is because the drivers he receives from above, at the system application level, and those he receives from below, from the component technologies, are never sufficiently well defined to make choices easy. Nevertheless, designing a system requires keeping both sets of requirements and constraints in mind simultaneously while integrating the full system.

2.2   Looking at Systems

The hope of the system designer is that the problem can be subdivided in such a way that the parts can be handled somewhat separately. Of course, a system can be described in many ways. For example, one can look at a system in terms of the phases of its chronological development and operation, a long timescale. Or one can look at only the short timescale of operation, the time sequence of system functioning, focusing on time evolution of quantities. Here we choose to describe systems in terms of the subsystems.

The nomenclature we will use is shown in Figure 2.2. The entire *system* is made up of *subsystems,* by which we mean the technological subassemblies that make it up. A common set of subsystems are prime power, pulsed power, etc., as in Figure 2.1. Each subsystem is composed of *components.* For example, a microwave source can have a slow wave structure and may require a magnet. In turn, components are made up of individual *parts;* a Marx generator's parts would be capacitors, inductors, and resistors. In an antenna, the components are a pedestal, a dish, and a feed horn.

Existing system design methods and expert systems use this system–subsystem–component–part breakdown point of view. An excellent example of this is HEIMDALL, the expert system for HPM weapons,[1] which does system concepts, meaning that it includes the subsystems but not all the ancillary equipment, such as vacuum pumps. In Figure 2.1 energy flows from prime power to higher-voltage, shorter-pulse-length pulsed power, then to the HPM source. Here the energy is converted into microwaves, is extracted in

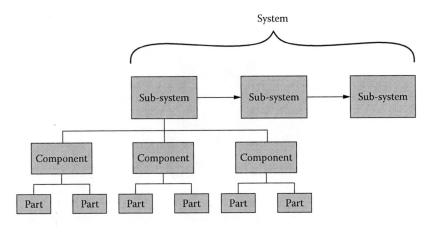

FIGURE 2.2
Nomenclature: Systems are made up of subsystems (pulsed power, etc.) that have components (Marx generator, etc.) composed of parts (capacitors, etc.).

a waveguide mode, which may be converted to another mode, and then radiates from an antenna. This diagram applies exactly to systems for power beaming and HPM weapons (see Chapter 3). In particle accelerators, the energy would not radiate, instead going directly to an accelerating cavity.

2.3   Linking Components into a System

Conceptually, think of systems as collections of subsystems in the way that, although an atom has a complicated internal structure, for the study of chemistry one needs to know only an atom's valence. The building blocks of a system, the subsystems, can be described as shown in Figure 2.3. Input parameters combine with features of the subsystem to produce the output parameters, which in turn become input parameters for the next subsystem. The output parameters flow to the next subsystem, providing it essential input information. System features such as the total weight, total size, and total electrical energy consumed can be calculated by summing those quantities for each subsystem.

Each subsystem in Figure 2.1 has many parameters. They vary from simple physical parameters such as sizes and weights, to electrical parameters such as current, voltage, and impedance, to microwave parameters such as frequency, bandwidth, and waveguide mode. To assemble components into a system, some choice of sets of parameters must be made. In this chapter, we describe a particular *linking parameter set* of input and output parameters and how to use them in designing conceptual systems.

Choosing to take some parameters as independent variables to describe components means that other parameters are then dependent, being derived

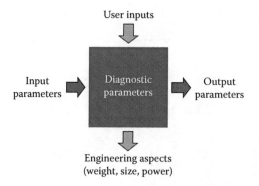

FIGURE 2.3
Interface parameters knit the components together. For examples of linking parameters, see Figure 2.4.

from independent parameters. (Choice of independent variables may not be the same for all systems; selecting which is part of the system art.) In order to capture a description of the components in terms of major parameters, the particular linking parameter set for each component class, such as pulsed power components, has its own particular characteristics. We show a well-tested set in Figure 2.4. These are one useful way to consistently connect component classes to operate with other component classes. This is the energy flow direction. However, for system design efforts, there is no *intrinsic* direction of flow for the exchange of information or parameters between subsystems.

Note that there usually are several stages of electrical pulse compression, called *power conditioning*. It is less common to use an additional stage of compression of the microwave pulse itself. Although it is quite common to

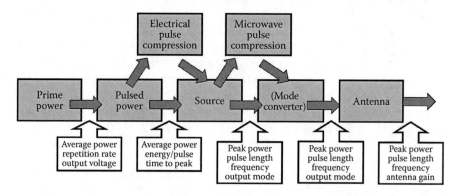

FIGURE 2.4
Energy flow between component classes in an HPM system. Quantities between component classes are those needed for linking them into a system. Optional branches are for an additional stage of electrical pulse compression (e.g., Marx to PFL) and pulse compression of microwaves after their generation.

see microwave pulse compression in conventional radar systems, it is less commonly used in the HPM world, although it was considered at the level of 100 MW for the Next Linear Collider (see Chapter 3 and Section 5.4). The advantage of microwave pulse compression allows use of lower-power, and therefore lower-voltage, pulsed power components.

Compatibility, i.e., which other technologies a subsystem can or cannot be mated to, is important to linking components together into a system. Beyond compatibility are *preferences,* the typical best design practice in linking subsystems. For example, each microwave source will produce radiation in a specific waveguide mode. That mode will influence the type of antenna with which a microwave source can operate. Whether available hardware can handle the mode via commercial availability or easy fabrication influences the design choice. Therefore, one must consider and specify which modes a source will produce and which modes an antenna can accept and radiate.

2.3.1 Prime Power

For linking prime power to pulsed power (Figure 2.5), the important parameters are the average power, the repetition rate — which together define the energy per pulse — and the output voltage. There are many technology choices for prime power. In Figure 2.6, we show a set of options that could be employed to provide continuous prime power output to a number of pulsed power options mentioned at the right of the figure, which would ultimately drive a microwave source. Our use of the term *continuous* is somewhat guarded, since the subsystem at the bottom of the figure, which involves an explosively driven magneto-cumulative generator, is purely a single-shot subsystem. For clarity, we emphasize that the diagram is not the connection diagram for a prime power subsystem, but rather a presentation of possible component choices within a prime power subsystem. A common choice for many systems is to use a generator powered by an internal combustion engine, such as the diesel alternator or the turbo alternator, but it is also possible to drive systems with batteries for long periods. In either case, the common feature of these subsystem options is that either an alternating current (AC) internal combustion generator or a direct current (DC) battery provides long-pulse or continuous power that is converted to a DC output that functions as input to the pulsed power.

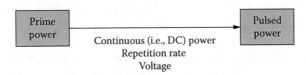

FIGURE 2.5
Linking parameters for connecting prime power to pulsed power.

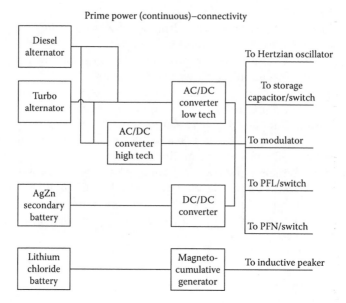

FIGURE 2.6
Possible prime power component connections for continuous power.

For firing a system in a burst mode so that the energy store is exhausted over the course of firing a burst of pulses, then replenished between bursts, other alternatives exist, as shown in Figure 2.7. Note that each option within this set of subsystem options has four components: an internal combustion prime mover producing an AC output; an interface component followed by an energy store and electrical source that, depending on the option, operates at either AC or DC; and a converter, the output of which provides a series of switched DC pulses to the pulsed power.

To better understand the operation, consider the timing diagram of Figure 2.8. The upper plot shows the power pulses delivered by the burst-mode prime power store to the input of a pulsed power system, with power pulses of duration τ_B separated by an interburst period of length τ_{IB}. Below, we see the timing for the individual pulses of duration τ_P delivered by the pulsed power during the period τ_B. The prime mover component delivers power continuously to the energy store and electrical source, which stores it over the interburst period and then delivers it at a higher power over the course of a burst. The pulsed power in turn stores the power during the burst and delivers it at much higher power during each pulse. For example, it is possible to store the energy in a flywheel or pulsed alternator and then convert it to AC or DC.

Thus, the key difference between the continuous and burst-mode systems is the added energy storage in the latter. Average power is much higher in continuous mode, so thermal management becomes a larger issue. In the burst mode, the system has four operational states: firing, charging, charged and waiting, or off.

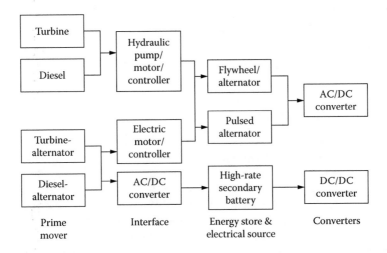

FIGURE 2.7
Possible prime power component connections for burst-mode power. High-tech converters use modern solid-state and digital technologies, and low-tech converters use older transformer/rectifier technologies.

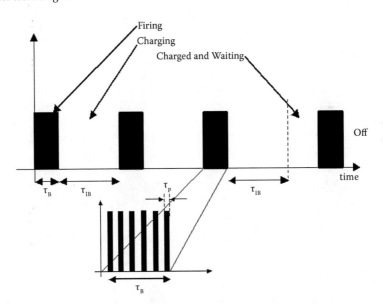

FIGURE 2.8
The four states of burst-mode operation.

The repetition rate for peak power selects the switching technologies available to the system. For example, pulsed power repetition rates above 150 Hz suggest a preference for spark gap, magnetic, or solid-state switches. For airborne systems with stringent requirements to limit size and weight, the prime power is a major driver because it is heavy, or the system must tap the platform power system, usually coming from the engine. Ground-based systems can use more conventional prime power, such as internal combustion engines. The designer of airbourne systems tries to avoid adding another internal combustion engine, so will prefer tapping the system power.

2.3.2 Pulsed Power

Pulsed power is used in most applications, with the major exceptions being continuous power beaming and plasma heating. The existing types of pulsed power (see Chapter 5) are modulators, Marx generators, pulse-forming lines (PFLs), pulse-forming networks (PFNs), and inductive energy storage in combination with opening switches. With the input voltage, average power, and repetition rate taken from prime power, the output of pulsed power components can be characterized by three classes of possible downstream elements: (1) another pulsed power component for further pulse compression, (2) directly to a microwave source, or (3) an impedance and voltage transformer. In the last case, the pulsed power component may produce the correct pulse length, but not the voltage or impedance, in which event one could use a pulse transformer, voltage adder (linear induction accelerator [LIA]; see Chapter 5), or a tapered transmission line, which is another kind of impedance transformer.

The most typical case is connecting a series of pulsed power components. A component, such as a Marx generator, provides the input to a further stage of electrical pulse compression, such as a PFL. The other, less typical, case is that the output of the first pulsed power component is connected to a microwave source, such as a Marx directly driving an antenna. For both cases, the output qualities that must be specified are output voltage, impedance (therefore current, the ratio of voltage to impedance), and pulse length (Figure 2.9). It is also important to specify rise and fall times because of resonance conditions that must be met for the source to generate microwaves. Electrical energy sent into the source before and after the resonance is met is useless, and unused electrical energy passed through the source must eventually be handled by the cooling system. A complex issue at the pulsed

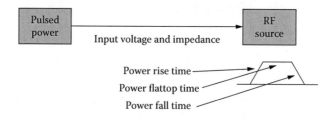

FIGURE 2.9
Linking parameters for connecting pulsed power to microwave source.

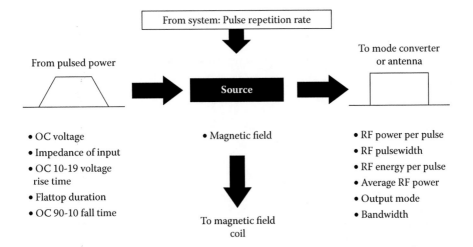

FIGURE 2.10
Linking parameters for connecting pulsed power to microwave source and on to the antenna. The source will have ancillary equipment: vacuum pumps, cooling. A master oscillator will be required for amplifiers. OC = open circuit.

power–microwave source interface comes from the dynamic impedance of the microwave source, making impedance matching a serious design issue. Source impedance almost always varies in time and is voltage dependent as well (see Sections 4.6 and 5.3).

2.3.3 Microwave Sources

Microwave sources are usually the most complex element of the system, as Figure 2.10 shows. Simply put, for purposes of connecting to downstream elements (most typically a direct connection to the antenna), the essential features of the source can be taken as the peak microwave power, frequency, pulse length, and output waveguide mode. The bandwidth can also influence the choice of downstream components. Sources also have substantial supporting equipment: a vacuum pump, a magnet in many cases, a cooling system (as do other subsystems), and a collector or "dump" to capture the beam. There may be a master oscillator if the source is an amplifier. Its power, frequency, pulse duration, and pulse width describe this oscillator. The magnetic field will be specified by the physics of the source. There may also be a need for an x-ray shield to protect personnel.

Perhaps the most important interface in an HPM system is between the pulsed power and microwave source. Electrically, it can be characterized as shown in Figure 2.11, where the pulsed power is represented by an equivalent circuit with a specified open-circuit voltage, V_{OC} (voltage produced in an open circuit by the pulsed power), and a pulsed power output impedance, Z_{PP}, while the microwave source is characterized by its impedance, Z_{SOURCE}.

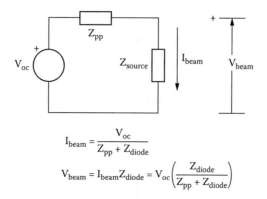

$$I_{beam} = \frac{V_{oc}}{Z_{pp} + Z_{diode}}$$

$$V_{beam} = I_{beam} Z_{diode} = V_{oc}\left(\frac{Z_{diode}}{Z_{pp} + Z_{diode}}\right)$$

FIGURE 2.11
Circuit for the vital pulsed power–microwave source interface. The combination of impedances determines electrical input to source.

The importance of this interface comes from the need for the impedances of the pulsed power and the microwave source to be well matched, or serious energy and power efficiency losses will occur (see Section 5.2). This in turn affects the size, mass, and cost of the system. Impedance matching is made complex by the fact that typically the source impedance varies with time. It usually falls from an initial value because of gap closure in the beam-generating diode. For a Child–Langmuir and foil-less diodes (see Section 4.6.1), which many sources have, the impedance varies with the anode–cathode gap d, but weakly with voltage. For Child–Langmuir diodes,

$$Z_{source} \sim \frac{d^2}{V^{1/2}} \tag{2.1}$$

so plasma generated at the cathode or anode and moving to close the gap can change impedance steadily. Therefore, the time when peak microwave power should occur (when the microwave resonance is reached) is the best time for impedance match to occur.

2.3.4 Mode Converter and Antenna

The output parameters of the source determine the connection to the antenna (Figure 2.12). The most crucial factor is a waveguide mode, which must be made compatible with the desired antenna. The characteristics of the output of the antenna are the power, frequency, and antenna gain or angular beam width. They determine the output beam propagation characteristics (Figure 2.13). Antennas are discussed in Section 5.5.

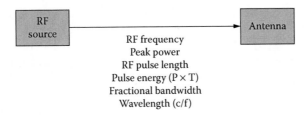

FIGURE 2.12
Linking parameters for connecting microwave source to the antenna.

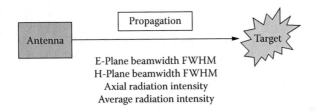

FIGURE 2.13
The antenna output parameters determine propagation of the microwave beam.

2.4 Systems Issues

It is frequently thought that the factor that drives selection of components for sources is efficient transfer of energy from electricity to microwaves. While this is a factor, it is actually a stalking horse for the truly important factors. Various applications have different driving factors. In Chapter 1 we described the early years of HPM, when peak power was the driving factor. But more recently, the compatibility of sources with the other elements in the system has become more important. There are also other factors, such as reliability, complexity, maintainability, and mobility. It has been said that we are now in a new era of "-ilities."

For military applications, which typically require a mobile system, the driving parameters for the system are usually the volume and on-board average power requirement of the system. Weight is less important, which may seem counterintuitive. For example, airborne usually means lightweight. An HPM weapon will usually be fitted onto an existing platform in the space previously reserved for another weapon. Such weapons are usually denser than HPM devices. For example, if an HPM system, with some vacuum volume, substitutes for an explosive warhead with high density, the weight actually drops. So, often the real constraint is volume.

Power-beaming applications are usually limited by the available antenna apertures at either the transmitting antenna or the receiving rectenna. For particle accelerators, on the other hand, one major driving factor is the cost of the source when manufactured in quantity. This is true because of very stringent performance requirements on lifetime, average power, and energy per pulse. Another is reliability, which stems from the complexity of the system: when there are so many microwave sources, frequent breakdowns would be intolerable. This requirement argues in favor of more conservative technology choices.

The key distinguishing electrical parameters of the system components are the voltage and current and their ratio, *impedance*. Impedance is a quite general quantity, as it also relates to appropriate components of E and B through the component geometry and to particle drifts in the source, which are proportional to local electric and magnetic fields, E/B. Impedance also relates to antennas through the complex impedance, including both the real part, resistance, and the imaginary part, reactance.[2] Antenna design frequently consists of reducing the reactive component, making impedance matching simpler. The voltage–impedance parameter space into which HPM source technologies fall is shown in Figure 5.2. With voltages of <100 kV and impedances of >1 kΩ, conventional microwave applications (radar, electronic countermeasures, communications) lie at the low power end (power = V^2/Z). Most, but not all, present-day HPM devices cluster at 10 to 100 Ω. Voltages generally span the range of 0.1 to 1 MV, but some sources have been operated up to several megavolts.

The implication is that efficient transfer of electrical energy to microwave sources that span a broad range of impedances will require pulsed power technologies that similarly span a substantial impedance range. When adjoining circuit elements do not have the same impedance, voltage, current, and energy transferred to the downstream element can be very different (Figure 5.3). Efficient energy transfer suffers if impedance mismatch exists, especially if the downstream component is of lower impedance than the circuit driving it. Nevertheless, there are exceptions: mismatching upward to higher-load impedance is sometimes used as an inefficient means of increasing load voltage.

As an example of the effect of system considerations, consider the case of the use of flux compression generators (FCGs, see Section 5.2.2), also known as explosive generators and magneto-cumulative generators (MCGs), to drive single-shot HPM weapons. These devices have been extensively developed in Russia and the U.S. and are now being widely studied. FCGs operate at time-varying impedances well below 1 Ω, producing very high currents at low voltages. As a consequence, workers use transformers, in part to get better rise times for inductive loads. Because of the very high currents, the preferred pulsed power system for such devices is inductive energy storage with an opening switch (see Section 5.2). The impedance transformation required to couple FCGs to ~100-Ω sources would be very inefficient. In the 1990s the search for high power, low-impedance microwave sources to better match FCGs led to interest in the Magnetically Insulated Line Oscillator (MILO) (Chapter 8), which has an impedance of ~10 Ω. But the MILO is not tunable, is heavy, and requires a more complex antenna to radiate in a useful pattern. Consequently, the Russians chose to use both a vircator (Chapter 10), which, although higher in impedance, is lower than other Russian sources, and the backward wave oscillator (BWO) (Chapter 8). The vircator's other advantages are its small size and simple, inexpensive construction. Its very low efficiency is made up for by the very high energies available from FCGs and low mass due to the very high energy density of high explosives as a source of energy. This shows why system considerations will determine what components are actually used rather than single parameter choices, such as impedance or efficiency.

2.5 Scoping an Advanced System

As we have said, designing a system is at least as much art as it is science, and creating a design concept is a highly iterative process. To show the thought processes, here we construct and work through a specific example at the state of the art in HPM systems. Our example system does not actually exist, but, as we shall see, it is similar to those either built or proposed by others. We will call it the SuperSystem, which we specify by the top-level system requirements shown in Table 2.1 (see Problem 1). All of those requirements govern the output of the system. There are no explicit constraints on size or mass; however, we assume the system will be ground mobile, so that the choice of a platform would ultimately limit the size and mass of the system.

Operating in the burst mode is essential if the SuperSystem is to have a transportable size and weight. Figure 2.14 shows a time plot with the four states for a system firing four bursts. Within each burst of duration τ_B, during which the system is in the firing state, individual shots of pulse length τ_{RF} are radiated from the antenna at a pulse repetition rate PRR of 500 Hz. Between bursts, during which the system is in the charging state, an inter-burst charging period of duration τ_{IB} is required to recharge the system for the next burst. If the period between bursts exceeds τ_{IB}, as it does between the third and fourth bursts in Figure 2.14, the system goes into the charged-

TABLE 2.1

Specifications for an Advanced HPM System Called SuperSystem

Parameter	Value	Comment
Radiated power	500 MW	ORION system level
Pulse length	10–20 nsec	Means smaller pulsed energy store
Pulse repetition rate	500 Hz	Has been done for BWOs and relativistic magnetrons
Frequency	X-band	Much BWO experience here
Antenna gain	45–50 dB	Sufficient for long range
Burst duration	1 sec	Consistent with defense missions
Interburst period	10 sec	Consistent with defense missions

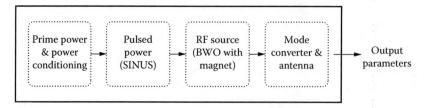

FIGURE 2.14
Internal structure of the SINUS pulsed power generator.

and-waiting state. When no more bursts are fired, the system is off. For this example, we choose to make the two timescales, τ_B and τ_{IB}, 1 and 10 sec, respectively. These timescales are consistent with those for defense missions. We will describe burst-mode operation and the effect of the parameter choices on system performance in detail below. First, we discuss technology choices.

The operating parameters of the SuperSystem are all compatible with a number of known systems that are based on the SINUS series of pulsed power generators originally developed at the Institute of High-Current Electronics (IHCE) in Tomsk, Russia. Examples of different versions of SINUS pulsed power generators are listed in Table 8.3, which shows a range of voltages, currents, pulse lengths, and repetition rates. The same basic technology is involved in the construction of each of these, as shown in Figure 2.14: capacitive input energy storage, an integrated pulse-forming line and resonant Tesla transformer, a high-repetition-rate gas switch, and a voltage- and current-transforming transmission to match the radio frequency (RF) source electrical requirements. We therefore assume that the SuperSystem utilizes SINUS technology, based particularly on the distinctive pulse length and repetition rate.

2.5.1   NAGIRA: Prototype for the SuperSystem

To model the SuperSystem, first consider a SINUS-based system produced by IHCE: the NAGIRA[3] high power microwave radar (acronym for Nanosecond Gigawatt Radar). It was delivered to the U.K. Ministry of Defense and the British company GEC-Marconi in 1995 and is now in the U.S. The radar system is housed in two containers, the transmitter cabin and the operations container. The transmitter cabin is shown in Figure 2.15, a photograph taken from the rear of the transporting trailer. The SINUS accelerator in the NAGIRA system produces an electron beam of 600 kV and 5 kA in a 10-nsec pulse with a 150-Hz repetition rate. This 3-GW electron beam is injected into a backward wave oscillator (BWO) to generate an RF output in the X-band with a power of nominally 500 MW, implying a *power efficiency* in the BWO of about 17%. The TM_{01} output mode of the BWO passes through a TM_{01}-to-TE_{11} mode converter, after which a quasi-optical coupler passes the output to a 1.2-m-diameter parabolic dish. The superconducting magnet for the BWO produces a 3-T magnetic field.

FIGURE 2.15
Transmitter cabin for the NAGIRA system. Most prominently, note the back end of the SINUS pulsed power generator. The motor generator supplying prime power is seen to the lower right. The lower dish on the roof is the transmitting antenna, while the upper dish is the receiving antenna.

The radiated output is about 400 MW in a 5-nsec pulse. Since the power from the BWO is 500 MW and the mode converter is said to have an efficiency of $\eta_M = 95\%$, we infer that coupling to the antenna has an efficiency of about $\eta_A = 85\%$. The electrical power into the BWO is, as stated previously, 3 GW in a 10-nsec electron beam, so we define an *energy efficiency* for the BWO of

$$\eta_{BWO} = \left(\frac{\tau_{RF}}{\tau_{BEAM}} \right)\left(\frac{P_{RF}}{V_0 I_b} \right) = \left(\frac{5}{10} \right)\left(\frac{500}{3000} \right) = 0.083 \qquad (2.2)$$

where τ_{RF} is the length of the microwave pulse, τ_{BEAM} is the length of the beam pulse, P_{RF} is the microwave output of the BWO (500 MW vs. the 400 MW radiated from the antenna), and V_0 and I_b are the beam voltage and current. To trace back and determine the prime power requirement, we also need the electrical efficiency of the SINUS pulsed power machine, η_S, and the efficiency of the power conditioning system, η_{PC}, which we take to be a transformer-rectifier with capacitive storage, which is essentially an AC-to-DC converter. The SINUS machines have a typical efficiency of $\eta_S = 0.69$, and a power conditioning efficiency of $\eta_{PC} = 88\%$ in the case of three-phase prime power input. The *average* power required from the prime power system is therefore given in terms of the radiated microwave power, $P_{RAD} = 400$ MW, the electron beam pulse length, and the pulse repetition rate (PRR) by the expression

$$\langle P_{PRIME} \rangle = \frac{P_{RAD}\tau_{RF}PRR}{\eta_A \eta_M \eta_{BWO} \eta_S \eta_{PC}} = \frac{\left(4 \times 10^8\right)\left(5 \times 10^{-9}\right)(150)}{0.85 \times 0.95 \times 0.083 \times 0.69 \times 0.88} = 7.4 \text{ kW} \quad (2.3)$$

The difference between the 5-nsec microwave pulse length and the 10-nsec electron beam pulse length is taken account of in Equation 2.3.

The layout of the NAGIRA system within the transmitter cabin is shown in Figure 2.16. The portion of the system on the right, pulse-forming line, high power switch, and impedance transformer, comprise the SINUS pulsed power generator. Note that it takes up a substantial fraction of the length of the container, so that the transmission line is bent upward to the BWO, which is surrounded by a cryostat for its superconducting magnet. The quasi-optic feed couples the BWO output to the transmitting dish antenna through open air. Although not labeled, it would appear that the motor generator supplying prime power is located at the back of the container, opposite the pulse former.

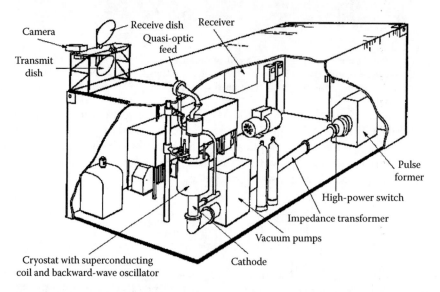

FIGURE 2.16
Schematic of the transmitter cabin of the NAGIRA system. (From Clunie, D. et al., The design, construction and testing of an experimental high power short-pulse radar, in *Strong Microwaves in Plasmas*, Litvak, A.G., Ed., Novgorod University Press, Nizhny Novgorod, Russia, 1997. With permission.)

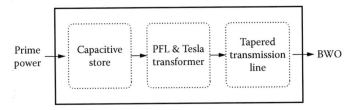

FIGURE 2.17
Top-level schematic of the SuperSystem. (From Clunie, D. et al., The design, construction and testing of an experimental high power short-pulse radar, in *Strong Microwaves in Plasmas*, Litvak, A.G., Ed., Novgorod University Press, Nizhny Novgorod, Russia, 1997. With permission.)

2.5.2 Constructing a SuperSystem

Using information for other NAGIRA-like systems, as well as models for the individual components of the NAGIRA system, we can construct a model of the SuperSystem itself. To define the performance required of the components of the SuperSystem, we work backward through the system, from the microwave output requirements to the prime power. A top-level schematic for the SuperSystem is shown in Figure 2.17. We include the effect of the parameters τ_B and τ_{IB} only on the prime power, which implies that we ignore the effect of these parameters on the mass and volume of the cooling system required for the pulsed power and BWO. Since both elements of the system will probably only just reach their equilibrium temperature during a burst, this assumption is justified.

In the following, we will rely on phenomenological models with assumed parameters. For example, we will assume an efficiency for conversion of electron beam power to microwave power at the output of the microwave source; no attempt will be made to compute or even estimate that efficiency from first principles. Rather, it is an independent input parameter; we will choose a value in line with those observed in actual experiments. We can vary it to uncover the effect on SuperSystem. This is one of the key elements of the process of scoping a system using a systems modeling approach.

2.5.3 Antenna and Mode Converter

Returning to Table 2.1, we see that the system is to operate in the X-band, so we choose a frequency of 10 GHz. The gain of the system antenna is to be 45 to 50 dB. Denote this gain, measured in decibels, G_{dB}. This value actually corresponds to an actual numerical value of (see Section 5.5)

$$G = 10^{G_{dB}/10} \tag{2.4}$$

For G_{dB} between 45 and 50 dB, G lies between about 3.2×10^4 and 10^5. In terms of the antenna diameter D and the frequency and wavelength of the microwave output, f and λ, G is given for a parabolic dish antenna by

$$G = 5.18\left(\frac{D}{\lambda}\right)^2 = 5.18\left(\frac{Df}{c}\right)^2 \tag{2.5}$$

For our X-band frequency, we choose f = 10 GHz, so that λ = 3 cm. Now, if we choose a gain of 45 dB, so that G = 3.2×10^4, we can use Equation 2.5 to find the antenna diameter: D = 2.4 m.

We can now compute several other output parameters of the system. The output beam width of the microwave beam (or the width of the central antenna lobe) from a parabolic dish antenna (see Chapter 5) is the following:

$$\theta\,(radians) = 2.44\left(\frac{\lambda}{D}\right) = 2.44\left(\frac{c}{Df}\right) \tag{2.6}$$

For our X-band frequency and computed diameter D, θ = 30 mrad. The *effective radiated power* (ERP), the product of the radiated power and the gain, is

$$ERP = GP_{RAD} \tag{2.7}$$

which, in the case at hand, has the value ERP = 16 TW for a radiated power of P_{RAD} = 500 MW. Finally, the peak angular radiation intensity, measured in watts/steradian, is given by

$$J\left(\frac{W}{steradian}\right) = \frac{ERP}{4\pi} \tag{2.8}$$

To move farther upstream to the BWO, we need only two other parameters: the power conversion efficiency of the antenna, η_A, and the efficiency of the mode converter that converts the TM_{01} output of the BWO to a TE_{11} mode required by the antenna, η_M. For our system, we choose η_A = 0.85 and η_M = 0.95. Note once again that we do not need to know the details of the coupling to the antenna or the mode conversion process. We simply use values that are consistent with past practice.

2.5.4   Backward Wave Oscillator

For the power level and repetition rate, the best choice for X-band is the BWO. At lower frequencies, magnetrons and klystrons become competitive. BWOs have reached powers of 5 GW, so 0.5 GW is well within their capability. Working backward from the 500-MW output radiated from the antenna, given the antenna and mode converter efficiencies of $\eta_A = 0.85$ and $\eta_M = 0.95$, we infer an output from the BWO of

$$P_{BWO} = \frac{P_{RAD}}{\eta_A \eta_M} = \frac{500\,MW}{(0.85)(0.95)} = 620 \text{ MW} \qquad (2.9)$$

in a pulse of length τ_{RP}.

Next, to determine the electrical input requirement for the BWO, we must take account of both the instantaneous power efficiency for converting electron beam energy to microwave power and the differences in pulse length between the microwave output and the electron beam input. In our earlier treatment of BWO efficiency in Equation 2.2, we combined both effects. To compute the electrical input requirement for the BWO, we make the following assumptions:

- To take account of the longer pulse length of the SuperSystem relative to the NAGIRA system, we assume a constant difference between the electron beam and output microwave pulse lengths, rather than a constant proportionality. Thus, we retain the 5-nsec difference for NAGIRA to arrive at an electron beam pulse width for the SuperSystem of $\tau_{BEAM} = \tau_{RF} + 5$ nsec. This 5 nsec is the time required for the electron beam current to rise to the level required to sustain oscillation in the BWO and for those oscillations to build up to their full amplitude.
- Factoring out pulse-length effects, we assume that the power efficiency for generating microwaves, the relationship between the electron beam power and the microwave output power of the BWO, is the same for the SuperSystem as it was for NAGIRA. In NAGIRA, a 3-GW electron beam produced a 500-MW BWO output, so that the power efficiency $\eta = 0.17$. Therefore, to produce a 620-MW BWO output, an electron beam power of $P_{BEAM} = 3720$ MW is required.
- We assume that the SuperSystem BWO will have the same 150-ohm impedance, Z_{BWO}, as the BWO in the IHCE system, as this is typical of BWOs (see Figure 5.2). Thus, with V_0 the anode–cathode voltage in the diode,

$$P_{BEAM} = V_0 I_b = \frac{V_0^2}{Z_{BWO}} = 3720 \text{ MW} \qquad (2.10a)$$

so

$$V_0 = \left(P_{BEAM} Z_{BWO}\right)^{1/2} = 747 \text{ kV} \qquad (2.10b)$$

$$I_b = \frac{V_0}{Z_{BWO}} = 4.98 \text{ kA} \qquad (2.11)$$

As indicated in the system-level diagram of Figure 2.17, the BWO will require a magnet. In the NAGIRA system, the magnet is superconducting, as is typical of fielded systems (such as ORION; see Chapters 5 and 7). This magnet adds mass to the system, and it requires its own small power supply and cooling system. Field systems use superconducting magnets because, while they need cryogenics for cooling, their ancillary power supply and vacuum equipment are far smaller and lighter than the power supply of normal magnets and they are lighter than permanent magnets.

2.5.5   Pulsed Power Subsystem

The internal structure of the SINUS generator consists of three components, as indicated in Figure 2.14. From the end closest to the prime power, these are the low-voltage capacitive energy store, the integrated Tesla transformer/pulse-forming line (PFL), and the tapered output transmission line. A layout of a SINUS machine is shown in Figure 2.18. The low-voltage capacitive store charges up at the output voltage of the power conditioning end of the prime power subsystem. The Tesla transformer is integrated with a cylindrical PFL. A single-turn primary is wound around the inner wall of the outer conductor, while the secondary crosses the gap between the coaxial conductors in the PFL, so that its high-voltage end contacts the center conductor, driving it to a large negative voltage. The Tesla transformer steps up the voltage from the capacitive store to the required output voltage, and the PFL shapes the output pulse and defines the duration of the voltage pulse at the output.

Unfortunately, for reasons related to the volume efficiency of storing electrical energy, the output impedance of the oil-insulated PFL itself, Z_{PFL}, tends

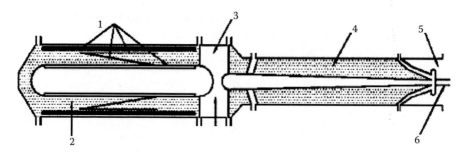

FIGURE 2.18
Physical configuration of SINUS pulsed power devices: 1, pulsed transformer; 2, coaxial oil-insulated PFL; 3, gas spark gap; 4, output tapered transmission line; 5, vacuum region; 6, cathode for beam generation, driving BWO. (From Rostov, V.V. With permission.)

to lie between 20 and 50 Ω (Section 5.2). The role of the tapered transmission line is to transform between the output impedance of the PFL and the impedance of the BWO, $Z_{BWO} = 150\ \Omega$. The length of the tapered transmission line is directly proportional to the length of the electron beam pulse, τ_{BEAM}:

$$L_{TL} = \frac{3}{2} c' \tau_{BEAM} \tag{2.12a}$$

Because we expect that the transmission line will be filled with insulating oil, the speed of light in the line is reduced to $c' = 2 \times 10^8$ m/sec, so that in units appropriate to the problem,

$$L_{TL}(m) = 0.3\tau_{BEAM}(nsec) \tag{2.12b}$$

Presuming that we wish to align the transmission line with the Tesla transformer/PFL, as was the case for NAGIRA in Figure 2.16, this length must be added to that of the PFL when one is determining the overall length of the platform that would carry this system. The length of the Tesla transformer/PFL, which will also be filled with insulating oil, is approximately the length of the PFL itself, which for a pulse length of τ_{BEAM} is given by

$$L_{PFL} \approx \frac{1}{2} c' \tau_{BEAM} \tag{2.13a}$$

or

$$L_{PFL}(m) \approx 0.1\tau_{BEAM}(nsec) \tag{2.13b}$$

Adding Equations 2.12b and 2.13b, we get

$$L_{TL}(m) + L_{PFL}(m) = 0.4\tau_{BEAM}(nsec) \tag{2.14}$$

From this relation, we can see that for a 10-nsec pulse, the length is about 4 m, while for a 20-nsec pulse, the length doubles to about 8 m. Because 8 m may be too long for some containers, we must consider the options with and without the transmission line, evaluating the trade-off between the added length of the tapered transmission line and the inefficiency that results if it is not used.

To treat the power flow, in Figure 2.11 we rename the voltage source and its output impedance V_{PFL} and Z_{PFL}. The beam voltage across Z_{BWO} and beam power through Z_{BWO} are

$$V_0 = V_{PFL}\left(\frac{Z_{BWO}}{Z_{PFL}+Z_{BWO}}\right) \tag{2.15}$$

$$P_{BEAM} = \frac{V_0^2}{Z_{BWO}} = V_{PFL}^2\,\frac{Z_{BWO}}{\left(Z_{PFL}+Z_{BWO}\right)^2} \tag{2.16}$$

The maximum value of P_{BEAM} for a given PFL output impedance is given by Equation 2.16 when $Z_{BWO} = Z_{PFL}$:

$$P_{BEAM,max} = \frac{V_{PFL}^2}{4Z_{PFL}} \tag{2.17}$$

We define the pulsed-power efficiency to be the fraction of the maximum available pulsed-power output that goes into the beam:

$$P_{BEAM} = \eta_{PP} P_{BEAM,max} \tag{2.18}$$

If a tapered transmission line is used, Z_{BWO} will appear to the PFL to be approximately equal to Z_{PFL}, no matter what its actual value, because the line is an impedance transformer. The line is not perfectly efficient, and our rule of thumb is that with tapering

$$\eta_{PP} = \eta_{TL} = 0.85 \tag{2.19}$$

If the tapering is not used, perhaps because it is too long for a given platform, we combine Equations 2.16, 2.17, and 2.18 to get

$$\eta_{PP} = \frac{P_{BEAM}}{P_{BEAM,max}} = \frac{4Z_{BWO}Z_{PFL}}{\left(Z_{PFL}+Z_{BWO}\right)^2} \tag{2.20}$$

The decision to use the tapered line or not will depend on overall system length. Two cases will arise; we will:

- Use the tapered line if it doesn't make the system too long for the platform. It will have optimum impedance from the standpoint of minimizing the volume in which its energy is stored, so that $Z_{PFL} = 20\ \Omega$ and $\eta_{PP} = 0.85$.
- Not use it if it makes the system too long. The impedance will be higher, $Z_{PFL} = 50\ \Omega$; from Equation 2.21b, $\eta_{PP} = 0.75$ for BWO impedance $Z_{BWO} = 150\ \Omega$.

The SINUS is a well-engineered pulsed power system, so about 90% of the energy stored in the PFL is available to deliver power to the beam, the remainder being lost to rise and fall times and other intrinsic losses. When the tapered line is used, Z_{BWO} appears to have the value Z_{PFL}, no matter what its actual value. With and without the tapered line, P_{PFL}, the power flowing out of the voltage source V_{PFL} into the series impedances Z_{PFL} and Z_{BWO}, is given by the two expressions (with $V_0 = 747$ kV, $Z_{PFL} = 20$ or $50\ \Omega$, depending on whether the tapered line is used, and $Z_{BWO} = 150\ \Omega$):

With the tapered line ($Z_{PFL} = 20\ \Omega$)

$$P_{PFL} = \frac{V_{PFL}^2}{2Z_{PFL}} = 2\frac{P_{BEAM}}{\eta_{PP}} = 8.75\ \text{GW} \tag{2.21}$$

Without the tapered line ($Z_{PFL} = 50\ \Omega$)

$$P_{PFL} = \frac{V_{PFL}^2}{Z_{PFL} + Z_{BWO}} = \frac{V_0^2}{Z_{BWO}}\left(\frac{Z_{PFL} + Z_{BWO}}{Z_{BWO}}\right) = 4.96\ \text{GW} \tag{2.22}$$

In either case, assuming a 5-nsec rise time for the microwaves, $\tau_{BEAM} = \tau_{RF} + 5$ ns, the energy stored in the PFL must be

With the tapered line ($Z_{PFL} = 20\ \Omega$)

$$E_{PFL}\left(J\right) = \frac{P_{PFL}\tau_{BEAM}}{0.90} = 9.72\tau_{RF}\left(ns\right) + 48.6 \tag{2.23a}$$

Without the tapered line ($Z_{PFL} = 50\ \Omega$)

$$E_{PFL}\left(J\right) = \frac{P_{PFL}\tau_{BEAM}}{0.90} = 4.69\tau_{RF}\left(ns\right) + 23.4 \tag{2.23b}$$

To find the values of V_{PFL} are

With the tapered line ($Z_{PFL} = 20\ \Omega$)

$$V_{PFL} = 2\left(\frac{P_{BEAM}Z_{PFL}}{\eta_{PP}}\right)^{1/2} = 592\ \text{kV} \tag{2.24a}$$

Without the tapered line ($Z_{PFL} = 50\ \Omega$)

$$V_{PFL} = V_0 \left(\frac{Z_{PFL} + Z_{BWO}}{Z_{BWO}} \right) = 966 \text{ kV} \tag{2.24b}$$

The energy in the fast energy store, E_{FAST}, is transferred to the PFL with 90% energy efficiency:

With the tapered line (Z_{PFL} = 20 Ω)

$$E_{FAST}(J) = \frac{E_{PFL}(J)}{0.90} = 10.8\tau_{RF}(ns) + 54.0 \tag{2.25a}$$

Input power required to recharge the capacitive energy store between shots is approximately given by

With the tapered line (Z_{PFL} = 20 Ω)

$$P_{FAST} \approx E_{FAST} PRR = PRR\left[10.8\tau_{RF}(ns) + 54.0\right] \tag{2.25b}$$

A comparison of the parameters required for SuperSystem with the ranges for the SINUS series is given in Table 2.2. One can see that the required input power to charge the capacitive energy store of the SINUS generator is about 53% above the level seen in previous versions of this device.

Moving upstream, we come to the burst-mode prime power options of Figure 2.7. Starting with the converters at the right of the figure, we note that AC/DC converters (transformer-rectifiers) and DC/DC converters (essentially switching power supplies) have typical power transfer efficiencies of 0.88. Therefore, the input power into the controllers is

$$P_C = \frac{P_{FAST}}{\eta_C} = \frac{P_{FAST}}{0.88} \tag{2.26}$$

Note that this is the power that must be transferred through either the AC/DC or DC/DC converters, depending on the option chosen, to recharge the capacitive store for each shot. The energy per shot that must be transferred into the controller is

$$E_C = \frac{P_C}{PRR} \tag{2.27}$$

At this point, we make the decision that the output voltage for the controllers, V_C, is the default input voltage to the SINUS generator: 300 V.

TABLE 2.2

Comparison of the SuperSystem Parameters to Those of the SINUS Series of Pulsed Power Machines

Parameter	SuperSystem Value	SINUS Range
Average input power, P_{FAST}	40.5–76.5 kW	0.1–50 kW
Output voltage, Φ_{BEAM}	747 kV	100–2000 kV
Output pulse width, τ_{BEAM}	15–25 nsec	3–50 nsec
Output impedance, Z_{BWO}	150 Ω	20–150 Ω
Pulse repetition rate, PRR	500 Hz	10–1000 Hz

TABLE 2.3

Eight Parameters Describing the Burst Mode of Operation

Mission Parameters		Burst Store Parameters	
PRR	Pulse repetition rate	E_S	Energy stored in the energy store and electrical source component to deliver one microwave pulse
N_S	Number of shots in a burst	E_B	Energy stored in the energy store and electrical source component to deliver one burst
τ_B	Time duration of one burst	P_{AVE}	Average output power delivered by the energy store and electrical source component during a burst
τ_{IB}	Interburst (recharge) interval	P_{RC}	Average power delivered into the energy store and electrical source component during the interburst recharge time

As we move upstream into the prime power system options captured in Figure 2.7, we note that burst-mode prime power options are defined by eight parameters, four *mission parameters* and four *burst store parameters*, which are shown in Table 2.3; we rename some of the parameters here for convenience in making them consistent with our SuperSystem model. Of the four mission parameters shown in the table, we are familiar with three already; the fourth, the number of shots in a burst, N_S, is given by

$$N_S = PRR \times \tau_B \tag{2.28}$$

Given the range of possible user choices for τ_B, N_S can range from 500 to 2500 shots per burst. The burst store parameters relate to stored energies and power flows within the prime power. The energy per shot, E_S, is defined as the amount of energy that must be stored in the energy store and electrical source component to ultimately deliver one 500-MW radiated microwave pulse with a pulse length of τ_{RF} at an efficiency η_{ESES}. Thus,

$$E_S = \frac{E_C}{\eta_{ESES}} \tag{2.29}$$

TABLE 2.4

Power Transfer Efficiencies of the Different
Energy Store and Electrical Source Options
of Figure 2.7

Energy Store and Electrical Source Option	h_{ESES}
Flywheel/alternator	0.96
Pulsed alternator	0.96
High-rate secondary battery	0.50

The efficiency η_{ESES} varies as shown in Table 2.4 for the three energy store and electrical source options of Figure 2.7. The energy per burst, E_B, is the amount of energy extracted from the energy store and electrical source component at an efficiency η_{ESES} in the course of a full burst:

$$E_B = N_S E_S \tag{2.30}$$

The average power, P_{AVE}, is defined as the average power flow *into* the converter component from the energy store and electrical source component over the course of a burst; in fact,

$$P_{AVE} = P_C = \eta_{ESES} E_S PRR = \eta_{ESES} \frac{E_B}{\tau_B} \tag{2.31}$$

The last step in this equation involved the use of Equations 2.23 and 2.25. Now, the average recharging power, P_{RC}, is the power that flows into the energy store and electrical source component to recharge it during the inter-burst recharging period. Thus,

$$P_{RC} = \frac{E_B}{\tau_{IB}} = \left(\frac{\tau_B}{\tau_{IB}} \right) \frac{P_{AVE}}{\eta_{ESES}} \tag{2.32}$$

The power from the prime mover components required to supply this power must pass through the interface components. From the literature, the efficiency with which this power passes through, η_I, varies with each interface component option, as shown in Table 2.5. Thus,

$$P_{PM} = \frac{P_{RC}}{\eta_I} \tag{2.33}$$

Estimation of the volumes and weights of the subsystems and the total SuperSystem can then proceed. Figure 2.19 shows a concept of the subsystem

TABLE 2.5

Efficiencies for the Different Interface Options
of Figure 2.7

Interface Option	Efficiency, η_I
Hydraulic pump motor controller	0.80
Electric motor controller	1.00
AC/DC converter	0.88

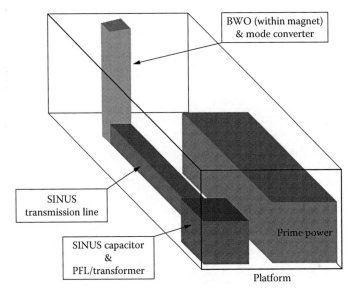

FIGURE 2.19
Internal configuration of the SuperSystem.

layout based on the layout of NAGIRA (Figure 2.16). For a higher power
system, see Problem 2.2.

2.6 Conclusion

As an art, not a science, only experience with choices and detailed analysis
can make systems design and the underlying issues clear. It is a school of
hard knocks, as optimistic initial assumptions are frequently followed by
detailed considerations that reduce performance. Nevertheless, it is the only
way to move forward to allow the building of advanced HPM systems that
the applications described in the next chapter need.

Problems

1. For SuperSystem, at a range of 10 km what is the power density and energy density for a single pulse?

2. Design a more advanced SuperSystem with increased peak power, longer pulse duration, but lower repetition rate and microwave frequency: 5 vs. 0.5 GW, 100 vs. 20 nsec, 100 vs. 500 Hz, 3 vs. 10 GHz. Keep the burst time and interburst time fixed at 1 and 10 sec. Keep the antenna gain fixed at 40 dB, to keep the diameter reasonable.

References

1. Butler, C.J., Benford, J., and Sitkin, E., Heimdall: An Expert System Code for Modelling RF Weapon Sources and Systems, paper presented at the Proceedings of the Seventh National Conference on High Power Microwave Technology, 1994.

2. Balanis, C.A., *Antenna Theory*, John Wiley & Sons, New York, 1997, p. 73.

3. Clunie, D. et al., The design, construction and testing of an experimental high power, short-pulse radar, in *Strong Microwaves in Plasmas*, Litvak, A.G., Ed., Novgorod University Press, Nizhny Novgorod, Russia, 1997.

4. Humphries, S., Jr., *Principles of Charged Particle Acceleration*, John Wiley & Sons, New York, 1986, p. 249.

3

High Power Microwave Applications

3.1 Introduction

There is a strong, symbiotic relationship between a developing technology and its applications. New technologies, or dramatic increases in the performance of an existing technology, can generate applications previously either unrealizable or impractical. Conversely, the demands posed by an existing application can spur the development of a new technological capability. Examples of both types can be found in the evolution of high power microwaves (HPM). The high power and energy output made possible by HPM research have created a technology-driven interest in HPM directed energy weapons. On the other hand, the requirements for electron cyclotron resonance heating of fusion plasmas have resulted in an application-driven program to develop high-frequency, high-average-power microwave sources. In this chapter, we will address the present state of such applications.

Compared to the first edition 15 years ago, some applications have grown, some have declined, and some are no more (for example, laser pumping using microwaves has disappeared). The most successful is HPM directed energy weapons, now at the point of deployment. Funding over the past few decades has been highest for defense-oriented work. Power beaming research is vibrant, as is microwave-driven plasma heating. Plasma heating is perhaps the longest studied application. High power radar is currently inactive, except for short-pulse mesoband ultrawideband. Following the rejection of high power klystrons for the International Linear Collider, particle acceleration has receded toward the distant horizon and laser pumping using microwaves has disappeared.

3.2 High Power Microwave Weapons

A time in the near future ...

We came in low toward the target. It was a black night and for now there was no sign that they had detected us as we came in from the sea. There were four other pilots making the same attack plan against nearby nodes at the same time. One of them would get painted for sure, and then the whole air defense system would light up and they'd start finding us. Best to get on with it.

We had a couple of these new model HPM systems hanging off hard points under the wing, and we were going to try using them tonight. We'd tried the first model last year and it didn't work out. Too much trouble to use out here in real life.

So I activated the HPM icon on the HUD* display and looked over the settings menu. Last time it'd been as long as my arm. This time they had a new interface and claimed it was "smart." We'd see. It sure was simpler.

Last time we'd had to set everything on it. First, we'd use the emissions detection gear to tell which systems they were using. These days everything is for sale, so it could be Russian, French, even ours.

So we had to figure out from its comparison library what we had looking at us. Plus what they were using for comm. Then we'd send that to the HPM system. Then we'd have to tell it what frequencies it should produce, how fast to shoot, all sorts of stuff. Stuff we didn't have time for.

The slow part was us, mere slo-mo humans.

Now it was different. I just picked the target set. The HPM's new operating mode identified what emissions to listen for. Then it looked up electronic vulnerability tables based on real field tests, tests to burnout, of assets captured in some Third World scrap or just bought on the international arms market. Then it set its own parameters: repetition frequency, burst length, and things I hadn't a clue about, like polarization.

My navigator said the GPS had us 10 miles out from the target area, so I took her down to 100 yards and we began the approach. Then they found us.

First the RWR† sound started up and our ECM‡ cut in, singing and flashing direction lights. A fiery stream of antiaircraft shells snaked toward us, looking like red bubbles in black champagne. We evaded. The ground lit up like a Christmas tree. Then they launched a surface-to-air *missile* somewhere and I got a big red "missile" on the HUD. The new HPM system had a self-protect mode, so I clicked that icon and it started an omni broadcast around us. And I jammed the stick hard over right, jinking to evade. But I needn't have bothered. The HPM caused the brains

* HUD: Heads-up display.
† RWR: Radar warning receiver.
‡ ECM: Electronic countermeasure.

of the thing to lose interest in us and it slammed into the sand below. But the shells still streamed through the night; we had to get on with it.

The HPM weapon was mounted as an expendable, so we just had to get it close enough to launch it. So I set up to release it, first presetting the "release" icon. By then the bird had listened to the electronic traffic around us and knew what to go after. I squeezed off the release. The HPM bird fell from the rack. We felt an upward jolt as the plane lightened. The "release complete" indicator on the display flashed and then began reading the telemetry from the launched bird. I jinked hard left and glanced outside. I could just barely see its low-observable engine plume as it snaked to the right after its first target. Before long I heard in my earphones its electronic chatter as it fired a burst of pulses into the first concentration it found. It would adaptively adjust all those parameters until it reached a set the tables said would put each target in the node out of action. Then it moved on. It would go ahead, burning out their electronic eyes and ears, clearing a path through the primary nodes of their defense, bringing their system down. Then the fighter-bombers could get through to the big targets.

As we set off on a course to drop the next smart bird, I remembered the first time we used HPM. We had to set up a slew of parameters for the specific target and then fire some test shots to make sure we were getting the right power and such. Now we just let her go and she tuned her parameters as she went. The guys who make these things have finally got the true meaning of "fire and forget" and "man out of loop." We don't have time to fiddle with their gadget when there's live fire around us!

Science fiction? A realistic projection? This section introduces you to the main proposed application for high power microwaves: electromagnetic nonlethal weapons.

Weapons that direct energy instead of matter on targets have undergone extensive research in the last two decades. They have two potential advantages over existing weapon systems. First, they use a power supply rather than a magazine of explosive munitions; this "deep magazine" is unlikely to be expended in battle. Second, they attack at the speed of light, 160,000 times faster than a bullet, thus making avoidance of the incoming bolt impossible and negating the advantage of increasingly swift tactical missiles.

Directed energy weapons (DEWs) generally fall into three categories: lasers, microwave or radio frequency (RF) energy weapons, and charged particle beam weapons. HPM has an advantage over the other DEWs in that microwaves do not face a serious propagation issue. Particle beams and lasers have difficulties in propagating through the atmosphere, and electron beams cannot propagate in space. Moreover, both are pinpoint weapons with small spot sizes requiring precise pointing to hit the target. Antenna directed microwaves, on the other hand, spread through diffraction and have spot sizes large enough to accommodate some lack of precision in pointing and tracking. Lasers and particle beams are also much less electrically efficient, more complex, and therefore more costly.

3.2.1 General Aspects of High Power Microwave Weapons

One of two outcomes is expected of a DEW attack: soft kill, in which mission-critical components are disabled while the target body remains largely undamaged, and hard kill, with large-scale physical destruction of a target. The military prefers a hard-kill capability, "smoking rubble," or at least an assurance that an adversary cannot circumvent the weapon effect; however, most DEW scenarios for HPM are soft-kill missions. This is similar to radar jamming and electronic warfare, which can be predicted with confidence, but not directly verified.

The concept of using powerful microwave pulses as a weapon dates at least as far back as British radar studies during World War II. The idea has had several reincarnations, including early ideas of thermal and structural damage, which require enormous powers. Now both antielectronics and nonlethal antipersonnel HPM weapons are deployed.

In the 1980s the prospect became more credible because of two converging technology developments: (1) the development of sources capable of producing peak powers in excess of a gigawatt and (2) the increasing miniaturization of, and dependence on, electronic components in military and consumer electronics. The small scale of today's electronic components makes them vulnerable to small amounts of microwave energy — thus the emergence of the "chip gun" and "E-bomb" concepts, a transmitter designed to upset or burn out integrated circuits in the electronic brains of modern systems.[4] The continuous trend toward miniaturization and lower operating voltages has made HPM weapons more attractive. Vulnerability has increased because the recent use of unhardened commercial equipment at lower cost has meant replacement of metal packaging by plastic and composite materials.

Some virtues of electronic attacks with HPM are:

- Electronic attacks produce little or no collateral damage, nonlethal to humans.
- Most defense systems are not hardened, so to counter, an entire system must be hardened. An HPM weapon made effective against a deployed system requires modification of a large inventory in order for confidence to be restored.
- Entry can be by front door or back door.
- Area attacks, with many targets, are possible.
- Little sensitivity to atmospheric conditions such as fog and rain.
- Few legal barriers to their development or use.
- Cost little relative to conventional munitions per target.

- Repair requires high level of expertise, so probably cannot be done at the site.
- Can provide additional rungs on an escalation ladder, expanding the range of options available to decision makers.

A limitation of electromagnetic weapons is the difficulty of kill assessment. Absence of emissions does not necessarily mean that an attack has been successful. Targets successfully attacked may appear to still operate. Radiating targets such as radars or communications equipment may continue to radiate after an HPM attack, even if their receivers and data processing have been damaged or destroyed. A deceptive response for a system coming under attack is to shut down.

In descending order of required power density on target, the hierarchy of HPM directed energy effects is the following:

1. *Burnout*: Physical damage to electronic systems.
2. *Upset*: Temporary disruption of memory or logics in electronic systems.
3. *Jamming*: Blinding of microwave or RF receivers or radar.
4. *Deception*: Spoofing of the system into mission failure.

This last category is akin to *electronic warfare* (EW) at higher power levels, and therefore at much greater effective ranges. The third level in the above list is an overpowering of enemy systems in the same manner as contemporary battlefield jamming systems, but HPM-based super-jammers would be capable of totally dominating the battlefield, allowing no chance for "burn-through" of their jamming signal. Between high power jammers and burnout devices lies a middle ground where electronics can be upset, e.g., lose information in digital systems so that a missile may become disoriented (break lock) or tactical communications confused. As the hierarchy above is ascended, the associated missions become increasingly generic; i.e., broader classes of targets can be attacked at the higher power levels, while electronic warfare techniques are very target specific.

The distinctions between HPM directed energy and electronic warfare can be viewed in terms of the trade between the sophistication and power level of an attack, as shown in Figure 3.1. Historically, the line between electronic warfare and HPM directed energy weapons has been fairly sharply drawn, at least in the West. On the one hand, electronic warfare uses sophisticated techniques at orders of magnitude lower power (~1 kW) to deny an opponent effective use of communications and weapon systems, while protecting one's own use of the electromagnetic spectrum. Electronic warfare has emerged as a vital element of military strategy and an effective way of neutralizing enemy forces and enhancing the power of friendly forces. It has also become extremely expensive in its increasing sophistication because of the diversity of threats and a continuing race between the techniques of generating *elec-*

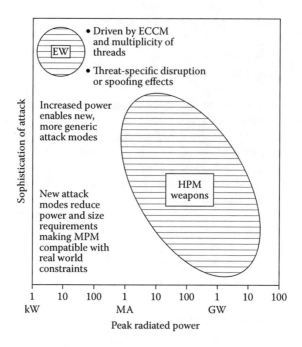

FIGURE 3.1
Domains and trends of HPM and electronic warfare.

tronic countermeasures (ECMs) and countering such interference with *electronic counter-countermeasures* (ECCMs). On the other hand, HPM DEWs, in their earliest embodiment, were seen as a means of attacking a multitude of targets, using simple pulses from a generic weapon at >1 GW to provide higher cost effectiveness than EW. Then efforts turned to developing intermediate alternatives, combining features of HPM and EW, known as *smart microwaves*, or medium power microwaves (MPM).[1] The emphasis is on employing more sophisticated waveforms at a power reduced relative to HPM. Whereas some applications of HPM have envisaged burnout of the target in a single pulse, smart microwave attacks would use repetitive pulsing or amplitude and frequency modulation and other forms of pulse shaping to lower the damage level of electronics.

Two types of attack modes against electronics have been proposed: point weapons and area weapons. In one, an HPM weapon fires intense pulses to disable a specific target at substantial range. A high power density is produced on that target by the weapon and upset or burnout is obtained. In the modern armory there are substantial numbers of target radars, semiactive homing missiles, and communications and control systems that could be vulnerable to such attacks. For example, several modern aircraft are marginally unstable and require sophisticated control electronics to fly. Burning out the chips of such an aircraft flight computer could destroy the aircraft itself. Civil aircraft use electronic systems for critical safety functions such as flight control, engine control, and cockpit display. Such systems may be susceptible to interference or damage from terrorists.

In the second class of HPM attack, large areas are swept with a radiating pulse in hopes of disabling a significant number of targets. For an area weapon to be effective, either the amount of radiated energy must be large or the threshold vulnerability level of the target must be very low. For example, the fluence to cause bit errors in unshielded computers is roughly 10^{-7} J/cm^2, so that an area weapon would be disruptive to many consumer electronic items. It is debatable that this is an effective military attack, but it could have substantial economic consequences. One method of delivering high fluences to large areas is the microwave bomb proposed in the 1980s for the Strategic Defense Initiative, in which a microwave pulse radiated in space from a nuclear explosion would damage electronic equipment over a large area.[2] Another example of an area weapon would be an RF system on a dedicated airframe, flying at low altitude and sweeping battlefield areas. Such an attack might blank out battle management systems and disable tactical communications.

The requirements of the services will differ substantially depending upon their mission and the platforms available. Since shipboard systems are larger, the Navy might be the first to use HPM because it would be less stringently limited by the size and weight of HPM systems. The threat to navies is clear: the sinking of the *Sheffield* and the near sinking of the *Stark* show that modern ships are particularly vulnerable to low-flying, sea-skimming cruise missiles. (Air defense destroyer HMS *Sheffield* burned to the waterline after being hit by a single Aerospatiale AM39 Exocet ASCM.) Such missiles have become cheaper, smarter, and longer in range, and are now being deployed in the Third World. Defense against a saturation attack by such missiles is disadvantageous since the fleet can carry only a limited number of missiles for defense.

Ground-based point attack HPM weapons would probably be mounted on large tracked vehicles with high gain antennas on a mast to provide targeting, such as the Ranets-E system shown in Figure 3.2. The size and weight requirements would be more restrictive than for the fleet, and the antenna sizes would be large to produce high directivity to avoid fratricide.

The air forces have a more difficult problem. Airborne systems need to be much smaller and operate at lower power levels because of limited on-board power. Airborne missions place severe requirements on antennas to be steerable (mechanically or by phase) and small but still avoid air breakdown around them. Aircraft reach their volume limitation long before their weight limitation, so long dimensions and size are the true issue. An example of this is the case of the forthcoming U.S. system UCAV. The HPM possibilities have been explored by L3 Communications, as shown in Figure 3.3 and Figure 3.4 below.[3]

Compactness is in fact a general problem of military applications of HPM. The size and weight limitations of military platforms require squeezing as much power into a specific volume as possible, and HPM systems to date

M R T I   &   R O S O B O R O N E X P O R T   P R E S E N T

RANETS-E Mobile Microwave Protection System

Purpose

A research and development project is proposed to create a mobile microwave protection system (MMPS) - Ranets-E.

The Ranets-E MMPS is intended for:

• evaluation of electromagnetic resistance of military electronic systems (stationary or moving) to high-power microwave radiation;

• microwave protection from high precision weapons.

FIGURE 3.2
Brochure for an air defense HPM weapon. Russian firm offers to co-develop a system with goals of 0.5 GW, 10 to 20 nsec, 500 Hz in X-band, with 45- to 50-dB gain antenna.

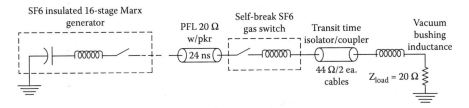

FIGURE 3.3
Circuit diagram of pulse power driving a magnetron. The layout is shown in Figure 3.4. (Reprinted from Price, D. et al., Compact pulsed power for directed energy weapons, *J. Directed Energy*, 1, 56, 59, 2003. By permission of DEPS.)

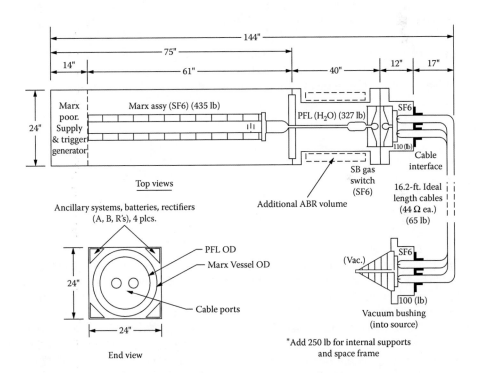

FIGURE 3.4
Top and end view of a compact pulse power system concept for mounting into the two pods of UCAV. Pulsed power (see Figure 3.3) fits into one pod and connects through cables to the second pod's vacuum bushing. Magnetron source is not shown. (Reprinted from Price, D. et al., Compact pulsed power for directed energy weapons, *J. Directed Energy*, 1, 56, 59, 2003. By permission of DEPS.)

have not been designed with this as a criterion, their purpose being HPM effects testing. In recent years, great strides have been made in compactness of capacitive energy stores, so that now the bulk of an HPM system is likely to be in other elements, such as insulating dielectrics and magnetic field coils. The specific microwave source used also has a big impact on size and weight. For example, free-electron lasers requiring a magnetic wiggler are frequently longer and heavier than other sources. The vircator and MILO (see Section 7.7), in contrast, require no magnetic field and are therefore quite simple and compact. Here efficiency comes into play. Low-efficiency sources, such as the vircator, require a large energy store to produce a given microwave power output, hence the emphasis in HPM circles on efficiency.

The relevant parameter is the *energy efficiency* of conversion from electricity to microwaves averaged over the entire electrical pulse. This is not the *power efficiency*, which can be high at some part of a pulse but low elsewhere. The difference is due to *pulse shortening*.[4] Let ε_p be the power efficiency, the ratio of instantaneous microwave power P_μ to beam power P_b, and ε_E be the energy efficiency, the ratio of microwave energy in the pulse E_μ to the pulsed electrical energy that drives the beam, E_b. For simplicity, assume powers are constant in time. Then, with t_μ the pulse duration of microwaves and t_b the pulse duration of the beam, the relations are

$$E_\mu = P_\mu t_\mu \tag{3.1}$$

$$E_b = P_b t_b \tag{3.2}$$

Therefore,

$$\varepsilon_E / \varepsilon_p = t_\mu / t_b \tag{3.3}$$

The pulse-shortening ratio t_μ/t_b is the ratio of the energy efficiency to the power efficiency. Present values are $\varepsilon_p \sim 0.2$ to 0.4, but pulse shortening makes $\varepsilon_E < 0.1$. The HPM system designer is concerned with the *energy* he must provide from prime power and store in pulsed power because energy drives HPM system volume and mass. Therefore, for designers, ε_E is more meaningful than ε_p. The distinction between them is the pulse-shortening ratio. For time-varying P_b and P_μ, one integrates over the power over the pulse length; details of P_μ, which tends to vary rapidly, can strongly influence results (see Problem 1).

Since most microwave devices operate with a resonance condition, which depends on current and voltage, and both typically vary during the pulse, a good deal of the electrical pulse may not be useful in producing microwaves. The solution is to satisfy the resonance condition (usually a stringent requirement on voltage constancy) during virtually the entire pulse, requiring sophisticated circuitry, which increases the cost if it can be done at all. Pulsed power technology that is particularly appropriate for compact HPM is the *linear induction accelerator,* also called a *voltage adder* (see Chapter 5), but its cost and weight are high.

A further practical limitation on military HPM is the constraint of on-board power. For example, if an airborne HPM device generates 1 GW for a microsecond, thus producing 1 kJ/shot, repetitive operation at 10 Hz will require an average power of 10 kW. Such powers are available on military aircraft, but this power is usually already allocated. Making room for an HPM weapon will require additional power or eliminating some other military system on board. Therefore HPM must show its utility to justify its power consumption. In this example, a higher repetition rate, for example, a kilohertz, would be beyond the available power on an airframe. Power constraints are not as demanding on ground vehicles or on Navy vessels.

There are tactical issues associated with HPM DEWs. The disadvantages inherent in an antenna directed weapon are sidelobes and strong local fields (see Chapter 5). These produce potential problems for friendly forces, which, in the HPM community, are termed *fratricide* and *suicide.* Fratricide is unintended damage to nearby electronics or personnel due to sidelobe emission. In the "fog of war" the potential for damage to friendly forces near at hand could be a serious limitation. Since the modern battlefield is so dependent upon electronics, damage or interference with electronic systems such as the extensive electromagnetic environment of fleet defense may prevent HPM from being deployed. The problem may be mitigated somewhat by sidelobe suppression (see Chapter 5). One solution is to operate HPM only in an area empty of friendly systems, such as behind enemy lines. The second problem, suicide, is unintended damage to the subsystems of the HPM platform itself due to its own emitted pulse. This problem may be soluble by shielding or shutting down subsystems while the HPM is operating.

Because they are high power radiators, HPM weapons will have their greatest potential vulnerability to antiradiation missiles (ARMs), which home in on microwave signals. ARMs are the natural enemies of HPM; therefore, HPM systems must be able to attack ARMs if they are to survive. Potential attack mechanisms include interference with the guidance system circuits, to make them break lock, or predetonation of the warhead.

3.2.2 E-Bombs

The use of HPM warheads on precision munitions is an attractive coupling of electronic attacks with precision guided munitions (PGM): accurate missiles, glidebombs, and unmanned aerial vehicles (UAVs). Carlo Kopp, who coined the term *E-bomb* in 1995, when the U.S. Air Force originally published his work, envisioned combining a smart bomb with a HPM warhead.[5]

The massed application of such electromagnetic weapons in the opening phase of an electronic battle can quickly command the electromagnetic spectrum. This would mean a major shift from physically lethal to electronically lethal attacks. Potential platforms for such weapons are the USAF-deployed

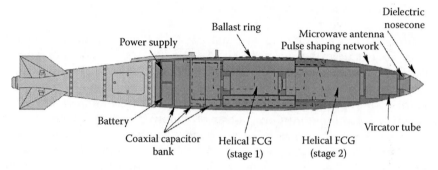

Mk.84 900 kg 3.84 m × 0.46 m dia

FIGURE 3.5

E-bomb concept, with two-stage flux compressor driving a vircator, fits into envelope of Mark 84 bomb. (Reprinted by permission of Dr. Carlo Kopp.)

global positioning system (GPS) aided munition on the B-2 bomber and the GPS/inertially guided GBU-29/30 JDAM (Joint Direct Attack Munition) and the AGM-154 JSOW (Joint Stand Off Weapon) glidebomb. Other countries are also developing this technology. The attractiveness of glidebombs delivering HPM warheads is that the weapon can be released from outside the effective radius of target air defenses, minimizing the risk to the launch aircraft, which can stay clear of the bomb's electromagnetic (EM) effects. In principle, all aircraft that can deliver a standard guided munition become potential delivery vehicles for an E-bomb. But, E-bombs must fit onto existing weapon platforms. Failure to do so, by requiring a dedicated new platform such as a specialized UAV, would make E-bombs unattractive due to cost and logistics. Fitting onto existing platforms can be difficult, as Figure 3.4 shows. This is a conceptual design, by L-3 Communications, of an HPM system fitted to the UCAV unmanned aerial vehicle. The UAV has two payload pods, so the challenge is to divide the system onto two portions, yet have efficient operation. The prime and pulsed power are in the first pod, with cables connecting to the other pod containing the vacuum interface, followed by the relativistic magnetron source and antenna, which are not shown. The circuit, in Figure 3.4, is Marx/pulse-forming line (PFL)/magnetron. Another approach by Kopp is use of a single-shot vircator driven by a flux compressor,[5] as shown in Figure 3.5 (see Chapter 5). Kopp also originated the idea of using circular antenna polarization to improve HPM power coupling into targets.

3.2.3 First-Generation High Power Microwave Weapons

HPM DEWs are not a panacea, a silver bullet, and must be used in missions where it is uniquely qualified. Below are several applications that are actually being developed.

3.2.3.1   Active Denial

Beamed microwave energy can inflict intense pain without actually injuring the people against whom it is directed.[6] This *pain gun* concept uses a continuous-wave beam at a frequency near 94 GHz. The radiation passes almost unattenuated through the atmosphere to be absorbed in the outer layer of a target individual's skin. The energy is deposited near the nerve endings, where water in the skin absorbs the radiation, creating a sensation like that of touching a flame. This less than lethal weapon, as it has been classified by the U.S. Pentagon, can be used to disperse an angry mob — which may offer cover for a more dangerous terrorist — or prevent unauthorized individuals from entering a prohibited area.

An Active Denial System (ADS) has been developed by Air Force researchers and built by Raytheon. It fires a 95-GHz continuous microwave beam produced by a gyrotron (see Chapter 10). The ADS weapon's beam causes pain within 2 to 3 sec, which becomes intolerable after less than 5 sec. People's involuntary reflex responses force them to move out of the beam before their skin can be burned. It exploits a natural defense mechanism that helps to protect the human body from damage. The heat-induced pain is not accompanied by actual burning, because of the shallow penetration of the beam, about 0.4 mm, and the low levels of energy used. Tests show that the effect ends the moment a person is out of the beam, and no lasting damage is done as long as the person exits the beam within a certain, rather long, duration. The transmitter needs only to be on for a few seconds to cause the sensation. The range of the beam is >750 m, farther than small arms fire, so an attacker could be repelled before he could pull a trigger (see Problem 2).

The system is compact enough to be fielded (Figure 3.6) and is dominated by the hardware for generating the high average power. Since gyrotrons are at best 50% efficient in conversion of electricity into microwaves, a 100-kW beam requires at least 200 kW of electron beam energy; inefficiencies in the power converter from the alternating current prime power generator to the direct current electron beam accelerator increase the prime power requirement to a somewhat greater number. Because the operating range would need to be much larger, making ADS airborne would require serious mass reductions; kW/kg would have to drop to make the higher-power system manageable in size and mass.

The antenna is the innovative Flat Parabolic Surface (FLAPS), which seems to be a contradiction, but it is in fact possible to design a geometrically flat surface to behave electromagnetically as though it were a parabolic reflector.[7] A FLAPS can directly replace a parabolic dish. It has advantages such as being easy to store and offering less wind resistance. The FLAPS consists of an array of dipole scatterers. The dipole scatterer unit consists of dipoles positioned approximately 1/8 wavelength above a ground plane. In Figure 3.7, a crossed shorted dipole configuration is shown; each dipole controls its corresponding polarization. Incident RF energy causes a stand-

FIGURE 3.6
Active Denial System, a nonlethal microwave weapon.

ing wave to be set up between the dipole and the ground plane. The dipole reactance is a function of its length and thickness. The combination of standing wave and dipole reactance causes the incident RF to be reradiated with a phase shift, which can be controlled by a variation of the dipole's length, thickness, its distance from the ground plane, the dielectric constant of the intervening layer, and the angle of the incident RF energy and adjacent dipoles. Typically, the dipole lengths vary to achieve a full 360° range of phase shifts.

3.2.3.2 Neutralizing Improvised Explosive Devices

In the War on Terror, a new equalizer has appeared, the improvised explosive device (IED). Made from plastic or other explosive and triggered frequently by cell phone, pager, or other radio command, IEDs are cobbled together from whatever the people that plant them can find, so there is no magic bullet that will universally apply to the IED threat. One hope is that a pulse of electromagnetic energy can fry the circuits, and there are several methods that might apply, from jamming the cell phone to setting off the fuse remotely. The NIRF system (neutralizing IEDs with RF) pro-

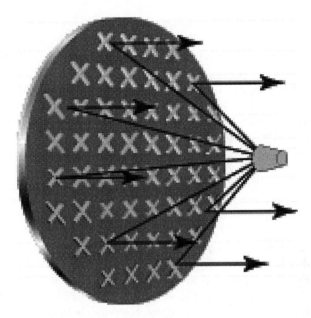

FIGURE 3.7
FLAPS planer antenna with crossed dipoles. Although flat, it converts a quasi-spherical wave from the feed into a focused beam, much as a parabolic dish would. (Reprinted by permission of Malibu Research, Camarillo, CA.)

duces high power, in the microwave range, at very short range to attack an IED's electronics.

3.2.3.3 Jamming or Predetonating Proximity-Fused Munitions

Proximity fuses in artillery or mortar shells use a radar signal to determine their range to a target so that the shell can be detonated at a preselected distance to the target, without actually having to hit it. A countermeasure is to silence the rounds by jamming or predetonation. The U.S. military uses *Warlock Green* and *Warlock Red* for this task, which are descended from an earlier antiartillery predetonation weapon system called the *Shortstop Electronic Protection System* (SEPS). It was introduced as a proximity fuse countermeasure for prematurely detonating incoming artillery and mortar rounds by receiving the fuse signal, modifying it, and sending a reply that makes the fuse think it is close to the ground. These weapons, already in the U.S. inventory, essentially instruct the target shell to detonate. Packaged in a suitcase-sized case and fitted with a small multidirectional antenna (in a way strikingly similar to that of the ultrawideband [UWB] short-range weapon DS100 described in Chapter 6), the Shortstop system can be activated and operational within seconds. Shortstop's passive electronics and operational features make it resistant to detection by enemy signal intelligence sensors. Shortstop weighs about 10 kg; versions designed as backpacks and vehicle-mounted units are on the near horizon.

3.2.3.4   *Vigilant Eagle*

Another credible mission moving toward deployment is Vigilant Eagle, a ground-based high-average-power microwave system providing protection of aircraft from shoulder-launched surface-to-air missiles (SAMs). The system is a network of sensors and weapons installed around an airport instead of on individual planes.[8] That is advantageous because planes are most vulnerable to such missiles at takeoff and landing because of the short range of the missiles. The system's estimated cost is much less than a system mounted directly on many potential target aircraft.

Vigilant Eagle has three interconnected subsystems: a distributed missile warning system (MWS), a command and control computer, and the high power amplifier-transmitter (HAT), which is a billboard-sized electronically steered array of highly efficient antennas linked to thousands of solid-state modular microwave amplifiers driving an array antenna. The MWS is a prepositioned grid of passive infrared sensors, mounted on towers or buildings much like cell phone towers (Figure 3.8). Missile detection by at least two sensors gives its position and launch point. The sensors track a heat-seeking SAM and trigger firing of HAT. The control system sends pointing commands to the HAT and can notify security forces to engage the terrorists

FIGURE 3.8
Vigilant Eagle (VE), a ground-based high-average-power microwave system protecting aircraft from shoulder-launched surface-to-air missiles. Missile detection by at least two sensors from a grid of passive infrared sensors on towers gives position and launch point. VE radiates a tailored electromagnetic waveform to confuse missile targeting. (Reprinted by permission of Raytheon Company, Tucson, AZ.)

who fired the missile. The HAT radiates a tailored electromagnetic waveform to confuse the targeting of the missile and can damage the targeting electronics, so that the missile loses track of the aircraft and deflects away from it. The beam is about 1° wide, with spot size much larger than the missile, so it has easier pointing requirements than a laser weapon would. EM fields are within safety standards for human exposure limits (Chapter 5) (see Problem 3).

3.2.4 Missions

The host of DEW missions proposed for HPM cover a broad parameter space. They vary from short-range missions, such as fighter self-defense against SAMs, to longer-range missions, such as the defense of fleets against missiles and the direct attack of ground systems from the air. The longest-range missions are antisatellite (ASAT; a constellation of satellites with HPM weapons with the capability of disrupting, disabling, or destroying wide varieties of electronics) and its inverse, the attack of ground systems from orbit. Most ASAT devices have been kinetic energy (KE) warheads, meaning that they collide with or explode near satellites. However, KE weapons would add to the growing problem of debris in orbit, so they are unlikely to be deployed by the U.S.[9] HPM, however, is an electronic kill and has no such deployment drawback.

In order to assess the practicality of such missions, engagement analysis must be performed, a process aided by the nomograph in Figure 3.9. This analysis assumes a parabolic antenna with 100% efficiency; a typical practical value is 50 to 80%. The radiated power is twice that calculated from the product of power density and spot sizes because about half the beam falls outside the diffraction spot. There will be propagation loss at the higher frequencies and longer ranges (see Chapter 5). One starts by choosing any one of the variables and then drawing a rectangle by choosing the remaining variables, determining the final variable.

For example, in Figure 3.9, choosing 10-GHz operating frequency and a 10-m antenna diameter gives a beam width of 3.7 mrad. Following a horizontal line and choosing a target range of 10 km gives a spot 37 m wide. Following a vertical line and choosing to irradiate the spot at 100 W/cm², the power density chosen by Velikhov,[10] gives a radiated power of 15.6 GW. A horizontal line leads to a point beyond the state of the art for single-pulse HPM sources. (The shaded area in Figure 3.9 is based on Figure 1.3.) A source requirement within the existing art could be obtained, for example, by choosing a larger antenna or shorter range.

The nomograph could be used to argue in either of two ways: (1) Assuming a required power density to produce the desired effect on a target, one can deduce that the effective radiated power (ERP) has to exceed a critical value. This places the burden of HPM weaponry on the technologist to find a source that is powerful and yet compact enough to fit onto the proposed platform.

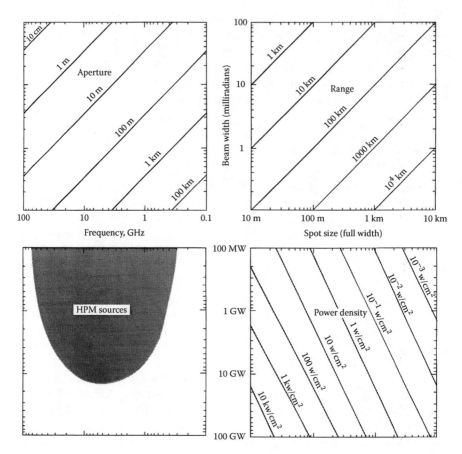

FIGURE 3.9
Nomograph for estimating mission parameters.

(2) Assuming that platform constraints limit ERP and antenna diameter, one deduces a power density that can be incident on a target at the required range. This puts the requirement on the effects (e.g., by tailoring waveforms or properly choosing frequencies) at the power density that technology can practically produce. This line of reasoning has led to the medium power microwave program described previously. This logic accepts the realities of technology and that new platforms are unlikely to be created exclusively for HPM, and places an emphasis on reducing the lethality threshold by increasing the sophistication of the attack.

Figure 3.10 shows another form of nomograph for DEW or power beaming.

3.2.5 Electromagnetic Terrorism

Complex systems such as government, industry, and the military cannot function without fast information flow. The West's data and communications

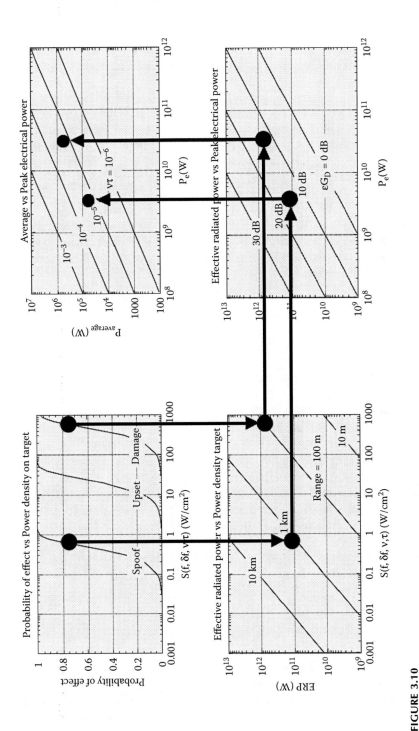

FIGURE 3.10

Another DEW nomograph. Choosing a high probability of effect for either spoofing or damage, and range for attack, gives ERP. Then choosing antenna gain gives peak power. Finally, choosing duty factor determines average power.

infrastructure will remain a soft electromagnetic target in the foreseeable future. That is because no one wants to pay for it. (*Soft kill* will inhibit or degrade the function of a target system leaving the target system electrically and physically intact. *Hard kill* will damage or destroy the target system.) This is a fundamental and growing EM infrastructure vulnerability; it can be exploited.

The technology is increasingly available and inexpensive. For example, see the DS110 in Figure 6.13. The vulnerability of infrastructure is suspected, but not widely known. Potential targets vary from small systems, such as individual automobiles, to the highly complex (aircraft on takeoff and landing, electrical power grid, oil and gas pipelines, banking, financial centers such as stock markets).

In Russia, former Soviet laboratories have suffered. Their extensive microwave and pulse power technology can be and is being sold to and copied by Third World nations or even terrorists. So far no counterproliferation regimes exist. Should treaties arise to limit proliferation of electromagnetic weapons, they would be virtually impossible to enforce, given the common availability of the technology.

3.2.6 Coupling

Whatever the mode of attack, microwave energy can propagate to the target system internal electronics through two generic types of coupling paths: *frontdoor coupling* and *backdoor coupling*.

- *Frontdoor* denotes coupling through intentional receptors for electromagnetic energy such as antennas and sensors; power flows through transmission lines designed for that purpose and terminates in a detector or receiver.

- *Backdoor* denotes coupling through apertures intended for other purposes or incidental to the construction of the target system. Backdoor coupling paths include seams, cracks, hatches, access panels, windows, doors, and unshielded or improperly shielded wires.

In general, the power coupled to internal circuitry P from an incident power density S is characterized by a coupling cross section S, with units of area:

$$P = S \sigma \tag{3.4}$$

For frontdoor paths, σ is usually the effective area of an aperture, such as an antenna or slot. The effective area peaks at the in-band frequency for the antenna and falls off sharply with frequency as roughly f^{-2} above the in-band frequency and f^4 below it. These are only very general relations, very dependent upon mismatch effects and construction details. Therefore, to gain entry

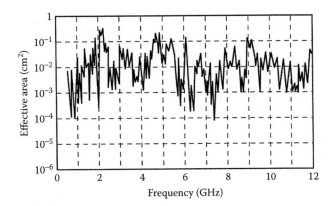

FIGURE 3.11
Typical cross section for coupling into the interior of an object through backdoor channels.

through an antenna, one wants to operate at the in-band frequency if it can be determined. The above discussion is for irradiation into the main lobe of the antenna. Radiating at random angles or in the sidelobes will reduce the coupled power substantially (see Problem 4).

Entering through a backdoor coupling path has many complexities. Figure 3.11 illustrates a general result: the rapid variation of coupling cross section as a function of frequency due to the overlapping sets of coupling paths. This occurs quite generally for backdoor attacks. Therefore, the cross section of coupling is difficult to predict for a specific object without detailed testing, although properties averaged over frequency bands can be predicted.

Microwaves directly couple into equipment through ventilation holes, gaps between panels, and poorly shielded interfaces because apertures into the equipment behave much like a slot in a microwave cavity, allowing microwaves to enter and excite the cavity. Electronic equipment cavity Q factors (see Equation 4.88) are in the range of Q ~ 10, so the time it takes to fill it with energy is about 10 cycles. This is one reason that mesoband pulses (see Chapter 6), which last only a few cycles, can be effective: their pulse durations allow cavities to fill, then end. A general approach for dealing with wiring- and cabling-related backdoor coupling is to determine a known lethal voltage level, and then use this to find the required field strength to generate this voltage. Once the field strength is known, the effective range for a weapon can be calculated.

One might think that one of the most attractive targets for HPM is radar. However, most radar sets are designed with *receiver protection devices* (RPDs), such as transmit/receive switches, to prevent near-field reflections of the radar transmitter from damaging the receiver. RPDs could also protect against HPM. The RPD triggers and strongly attenuates incoming signals that exceed the damage threshold of downstream elements, such as mixers. The typical RPD allows an early spike to propagate through the system and then shorts the remainder of the pulse. The duration and intensity of the spike leakage are crucial to being able to burn out the more sensitive diodes of the down-stream receiver. The technology of the devices used to limit the pulse semi-conductors — gas plasma discharge devices, multipacting devices, and ferrite limiters, as well as hybrids and combinations of these devices — is very well developed. However, much fielded hardware has far simpler, older RPD technology, and in some cases this technology is downgraded or removed in the field to facilitate system operation. Attacks against specific systems will require system testing. Short, high power pulses are better for this attack.

It is important in discussions of HPM DEW missions to understand effects nomenclature. *Susceptibility* occurs when a system or subsystem experiences degraded performance when exposed to an EM environment. Electromagnetic *vulnerability* is when this degradation is sufficient to compromise the mission. *Survivability* occurs when the system is able to perform a mission, even in a hostile environment, and *lethality* occurs when a target is incapable of performing its mission after being irradiated.

Most of the effort within the U.S. military research and development community is on lethality, while in the rest of the world it is on survivability.

3.2.7   Hardening

HPM is susceptible to shielding or *hardening* as a countermeasure. Electromagnetic hardening can reduce the vulnerability of a target substantially without the attacker realizing that hardening has been employed. Hardening is produced by control of the entry paths for radiation into the target system and by reduction of the susceptibility of subsystems and the components to which radiation can couple. The drawback of hardening is that it may degrade the target system response and even increase its weight and volume. Moreover, the retrofit cost may be an excessive penalty.

The most effective method is to wholly contain the equipment in a *Faraday cage*, an electrically conductive enclosure, so that HPM cannot penetrate into the equipment. However, most such equipment must communicate with and be fed with power from the outside world, and this can provide entry points via which electrical transients may enter the enclosure and effect damage. While optical fibers address this requirement for transferring data in and out, electrical power feeds remain an ongoing vulnerability.

Shielding electronics by equipment chassis provides only limited protection, as any cables running in and out of the equipment will behave very much like antennas, in effect guiding the high-voltage transients into the equipment.

A low-frequency weapon will couple well into a typical wiring infrastructure, as most telephone lines, networking cables, and power lines follow streets, building risers, and corridors. In most instances, any particular cable run will comprise multiple linear segments joined at approximately right angles. Whatever the relative orientation of the weapons field, more than one linear segment of the cable run is likely to be oriented such that a good coupling efficiency can be achieved.

It is clearly evident that once the defense provided by a transformer, cable pulse arrestor, or shielding is breached, voltages even as low as 50 V can inflict substantial damage upon computer and communications equipment. Items (computers, consumer electronics) exposed to low-frequency, high-voltage spikes (near lightning strikes, electrical power transients) in many instances experience extensive damage, often requiring replacement of most semiconductors in the equipment.

3.2.8   High Power Microwave Effects on Electronics

Once radiation penetrates into the interior of a target system, the susceptibility of the small-scale semiconductor devices, which make up the interior electronics, becomes the key issue in DEW utility. Failures in semiconductor devices due to thermal effects occur when the temperature at the critical junctions is raised above 600 to 800° Kelvin, resulting in changes in the semiconductor up to and including melting. Because the thermal energy diffuses through the semiconductor material, there are several failure regimes, depending upon the duration of the microwave pulse. If the time-scale is short compared to thermal diffusion times, the temperature increases in proportion to the deposited energy. It has been established experimentally that semiconductor junction damage depends only upon energy for pulse durations less than 100 nsec, when thermal diffusion can be neglected. Therefore, in this regime the threshold power for damage varies as t^{-1}, as shown in Figure 3.12. For pulses greater than 100 nsec, thermal diffusion carries energy away from the junction. The general result is the *Wunsch–Bell relation*[11] for the power to induce failure:

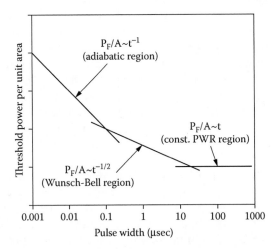

FIGURE 3.12
Power density for burnout vs. pulse width. UWB HPM sources operate in the adiabatic region and narrowband sources in the Wunsch–Bell or constant power regions.

$$P \sim \frac{1}{\sqrt{t}} \qquad\qquad (3.5)$$

The Wunsch–Bell relation is generally applicable because, even though thermal conductivity and specific heat vary with temperature, these effects cancel out. Therefore, *in the domain between 100 nsec and ~1 msec, the energy required to cause semiconductor junction failure scales as $t^{1/2}$ and the power requirement scales at $t^{-1/2}$.* For pulses longer than ~10 μsec, a steady state occurs in which the rate of thermal diffusion equals the rate of energy deposition. Therefore, the temperature is proportional to power, resulting in a constant power requirement for failure. The energy requirement then scales as t. The consequence of these scaling relations is that the shortest pulses require the highest powers but the least energy. Conversely, the highest energy and lowest power are required for long pulses. UWB HPM sources (Chapter 6) operate in the adiabatic, narrowband sources in the Wunsch–Bell or constant power regions. If energy is to be minimized in deployed weapons, the shorter pulses will be used, and if power is the limiting requirement, then longer pulse durations are indicated.

The above relations apply to single-pulse damage. If there is insufficient time between successive pulses for heat to diffuse, then accumulation of energy or *thermal stacking* will occur. From Figure 3.12, thermal equilibrium occurs in less than 1 msec; therefore, $\geq$1-kHz repetition rates are required for thermal stacking. The required value of repetition rate will vary with specific targets. There is also the possibility that permanent damage may result from the accumulation of multiple pulses. Some data suggest that gradual degradation occurs even at repetition rates much lower than the rate required for thermal stacking, possibly due to incremental damage. An analogy would be insulator flashover. This effect may allow reductions in the threshold power requirement for electronic damage. Some data indicate that when hundreds to thousands of pulses are used, the damage threshold can be reduced an order of magnitude. Such effects have not been quantified sufficiently. The repetition rate requirement can be relaxed, relative to thermal stacking, and will be determined by mission constraints, such as time on target. In any case, repetitive operation is required for any real engagement, and therefore accumulating damage effects may be inherent in HPM DEW engagements.

Damage is not the only mechanism to consider. *Upset* of digital circuits occurs when HPM couples to the circuit, is rectified at, for example, p-n junctions, and produces voltages equivalent to normal circuit operating voltage.

The development trends of electronic processors have led them to be more susceptible and vulnerable to microwaves. The basic component is the transistor gate, shown in Figure 3.13, which has a working voltage V and junction parasitic capacitance C_p, proportional to the area of the gate. Switching of the transistor occurs by the flow of a current flow of charge $Q = C_p V$ in a

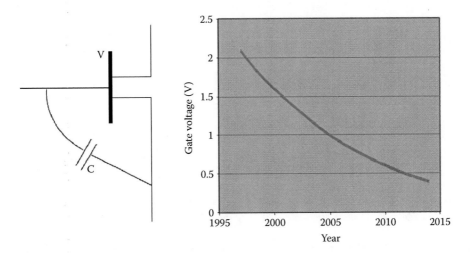

FIGURE 3.13
Circuit of transistor gate and historical trend of gate voltage.

time Δt. The rate of processing, or clock rate, is given by $f = 1/\Delta t$. Power flowing to the transistor is

$$P = VI = VQ/\Delta t = C_p V^2 f \tag{3.6}$$

Commercial pressure speeds clock rates, with the result that dissipated power in the gate increases, heating the junction. Such power must be extracted, limiting how dense the gates can be made, which limits processor compactness. Some improvements can be realized by reducing the gate area and hence C_p, but this shrinking of the gates is offset by the need to increase the total number of transistors, and so the dissipated power density remains constant. The attractive way out is to reduce voltage, as $P \sim V^2$. Figure 3.13 shows the continuing fall of voltage, with future declines projected to come. The lowered voltage reduces operating margins, making the circuit more susceptible to noise. This trend makes the chip containing the gates more vulnerable to microwave attack. If a microwave beam falling on the device housing the processor causes high enough voltage at the junction, effects up to burnout of the transistor will occur. Figure 3.14 shows a trend toward lower thresholds for susceptibility as the scale size of integrated circuits is reduced to fit more electronics on single chips.

For classes for commercial off-the-shelf (COTS) equipment, the most common system is the computer. Clock rate as well as the number of transistors is increasing, as the minimum structure size is decreasing with successive computer generations. Figure 3.15 shows effects testing on Pentium III processor-based PCs.[12] The electric field for the onset of effects is given. There is substantial scatter in effects data because of the complex sequence of penetrating the system and computer housing, and the various chassis prop-

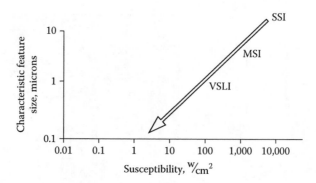

FIGURE 3.14
Feature size of chip technologies falls, lowering the threshold for HPM effects.

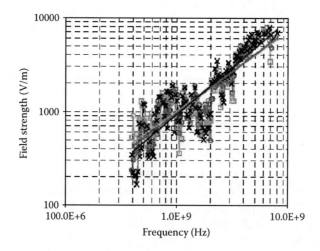

FIGURE 3.15
Electric field strength at which first breakdown effects occur in three Pentium III 667-MHz computers. (Reprinted from Hoad, R., *IEEE Trans. Electromag. Compatibility*, 46, 392, 2004 [Figure 3]. By permission of IEEE.)

agation paths combined with the inherent statistical nature of electronics effects. The system is more vulnerable at lower frequencies, which is a general result for COTS electronics. It is easier to produce effects at lower frequencies when the system is relatively open and has lots of wires. So cable effects dominate below 1 GHz for computers with power cords and wiring. Openings such as slots are more important at higher frequencies. In general, in-band radiation is highly effective at entering and causing severe problems if there is an operating frequency in the microwave. However, the ease of backdoor penetration and the susceptibility of modern digital electronics make this mechanism more likely in COTS systems. However, modern high-speed computers are harder than their predecessors. But when they do fail, the effects are more severe. The most common failure mode is computer

TABLE 3.1

Categories and Consequences of Electronic Effects

Failure Mode	Power Required	Wave Shape	Recovery Process	Recovery Time
Interference/disturbance	Low	Repetitive pulse or continuous	Self-recovery	Seconds
Digital upset	Medium	Short pulse, single or repetitive	Operator intervention	Minutes
Damage	High	UWB or narrowband	Maintenance	Days

crash with manual restart required. This might be because manufacturers are getting smarter about shielding for emissions and interference.

Table 3.1 shows in general the categories of consequences of effects. It demonstrates that the recovery time can vary from seconds to hours. Some think this a major advantage of HPM, obtaining a variety of effects from a single source.

In Europe, effects work concentrates on understanding electronics failure mechanisms to harden their electronics. It is well coordinated between the universities, the electromagnetic interference/compatibility (EMI/EMC) communities, and government/contractor laboratories. The U.S. concentrates on lethality assessment; the effects program is spread between the services, and there is little involvement of the EMI/EMC community.

3.2.9   Conclusion

If readers gather from this survey that HPM weapons face a complex set of factors that influence their true military utility, they are correct. The complexity of HPM DEWs requires a systematic approach with careful consideration of utility as a weapon, meaning an engagement analysis, coupling analysis and testing, electronic component testing, system testing, and a countermeasure (hardening) study.

As in the episode at the beginning of this section, easy, speedy battlefield use is critical. Only systematic integration with existing network-centric systems can give the warfighter quick, useful weapons. Just as "fire and forget" became a paradigm, so could "HPM kill."

3.3   High Power Radar

Radar applications for HPM depend not on the higher power directly, but rather on very short pulses at high repetition rate at high power. Here we cover short pulses (<10 nsec) at high power: ultrashort pulse radar, called ultrawideband (UWB) radar.

The most obvious application of HPM to radar is simply to increase the transmitter power P_T, thus increasing the maximum detection range for a target with a specified radar cross section. In radar, all other things being fixed, the maximum detection range scales as $P_T^{1/4}$; therefore, increasing a high power radar from 1 MW to 10 GW will increase the detection range by an order of magnitude. However, for conventional radar the limitation is not power, but signal-to-noise ratio produced by ground clutter and other interfering effects. Therefore, simple power scaling does not address the real limitations of conventional radar. The most fruitful use of HPM in radar is to address some of conventional radar's basic limitations.[13]

A basic issue is *dead time*, the time when the receiver is turned off while the transmitter operates. Using shorter pulses minimizes this, but the radar must also have a good return signal at the desired range. The solution is to operate at GW power levels with pulses just long enough to have a bandwidth about that of a wideband receiver. A shorter pulse would give a reduced receiver signal due to some of the power being out of band.

The most developed HPM radar is the[14] NAGIRA system (Nanosecond Gigahertz Radar) built in Russia and comprehensively tested in the U.K. The system, shown in Figure 2.15 and Figure 2.16, is based on a SINUS-6, in which a Tesla transformer charges a coaxial oil-filled pulse-forming line (PFL) to 660 kV. The line is discharged by a triggered gas switch into a transmission line that tapers to 120 Ω impedance at 660 kV to match that of a backward wave oscillator. This produces 7-nsec pulses and operates up to 150 Hz in X-band (10 GHz, so 70 cycles and ~2% bandwidth). After TE_{01}-to-TM_{11} mode conversion, the extracting waveguide terminates at a feed horn with a vacuum window through which the Gaussian-shaped excitation illuminates an offset, 1.2-m parabolic reflector antenna, forming a pencil beam with a beam width of 3° (see Problem 5).

Range resolution of a radar is given by $\delta = c\tau/2$, where τ is the duration of the pulse, so that a 10-nsec radar signal will give a range resolution of about 1.5 m. Each radar range cell in the signal processor is only 1 m long. Therefore, large extended targets can have their "portraits painted" by a series of signals coming from different parts of the target.[15] This allows some degree of target identification in principle, and experiments show that it works in practice. NAGIRA can detect a small airplane at 100 km. Range resolution is ~1 m, so targets can be identified.

An important factor is pulse-to-pulse frequency stability because the moving target motion is detected by separating from clutter by subtracting subsequent pulses. Therefore, voltage and current into the BWO must be controlled. Using short high power pulses makes possible precise range resolution for detection and tracking of low-cross-section moving targets in the presence of strong local reflections.

Gyrotrons and relativistic magnetrons would also be good choices for this radar because they too can produce high power at higher frequencies. Outstanding technical issues for HPM radar remain. Principal among them are cathode life and scanning antennas at high power. The cathode life of the

NAGIRA graphite cathodes, as well as other explosive emission materials, is 10^8 shots; a much longer lifetime would be needed in a practical device. To enable scanning of the beam on azimuth and elevation will require a high power rotating joint in the antenna feed.

3.4 Power Beaming

A variety of schemes have been suggested for transferring energy from Earth to space, space to Earth, space to space, and between points on Earth using HPM beams. The purpose is to beam large amounts of energy; the peak and average power requirement is set by the duration of the application.[16] Some of the applications are solar-powered satellite-to-ground transmission, inter-satellite power transmission, and short-range applications such as between parts of a large satellite or from a lander to a rover in interplanetary exploration.[17] Another class of applications of directed power beams where the receiving area is a target includes microwave weapons and ionosphere modification by artificial ionization of the upper atmosphere for radar or atmospheric chemistry.[18] Free-space transmission is desirable because it is far more efficient over long distances (>100 km) than competitor techniques, such as coax cable, circular waveguide, or fiber-optic cable.[19]

All such applications depend on far-field transfer, where the separation between transmitting antenna and receiving area is greater than or equal to the far-field distance $2D^2/\lambda$, where D is the diameter of the largest area and λ is the wavelength. (However, the method described here can also be applied in the near field with a focused transmit antenna. This is best done with a phased-array antenna to produce a focused pattern with spot size less than D_t.) The electromagnetics are quite general, depending on only the basic parameters of the system.

The schematic is shown in Figure 3.16. The transmitting antenna of area A_t places its beam on the receiving area A_r, which can be another antenna, a rectenna, or a target, such as a spacecraft. At the range R, the beam, defined by its half-power beam width, will typically be larger than A_r.

In an exact general treatment,[20] the power transfer efficiency for circular apertures is governed by the power–link parameter

$$Z = \frac{D_t D_r}{\lambda R} \tag{3.7}$$

where D_t and D_r are the diameters of the transmitting and receiving antennas, respectively, and R the distance between transmitter and receiver. For efficient transfer, the spot size must be ~D_r and $Z \geq 1$.

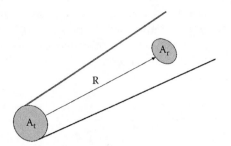

FIGURE 3.16
Power beaming schematic; transmitting and receiving areas separated by distance R.

To understand the power–link relation, note that the power received on a target is the power density integrated over the target area. If the transmitting antenna has an isotropic radiation pattern, then the power density S is just the power P distributed over the surface area of a sphere through which the power passes:

$$S = \frac{P}{4\pi R^2} \tag{3.8}$$

The typical case is that the beam is focused by the antenna (see Chapter 5). If it has a gain G, then the power density is enhanced to

$$S = \frac{PG}{4\pi R^2} \tag{3.9}$$

The antenna gain is

$$G = \frac{4\pi \varepsilon A_t}{\lambda^2} \tag{3.10}$$

where ε is the antenna efficiency, also called the aperture efficiency, and A_t is the transmitting antenna area. The efficiency with which power is received is then roughly

$$\frac{P_r}{P_t} = \left(\frac{\pi}{4}\right)^2 \varepsilon \left[\frac{D_t D_r}{\lambda R}\right]^2 = \left(\frac{\pi}{4}\right)^2 \varepsilon Z^2 \tag{3.11}$$

The efficiency of transfer varies roughly with the square of Z. This is approximate because this simple analysis does not average over the target area or take into account the antenna pattern and is invalid above Z ~ 1.

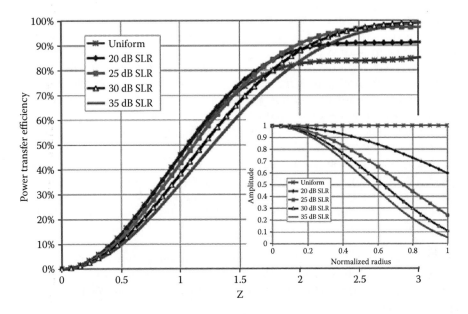

FIGURE 3.17
Power transfer between distant apertures as a function of principal parameters, $Z = D_t D_r / \lambda R$. Controlling sidelobe ratio (SLR) by tapering the power across the antenna (inset, see Ref. 20) reduces side radiation but broadens beam width.

An exact treatment gives the curve of Figure 3.17. Power transfer can be made efficient by proper choice of diameters and wavelength for a specific range. Choosing $Z = 2$ gives high efficiency, but $Z = 1.5$ may well be a more practical compromise to reduce aperture areas. An approximate analytic expression, suitable as a rule of thumb, for the exact solution in Figure 3.17 is

$$\frac{P_r}{P_t} = 1 - e^{-Z^2} \tag{3.12}$$

For comparison, for $Z = 1$, and uniform aperture amplitude distribution, Equations 3.11 and 3.12 give 62% efficiency and the exact solution gives 48%.

These efficiencies are for circular polarization-matched antennas; linear or dual-linear polarization can lead to additional losses. They also do not include losses due to propagation effects and intervening attenuations, such as that from atmosphere.

Note in the inset in Figure 3.17 that the transmitting antenna *taper*, defined as gradual reduction of amplitude toward the edge of the transmitting antenna of the array, has a great influence on efficiency. The abruptness with which the amplitude ends at the edge of the antenna is closely related to how much energy is found in sidelobes. In many cases, the power transfer efficiency is regulated by the desired sidelobe ratio (SLR), to avoid interference with other platforms, for example. A 25-dB SLR means the first sidelobe is 25 dB down (reduction factor of 316) from the main beam, and it requires that a specific amplitude taper be applied to the aperture. Higher SLR values reduce the amount of power radiated in the sidelobes at the expense of a broader beam width. A 25-dB SLR taper provides a good compromise between low sidelobes and high power transfer efficiency.

The power beaming relation scaling shown in Figure 3.17 has been demonstrated many times and is well documented.[2] Efficiencies of aperture-to-aperture transfer from transmitter to receiver of >80% have been demonstrated, although designs typically compromise efficiency somewhat to improve other factors, such as cost and ease of assembly and maintenance. As an example, consider an aircraft at 1-km altitude receiving power for its propulsion from an antenna radiating at 12 GHz from a 4-m-diameter aperture. If the receiving rectenna has a diameter of 12.9 m, from Figure 3.17, Z = 1.98 and collection efficiency is 90%. However, if the rectenna on the aircraft were elliptical, perhaps to provide conformal fit, but with the same area as the circular rectenna, the efficiency would be reduced by the differing power distributions.

The diffraction-limited beam width is

$$\theta = 2.44\ \lambda/D_t \tag{3.13}$$

This is the diffraction-limited divergence, giving the first null point of the Bessel function for a circular aperture and including 84% of the beam power. The spot size is just $D_s = \theta\ R = 2.44\ \lambda\ R/D_t$. For example, a 2-m-diameter ground-based transmitter at 35 GHz radiates to an aircraft 1 km away. For a 7.5-m receiving aperture (or target size), Z = 1.75 and 83% of the beam is received. The beam width is 0.6° and spot size is 10.5 m.

At the receiving end, the *rectenna*, a rectifying antenna, a component unique to power beaming systems, converts microwave power to DC power. It consists of an array of antenna elements, usually simple dipoles, each terminated with a rectifying diode. The circuit elements are shown in Figure 3.18 and many are shown in Figure 3.19. Diodes rectify most efficiently when the DC output voltage is operating at half the diode's breakdown voltage. Efficiency varies with power density on the element; most work to date has been at densities on the rectenna of 10 to 100 mW/cm². Because a rectenna element is terminated by a diode and RF shorting filter, the array radiation pattern resembles the cosine pattern of a single dipole element. No sidelobes are created by the rectenna array (see Problem 6). But there can be radiation of self-generated harmonics due to circuit nonlinearities. High rectenna efficiencies can be achieved, but only with proper selection of diode parameters. The chief limiting factor is high cutoff frequency, meaning that frequency is constrained by achieving low-diode junction capacitance and low series resistance.

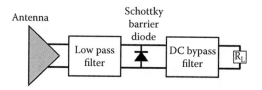

FIGURE 3.18
Rectenna converts microwaves to DC electrical energy. The antenna is typically a horizontal dipole placed 0.2λ above a reflecting ground plane. Filters contain the self-generated harmonics produced by the diode and provide the proper impedance match to the diode at the power beaming frequency. (Reprinted from McSpadden, J.O. and Mankins, J.C., *IEEE Microwave Mag.*, 3, 46–57, 2002. By permission of IEEE.)

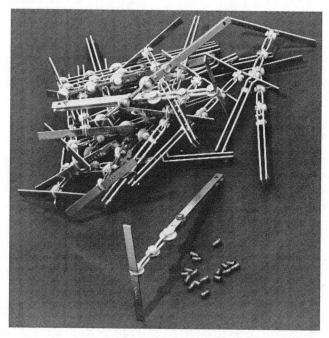

FIGURE 3.19
Group of typical rectenna elements, dipoles, diodes, and circuitry. (Reprinted from McSpadden, J.O. and Mankins, J.C., *IEEE Microwave Mag.*, 3, 46–57, 2002. By permission of IEEE.)

Rectenna experiments have demonstrated up to 54% DC-to-DC power transfer efficiency at 2.45 GHz, and there is some work on higher frequency rectennas.[21] Figure 3.20 shows the highest recorded rectenna efficiencies at 2.45, 5.8, and 35 GHz. Aircraft and blimps have been flown, sometimes for considerable times, on power beams.[22,23] A demonstration of power beaming over a 1.5-km distance produced a received power of 34 kW at 2.388 GHz by Dickinson[24] and Brown.[25]

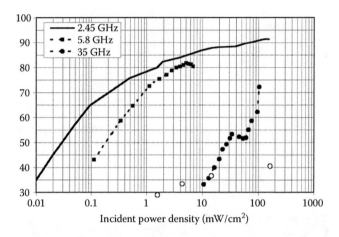

FIGURE 3.20
Highest measured rectenna conversion efficiencies. The numbers on the 2.45-GHz curve indicate the efficiency using GaAs-W Schottky diodes with varying breakdown voltages. (Reprinted from McSpadden, J.O. and Mankins, J.C., *IEEE Microwave Mag.*, 3, 46–57, 2002. By permission of IEEE.)

William Brown, who invented the rectenna, conducted detailed development for space applications of it, most especially an interorbital ferry for transporting from low Earth orbit to geosynchronous orbit.[25] It would use beamed power to a large array (225 m^2) of rectennas, which drive ion thrusters continuously. For a cargo mass fraction of ~50%, orbit-raising transit times are about 6 months. Power beamed to the ferry is of the order 10 MW, with a low duty factor.

The use of a microwave beam to efficiently transport power from solar cells in space to the Earth's surface was proposed in the 1960s and reinvestigated by NASA in the 1990s.[26] The reference system concept, a large capital-intensive no-pollution energy system for the 21st century, is ~10 GW continuous power available at an efficiency of >50%. As shown in Figure 3.21 and Figure 3.22, a station in geosynchronous orbit (22,300 miles altitude) emits a narrow beam (0.3 mrad) at frequencies in S-band (2.45 GHz). Basic issues are efficient conversion of the DC power to microwave power and preeminently the construction of extremely large massive antennas (apertures in space typically 1 km across and a rectenna array about 7 km on the Earth).[27]

The microwave tube requirement could be met by a vast distributed array of crossed-field amplifiers, magnetrons, gyrotrons, or klystrons operating at average powers of ~10 kW each. Magnetrons for continuous operation are now available at 100 kW. Therefore, ~10^5 tubes would be required and low-cost production techniques would need to be developed to keep the cost practical. Higher-average-power MW-class gyrotrons (Chapter 10) could be competitive, but they operate at higher frequencies. For example, commercially available continuous-wave (CW) gyrotrons are capable of

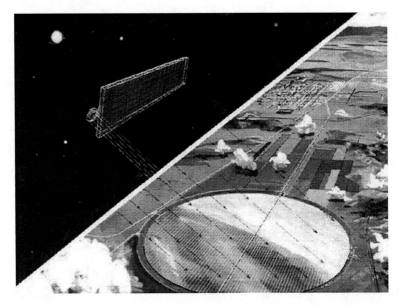

FIGURE 3.21
Solar power satellite beams enormous microwave power to rectenna farm for distribution.

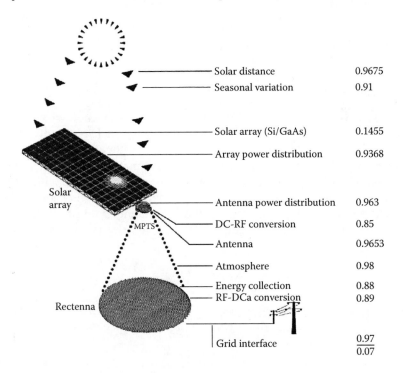

Solar distance	0.9675
Seasonal variation	0.91
Solar array (Si/GaAs)	0.1455
Array power distribution	0.9368
Antenna power distribution	0.963
DC-RF conversion	0.85
Antenna	0.9653
Atmosphere	0.98
Energy collection	0.88
RF-DCa conversion	0.89
Grid interface	$\dfrac{0.97}{0.07}$

FIGURE 3.22
Solar power satellite estimated efficiencies.

MW-class powers in the 35- and 94-GHz atmospheric window (see Chapters 5 and 10). An alternative is a much larger number of solid-state sources at much lower power.

The basic issue for all space-to-Earth power transmission schemes is not technical, but economic. The costs are strongly contested[28] and will determine the practicality of large-scale CW HPM systems in space. It is at best a distant prospect. An even larger-scale proposal is to deliver energy from solar power stations on the moon built from all lunar materials, beaming power directly to receiving sites on Earth.[29] A reciprocal application wherein the power is beamed upward from Earth to low Earth orbit satellites requires ~10- to 100-MW power levels for tens of seconds.[30]

3.5   Space Propulsion

Microwave beams have been studied for moving spacecraft by a variety of means:

- Launch to orbit
- Launch from orbit into interplanetary and interstellar space
- Deployment of large space structures

3.5.1   Launch to Orbit

Today, payloads are launched into orbit the same way they were 50 years ago — by chemical rockets. Expendable multistage rockets usually achieve payload fractions of less than 5%. As described by the rocket equation, this is due partly to the structural limits of existing materials and partly to the limited efficiency of chemical propellants. The rocket equation, which drives the advance of space exploration, is simply derived.[31] If a rocket of mass M produces momentum change by expelling a small amount of reaction mass dM at high velocity V_e, the change in rocket velocity is, from momentum conservation,

$$M dV - V_e dM \qquad (3.14)$$

By integration, when the fuel is spent so that the initial mass M_0 is reduced to a final mass M_f, the resulting change in rocket velocity,

$$\Delta V = V_e \ln M_0/M_f \qquad (3.15)$$

Since the natural logarithm increases very slowly with the mass ratio, large mass ratios do not produce large ΔV. Another way to see this is

$$\frac{M_o}{M_f} = e^{\frac{\Delta V}{V_e}} \tag{3.16}$$

which means that getting the high velocities needed for orbit or interplanetary probes with a single rocket is almost impossible because the rocket structural mass is too high relative to the fuel mass (see Problem 7). The structural economies made to preserve these minute payload fractions result in fragile, expensive-to-build rockets. Despite 50 years of incremental rocket development, better materials, novel propellants, and modestly improved reliability have left the basic economics of launch unchanged at a payload cost of around $5000/kg of delivered payload.[32] This makes space development and exploration very expensive. If price can be lowered, an increase in demand great enough to offset lower prices and increase the overall market size is thought to exist only at a price point below $1000/kg. It is doubtful that conventional chemical rocket launches will ever be able to reach such a price.

This led Parkin[33] to propose new HPM-based launch schemes, which offer greater I_{sp} than possible with conventional systems. (The efficiency of a rocket, how many seconds a pound of fuel can produce a pound of thrust, is termed specific impulse, I_{sp}. Chemical rockets have reached a practical limit of $I_{sp} < 500$ sec.)

The physics of HPM propulsion are different, and in some ways more complex, than conventional propulsion. The microwave thermal rocket, shown in Figure 3.23, is a reusable single-stage vehicle that can afford the mass penalty of robust, low-cost construction because it a uses a high-performance microwave thermal propulsion system with double the I_{sp} of conventional rockets.[34,35] It is premature to quantify the reduction in transportation costs that these simplifications will bring, but clearly the cost reduction could be dramatic, transforming the economics of launch to space.

Microwave thermal thrusters operate on an analogous principle to nuclear thermal thrusters, which have experimentally demonstrated specific impulses exceeding 850 sec. (The large weight of the nuclear reactors relative to the thrust the propulsion systems produce precludes their use in practical rockets.) It is a reusable single-stage vehicle that uses a HPM beam to provide power to a heat exchanger-type propulsion system, called the *microwave thermal thruster*. By using HPM, the energy source and all the complexity that it entails are moved onto the ground, and a wireless power transmission system is used between the two. When Parkin and Culick proposed this system[36] and analyzed its performance, they found that beam divergence and aperture area are key constraints for such a system and devised a high-acceleration ascent trajectory, which provides most of the transfer of energy at short range, in order to minimize the size of the radiating aperture required. For such a system, the power–aperture–efficiency trade given in Equation 3.17 is key to minimizing the cost, which could potentially fall to

FIGURE 3.23
Microwave thermal rocket in flight. (Reprinted by permission of Dr. Kevin Parkin.)

as low as a few times the *energy* cost of launch (as opposed to capital cost of the system), which is around $50/kg for such a system.

Ascent trajectory analysis predicts a 10% payload fraction for a 1-ton payload on a single stage to orbit a microwave thermal rocket. By using a single reusable stage with high specific impulse, low-cost and reusable launchers are possible. For comparison, the Saturn V had three nonreusable stages and achieved a payload fraction of 4%. In Figure 3.24, the spacecraft flat aeroshell underside is covered by a thin microwave absorbent heat exchanger that forms part of the thruster. The exchanger is likely to be of silicon carbide and consists of ~1000 small channels carrying fuel to the motor. The rocket can be launched by a conventional first stage, by airdrop into the vicinity of the beam, or by a physically small but powerful microwave source located at the launch site, which powers the first few kilometers of atmospheric ascent before it is acquired by the main beam.

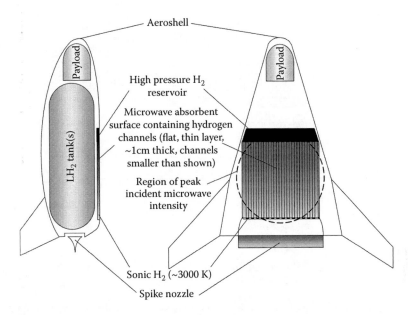

FIGURE 3.24
Major elements of the microwave thermal rocket. Vehicle length is 6 m, 5 m width at base. Payload is 100 kg; structure, 180 kg; LH_2 fuel, 720 kg. (Reprinted by permission of Dr. Kevin Parkin.)

During the ~3-min powered ascent, acceleration varies from 2 to 20 g, so as to stay in range of the beam source. The vehicle length is 6 m, 5 m wide at base. During ascent, the heat exchanger faces the microwave beam. Using a phased-array source, a high-intensity 3-m-diameter microwave spot can be projected onto the underside of a launcher over 100 km away.

The power density of 300 MW impinging on a 7-m^2 converter on the underside drives the hydrogen fuel to temperatures of 2400 K. Microwave frequency determines the maximum beam energy density via the constraint of atmospheric breakdown. Ionizing air into plasma can distort and reflect the incoming beam (see Chapter 5). Atmospheric breakdown occurs more easily at low frequencies, because fields have more time to accelerate background electrons. The continuous-wave (CW) atmospheric breakdown intensity, shown in Figure 3.25, demonstrates that a 300-GHz beam can achieve 1000 times the power density of a 3-GHz beam. A microwave beam of 300 MW on a 3-m spot means that the power density will average $\sim 40\ MW/m^2$, so frequencies above ~25 GHz are needed to prevent breakdown. High-frequency operation above ~100 GHz is preferable because the array area needed decreases substantially as beam frequency increases.

The feasibility and design of a 30-MW, 245-GHz, ground-based beam launcher facility using an array of parabolic dishes were described in papers by Benford, Myrabo, and Dickinson[37,38] in the 1990s. The average power output of gyrotron sources used has significantly increased to commercially available 1-MW gyrotron sources operating at ~100 GHz. Three hundred

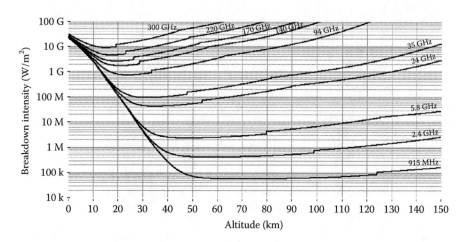

FIGURE 3.25
Breakdown variation with altitude and frequency. Launch to orbit requires higher frequencies.
(Reprinted by permission of Dr. Kevin Parkin.)

such sources would be sufficient to propel a 1-ton craft into Low Earth Orbit
(LEO). The gyrotron can be either an amplifier or a phase-injection-locked
oscillator.[39] In the former unit, the array beam steering is done simply by a
combination of mechanical pointing and phase shifting the module drive
signal. In the latter case, the beam steering is done by a combination of
mechanical pointing and phase shifting, but must also include means such
as a magnetic field strength trim to set the rest frequency of the oscillator
(Chapter 10).

Figure 3.26 gives the layout of a 245-GHz, 30-MW ground-based power
beaming station.[38] It has a span of 550 m and contains 3000 gyrotrons at 10
kW of power. A single radiating element is the 9-m dish shown in Figure 3.27.

Augmentation of rocket engines for launch to orbit have also been suggested by beaming to rectennas.[40] A crucial limitation of such schemes is the
weight of the rectenna, 1 to 4 W/g, which needs improvement of an order
of magnitude for space applications. Supporting a platform in the upper
atmosphere at ~70 km by means of the radiometric effect using a ~1-MW
microwave beam to heat the underside of the vehicle has also been analyzed
and seems practical.[41]

The cost of such large HPM systems described here is driven by two
elements: the *capital cost*, including the cost of the microwave source and the
cost of the radiating aperture, and the *operating cost*, the cost of the electricity
to drive the system:

$$C = C_C + C_O$$

$$C_C = C_S + C_A \tag{3.17}$$

$$C_O = C_E$$

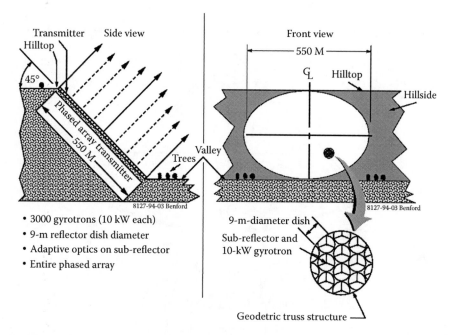

FIGURE 3.26
Ground-based power beaming station.

A simple approach is to assume linear scaling with coefficients describing the dependence of cost on area $B_1(\$/m^2)$, microwave power $B_2(\$/W)$, and prime power $B_3(\$/W)$:

$$C_A = B_1 D^2_t \frac{\pi}{4}$$

$$C_S = B_2 P$$

$$C_E = B_3 P \tag{3.18}$$

$$C_C = B_2 P + B_1 D^2_t \frac{\pi}{4} + B_3 P$$

An empirical observation made independently by Dickinson[38] and O'Loughlin is that, for a fixed effective isotropic radiated power (EIRP; the product of radiated power and aperture gain), *minimum capital cost is achieved when the cost is equally divided between antenna gain and radiated power*. Here, the gain includes the antenna, its supports, and subsystems for pointing and tracking. Power cost includes the source, power supply, and cooling equipment. If this rule of thumb holds,

$$C_S = C_A, \ C = [2B_2 + B_3]P \tag{3.19}$$

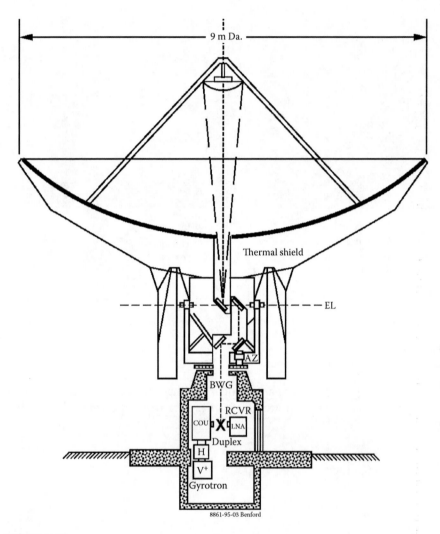

FIGURE 3.27
Single 9-m dish element for the ground-based power beaming station. Beam Waveguide (BWG)
feeds dish from gyrotron, reflected signal received (RCVR) by low noise amplifier (LNA).

Only rough estimates of the capital costs are possible with such a simple
model. However, when mission specifics are introduced, giving additional
constraints, useful scalings result. Kare and Parkin have cost modeled a
microwave thermal rocket.[42] They also account for the *learning curve*, the
decrease in unit cost of hardware with increasing production. This is
expressed as a learning curve factor f, the cost reduction for each doubling
of the number of units. The cost of N units is

$$C_N = C_1 N^x \qquad (3.20)$$

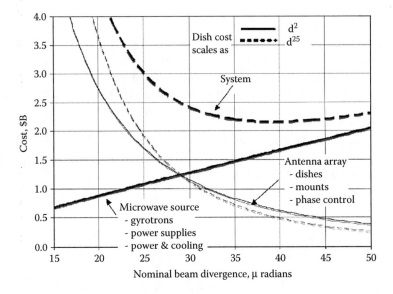

FIGURE 3.28
Scaling of aperture and source cost (including prime power) for the microwave launcher system as a function of the beam divergence.

For cost scaling, exponent x = 1 + ln(1 − f)/ln2, e.g., a 15% learning curve corresponds to an exponent of 0.76. For example, if gyrotron cost scales as the power P, but if there is a 15% learning curve, cost scales as $P^{0.76}$. MW units costing $1/W in single units will cost $0.28/W for 200 units.

For antenna costs, data exist that scale cost with diameter D^δ to an exponent ranging from δ = 2 to 2.5. Figure 3.28 shows the scaling of the aperture cost and source cost (including prime power) for the microwave system as a function of the beam divergence, for δ = 2.0 and 2.5. (Both the number and size of the antennas enter into the calculation, so the total array cost varies as $D_t^{\delta-2x}$.) Note that the Dickinson observation holds approximately: the cost minimum is close to where the source and antenna costs are equal.

3.5.2 Launch from Orbit into Interplanetary and Interstellar Space

Microwave-propelled sails are a new class of spacecraft that promises to revolutionize future space travel. For a general introduction to solar and beam-propelled sails, refer to McInnes.[43]

The microwave-driven sail spacecraft was first proposed by Robert Forward as an extension of his laser-driven sail concept.[44] The acceleration from *photon momentum* produced by a power P on a thin sail of mass m and area A is

$$a = [\eta + 1]P/m\,c \tag{3.21}$$

where η is the reflectivity of the film of absorbtivity α and c is the speed of light. The force from photon acceleration is weak, but is observed in the trajectory changes in interplanetary spacecraft, because the solar photons act on the craft for years. For solar photons, with a power density of ~1 kW/m^2 at Earth's orbit, current solar sail construction gives very low accelerations of ~1 mm/sec^2. Shortening mission time means using much higher power densities. In this case, to accelerate at one Earth gravity requires ~10 MW/m^2 (see Problems 8 and 9).

Of the power incident on the sail, a fraction αP will be absorbed. In steady state, this must be radiated away from both sides of the film, with an average temperature T, by the Stefan–Boltzmann law:

$$\alpha P = 2A \, \varepsilon \, \sigma \, T^4 \qquad (3.22)$$

where σ is the Stefan–Boltzmann constant and ε is emissivity. Eliminating P and A, the sail acceleration is

$$a = 2 \, \sigma/c[\varepsilon \, (\eta + 1)/\alpha](T^4/m) = 2.27 \times 10^{-15} \, [\varepsilon(\eta + 1)/\alpha](T^4/m) \qquad (3.23)$$

where we have grouped constants and material radiative properties separately. Clearly, the acceleration is strongly temperature limited. This fact means that materials with low melt temperatures (Al, Be, Nb, etc.) cannot be used for fast beam-driven missions. Aluminum has a limiting acceleration of 0.36 m/sec^2. The invention of strong and light carbon mesh materials has made laboratory sail flight possible[45] because carbon has no liquid phase and sublimes instead of melting. Carbon can operate at very high temperature, up to 3000°C, and limiting acceleration is in the range of 10 to 100 m/sec^2, sufficient to launch in vacuum (to avoid burning) in Earth-bound laboratories.

Recently, beam-driven sail flights have demonstrated the basic features of the beam-driven propulsion. This work was enabled by invention of strong, light carbon material, which operates at high enough temperatures to allow liftoff under one Earth gravity. Experiments with carbon–carbon microtruss material driven by microwave and laser beams have observed flight of ultralight sails at several gee acceleration.[46] In the microwave experiments, propulsion was from a 10-kW, 7-GHz beam onto sails of mass density of 5 g/m^2 in 10^{-6} Torr vacuum. At microwave power densities of ~kW/cm^2, accelerations of several gravities were observed (Figure 3.29). Sails so accelerated reached >2000 K from microwave absorption and remained intact. Photonic pressure accounts for up to 30% of the observed acceleration. The most plausible explanation for the remainder is desorption pressure and evaporation of absorbed molecules (CO_2, hydrocarbons, and hydrogen) that are very difficult to entirely remove by preheating. The implication is that this effect will always occur in real sails, even solar sails at low temperature, and can be used to achieve far higher accelerations in the initial phase of future sail missions.

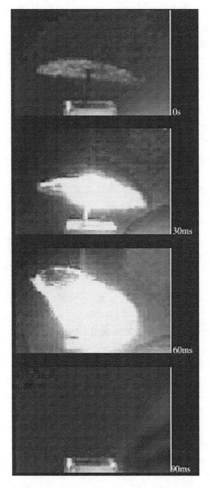

FIGURE 3.29
Carbon sail lifting off of a rectangular waveguide under 10 kW of microwave power (four frames, first at top). Frame interval is 30 msec.

This other acceleration mechanism, *desorption pressure*, is due to mass ejected from the material downward to force the sail upward:

$$F = v_T \, dm/dt \tag{3.24}$$

where v_T is the thermal speed of the evaporated material — carbon or molecules absorbed into the carbon substrate. The magnitude of this effect can vastly exceed the microwave photon pressure if the temperature is high enough because it couples the energy of photons, not the momentum, as in photon pressure. The essential factor is that this force must be asymmetric; if the sail is isothermal, there is no net thrust. Thermal analysis shows that, although for thin material thermal conduction reduces such differences greatly, the skin effect produces a substantial temperature difference between the front and back of sails.

A variety of compounds not typically thought of as fuels can be "painted" on sails and, depending on which physical process occurs, sublimed, evaporated, or desorbed. Atoms embedded in a substrate can be liberated by heating, an effect long studied in the pursuit of ultraclean laboratory experiments. This effect is called *thermal desorption* and dominates all other processes for mass loss above temperatures of 300 to 500°C.

A molecule is *physisorbed* when it is *adsorbed* without undergoing significant change in electronic structure, and *chemisorbed* when it does. Physisorbed binding energies (~2 to 10 kcal/mole) are typically much less than chemisorbed energies (~15 to 100 kcal mole), by as much as an order of magnitude. This implies that two different regimes of mass liberation can be used, with physisorbed molecules coming off at lower temperatures, and hence lower thrust per mass, while chemisorbed molecules can provide higher thrust per mass. Cuneo[47] offers this general schematic for desorption in layers from bulk substrates: the rate of mass loss under heating is

$$dn/dt = an \ e^{-Q/kT} \qquad (3.25)$$

where a is $\sim 10^{13}$ sec^{-1}, Q is the required liberation energy (usually <1 eV), and n is the area density in atoms/m^2, so that dn/dt is the desorbed flux under heating in atoms/m^2-sec. (We neglect readsorption, which is tiny in a space environment.) The exponential factor means that thermal desorption of molecules from a sail lattice will have a sudden onset as the sail warms. When temperature T varies with time, the equation can be formally solved:

$$n(t)/n° = exp\{-a \ exp[-Q/kT(t)] \ dt\} \qquad (3.26)$$

As the binding energy Q increases, the time to desorb gets longer. The relationship between Q and T*, the temperature at the peak in the desorption rate dn/dt, is, for heating rate dT/dt,

$$Q/kT^{*2} = a \ exp(-Q/kT^*)/(dT/dt) \qquad (3.27)$$

At the peak desorption rate, 63% of the mass inventory has been lost, so this is a good estimate of when the effect is largest for a given molecule of binding energy Q.

Hydrogen is often easiest to liberate, with a Q of 0.43 to 1.5 eV, depending on the substrate. Water has Q = 0.61 eV. Generally, candidate chemisorbed compounds like hydrocarbons have a Q of around 1 eV (11,605 K). CO is more strongly bound and may be the most tightly held in a carbon sail lattice. Sails experiencing strong, sudden-onset lift may be desorbing CO at a critical temperature onset of >2300 K (see Problem 10).

An early mission for microwave space propulsion is dramatically shortening the time needed for solar sails to escape earth orbit. By sunlight alone,

(a)

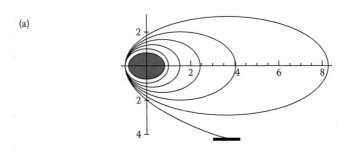

Normalized radius vs time for solar sail receiving approximately $1/5\ E_{TOTAL}$ per boost

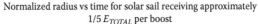

(b)

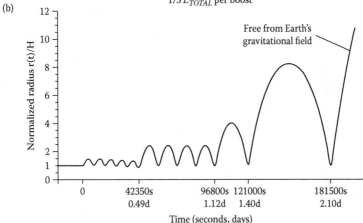

FIGURE 3.30

(a) Beam-driven sail trajectory out of Earth orbit. Units are the radius normalized to the initial height H. Solar sail is not to scale. (b) Radius normalized to Earth radius vs. time for solar sail. (Reprinted by permission of Dr. Gregory Benford, University of California–Irvine.)

sails take about a year to climb out of the Earth's gravity well. Computations show that a ground-based or orbiting transmitter can impart energy to a sail if it has resonant paths — that is, the beamer and sail come near each other (with either the sail overhead an Earth-based transmitter or the sail nearby orbits in space) after a certain number of orbital periods. For resonance to occur relatively quickly, specific energies must be given to the sail at each boost. If the sail is coated with a substance that sublimes under irradiation, much higher momentum transfers are possible, leading to further reductions in sail escape time. Simulations of sail trajectories and escape time are shown in Figure 3.30. In general, resonance methods can reduce escape times from Earth orbit by over two orders of magnitude vs. using sunlight alone on the sail.[48]

While microwave transmitters have the advantage that they have been under development much longer than lasers and are currently much more efficient and inexpensive to build, they have the disadvantage of requiring much larger apertures for the same focusing distance. This is a significant disadvantage in missions that require long acceleration times with correspondingly high velocities. However, it can be compensated for with higher acceleration. The ability to operate carbon sails at high temperature enables much higher acceleration, producing large velocities in short distances, thus reducing aperture size. Very low mass probes could be launched from Earth-based microwave transmitters with maximum acceleration achieved over a few hours using apertures only a few hundred meters across.

A number of such missions have been quantified.[49] These missions are for high-velocity mapping of the outer solar system, Kuiper Belt, Plutinos, Pluto and the Heliopause, and the interstellar medium. The penultimate is the interstellar precursor mission. For this mission class, operating at high acceleration, the sail size can be reduced to less than 100 m and an accelerating power of ~100 MW focused on the sail.[50] At 1 GW, sail size extends to 200 m and super-light probes reach velocities of 250 km/sec for very fast missions (see Problem 11).

3.5.3   Deployment of Large Space Structures

Will sails riding beams be stable? The notion of *beam riding*, the stable flight of a sail propelled by a beam, places considerable demands on the sail shape. Even for a steady beam, the sail can wander off if its shape becomes deformed or if it does not have enough spin to keep its angular momentum aligned with the beam direction in the face of perturbations. Beam pressure will keep a concave shape sail in tension, and it will resist side-wise motion if the beam moves off center, as a side-wise force restores it to its position.[51] Experiments have verified that beam riding does occur.[52] Positive feedback stabilization seems effective when the side-wise gradient scale of the beam is the same as the sail concave slope. A broad conical sail shape appears to work best.

Recently, stable beam riding has led to a proposed new type of aircraft/ spacecraft, supported from below at altitudes of ~70 km by a radiometric temperature difference effect.[53] An ultralight, mostly carbon fiber "Lifter" is feasible using the radiometric force, an effect known since the 19th century. A powerful microwave beam illuminates the Lifter underside to provide the required temperature difference. Optimally, the full ambient atmospheric pressure can be delivered to one side of the Lifter area by heating it well above the ambient air at 200 K. Beam powers of ~1 MW can support masses of ~100 kg, riding on the narrow microwave beam, and active stabilization is available through beam manipulation. The effect occurs at pressures corresponding to 30 to 150 km altitude, where no aircraft or spacecraft can sustain flight. It can perform communications relay, environmental monitoring, data telemetry, and high-quality optical imaging.

Finally, beams can also carry angular momentum and communicate it to a sail to help control it in flight. Circularly polarized electromagnetic fields carry both energy and angular momentum, which acts to produce a torque through an effective moment arm of a wavelength, so longer wavelengths are more efficient in producing spin. A variety of conducting sail shapes can be spun if they are *not* figures of revolution.

The wave angular momentum imparted to the conducting object scales as wavelength λ. For efficient coupling, λ should be of the order of the sail D_r. This arises because a wave focused to a finite lateral size D generates a component of electric field along the direction of propagation of magnitude λ/D times the transverse electric field, as long as D exceeds the wavelength. The spin frequency S of an object with moment of inertia I grows according to

$$\frac{dS}{dt} = \frac{2P(t)}{I}\frac{\alpha}{\omega} \tag{3.28}$$

where P is the power intercepted by the conductor and α is the absorption coefficient.

Konz and Benford[54] generalized the idea of absorption to include effects arising from the geometric reflection and diffraction of waves, independent of the usual material absorption. They showed that axisymmetric perfect conductors cannot absorb or radiate angular momentum when illuminated. However, any asymmetry allows wave interference effects to produce absorption, even for perfect reflectors, so $\alpha < 0$. Microwaves convey angular momentum at the *edges* of asymmetries because of boundary currents. Such absorption or radiation depends solely on the specific geometry of the conductor. They termed this *geometric absorption*. Conductors can also radiate angular momentum, so their geometric absorption coefficient for angular momentum can even be negative. The geometric absorption coefficient can be as high as 0.5, much larger than typical simple material absorption coefficients, which are typically ~0.1 for absorbers such as carbon and ~0 for conductors. Experiments show the effect is efficient and occurs at very low microwave powers.[55] This effect can be used to stabilize the sail against the drift and yaw, which can cause loss of beam riding, and allows hands-off unfurling deployment through control of the sail spin at a distance.[56]

3.6   Plasma Heating

Controlled thermonuclear fusion promises much as a clean energy source. It has been 50 years in the future for some 50 years now. Nevertheless, progress in fusion parameters of temperature, confinement time, and energy production is slowly moving forward. In the last decade, use of high-average-power microwave beams to heat plasmas to thermonuclear temperatures has expanded greatly. Here we concentrate on electron cyclotron resonance heating (ECRH) because it is the method receiving the most development. For a detailed description of ion cyclotron resonance heating and lower hybrid heating, consult the first edition. Microwave beams are also used to generate currents in fusion devices (electron cyclotron current drive [ECCD]); we briefly summarize that application.

Both ECRH and ECCD can be highly localized, and therefore are used for toroidal plasma confinement. The applications are stabilization of magneto-hydromagnetic instabilities such as tearing notes, production of desirable current density and plasma pressure profiles, and detailed control of plasma energy transport.

For ECCD, electron cyclotron waves drive current in toroidal plasma. A microwave beam launched obliquely into the toroidal plasma excites electrons traveling in one direction. Strong damping of the wave depends upon the velocity direction of the absorbing electrons. These electrons are accelerated to higher energy in that direction, and therefore the current is generated. This method has been widely applied, giving currents of 10 to 100 kA and aiding stabilization of tearing modes in localized regions. For a detailed account of ERCH and ECCD, see Prather's review.[57]

There are three basic microwave heating variants, differing in the frequency of the microwaves that are radiated into the plasma to be resonantly absorbed by the electrons or ions. The three types of heating that have received the most attention and the associated resonant frequencies are as follows:

- Ion cyclotron resonance heating (ICRH)

$$f = f_{ci} = \frac{eB}{2\pi m_i} \tag{3.29}$$

- Lower hybrid heating (LHH)

$$f = f_{LH} \cong \frac{f_{pi}}{\left(1 - \frac{f_{pe}^2}{f_{ce}^2}\right)^{1/2}} \tag{3.30}$$

- Electron cyclotron resonance heating (ECRH)

$$f = f_{ce} = \frac{eB}{2\pi m_e}, \text{ or } f = 2f_{ce} \tag{3.31}$$

where e is the magnitude of the charge on an electron or an ionized isotope of hydrogen, m_i and m_e are the ion and electron masses, B is the magnitude of the local magnetic field, $f_{pe} = (n_e e^2 / \varepsilon_0 m_e)^{1/2} / 2$, with n_e the volume density of electrons and ε_0 the permittivity of vacuum, and $f_{pi} = (n_i Z_i^2 e^2 / \varepsilon_0 m_i)^{1/2}/2$, with $Z_i = 1$ the charge state of the ions in the fusion plasma. At the densities and field values necessary for fusion, $f_{ci} \simeq 100$ MHz, $f_{LH} \simeq 5$ GHz, and $f_{ce} \simeq 140$ to 250 GHz.

The disadvantage of IRCH and LHH frequencies is that it is difficult to design an antenna that efficiently couples the microwaves into a plasma containment vessel with dimensions of the order of the wavelength. In the case of ICRH, there is an added difficulty with coupling the microwaves into the plasma itself, since they must pass through an outer region in the plasma where the waves are attenuated. Overcoming these problems sometimes involves inserting the input antenna into the plasma, which can damage the antenna and inject undesirable impurities into the plasma.

ECRH is now the auxiliary heating method of choice for major fusion devices such as tokamaks and stellerators. Compared to neutral beam heating, ERCH has the advantage of sufficiently high average power to heat plasma in localized regions because of the magnetic field dependence of the mechanism. The other virtue is the ability to place the launching antenna far from the plasma to not perturb or contaminate the fusion experiment. In typical ERCH fusion heating experiments, the gyrotron can be located tens to hundreds of meters from the fusion device and injects into it through small cross-section ports. Both of these virtues are particularly useful in the coming era of major fusion experiments when significant radiation shielding will be required, meaning that small ports are essential and all ancillary equipment must be at some distance.

At the high frequencies and short wavelengths appropriate to ECRH, waves coupling into the plasma and antenna design are much more straightforward. The electron cyclotron wave propagates well in vacuum and couples efficiently to the plasma boundary, in contrast to the other techniques. Additionally, because the magnetic field varies across the plasma, a tunable source could allow selective heating of the plasma at varying depths, and so tailor the density profile of the plasma by taking advantage of the expansion that accompanies a local rise in temperature.

Equation 3.31 shows that ECRH is accomplished at either f_{ce} or $2f_{ce}$. In fact, the choice of frequency depends on the polarization of the microwaves and the point at which they are injected. This can be explained with a rather simple model for the propagation of high-frequency waves in infinite plasma (Figure 3.31); for a more sophisticated treatment, see Ott et al.[58] A uniform magnetic field along the z axis represents the magnetic field confining the plasma. We are interested in waves that will penetrate the confined plasma, so we look at waves propagating across the magnetic field; we designate this direction the x axis, which is the direction of the wavevector, $\mathbf{k} = \hat{x}k$. The microwave fields are represented as plane electromagnetic waves, e.g., $\mathbf{E}(x, t) = Ee^{i(kx-\omega t)}$. There are two natural modes of oscillation for the electromagnetic waves in the plasma, depending on the polarization of the electric field vector relative to the magnetic field:

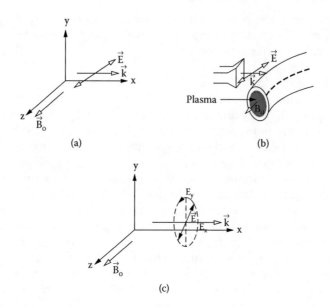

FIGURE 3.31
Models for propagation of modes in plasma: (a) O-mode, showing polarization of E; (b) geometry for O-mode plasma heating; (c) polarization for X-mode — note elliptical polarization of E.

- The *ordinary*, or O-, *mode*, with an electric field vector that is parallel to the magnetic field
- The *extraordinary*, or X-, *mode*, with electric field components in the x-y plane, i.e., perpendicular to the magnetic field, but with a component along the wavevector

In the case of the O-modes, because **E** and **k** are perpendicular to one another, $\nabla \cdot \mathbf{E} = 0$, so that there is no microwave space-charge associated with the wave. On the other hand, $\nabla \cdot \mathbf{E}$ does not vanish for the X-mode, so that there is a space-charge component to these waves. To describe the propagation of waves in the plasma, we turn to the concept of the dispersion relation, which is discussed in Chapter 4. Briefly, a dispersion relation is a mathematical relationship between $\omega = 2f$ and $k = 2/\lambda$ (λ is the wavelength along x), which for our purposes will be used to determine where waves may and may not propagate within the plasma and, consequently, where microwaves of a particular frequency and polarization should be launched. Here, we state the O- and X-mode dispersion relations without derivation (the interested reader is directed to Chen[59]):

- O-mode

$$k^2 c^2 = \omega^2 - \omega^2_{pe} \tag{3.32}$$

where $\omega_{pe} = f_{pe}/2$

- X-mode

$$k^2 c^2 = \frac{\left(\omega^2 - \omega_-^2\right)\left(\omega^2 - \omega_+^2\right)}{\omega^2 - \omega_{uh}^2} \tag{3.33}$$

with

$$\omega_\pm = \frac{1}{2}\left[\left(\omega_{ce}^2 + 4\omega_{pe}^2\right)^{1/2} \pm \omega_{ce}\right] \tag{3.34}$$

$$\omega_{uh}^2 = \omega_{ce}^2 + \omega_{pe}^2 \tag{3.35}$$

To demonstrate the relationship between the dispersion relations in Equations 3.32 and 3.33 and the ECRH frequency requirements, we consider first the O-mode waves with the dispersion relation given in Equation 3.32. There are some frequencies for which there are no real solutions for the wavenumber k; specifically, for frequencies below the plasma frequency, k is imaginary. The implication is that waves at those frequencies exponentially diminish in magnitude with distance into the plasma (causality rules out the growing wave solution in this case). Thus, waves with $\omega < \omega_{pe}$ cannot penetrate the plasma. In actual fusion plasma confined in the toroidal geometry of a tokamak, the plasma density increases from zero at the edges of the torus to a maximum somewhere near the center. On the other hand, the magnetic field, which is proportional to the cyclotron frequency ω_{ce}, is largest on the inner radius of the torus and decreases toward the outside radius. Applying our knowledge of wave dispersion to this inhomogeneous case, we construct the wave accessibility diagram shown in Figure 3.32a. There we have shaded those regions where locally the frequency is such that we cannot solve the dispersion relation for real values of k (i.e., where $\omega < \omega_{pe}$), on the assumption that waves will not penetrate those regions of the plasma. The electron cyclotron frequency will usually lie above the plasma frequency everywhere in such a plasma, so that waves with O-mode polarization should always be able to penetrate to the electron cyclotron resonance layer.

The X-mode situation is somewhat more complicated. Figure 3.32b shows the wave accessibility diagram for this polarization. We can see that waves are cut off for frequencies below ω_-, which is inconsequential because this frequency is less than ω_{ce} everywhere in the plasma. Additionally, however, no real solutions to the X-mode dispersion relation exist for values of ω between ω_{uh} and ω_+. Since one would not expect waves with frequencies in this cutoff range to pass through the cutoff layer, one cannot achieve electron cyclotron resonance for X-mode waves launched from the outside of the plasma; however, resonance at the electron cyclotron fundamental can be reached by launching these waves from the inner side of the torus. As a

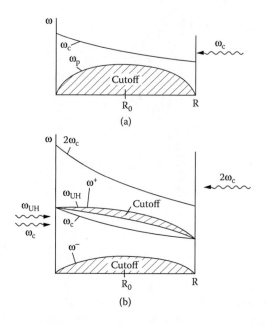

FIGURE 3.32
Wave accessibility diagrams for (a) O-mode and (b) X-mode propagation into plasma. R_0 is the plasma center and R is the edge. The torodial confining magnetic field decreases from inner to outer radius.

practical matter, this method of wave launching is usually undesirable because the inner surface of the torus is typically quite small and crowded. From this standpoint, it becomes desirable to generate X-mode waves at $2\omega_{ce}$, so that they can propagate to this resonant surface after launching from the outer surface of the plasma.

Other factors enter the process of selecting the optimal polarization and frequency for plasma heating. One of these is the strength of absorption at one of the resonant layers ($\omega = \omega_{ce}$ or $2\omega_{ce}$) for a wave with a given polarization. It turns out that the O-mode at $2\omega_{ce}$ is poorly absorbed at the early stages of plasma heating.[58] Therefore, the attractive choices are the O-mode at ω_{ce} and the X-mode at $2\omega_{ce}$, which have roughly equivalent absorption within the plasma. An additional practical consideration is that the production of a CW source at ω_{ce} (roughly 250 GHz at projected-fusion-reactor magnetic fields) with an average power of 1 MW is now a reality (Chapter 10), while the production of a 1-MW CW source at twice this frequency is somewhat more difficult. We therefore are left with O-mode waves at $\omega = \omega_{ce}$ as the primary immediate candidate for ECRH.

3.6.1 Sources for Electron Cyclotron Resonance Heating

Ten years ago, there were three candidate sources for ECRH. The survivor is the gyrotron, usually in a standard arrangement with the microwaves generated in a cavity aligned co-linearly with the beam (see Chapter 10). The other candidates were the cyclotron autoresonant maser (CARM), which promised high-frequency operation at high powers using more modest values of the guiding magnetic field, and the free-electron laser (FEL), which can operate continuously or in a repetitively pulsed, high-peak and average power mode. They were dropped on performance and cost grounds.

Although early heating experiments for fusion used gyroklystrons, and a free-electron laser was used in one experiment, all major ERCH experiments now employ gyro-oscillators. This is because the average power capabilities of gyro-oscillators make them far more desirable.

In addition, gyrotron oscillators have countered the extensive frequency tunability capabilities of gyro-amplifiers by using step tuning by varying the magnetic field in the gyrotron. High power, step-tunable gyrotrons are desirable for position-selective heating of fusion plasmas. Recent results have shown that ECRH can modify temperature profiles to suppress internal tearing modes. The frequency is adjusted to move the absorption region in the torus. This is achieved by using slow step tuning in which, as the magnetic field is varied over a broad range, the oscillation jumps from mode to mode, thus generating a broad and usually discrete radiation spectrum.

Gyrotron oscillators are now commercially available at average powers above 1 MW for times measured in seconds at frequencies of 75 to 170 GHz.[60] This is more than three orders of magnitude beyond the capability of other sources at these frequencies. The most ambitious experiment planned is the International Thermonuclear Experimental Reactor (ITER), which is to produce fusion power of 1 GW. With ECRH, a gyrotron pulse length of $\sim 10^3$ sec and power of ~ 100 MW will be required, meaning groups of gyrotrons operating simultaneously. Because cost is a driving factor, efficiencies greater than 50% will be necessary, and therefore depressed collectors will be essential. The present state of the art in the ITER frequency regime of 165 GHz is a coaxial gyrotron with 48% efficiency producing 2.2 MW.

3.7 Particle Accelerators

We test the limits of our knowledge of high-energy physics with the aid of high-energy accelerators. The triumphant standard model of particles and fields explains much, but is incomplete in not describing the necessary but elusive Higgs particle, as well as the particle masses themselves. The model also has no explanation for the dark matter that makes up 25% of the mass–energy of the universe, nor the dark energy, which is even larger. There are theories (e.g., supersymmetry, strings) of what lies at higher energies, but no accelerators are energetic enough to explore there. So, one line of accelerator development is aimed at the production of beams of electrons with energy of 1 TeV (1000 GeV). The accelerator producing such a beam

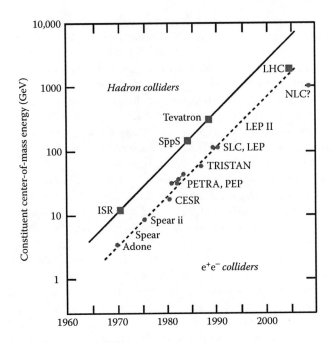

FIGURE 3.33
Evolution of accelerator beam energy for different accelerator types; circular accelerators of heavy particles on the left and linear electron accelerators (linacs) on the right.

will be an *RF accelerator*, in which radio frequency sources (actually microwave, as all approaches operate above the 0.3-GHz boundary between radio and microwave frequencies) produce the electric fields that accelerate the electrons. The evolution of particle accelerators in general, both electron and hadron (heavy particles such as protons) accelerators, is shown in Figure 3.33. Ion accelerators for protons and heavier nuclei use low frequencies (<1 GHz) for acceleration, basically because they are slow compared to the much lighter electrons. The state-of-the-art electron accelerators are the Stanford Linear Collider (SLC; 2.5 GHz, 45 GeV) at the Stanford Linear Accelerator Center (SLAC) in Palo Alto, CA, and the Large Electron-Positron (LEP; 209 GeV) Collider at the European Center for Nuclear Research (CERN) near Geneva, Switzerland. Both produce beams of high-energy electrons and positrons that collide and annihilate each other, generating as a result subatomic particles that provide information about the basic nature of matter and the elementary forces. The SLC, completed in 1987 and soon to be closed, has a linear accelerator, or *linac*, 3 km in length, producing 45-GeV particles. LEP is a large ring-shaped accelerator with a circumference of 26.7 km, initially producing oppositely directed beams of 55-GeV electrons and positrons, now with collision energy to 209 GeV.[61]

Because the synchrotron radiation losses of centrifugally accelerated electrons scale as E^4/R^2, with E the electron energy and R the radius of curvature of the electron orbit,[62] a 1-TeV electron accelerator will be linear. The power needed to drive electrons to such energies is far greater than a single source can supply, so gangs of sources are made to operate in phase synchronism. Amplifiers are used, though phase-locked oscillators can in principle work as well, especially for the smaller energies needed for industrial uses such as radiography and cancer therapy.

A key performance parameter for RF linacs is the accelerating gradient, the rate at which the energy of the electrons increases, measured in MeV/m. The higher the gradient, the shorter a linac can be made for a given final energy. This translates to lower costs for an accelerator, in terms of both hardware and site acquisition costs. The average SLC gradient is 17 MeV/m (the gradient within the accelerating sections is somewhat higher). The goal for a TeV machine is an accelerating gradient of the order of 100 MeV/m or more. As we will discuss shortly, meeting this goal is not possible at present and will require serious improvement in accelerator technology. Gradient increases with frequency because breakdown strengths also increase with frequency. On the other hand, the resonant cavities along the accelerator also become smaller with increased frequency, which reduces margins and tolerances, so much effort has been dedicated to accelerating efficiently through the smaller structures.

Achievement of higher-accelerating gradients would have the added benefit of making lower-energy RF linacs used in medical and industrial applications more compact. We mention two of these applications to illustrate their importance. If electrons are allowed to strike a metallic bremsstrahlung converter, in which their kinetic energy is converted to x-rays, they can be used for industrial radiography or cancer therapy. The energetic electron beams can also be passed through magnetic *undulators* or *wigglers* to produce synchrotron light sources or free-electron lasers (described in Chapter 10).

Much advanced HPM source research has been motivated by the need for higher-energy accelerators. Before we discuss the role HPM sources will play in meeting the goals of higher energy and very high accelerating gradient, we briefly review the basic principles of linacs. Figure 3.34 contains a block diagram for an RF linac. Within the accelerating sections, bunches of electrons are accelerated by the axial electric field component of HPM, as shown in Figure 3.35. The microwaves are produced in pulses of finite duration known as the *macropulse*, while an individual electron bunch is known as a *micropulse*. The number of RF wavelengths between micropulses is known as the *subharmonic number*. The bunched beam for the accelerator is produced in an injector. Of the two main types of injectors, the more common is the lower one shown in the figure, in which electrons from a source are initially accelerated into a bunching section. The bunching section is much like an accelerating section, except that in the *buncher*, electrons that may initially be uniformly distributed along the axis are either accelerated or decelerated according to the phase of the RF signal that they experience. This process, known as *autophasing*, will bunch the electrons in the potential wells of the RF. A second type of injector under development, known as a *photoinjector*, works by pulsed-laser illumination of a cathode made of a

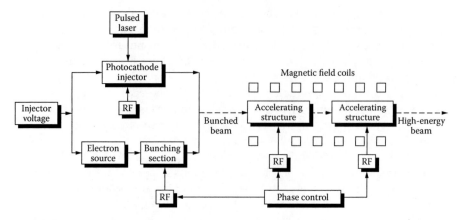

FIGURE 3.34
Diagram of microwave-driven electron linear accelerator. Photocathode injector or conventional electron source produces bunched beam that is injected into the accelerating cavities.

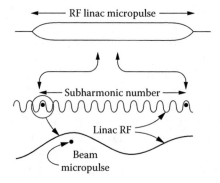

FIGURE 3.35
Electron beam micropulses within the macropulse in an RF wave of a microwave-driven linac.

photoemitting material. With the laser pulsed at the proper frequency, bunches with the proper spacing for the accelerating section can be made in this way.

The accelerating sections tailor properties of the microwave fields so that they can effectively interact with the electron bunches. An example of an accelerating section is shown in Figure 3.36. These sections work in one of two basic fashions. In the *traveling wave accelerator*, the RF is fed into the section at one end and travels through the section with its phase in synchronism with the bunches so that they feel a constant accelerating force; at the exit end of the section, the RF is absorbed. In a *standing wave accelerator*, the RF fields are arranged in a standing wave pattern, with zero-phase velocity along the section axis; the frequency and phase of the fields will be such that the bunch always feels an accelerating polarity of the fields as it passes from cavity to cavity in the accelerating section.

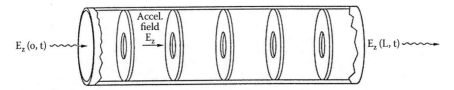

FIGURE 3.36
Accelerating section of a microwave-driven linac using disk-loaded waveguide of length L.

For decades, three accelerating technologies have been under development: copper RF cavities, superconducting RF cavities, and two-beam (also called driving-beam) acceleration. The copper cavity (warm) approach is an evolution of SLC and uses higher-frequency high power (~100-MW) relativistic klystrons, as described in Chapter 9. At SLC this method achieved a 50-MV/m gradient. The superconducting cavities (cold) approach offers higher efficiency, but lower accelerating gradient (~30 MV/m vs. ~100 MV/m projected for copper in future development).

A number of more speculative options are being explored to reach TeV energy levels with accelerating gradients of the order of 100 MeV/m. Some of these are quite speculative and lie outside the scope of our discussion: *wake field* accelerators, switched power accelerators, and laser *beat wave* and wake field accelerators using a plasma medium.[63]

Less speculative, but still unproven on a large scale, is the *two-beam accelerator.*[64] An electron beam at low energy and high current is used to produce HPM, which then accelerate electron bunches in an RF accelerator to high energies. The idea is not unlike that of a transformer from low voltage and high current to high voltage and low current. A possible configuration for such an accelerator using a free-electron laser to generate the RF is shown in Figure 3.37. An intense electron beam with an energy and current of the order of several MeV and several kA is generated in an induction linac, and microwaves are produced in the FEL wiggler sections and transmitted to the high-energy RF linac through the microwave power feeds; the energy tapped from the drive beam in the FEL is replenished in additional induction sections.

The decision has been made to build the International Linear Collider with the cold option, superconducting RF cavities. The decision was on cost/risk grounds and the fact that Germany is already building a coherent light source using the cold technology.[65] The sources required for the cold method are not high power, in contrast to the ~100-MW klystrons of the hot method. The candidates are ~10-MW, 1-msec multibeam klystrons.

The two-beam approach is being carried forward for multi-TeV energies by CERN, in a version different from that of Figure 3.37 called CLIC (Compact Linear Collider),[66] aiming for 3 to 5 TeV. The method is complex, with the ultimate acceleration driven by 30-GHz pulses at powers of several hundred MW. In the present concept, the CLIC Power Extraction and Transfer Structure (PETS) is a passive microwave device where bunches of the

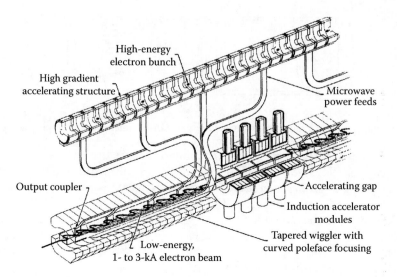

FIGURE 3.37

Two-beam accelerator concept. This type uses a free-electron laser, driven by an electron beam of several kA — several MeV to produce microwaves for the high-gradient accelerating structure of a linac. Induction modules (see Chapter 5) resupply the beam energy lost to microwaves.

drive beam interact with the impedance of a periodically corrugated waveguide and preferentially excite the synchronous traveling wave TM01 mode at 30 GHz.[67] The microwave power produced is collected at the downstream end of the structure by means of the power extractor and is conveyed to the main linac structure in rectangular waveguides. Principal issues are high surface electric and magnetic fields and simply how to build such a complex and carefully aligned system.

CLIC needs high power (~50-MW), long-pulse (~100-sec), 937-MHz multi-beam klystrons (see Chapter 9) to supply the RF power needed to accelerate the low-energy, high-intensity drive beam. The CLIC technology is not yet mature; challenging R&D is required.

The 30-GHz CLIC two-beam is the only method that promises to reach the >100-MV/m regime, making >1-TeV electron colliders possible. The sub-TeV energy range will be explored initially by the Large Hydron Collider (LHC) by 2010. If that range is not found to give physics results attractive enough to build ILC with cold technology, and if particle physicists feel multi-TeV energies are needed to access physics interesting enough to justify the cost of such a large project, CLIC is the alternative to the ILC.

A limiting factor in accelerating gradient is the electric field breakdown limit in the accelerating structure. Breakdown in normal conducting traveling wave structures has been observed to variously depend on a number of parameters, including surface electric field, pulse length, power flow, and group velocity. Current research gives a limit that is determined over the frequency range of 3 to 30 GHz by the quantity $P\tau^{1/3}/C$, where P is the microwave power flow through the structure, τ is the pulse length, and C

is the minimum circumference of the structure. This result comes from experiment.[68] The physical interpretation is that division of power by circumference gives the power flow above a unit transverse width of structure surface. This linear power density is the local power available to feed a discharge, which determines local heating of the structure surface. A limit is reached when the heating causes ablation of the surface material. When the temperature rise due to heat diffusion is caused solely by a constant deposited power, the ablation limit has a square-root time dependence. When an additional cooling mechanism, such as boiling, evaporation, radiation, or two-dimensional heat conduction, is included though, the time dependence of the limit is weaker and $\tau^{1/3}$ becomes plausible.

Breakdown strength also scales up with frequency, going approximately as $f^{7/8}$, experiments show. So higher frequencies can be more attractive, which is one reason CLIC chooses to operate at 30 GHz. As the frequency increases, certain other desirable features are seen: energy stored in the accelerating section and the average RF power required decrease as f^{-2}, dissipation within the structure and the peak RF power required decrease as $f^{-1/2}$, and the section length diminishes as $f^{-3/2}$.

Not all scalings are favorable, however. The dimensions of the accelerating structures scale as the wavelength of the microwaves $\sim f^{-1}$. This means that the size of a structure designed for operation at 30 GHz is measured in millimeters. Hence, tolerances on structure fabrication and beam location become more difficult to meet. Beam quality problems are also exacerbated at higher frequencies by the *wake fields* that are excited by the passage of the electron bunch through the accelerating structure. Roughly speaking, the bunch acts as an impulse that can excite higher-order, undesired modes of the structure. The wake fields associated with these modes can be longitudinal or transverse. The longitudinal wake fields increase as f^2 and cause longitudinal energy spread and a resultant bunch elongation. The transverse wake fields increase as f^3 and tend to push the beam off axis. Overall, at the currents needed in linear colliders, the reduced structure dimensions and increased wake field magnitudes accompanying the higher RF frequencies limit the upper accelerating frequency to about 30 GHz at present.

Problems

1. An HPM technologist, reporting on a device he has built, claims the efficiency of producing microwaves is 56%. The device produces 3 GW in a 100-nsec microwave pulse. The electrical input from a Marx/pulse-forming network (PFN) (see Section 5.2) is at 5.35 GW. This looks good to you for your application. However, you notice that the weight of the contractor's pulsed power system is half a ton

(about 500 kg), much too heavy for your platform. You know that Marx/PFNs typically are available at 300 g/J. What is the contractor not telling you?

2. To scope the effect that Active Denial relies on, suppose a 94-GHz signal is incident on salt water with conductivity at a frequency of 4.3 S/m. What is the skin depth (skin depth of a lossy dielectric material is given by Equation 4.51)? If the conductivity for human skin is 17 S/m, compute the skin depth.

3. If the Vigilant Eagle system attacks a missile with a lethality threshold of 1 mW/cm^2, at a range of 5 km from the airfield to the missile and a beam width of 1°, what transmitter power is required?

4. A missile target is measured to have a power coupled to circuitry inside of 1 W with 100 W/cm^2 incident power. Your tests show that 1 W coupled inside will produce mission failure. To get high P_k you must operate in X-band. Your mission analysis shows that for ensured kill you must defeat the missile at a range of at least 5 km. Your platform allows an antenna aperture no greater than 1 m, and you estimate the antenna efficiency as 60%. What is the coupling cross section? What is the minimum power radiated from the antenna required for this mission? What is the beam size at the target?

5. For NAGIRA, the BWO is reported to have an output of 0.5 GW. What is the microwave production efficiency of the BWO? If frequency stability requires that pulse-to-pulse microwave power be regulated to within 1%, and efficiency of microwave generation is independent of voltage, and the BWO diode is described by Child's law, what pulse-to-pulse voltage variation is allowed?

6. There is an optimum range of power density to operate a rectenna, set by the rectification efficiency of diodes (Figure 3.20). You want to change the operating frequency on a rectenna, but keep both the power beaming antenna aperture and the range to the rectenna fixed. You will need a new rectenna, and to keep efficiency up, will keep its area just large enough to fill the half-power beam width. You know that the area served by a single dipole in the rectenna (Figure 2.19) is proportional to the square of the wavelength. How does the number of dipole/diode elements you need vary with frequency? How does the power per diode vary?

7. Using the rocket equation, for mass ratios of 10, 100, and 1000, what multiples of exhaust velocity can a rocket achieve?

8. What force occurs on a 100-kg perfectly reflecting sail for 1-GW incident beam power? How long would the beam have to accelerate the sail to reach an interplanetary-scale speed of 1 km/sec?

9. For a beam-driven sail, at what range does the beam width size exceed the sail size? If power is constant in time, what speed is

attained at this point? How much more speed results if the beam remains on beyond this point?

10. In a sail using desorption-driven propulsion we measure the temperature T^*, at the peak in the desorption rate dn/dt, as 2000 K. At this time, dn/dt is given by Equation 3.15 and $dT^*/dt = bT^*$. If $b/a = 0.5$, what value of Q did the desorbing material have?

11. Model the capital cost C_C of an interstellar beamer as the cost of aperture with an areal cost coefficient of $B_1(\$/m^2)$, plus the cost of power with coefficient $B_2(\$/W)$. Find the value of beamer diameter D_t and power P that minimizes the cost for accelerating a mass m to a fraction k of the speed of light. Take r as the ratio of beamer diameter D_t to the diameter of the sail riding the beam, d. First, note that at maximum useful beaming range R, the sail diameter d is approximately the diffraction-limited beam width. Find an expression for the maximum useful accelerating time. Find P in terms of D and use Dickinson's observation that minimum cost for this two-parameter model occurs when the cost of power is equal to the cost of aperture. For a light 100-kg sail probe and r = 275, what diameter beamer and power are optimum at 100 GHz?

References

1. Rawles, J., Directed energy weapons: battlefield beams, *Defense Electronics*, 22, 47, 1989.
2. Taylor, T., Third generation nuclear weapons, *Sci. Am.*, 256, 4, 1986.
3. Price, D. et al., Compact pulsed power for directed energy weapons, *J. Directed Energy*, 1, 48, 2003.
4. Benford, J. and Benford, G., Survey of pulse shortening in high power microwave sources, *IEEE Trans. Plas. Sci.*, 25, 311, 1997.
5. Kopp, C., Electromagnetic Bomb, a Weapon of Electromagnetic Mass Destruction, http://abovetopsecret.com/pages/ebomb.html, 2003.
6. Beason, D., *The E-Bomb*, Da Capo Press, Cambridge, MA, 2005.
7. Introduction to FLAPS, http://www.maliburesearch.com/advancedAnt21_IntroFlaps.html.
8. Fulghum, D., Microwave weapons emerge, in *Aviation Week & Space Technology*, www.AviationNow.com/awst, 2005.
9. Singer, J., USAF interest in lasers triggers concerns about anti-satellite weapons, *Space News*, 17, A4, 2006.
10. Velikhov, V., Sagdeev, R., and Kokashin, A., *Weaponry in Space: The Dilemma of Security*, MIR Press, Moscow, 1986.
11. Wunsch, D.C. and Bell, R.R., Determination of threshold failure levels of semiconductor diodes and transistors due to pulsed power voltages, *IEEE Trans. Nucl. Sci.*, NS-15, 244, 1968.
12. Hoad, R. et al., Trends in susceptibility of IT equipment, *IEEE Trans. Electromag. Compatibility*, 46, 390, 2004.

13. Mannheimer, W., Applications of high-power microwave sources to enhanced radar systems, in *Applications of High Power Microwaves,* Gaponov-Grekov, A.V. and Granatstein, V., Eds., Artech House, Boston, 1994, chap. 5, p. 169.
14. Clunie, D. et al., The design, construction and testing of an experimental high power, short-pulse radar, strong microwave, in *Plasmas,* Litvak, A.G., Ed., Novgorod University Press, Nizhny Novgorod, Russia, 1997.
15. Baum, C. et al., The singularity expansion method and its application to target identification, *Proc. IEEE,* 79, 1481, 1991.
16. Nalos, E., New developments in electromagnetic energy beaming, *Proc. IEEE,* 55, 276, 1978.
17. McSpadden, J. and Chang, K., *Microwave Power Transmission,* Wiley Inter-Science, New York, in press.
18. Benford, J., Modification and measurement of the atmosphere by high power microwaves, in *Applications of High Power Microwaves,* Gaponov-Grekov, A.V. and Granatstein, V., Eds., Artech House, Boston, 1994, chap. 12, p. 209.
19. Pozar, D.M., *Microwave Engineering,* 2nd ed., John Wiley & Sons, New York, 1998, p. 665.
20. Hansen, R.C., McSpadden, J., and Benford, J., A universal power transfer curve, *IEEE Microwave Wireless Components Lett.,* 15, 369, 2005.
21. Koert, P. and Cha, J., Millimeter wave technology for space power beaming, *IEEE Trans. Microwave Theory Tech.,* 40, 1251, 1992.
22. Dickinson, R.M., Wireless power transmission technology state-of-the-art, in *Proceedings of the 53rd International Astronautical Congress,* Houston, 2002.
23. McSpadden, J. and Mankins, J., Space solar power programs and wireless power transmission technology, *IEEE Microwave Mag.,* 3, 46, 2002.
24. Dickinson, R.M., Performance of a high-power, 2.388 GHz receiving array in wireless power transmission over 1.54 km, in *1976 IEEE MTT-S International Microwave Symposium,* Cherry Hill, NJ, 1976, p. 139.
25. Brown, W.C., Beamed microwave power transmission and its application to space, *IEEE Trans. Microwave Theory Tech.,* 40, 123, 1992.
26. National Research Council, *Laying the Foundation for Space Solar Power: An Assessment of NASA's Space Solar Power Investment Strategy,* National Academy Press, Washington, DC, 2001.
27. Glaser, P., Davidso, F., and Csigi, K., *Space Solar Satellites,* Praxis Publishing, Chichester, U.K., 1998.
28. Fetter, S., Space solar power: an idea whose time will never come? *Physics and Society,* 33, 10, 2004. See also reply in Smith, A., Earth vs. space for solar energy, round two, *Physics Soc.,* 33, ?, 2002.
29. Criswell, D.R., Energy prosperity within the 21st century and beyond: options and the unique roles of the sun and the moon, in *Innovative Solutions to CO_2 Stabilization,* Watts, R., Ed., Cambridge University Press, Cambridge, U.K., 2002, p. 345.
30. Hoffert, M., Miller, G., Kadiramangalam, M., and Ziegler, W., Earth-to-satellite microwave power transmission, *J. Propulsion Power,* 5, 750, 1989.
31. Matloff, G., *Deep-Space Probes,* 2nd ed., Springer-Verlag, New York, 2005.
32. Zaehringer, A.J., *Rocket Science,* Apogee Books, Burlington, Ontario, 2004.
33. Parkin, K.L.G., DiDomenico, L.D., and Culick, F.E.C., The microwave thermal thruster concept, in *AIP Conference Proceedings 702: Second International Symposium on Beamed-Energy Propulsion,* Komurasaki, K., Ed., Melville, NY, 2004, p. 418.

34. Parkin, K.L.G. and Culick, F.E.C., Feasibility and performance of the microwave thermal rocket launcher, in *AIP Conference Proceedings 702: Second International Symposium on Beamed-Energy Propulsion*, Komurasaki, K., Ed., Melville, NY, 2004, p. 407.

35. Parkin, K.L.G., DiDomenico, L.D., and Culick, F.E.C., The microwave thermal thruster concept, in *AIP Conference Proceedings 702: Second International Symposium on Beamed-Energy Propulsion*, Komurasaki, K., Ed., Melville, NY, 2004, p. 418.

36. Parkin, K.L.G. and Culick, F.E.C., Feasibility and performance of the microwave thermal rocket launcher, in *AIP Conference Proceedings 664: Second International Symposium on Beamed-Energy Propulsion*, Komurasaki, K., Ed., Melville, NY, 2003, p. 418.

37. Benford, J. and Myrabo, L., Propulsion of small launch vehicles using high power millimeter waves, *Proc. SPIE*, 2154, 198, 1994.

38. Benford, J. and Dickinson, R., Space propulsion and power beaming using millimeter systems, *Proc. SPIE*, 2557, 179, 1995. Also published in *Space Energy Transp.*, 1, 211, 1996.

39. Hoppe, D.J. et al., Phase locking of a second harmonic gyrotron using a quasi-optical circulator, *IEEE Trans. Plas. Sci.*, 23, 822, 1995.

40. Lineberry, J. et al., MHD augmentation of rocket engines using beamed energy, in *AIP Conference Proceedings 608: Space Technology and Applications International Forum*, Huntsville, AL, 2003, p. 280.

41. Benford, G. and Benford, J., An aero-spacecraft for the far upper atmosphere supported by microwaves, *Acta Astronautica*, 56, 529, 2005.

42. Kare, J.T. and Parkin, K.L.G., Comparison of laser and microwave approaches to CW beamed energy launch, beamed energy propulsion — 2005, in *AIP Conference Proceedings 830: American Institute of Physics*, Komurasaki, K., Ed., Melville, NY, 2006, p. 388.

43. McInnes, C., *Solar Sailing: Technology, Dynamics, and Mission Applications*, Springer-Verlag, New York, 1999.

44. Forward, R.L., Starwisp: an ultra-light interstellar probe, *J. Spacecraft*, 22, 345, 1985.

45. Landis, G.A., Microwave-Pushed Interstellar Sail: Starwisp Revisited, paper AIAA-2000-3337, presented at the 36th Joint Propulsion Conference, Huntsville, AL, 2000.

46. Benford, J. and Benford, G., Flight of microwave-driven sails: experiments and applications, in *AIP Conference Proceedings 664: Beamed Energy Propulsion*, Pakhomov, A., Ed., Huntsville, AL, 2003, p. 303.

47. Cuneo, M.E., The effect of electrode contamination, cleaning and conditioning on high-energy pulsed-power device performance, *IEEE Trans. Dielectrics Electrical Insulation*, 6, 469, 1999. See also Cuneo, M.E. et al., Results of vacuum cleaning techniques on the performance of LiF field-threshold ion sources on extraction applied-B ion diodes at 1–10 TW, *IEEE Trans. Plas. Sci.*, 25, 229, 1997.

48. Benford, G. and Nissenson, P., Reducing solar sail escape times from earth orbit using beamed energy, *JBIS*, 59, 108, 2006.

49. Landis, G., Beamed energy propulsion for practical interstellar flight, *JBIS*, 52, 420, 1999.

50. Benford, G. and Benford, J., Power-beaming concepts for future deep space exploration, *JBIS*, 59, 104, 2006.

51. Schamiloglu, E. et al., 3-D simulations of rigid microwave propelled sails including spin, *Proc. Space Technol. Appl. Int. Forum AIP Conf. Proc.*, 552, 559, 2001.

52. Benford, G., Goronostavea, O., and Benford, J., Experimental tests of beam-riding sail dynamics, in *AIP Conference Proceedings 664: Beamed Energy Propulsion*, Pakhomov, A., Ed., Huntsville, AL, 2003, p. 325.

53. Benford, G. and Benford, J., An aero-spacecraft for the far upper atmosphere supported by microwaves, *Acta Astronautica*, 56, 529, 2005.

54. Konz, C. and Benford, G., Geometric absorption of electromagnetic angular momentum, *Optics Commun.*, 226, 249, 2003.

55. Benford, G., Goronostavea, O., and Benford, J., Spin of microwave propelled sails, in *AIP Conference Proceedings 664: Beamed Energy Propulsion*, Pakhomov, A., Ed., Huntsville, AL, 2003, p. 313.

56. Benford, J. and Benford, G., Elastic, electrostatic and spin deployment of ultralight sails, *JBIS*, 59, 76, 2006.

57. Prather, R., Heating and current drive by electron cyclotron waves, *Phys. Plas.*, 11, 2349, 2004.

58. Ott, E., Hui, B., and Chu, K.R., Theory of electron cyclotron resonance heating of plasmas, *Phys. Fluids*, 23, 1031, 1980.

59. Chen, F., *Introduction to Plasma Physics*, Plenum Press, New York, 1974, chap. 4.

60. Felch, K.L. et al., Characteristics and applications of fast-wave gyrodevices, *Proc. IEEE*, 87, 752, 1999.

61. Hinchliffe, I. and Battaglia, M., A TeV linear collider, *Physics Today*, 57, 49, 2004.

62. Humphries, S., *Principles of Charged Particle Acceleration*, John Wiley & Sons, New York, 1986.

63. Chandrashekhar, J., Plasma accelerators, *Sci. Am.*, 294, 40, 2006.

64. Sessler, A. and Yu, S., Relativistic klystron two-beam accelerator, *Phys. Rev. Lett.*, 58, 243, 1987.

65. Gamp, G., On the preference of cold RF technology for the International Linear Collider, in *AIP Conference Proceedings 807: High Energy Density and High Power RF, 7th Workshop*, Kalamata, Greece, 2006, p. 1.

66. Van Der Mear, S., The CLIC approach to linear colliders, *Part. Accel.*, 30, 127, 1990.

67. Battaglia, M. et al., Ed., Physics at the CLIC Multi-TeV Linear Collider, report CERN 2004-005 of the CLIC Physics Working Group, http://documents.cern.ch/cernrep/2004/2004-005/2004005.html.

68. Wuensch, W., The Scaling Limits of the Traveling-Wave RF Breakdown Limit, report CERN-AB-2006-013, CLIC Note 649, 2006.

4

Microwave Fundamentals

4.1 Introduction

High power microwaves are generated by transferring the kinetic energy of moving electrons to the electromagnetic energy of the microwave fields.* This process typically occurs in a waveguide or cavity, the role of which is to tailor the frequency and spatial structure of the fields in a way that optimizes the energy extraction from certain natural modes of oscillation of the electrons. In analyzing this process, we deal with the interactions between two conceptual entities: the normal electromagnetic modes of the waveguides and cavities and the natural modes of oscillation of electron beams and layers. The two exist almost independently of one another except for certain values of the frequency and wavelength, for which they exchange energy resonantly. We will therefore begin the chapter by reviewing the basic concepts of electromagnetics and considering the fields within waveguides in the absence of electrons. Our emphasis will be on two key properties of the electromagnetic fields within the waveguide: the spatial configuration of the fields and the relationship between the oscillation frequency and wavelength measured along the system axis. Our treatment will include both smooth-walled waveguides and periodic slow-wave structures, the treatment of the latter requiring a discussion of Floquet's theorem and Rayleigh's hypothesis. We will also touch on two features that play a role in determining the power-handling capability of high power devices: the relationship between the power in a waveguide or cavity and the peak perpendicular field at the wall, a key factor in breakdown, and resistive wall heating in high-average-power devices, either continuous or rapidly pulsed. From waveguides, we will graduate to cavities, which have normal modes of their own that can be treated largely by extension from the treatment of waveguides. The important cavity parameter Q, the so-called *quality factor*, will be a focus of the discussion.

---

* An exception to this is Hertzian oscillators, or wideband and ultrawideband sources such as those described in Chapter 6, in which microwaves are generated by direct coupling between a pulsed power source and an antenna.

Having introduced the key electromagnetic field concepts, we next consider space-charge-limited flow of electrons, both in a diode and for a beam in a drift tube, concepts that are key to the discussion of virtual cathodes. We follow this with a brief discussion of beam propagation, beam equilibria, and the formation of electron layers. We then proceed to a discussion of the natural modes of oscillation on electron beams and layers, and classify the different microwave sources according to the nature of the electron oscillations that facilitate the transfer of electron energy to the fields. Following that, we will take two alternative views of source classification, first by differentiating between oscillators and amplifiers, and then by considering the high- and low-current regimes of operation. We then proceed to address the topic of phase control between multiple sources being made to act in concert. Finally, we conclude with a discussion of the disposition of the spent electron beam downstream of the electromagnetic interaction region.

4.2 Basic Concepts in Electromagnetics

The electric and magnetic fields $\mathbf{E}$ and $\mathbf{B}$ in a microwave device are driven by the charge and current densities ρ and $\mathbf{j}$ according to Maxwell's equations, which are the following:

$$\nabla \times \mathbf{B} = \mu_0 \mathbf{j} + \frac{1}{c^2} \frac{\partial \mathbf{E}}{\partial t} \tag{4.1}$$

$$\nabla \times \mathbf{E} = -\frac{\partial \mathbf{B}}{\partial t} \tag{4.2}$$

$$\nabla \bullet \mathbf{B} = 0 \tag{4.3}$$

$$\nabla \bullet \mathbf{E} = \frac{\rho}{\varepsilon_0} \tag{4.4}$$

On occasion, some use the alternate field variables $\mathbf{D} = \varepsilon \mathbf{E}$ and $\mathbf{H} = \mathbf{B}/\mu$, where ε and μ are generalized values of the permittivity and permeability, equal to ε_0 and μ_0 in a vacuum. The solutions to these equations must conform to certain boundary conditions, which play an important role in determining the mode parameters in a waveguide or cavity, particularly when the mode wavelengths are comparable to the geometrical dimensions of the waveguide or cavity involved. The most straightforward boundary conditions are those that apply at the surface of a perfect conductor. At the surface of a perfect conductor, the tangential component of the electric field and the perpendicular component of the magnetic field must vanish. Therefore, at a point $\mathbf{x}$ on the surface of the conductor, with $\mathbf{n}_t$ a unit vector* tangential to the surface and $\mathbf{n}_p$ a unit vector perpendicular to the surface,

$$\mathbf{n}_t \bullet \mathbf{E}(\mathbf{x}) = 0 \tag{4.5}$$

$$\mathbf{n}_p \bullet \mathbf{B}(\mathbf{x}) = 0 \tag{4.6}$$

The perpendicular component of $\mathbf{E}$ and the tangential component of $\mathbf{B}$, however, can be discontinuous at the surface by an amount equal to the surface charge density and surface current density at the surface; these can be determined by integrating Equations 4.1 and 4.4 across the surface, which we shall discuss in context later in the chapter. We will also defer a discussion of the effect of a finite electrical conductivity σ at the surface.

The source terms ρ and $\mathbf{j}$ arise as a consequence of motion by individual electrons in response to the fields according to Newton's law,

$$\frac{d\mathbf{p}}{dt} = -e(\mathbf{E} + \mathbf{v} \times \mathbf{B}) \tag{4.7}$$

in which the momentum $\mathbf{p} = m\gamma\mathbf{v}$, where m is the mass of an electron, $\mathbf{v}$ is the velocity of an electron, and the relativistic factor $\gamma = (1 - |\mathbf{v}|^2/c^2)^{-1/2}$. Calculation of ρ and $\mathbf{j}$ involves some sort of summing process on the electrons. This closure process of relating electron motions back to the source terms in Maxwell's equations is an important element of any theoretical treatment. A wide variety of techniques are available for this task. For example, if the electrons in a beam or layer can be treated as a fluid, the momentum and velocity in Equation 4.7 can be treated as the momentum and velocity of a fluid element obeying the same equation, and the current density can be written

$$\mathbf{j} = \rho\mathbf{v} \tag{4.8}$$

Alternatively, if there is a considerable variation of the electron velocities about some mean — as would be the case for a beam with substantial divergence or axial velocity spread — then a classical approach based on the Vlasov or Boltzmann equations might be more appropriate. With the wide availability of computers, numerical techniques are increasingly applied. In any event, those who deal with theoretical treatments of microwave systems must be careful to evaluate the validity of an analysis on the basis of whether

---

* A unit vector $\mathbf{n}$ is defined by $|\mathbf{n}| = 1$.

certain common approximations are made. First, one must look for the degree of *self-consistency*, that is, the degree to which the fields, sources, and individual electrons mutually interact with one another. As an example, in systems operating at low currents, certain effects due to the space charge of the electrons can be neglected that might otherwise be significant in high-current devices. Second, one must determine if the approximation of *linearity* has been made. This is a perturbative approach in which functional variables are written as a large zero-order term, usually an equilibrium initial value plus a small time- and spatially-varying perturbation. Such analyses are adequate for determining the small-signal or early-time behavior of systems, but can fail for large-signal amplitudes. Third, one might see *multiple time-scale expansions*, in which time averaging is used to separate rapidly, but periodically, varying phenomena from the slow evolution of average system properties. Such analyses can fail when large-scale system features vary rapidly. Fourth, in numerical simulations, it is not uncommon to see reduced treatments of some sort, to limit the computing time and memory demand to manageable levels. Again, to give an example, in particle code simulations, many times the calculation is limited to one or two spatial dimensions, with variations in the remaining variables neglected.

4.3 Waveguides

Waveguides act as ducts for propagating microwave radiation and, under the proper circumstances in the presence of an electron beam, the interaction region within which microwaves are generated. They come in a wide variety of shapes and sizes, depending on the needs of the user. Let us first consider the case of waveguides with perfectly conducting walls, with no variation of the cross section along the axis and no electrons or dielectric within the waveguide (i.e., a vacuum waveguide). If we take the curl of Equation 4.1 and use Equations 4.2 and 4.3, we find that **B** obeys the wave equation within the waveguide when the source terms **j** and ρ vanish on the right-hand sides of Equations 4.1 and 4.4:

$$\nabla^2 \mathbf{B} - \frac{1}{c^2}\frac{\partial^2 \mathbf{B}}{\partial t^2} = 0 \qquad (4.9)$$

Similarly, taking the curl of Equation 4.2 and using Equations 4.1 and 4.4, we find that **E** obeys the same equation:

$$\nabla^2 \mathbf{E} - \frac{1}{c^2}\frac{\partial^2 \mathbf{E}}{\partial t^2} = 0 \qquad (4.10)$$

Note that both of these equations are linear in the fields, **B** and **E**, so that the solutions are valid for any magnitude of the field.

To proceed, let us consider the two independent classes of solutions:*

- *Transverse magnetic* modes (TM, or E-modes) with no axial component of **B** (i.e., $B_z = 0$)
- *Transverse electric* modes (TE, or H-modes) with $E_z = 0$

In the former case, we express the transverse field components in terms of E_z, while in the latter, we express the transverse field components in terms of B_z. Because we have assumed an axial symmetry for our waveguide, we can write both **B** and **E** in the form

$$\mathbf{E}(\mathbf{x},t) = \mathbf{E}(\mathbf{x}_\perp)\exp\left[i\left(k_z z - \omega t\right)\right] \tag{4.11}$$

where $\mathbf{x}_\perp$ is a vector in the plane perpendicular to the z axis. The simplicity of this relation depends on the fact that the shape of the cross section has no variation along the axis or in time. In Equation 4.11, the *wavenumber* k_z is related to the wavelength along the axis of the waveguide, λ_w (which is not, in general, equal to the free-space wavelength $\lambda = c/f$), by

$$k_z = \frac{2\pi}{\lambda_w} \tag{4.12}$$

and the radian frequency ω is related to the frequency f by

$$\omega = 2\pi f \tag{4.13}$$

Now, for TM modes, using Equation 4.10 and the form in Equation 4.11, E_z obeys

$$\nabla_\perp^2 E_z - k_z^2 E_z + \frac{\omega^2}{c^2} E_z = 0 \tag{4.14}$$

where $\nabla_\perp^2$ is the portion of the Laplacian operator that describes variations in the transverse plane of the waveguide. Similarly, for TE modes, using Equation 4.9, B_z obeys

---

* Another possible solution, the *transverse electromagnetic mode*, with no axial electric or magnetic field, is permitted in certain geometries, for example, a coaxial cylindrical waveguide or a parallel-plate configuration; however, it is not supported by a hollow cylindrical waveguide or a rectangular waveguide.

$$\nabla_\perp^2 B_z - k_z^2 B_z + \frac{\omega^2}{c^2} B_z = 0 \tag{4.15}$$

Both Equations 4.14 and 4.15 must be solved subject to the boundary conditions in Equations 4.5 and 4.6 for perfectly conducting walls, that the electric field components tangential to the wall and the magnetic field components perpendicular to the wall all vanish.

In general, for waveguides without axial variation in the cross section, we expect that we can separate the cross-sectional variation in the fields from that along the axis and in time, so that we can write

$$\nabla_\perp^2 E_z = -k_{\perp,TM}^2 E_z \tag{4.16}$$

$$\nabla_\perp^2 B_z = -k_{\perp,TE}^2 B_z \tag{4.17}$$

The eigenvalues on the right-hand side of these equations, $k_{\perp,TM}$ and $k_{\perp,TE}$, will depend on the cross-sectional shape of the waveguide; we will give examples shortly. Thus, we can rewrite Equations 4.14 and 4.15 in the form

$$\left(\frac{\omega^2}{c^2} - k_{\perp,TM}^2 - k_z^2 \right) E_z = 0 \tag{4.18}$$

$$\left(\frac{\omega^2}{c^2} - k_{\perp,TE}^2 - k_z^2 \right) B_z = 0 \tag{4.19}$$

Because **E** and **B** cannot vanish everywhere within the waveguide interior, the terms in parentheses in each of these equations must vanish, conditions we rewrite as:

$$\omega^2 = k_{\perp,TM}^2 c^2 + k_z^2 c^2 + \equiv \omega_{co}^2 + k_z^2 c^2 \tag{4.20}$$

$$\omega^2 = k_{\perp,TE}^2 c^2 + k_z^2 c^2 + \equiv \omega_{co}^2 + k_z^2 c^2 \tag{4.21}$$

where we have defined the *cutoff frequency* ω_{co} (without the subscripted TM or TE) as

$$\omega_{co} = k_{\perp,TM} c \quad \text{or} \quad k_{\perp,TE} c \tag{4.22}$$

We call ω_{co} the *cutoff frequency* because it is the minimum frequency at which a given mode can propagate along the waveguide; for lower frequencies, the mode is cut off. To see this, reorder the equation to see that $k_z^2 = (\omega^2 - \omega_{co}^2)/c^2$. Thus, if $\omega^2 < \omega_{co}^2$, $k_z^2 < 0$. Since there is no energy supplied to a passive waveguide (our beams come later), this means that k_z must be imaginary, and the waves decline exponentially from the point at which they are launched.

Equations 4.20 and 4.21 are examples of *dispersion relations*, the relationship between k_z and ω, or equivalently between λ_w and f, that must be obeyed within a device. Note here that only the eigenvalue $k_\perp$ — or the cutoff frequency ω_{co} — varies from one system to the next, depending only on the geometry of the waveguide. We will show that there are in fact an infinite number of cutoff frequencies within a waveguide, each corresponding to a *normal mode* of oscillation for the waveguide. These normal modes are the natural modes of electromagnetic oscillation in the waveguide, analogous to the waves on a string fixed at both ends, where higher-order modes correspond to shorter and shorter wavelengths along the string. They are distinguished by several properties. If we excite one of the normal modes, the oscillation in this mode will persist in time. If we drive the system with an impulse, then at long times after any initial transients have decayed the wave pattern within the system will be a linear sum, or *superposition*, of a set of normal modes (another consequence of the linearity of Maxwell's equations in the absence of charges and currents). Alternatively, this sum pattern of normal modes will be found at one end of a long system if we excite it at the other end with an impulse. In the following sections, we consider two of the most common types of waveguides, those with rectangular and circular cross sections.

4.3.1 Rectangular Waveguide Modes

Let us first consider a waveguide with rectangular cross section as shown in Figure 4.1, oriented so that a is the long dimension, $a \geq b$. The walls extend parallel to the x axis from $x = 0$ to $x = a$, and parallel to the y axis from $y = 0$ to $y = b$. In this geometry, $x_\perp$ is a vector in the x-y plane, and the eigenvalue equations in Equations 4.16 and 4.17 become

$$\nabla_\perp^2 E_z = \left(\frac{\partial^2}{\partial x^2} + \frac{\partial^2}{\partial y^2} \right) E_z = -k_{\perp,TM}^2 E_z \qquad (4.23)$$

$$\nabla_\perp^2 B_z = \left(\frac{\partial^2}{\partial x^2} + \frac{\partial^2}{\partial y^2} \right) B_z = -k_{\perp,TE}^2 B_z \qquad (4.24)$$

The boundary conditions are the following:

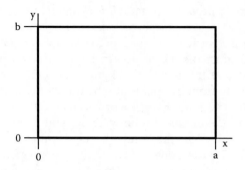

FIGURE 4.1
Cross section of a rectangular waveguide.

$$B_x\left(x=0,y\right)=B_x\left(x=a,y\right)=0\ ,0\le y\le b \tag{4.25a}$$

$$B_y\left(x,y=0\right)=B_y\left(x,y=b\right)=0\ ,0\le x\le a \tag{4.25b}$$

$$E_y\left(x=0,y\right)=E_z\left(x=0,y\right)=E_y\left(x=a,y\right)=E_z\left(x=a,y\right)=0\ ,0\le y\le b \tag{4.25c}$$

$$E_x\left(x,y=0\right)=E_z\left(x,y=0\right)=E_y\left(x,y=b\right)=E_z\left(x,y=b\right)=0\ ,0\le x\le a \tag{4.25d}$$

One can easily show that the solution to Equation 4.23 for the TM wave subject to the boundary conditions in Equation 4.25 is

$$E_Z = D\sin\left(\frac{n\pi}{a}x\right)\sin\left(\frac{p\pi}{b}y\right) \tag{4.26}$$

where D is a constant related to the amplitude of the wave. We can see from Equation 4.26 that for TM waves, neither n nor p can be zero. The eigenvalue and cutoff frequency are thus

$$k_{\perp,TM}\left(n,p\right)=\frac{\omega_{co}\left(n,p\right)}{c}=\left[\left(\frac{n\pi}{a}\right)^2+\left(\frac{p\pi}{b}\right)^2\right]^{1/2} \tag{4.27}$$

Similarly, one can show that the TE mode solutions are

$$B_z = A\cos\left(\frac{n\pi}{a}x\right)\cos\left(\frac{p\pi}{b}y\right) \tag{4.28}$$

TABLE 4.1

Expressions for the TM and TE Field Quantities for a
Rectangular Waveguide in Terms of the Axial Field
Component, Derived from Equations 4.1 to 4.4

Transverse Magnetic, $TM_{n,p}$	Transverse Electric, $TE_{n,p}$
$E_z = D\sin\left(\dfrac{n\pi}{a}x\right)\sin\left(\dfrac{p\pi}{b}y\right)$	$B_z = A\cos\left(\dfrac{n\pi}{a}x\right)\cos\left(\dfrac{p\pi}{b}y\right)$
$B_z \equiv 0$	$E_z \equiv 0$
$E_x = i\dfrac{k_z}{k_\perp^2}\dfrac{\partial E_z}{\partial x}$	$E_x = i\dfrac{\omega}{k_\perp^2}\dfrac{\partial B_z}{\partial y}$
$E_y = -i\dfrac{k_z}{k_\perp^2}\dfrac{\partial E_z}{\partial y}$	$E_y = -i\dfrac{\omega}{k_\perp^2}\dfrac{\partial B_z}{\partial x}$
$B_x = -i\dfrac{\omega}{\omega_{co}^2}\dfrac{\partial E_z}{\partial y}$	$B_x = i\dfrac{k_z}{k_\perp^2}\dfrac{\partial B_z}{\partial x}$
$B_y = i\dfrac{\omega}{\omega_{co}^2}\dfrac{\partial E_z}{\partial x}$	$B_y = i\dfrac{k_z}{k_\perp^2}\dfrac{\partial B_z}{\partial y}$

$$\omega_{co} = k_\perp c = \left[\left(\frac{n\pi c}{a}\right)^2 + \left(\frac{p\pi c}{b}\right)^2\right]^{1/2}$$

$$k_{\perp,TE}(n,p) = \frac{\omega_{co}(n,p)}{c} = \left[\left(\frac{n\pi}{a}\right)^2 + \left(\frac{p\pi}{b}\right)^2\right]^{1/2} \qquad (4.29)$$

In this case, either, but not both, n or p can vanish. Table 4.1 contains the
mathematical expressions for the other field components in terms of the axial
field component. We see that the eigenvalues for the TE and TM modes are
exactly the same, a situation that we will see does *not* hold for the normal
modes of a cylindrical waveguide.

Note that the cutoff wavelength for the mode with the lowest cutoff frequency, the TE_{10} mode (known as the *fundamental mode* for a rectangular
waveguide), is twice the long dimension of the waveguide:

$$\lambda_{co}(1,0) = \frac{2\pi c}{\omega_{co}(1,0)} = 2a$$

This feature has practical significance, based on a historical precedent. The
standard designation for the many sizes of waveguide that are commonly
built is the *WR number*, which stands for "waveguide, rectangular." The long

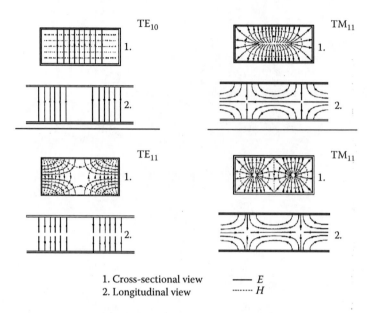

1. Cross-sectional view   —— E
2. Longitudinal view   ------ H

FIGURE 4.2
Field patterns for four of the lowest-order modes of a rectangular waveguide. In each case, the upper plot shows the fields in the cross section of the waveguide, the x-y plane, while the lower plot shows the fields in the y-z plane. (From Saad, T.S. et al., *Microwave Engineers Handbook*, Vol. 1, Artech House, Norwood, MA, 1971. With permission.)

dimension, commonly called a, measured *in inches*, is multiplied by 100 to give the number of the guide. So, a rectangular waveguide with a = 2.84 inches is called WR284. The guide has a cutoff in fundamental of 2.08 GHz and is optimal if operated in the 2.6- to 3.95-GHz band. A table of standard rectangular waveguide properties appears in the formulary (see Problem 1).

Field plots for four of the lowest-order modes in a rectangular waveguide are shown in Figure 4.2. Let us return to the dispersion relations in Equations 4.20 and 4.21. Dispersion relations are an essential tool in understanding waves in waveguides and cavities and how they interact with electrons to generate microwaves. We rewrite these equations in the form of an equation for ω (dropping the subscripts on $k_\perp$ for convenience):

$$\omega = \left(k_\perp^2 + k_z^2\right)^{1/2} c = \left(k_z^2 c^2 + \omega_{co}^2\right)^{1/2} \qquad (4.30)$$

There is a different cutoff frequency for each mode, $TM_{n,p}$ or $TE_{n,p}$, so that there are, in fact, a hierarchy of such dispersion curves. Remember that we have oriented the waveguide so that $a \geq b$, so that the lowest cutoff frequency is that for the $TE_{1,0}$ mode. To understand the physical significance of the cutoff frequency, note that if we establish the correspondence that $k_\perp^2 \sim k_x^2 + k_y^2$, with $k_x \sim n\pi/a$ and $k_y \sim p\pi/b$, then Equation 4.30 is essen-

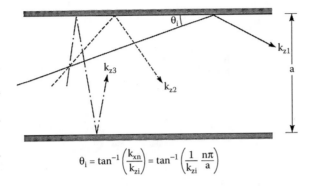

$$\theta_i = \tan^{-1}\left(\frac{k_{xn}}{k_{zi}}\right) = \tan^{-1}\left(\frac{1}{k_{zi}}\frac{n\pi}{a}\right)$$

FIGURE 4.3
Ray diagrams for waves traveling along a rectangular waveguide. As i increases from 1 to 3, k_{zi} approaches 0 and the wave approaches cutoff.

tially the dispersion relation for an electromagnetic wave propagating in free space, $\omega = kc = (k_x^2 + k_y^2 + k_z^2)^{1/2}c$. From a ray-optic viewpoint, we can view wave propagation within a rectangular waveguide as if a plane wave were reflecting from the walls with a wavefront perpendicular to the wavevector $\mathbf{k} = \hat{x}k_x + \hat{y}k_y + \hat{z}k_z$. Because the perpendicular component of **B** and the tangential component of **E** must vanish at the walls of this bounded geometry, the values of k_x and k_y are quantized. In Table 4.1, we see that n and p cannot both simultaneously vanish, or the field quantities would become infinite because $k_\perp$ would vanish.

In Figure 4.3, we see that as $k_z \to 0$, the direction of wave propagation along **k** becomes perpendicular to the axis, so that wave energy no longer propagates along the axis. Figure 4.4 is a plot of Equation 4.20 or Equation 4.21 for a single mode. For a given mode, we can see that there is no solution for ω in the range $0 \le \omega < \omega_{co}$, and there are no real values of ω for any mode below the lowest cutoff frequency for $n = 1$ and $p = 0$. On the other hand, for waves at a given frequency f, a waveguide can support every mode that has a cutoff frequency below $\omega = 2\pi f$. For example, imagine that we excite a waveguide with dimensions 7.214×3.404 cm, the dimensions for the WR284 waveguide, at $f = 5$ GHz. The lowest cutoff frequencies for the TE and TM modes are shown in Table 4.2. We can see that as many as five modes could be excited by a 5-GHz signal — $TE_{1,0}$, $TE_{2,0}$, $TE_{0,1}$, $TE_{1,1}$, and $TM_{1,1}$ — the cutoff frequency for the last two being the same, since the TE and TM modes have the same cutoff frequencies in rectangular waveguides (although there are no $TM_{0,p}$ or $TM_{n,0}$ modes). The cutoff frequencies for higher-order modes lie above the 5-GHz signal, so they will not be excited.

As a last point in this discussion of rectangular waveguides, we introduce the two important velocities associated with wave propagation: the *phase velocity*, v_ϕ, which is the speed at which the phase fronts advance along the axis,

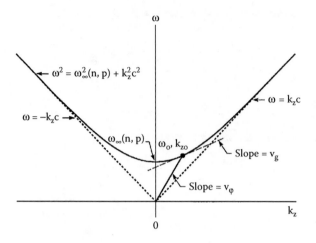

FIGURE 4.4
The dispersion relation between ω and k_z in Equation 4.30. The slopes of the two lines shown are defined in Equations 4.31 and 4.32.

TABLE 4.2

Cutoff Frequencies, Measured in GHz, for the Lowest-Order Modes of the WR284 Waveguide

n	p	$f_{co} = \omega_{co}/2\pi$ (GHz) a = 7.214 cm, b = 3.404 cm
1	0	2.079
0	1	4.407
1	1	4.873
1	2	9.055
2	0	4.159
2	1	6.059

$$v_\varphi = \frac{\omega}{k_z} \tag{4.31}$$

and the *group velocity*, v_g, which is the speed at which energy is transported along the axis,

$$v_g = \frac{\partial \omega}{\partial k_z} \tag{4.32}$$

In Figure 4.4, v_ϕ varies from ∞ at $k_z = 0$ to c for large values of the axial wavenumber, while v_g varies from 0 to c. The variation near $k_z = 0$ is consistent with our model of phase front motion in Figure 4.3; as the phase fronts becomes parallel to the walls of the guide, the phase velocity tends toward infinity, and as the waves bounce back and forth from one wall to the other, with no axially directed motion, the group velocity vanishes. For large values of k_z, the wavelength along the axis, $\lambda_w = 2\pi/k_z$, becomes much smaller than the cross-sectional dimensions of the guide, a and b, so that a mode propagates quasi-optically, with little effect from the waveguide.

4.3.2 Circular Waveguide Modes

Next consider a waveguide with circular cross section, with a perfectly conducting wall located at $r = r_0$. In this case, $x_\perp$ lies in the r-θ plane. Further, we can write **B** and **E** in the form

$$E(r,\theta,z,t) = E(r,\theta)\exp\left[i\left(k_z z - \omega t\right)\right] \tag{4.33}$$

Now, Equations 4.16 and 4.17 take the forms

$$\nabla_\perp^2 E_z = \frac{1}{r}\frac{\partial}{\partial r}\left(r\frac{\partial E_z}{\partial r}\right) + \frac{1}{r^2}\frac{\partial^2 E_z}{\partial \theta^2} = -k_{\perp,TM}^2 E_z \tag{4.34}$$

$$\nabla_\perp^2 B_z = \frac{1}{r}\frac{\partial}{\partial r}\left(r\frac{\partial B_z}{\partial r}\right) + \frac{1}{r^2}\frac{\partial^2 B_z}{\partial \theta^2} = -k_{\perp,TE}^2 B_z \tag{4.35}$$

The boundary conditions in cylindrical geometry are the following:

$$B_r\left(r = r_0\right) = E_\theta\left(r = r_0\right) = E_z\left(r = r_0\right) = 0 \tag{4.36}$$

Expressions for the transverse field components in terms of the axial field components for both the TM and TE modes are given in Table 4.3. There, we also show the axial field solutions to Equations 4.34 and 4.35, subject to the boundary conditions of Equation 4.36. Note first that the solutions for the axial field terms involve Bessel functions of the first kind, J_p, several examples of which are plotted in Figure 4.5. These functions are oscillatory, although the period of oscillation in the radial direction is not fixed. Second, note that the eigenvalues for the TM and TE modes differ, unlike those for the rectangular waveguide, which are the same for both modes. As we show in the table, the eigenvalues for the TM modes involve the roots of J_p:

$$J_p\left(\mu_{pn}\right) = 0 \tag{4.37}$$

TABLE 4.3

Expressions for the TM and TE Field Quantities for a Circular Waveguide in Terms of the Axial Field Components, Derived from Equations 4.1 to 4.4

Transverse Magnetic, $TM_{p,n}$ $(B_z = 0)$	Transverse Electric, $TE_{p,n}$ $(E_z = 0)$
$E_z = DJ_p(k_\perp r)\sin(p\theta)$	$B_z = AJ_p(k_\perp r)\sin(p\theta)$
$E_r = i\dfrac{k_z}{k_\perp^2}\dfrac{\partial E_z}{\partial r}$	$E_r = i\dfrac{\omega}{k_\perp^2}\dfrac{1}{r}\dfrac{\partial B_z}{\partial\theta}$
$E_\theta = i\dfrac{k_z}{k_\perp^2}\dfrac{1}{r}\dfrac{\partial E_z}{\partial\theta}$	$E_\theta = -i\dfrac{\omega}{k_\perp^2}\dfrac{\partial B_z}{\partial r}$
$B_r = -i\dfrac{\omega}{\omega_{co}^2}\dfrac{1}{r}\dfrac{\partial E_z}{\partial\theta}$	$B_r = i\dfrac{k_z}{k_\perp^2}\dfrac{\partial B_z}{\partial r}$
$B_\theta = i\dfrac{\omega}{\omega_{co}^2}\dfrac{\partial E_z}{\partial r}$	$B_\theta = i\dfrac{k_z}{k_\perp^2}\dfrac{1}{r}\dfrac{\partial B_z}{\partial\theta}$
$k_\perp = \dfrac{\omega_{co}}{c} = \dfrac{\mu_{pn}}{r_0}$	$k_\perp = \dfrac{\omega_{co}}{c} = \dfrac{\nu_{pn}}{r_0}$
$J_p(\mu_{pn}) = 0$	$J_p{'}(\nu_{pn}) = 0$

Note: A specific choice of sine functions has been made for convenience in the solutions for E_z and B_z, although both sines and cosines could be used. Note the ordering on the subscripts for the TE and TM modes; p and n are reversed, to be consistent with common practice.

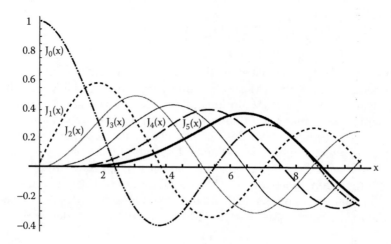

FIGURE 4.5
The first five Bessel functions of the first kind.

which for J_0 occur at $x = 2.40, 5.52, 8.65,$ and so on. Similarly, the eigenvalues for the TE modes involve the roots of the derivatives of J_p:

$$\frac{dJ_p\left(x = v_{pn}\right)}{dx} = 0 \qquad (4.38)$$

In Figure 4.5, we can see, for example, that the derivative of J_0 vanishes at $x = 3.83$ (n = 1), 7.02 (n = 2), 10.2 (n = 3), and so on; the solution at $x = 0$ is not valid. Third, we note that we have made a choice with regard to the θ dependence of the solutions; both the sine and cosine are valid solutions for the axial field components of the TE and TM modes, and we have made specific choices in the absence of any wall perturbation that might disturb the symmetry in a perfectly cylindrical waveguide. Fourth, note that by historical convention, the TE and TM modes in circular waveguides have the indices ordered in a manner consistent with the ordering of indices on the roots of the Bessel functions in Equation 4.37 and the roots of the derivatives in Equation 4.38, so that for cylindrical modes we write TE_{pn} and TM_{pn}, with the azimuthal index first; *to emphasize, the first index is always the azimuthal index for cylindrical modes.*

In the rectangular waveguides, as the indices n and p vary, the spacing between eigenmodes depends on both dimensions, a and b. In circular waveguides, on the other hand, as n and p vary, there is only one dimensional parameter to consider, r_0, so that relative spacing between modes, after we normalize to the dimensional parameter c/r_0, is fixed. In Figure 4.6, we plot the normalized values of the cutoff frequencies for the lowest-order normal modes of a circular waveguide for positive values of p. The index n must be greater than 0, as we noted; however, the index p can range over positive and negative. In the absence of a beam, $\omega(p, n) = \omega(-p, n)$; as we shall see later, though, this symmetry can be broken in the presence of a beam with rotating electrons. We see that the $TE_{1,1}$ mode has the lowest cutoff frequency of all modes in the circular waveguide. The figure also emphasizes the difference between the cutoff frequencies for the TE and TM modes. The dispersion relation for circular waveguides has exactly the same form as that shown in Figure 4.4, taking account of the difference in cutoff frequencies (see Problems 2 and 3).

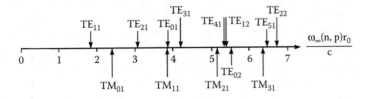

FIGURE 4.6
The cutoff frequencies for a circular waveguide, normalized to the factor c/r_0. The mathematical expressions for the cutoff frequencies are given in Table 4.3.

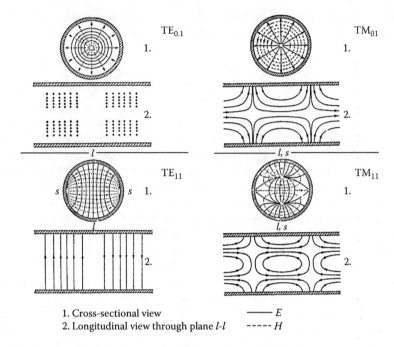

1. Cross-sectional view   ——— E
2. Longitudinal view through plane l-l ----- H

FIGURE 4.7
Field patterns for four normal modes of a circular waveguide. (From Saad, T.S. et al., *Microwave Engineers Handbook*, Vol. 1, Artech House, Norwood, MA, 1971. With permission.)

Figure 4.7 shows the field patterns for four of the circular waveguide modes. The $TE_{1,1}$ mode is commonly used because of the nearly plane-wave-like structure of the fields near the center of the waveguide. The $TM_{0,1}$ mode is also commonly seen, although it is less desirable in a number of applications because its intensity pattern has a minimum on the axis.

4.3.3 Power Handling in Waveguides and Cavities

The power-handling capability of a waveguide or cavity depends on two factors. At shorter pulses, roughly less than about 1 μsec, a key issue is breakdown, while at longer pulses, or at high duty factors so that the average power becomes significant, wall heating predominates. High-field breakdown of metallic waveguides and cavities is a complex subject that is reviewed elsewhere in greater detail.[1] Cleaning and conditioning of surfaces, maintaining stringent vacuum requirements, and applying surface coatings are elements necessary to maximize the power level at which breakdown occurs. The magnitude of the electric field at the wall is a key parameter determining when breakdown will occur. In a recent review, metal surfaces coated with a thin layer of titanium nitride (TiN) were found to break down when these fields exceeded one to several hundred MV/m. We therefore first consider the relationship between the power in a waveguide and the electric field at the wall. We will discuss how to determine this relationship for TM modes in a rectangular waveguide, and then provide the relationships for the TE mode in a rectangular waveguide, and the TM and TE modes in circular waveguides.

The *Poynting vector* $\mathbf{S} = \mathbf{E} \times \mathbf{H}$ measures the direction and magnitude of the power flux carried by an electromagnetic wave. The integral of this vector over a surface Σ is interpreted as the power flowing through that surface. For oscillating fields of the form in Equation 4.11, the average power through that surface is given by

$$P = \int_{\Sigma} \langle \mathbf{S} \rangle \bullet d\mathbf{A} \equiv \frac{1}{2} \int_{\Sigma} \mathbf{E} \times \mathbf{H}^* \bullet d\mathbf{A} = \frac{1}{2\mu_0} \int_{\Sigma} \mathbf{E} \times \mathbf{B}^* \bullet d\mathbf{A} \quad (4.39)$$

In a rectangular waveguide, this expression becomes

$$P = \frac{1}{2\mu_0} \int_0^b \int_0^a \left(E_x B_y^* - E_y B_x^* \right) dx dy \quad (4.40)$$

For TM waves, taking the form of the fields given in Table 4.1, integrating by parts, and using Equation 4.23 and the form for E_z in the table, we find

$$P_{TM} = \frac{1}{2\mu_0} \left(\frac{\omega k_z}{\omega_{co}^2} \right) \int_0^b \int_0^a \left| E_z \right|^2 dx dy = \frac{ab}{8\mu_0} \left(\frac{\omega k_z}{\omega_{co}^2} \right) |D|^2 \quad (4.41)$$

Here, D is the field magnitude. We express this in terms of the wall field using the expression for the maximum value of the perpendicular field component at the wall, which is the maximum value of either E_x at the surfaces $x = 0$ and $x = a$ or E_y at the surfaces $y = 0$ and $y = b$. Defining $E_{wall,max}$ to be the larger of the two magnitudes, we use the expressions for E_x and E_y in Table 4.1 to eliminate $|D|^2$, finding

$$P_{TM} = \frac{ab}{8\pi^2 Z_0} k_{\perp TM}^2 \left(\frac{\omega}{k_z} \right) \min \left[\left(\frac{a}{n} \right)^2, \left(\frac{b}{p} \right)^2 \right] E_{wall,max}^2 \quad (4.42)$$

where min(x, y) is the smaller of x or y, since the larger wall field, E_x or E_y, corresponds to the smaller of a/n or b/p (i.e., if a/n < b/p, $E_{wall,max}$ is the maximum value of E_x at the wall), and

$$Z_0 = \mu_0 c = \left(\frac{\mu_0}{\varepsilon_0} \right)^{1/2} = 377 \ \Omega \quad (4.43)$$

To complete this derivation, we use Equation 4.20 to find k_z in terms of ω and $k_{\perp,TM} = \omega_{co}/c = f_{co}/(2\pi c)$:

$$P_{TM} = \frac{ab}{8Z_0}\left[\left(\frac{n}{a}\right)^2 + \left(\frac{p}{b}\right)^2\right]\frac{1}{\left[1-\left(\frac{f_{co}}{f}\right)^2\right]^{1/2}}\min\left[\left(\frac{a}{n}\right)^2,\left(\frac{b}{p}\right)^2\right]E_{wall,max}^2 \qquad (4.44)$$

We see that as the waveguide dimensions increase, the power-handling capability increases for a constant $E_{wall,max}$. Also, the wall field for constant P decreases as $f \rightarrow f_{co}$.

In Table 4.4, we collect the expressions for the relationships between P and $E_{wall,max}$ for TM and TE modes in rectangular and circular waveguides. Separate expressions are required for TE modes in rectangular waveguides, depending on whether one of the field indices n or p vanishes, although both cannot vanish simultaneously. The factor of two difference stems from the fact that one of the cosine terms in the expression for B_z has a vanishing argument. One of the most important modes in a rectangular waveguide is the lowest-frequency $TE_{1,0}$ mode, for which Equation 4.42 reduces to

$$P_{TE}\left(TE_{1,0}\right) = \frac{ab}{4Z_0}\left[1-\left(\frac{c}{2af}\right)^2\right]^{1/2}E_{wall,max}^2 \qquad (4.45)$$

(see Problem 4).

In the case of circular waveguides, $E_{wall,max}$ is the maximum value of $E_r(r_0)$. The situation for TE modes is somewhat different, since E_r vanishes identically for p = 0. Therefore, the relation in the table holds only for p > 0. Two scaling relationships stand out in all of these expressions. First, the scaling with frequency is quite different for TM and TE modes. As $f \rightarrow f_{co}$ for a fixed value of the wall field, the power in the waveguide increases in the case of TM modes, while it must decrease for TE modes. Second, in the case of both TM and TE modes, for a fixed value of the wall field, the power increases roughly as the area of the cross section (see Problem 5).

The relationship between the electric field component perpendicular to the wall and the power flowing through a waveguide is important in the case of high-field breakdown on short timescales. When the length of the microwave pulse is long enough, or when a high power signal is pulsed rapidly enough at a high duty factor (defined as the product of the pulse repetition rate and the length of an individual pulse), the walls come into thermal equilibrium, and wall heating becomes an important operational factor for waveguides and cavities. To understand wall heating, we return to the analysis of the normal-mode fields. In our previous derivations, we assumed that the waveguide walls were perfectly conducting. In such ideal guides, a

TABLE 4.4

Relationship between the Power in a Given Mode, P, and the Maximum Value of the Electric Field at the Wall, $E_{wall,max}$

Mode	

Rectangular

TM

$$P_{TM} = \frac{ab}{8Z_0}\left[\left(\frac{n}{a}\right)^2 + \left(\frac{p}{b}\right)^2\right]\frac{1}{\left[1-\left(\frac{f_{co}}{f}\right)^2\right]^{1/2}}\min\left[\left(\frac{a}{n}\right)^2,\left(\frac{b}{p}\right)^2\right]E_{wall,max}^2$$

TE
n, p > 0

$$P_{TE} = \frac{ab}{8Z_0}\left[\left(\frac{n}{a}\right)^2 + \left(\frac{p}{b}\right)^2\right]\left[1-\left(\frac{f_{co}}{f}\right)^2\right]^{1/2}\min\left[\left(\frac{a}{n}\right)^2,\left(\frac{b}{p}\right)^2\right]E_{wall,max}^2$$

TE
n or p = 0

$$P_{TE} = \frac{ab}{4Z_0}\left[\left(\frac{n}{a}\right)^2 + \left(\frac{p}{b}\right)^2\right]\left[1-\left(\frac{f_{co}}{f}\right)^2\right]^{1/2}\min\left[\left(\frac{a}{n}\right)^2,\left(\frac{b}{p}\right)^2\right]E_{wall,max}^2$$

Circular

TM

$$P_{TM} = \frac{\pi r_0^2}{4Z_0}(1+\delta_{p,0})\frac{E_{wall,max}^2}{\left[1-\left(\frac{f_{co}}{f}\right)^2\right]^{1/2}}$$

TE
p > 0

$$P_{TE} = \frac{\pi r_0^2}{2Z_0}\left[1-\left(\frac{f_{co}}{f}\right)^2\right]^{1/2}\left[1+\left(\frac{v_{pn}^2}{p}\right)^2\right]E_{wall,max}^2$$

Note: Min(x, y) is the smaller of x and y, $Z_0 = 377\ \Omega$, and $\delta_{p,0}$ is a Kronecker delta.

boundary current flows in an infinitely thin layer at the conductor surface to prevent the tangential component of **B** from penetrating the wall. When the wall conductivity σ is finite, the situation changes. In general, one would have to re-solve the problem, taking account of the penetration of the fields and surface currents into the walls. Fortunately, waveguide materials of interest have high-enough conductivity that we can treat the effect of finite σ as a perturbation to the infinite σ case, using the fields that we have already determined for that circumstance.

In considering wall heating, we confine our attention to circular waveguides, since this geometry is applicable to circular cross-section cavities, where this issue is perhaps most important. Further, as we did when we determined the relationship between wall fields and power, we treat the case of TM modes in the most detail, presenting the results for TE modes without derivation. First, consider the effect of finite σ at the wall. For convenience, consider $TM_{0,p}$ modes with only an E_z component tangential to the wall. A nonzero value of E_z at the wall will drive a wall current according to Ohm's law:

$$j_z = \sigma E_z \tag{4.46}$$

Let us use this in Ampere's law (Equation 4.1), applied within the wall, assuming a wave with time dependence $e^{-i\omega t}$:

$$(\nabla \times B) \bullet \hat{z} = \mu_0 \sigma E_z - i \frac{\omega}{c^2} E_z \tag{4.47}$$

For walls of high conductivity, so that $\mu_0 \sigma \gg \omega/c^2 = \omega \mu_0 \varepsilon_0$, we ignore the second term on the right side. Now, take the curl of Faraday's law (Equation 4.2) and use Equation 4.47 and the fact that there are no free charges, so that $\nabla \bullet E = 0$:

$$\left[\nabla \times (\nabla \times E)\right] \bullet \hat{z} = -\nabla^2 E_z = i\omega (\nabla \times B) \bullet \hat{z} = i\omega \mu_0 \sigma E_z \tag{4.48}$$

For convenience, model this as a one-dimensional planar situation, so that Equation 4.48 takes the form (with α to be determined shortly)

$$\frac{d^2 E_z}{dx^2} = -\alpha^2 E_z \tag{4.49}$$

with x measured from the wall, at x = 0, and increasing into the wall; this has the solution

$$E_z = E_0 e^{-\alpha x} \tag{4.50}$$

with E_0 the axial electric field at the wall, x = 0, and

$$\alpha = \frac{(1+i)}{2^{1/2}} (\omega \mu_0 \sigma)^{1/2} \tag{4.51}$$

What we see is that the electric field drops off exponentially within the wall, and the phase is shifted as one moves into the wall. The distance scale for exponential field drop-off in the wall is known as the *skin depth*, δ, which is given by

$$\delta = \left(\frac{2}{\omega\mu_0\sigma}\right)^{1/2} = \frac{1}{\left(\pi\sigma\mu_0 f\right)^{1/2}} \tag{4.52}$$

As an example, in copper, $\sigma = 5.80 \times 10^7$ $(\Omega\text{-m})^{-1}$ so that the skin depth at 1 GHz is 2.1 μm; scaling to other frequencies is as $f^{-1/2}$ (see Problem 6).

Next, consider the current per unit width flowing in the wall, which we define in our simplified one-dimensional model by

$$J_z = \int_0^\infty j_z dx = \int_0^\infty j_0 e^{-(1+i)x/\delta} dx = \frac{\delta}{1+i} j_0 \tag{4.53}$$

Now define a complex surface impedance of the conducting wall by

$$Z_s = \frac{E_0}{J_z} = \frac{E_0}{j_0}\left(\frac{1+i}{\delta}\right) = \frac{1}{\sigma\delta}(1+i) \tag{4.54}$$

The real part of this impedance is the *surface resistance*:

$$R_s = \frac{1}{\sigma\delta} = \left(\frac{\pi f \mu_0}{\sigma}\right)^{1/2} \tag{4.55}$$

Note that R_s has the dimensions of a resistance, although strictly speaking, it is not a resistance in the classical sense.

Next, consider the relationship between J_z and the tangential component of **B** at the wall. Starting with Ampere's law again, neglecting the so-called displacement current term, $(1/c^2) \partial E / \partial t$, we integrate over the area perpendicular to the z axis that excludes the empty interior of the waveguide:

$$\iint \nabla \times B \bullet d = \oint B \bullet dl = \mu_0 \iint j \bullet dA \tag{4.56}$$

The transformation of the area integral to a closed line integral over the boundary of the area of integration is a direct consequence of Stokes' theorem. Assuming that the magnetic field drops to zero deep in the conductor and that symmetry causes the portions of the line integral perpendicular to the surface of the wall to vanish, the line integral becomes an integral over a path along the surface of the waveguide. Now for most cases of interest to us, the area integration on the far right-hand side of Equation 4.56 decomposes into separable integrations along the surface of the waveguide and perpendicular to the surface of the waveguide, into the wall. This is clearly at least approximately the case for our rectangular and circular waveguides, where one can separate the dependences on x and y, or r and θ, into products of functions that depend only on x or y, or r or θ. Therefore, in a cylindrical waveguide, for example, for a TM mode, we can rewrite Equation 4.56 as

$$J_z = \frac{1}{\mu_0} B_\theta \tag{4.57}$$

where J_z is defined analogously to Equation 4.53 as the current per unit circumference at the waveguide wall.

Finally, we use the Poynting vector, $\mathbf{S} = \mathbf{E} \times \mathbf{H}$, to determine the power flow per unit length from the electromagnetic field into the wall per unit length, which is power lost to the resistivity of the wall. We denote this quantity W_L, and it is given by

$$W_{L,TM} = \frac{1}{2\mu_0} \mathrm{Re} \oint E_z\left(r_0\right) B_\theta\left(r_0\right) r_0 d\theta \tag{4.58}$$

where the integral is over the surface of the waveguide at $r = r_0$. Note that when the wall is perfectly conducting, $E_z(r_0) = 0$ and there is no power loss. As we found earlier, though, E_z no longer vanishes at the wall for finite σ, so that we can see how the power loss in the wall arises. Making use of Equations 4.54 and 4.57, remembering that we take the real part of the integral in Equation 4.58, we find

$$W_{L,TM} = \frac{1}{2} r_0 R_S \int_0^{2\pi} \left|J_z\right|^2 d\theta \tag{4.59}$$

which emphasizes the significance of the surface resistivity, R_s, in determining power losses to the wall. In finding J_z, we note that because σ is a large quantity, E_z in Equation 4.46 is a small quantity, of order σ^{-1}, while J_z and B_θ are of order unity, so that we find J_z from Equation 4.57. For TM modes in a circular waveguide, B_θ is given in Table 4.3. From the table and Equation 4.57, we find

$$\left|J_z\right|^2 = \frac{|D|^2}{Z_0^2} \left(\frac{f}{f_{co}}\right)^2 \left[J_p'\left(\mu_{pn}\right)\right]^2 \sin^2\left(p\theta\right) \tag{4.60}$$

where, as usual, $Z_0 = 377\ \Omega$. Putting this expression into Equation 4.59 and using the identity $J_p'(\mu_{pn}) = J_{p-1}(\mu_{pn})$, we get

$$W_{L,TM} = \frac{\pi}{2Z_0^2} r_0 R_s \left(\frac{f}{f_{co}}\right)^2 |D|^2 \left[J_{p-1}(\mu_{pn})\right]^2 \qquad (4.61)$$

At this point, we want to eliminate D in favor of the power carried by the waveguide. To do so, we return to Equation 4.39:

$$P_{TM} = \frac{1}{2\mu_0} \int_0^{r_0} rdr \int_0^{2\pi} d\theta \left(\mathbf{E}\times\mathbf{B}^*\right)\bullet \hat{z} = \frac{1}{2\mu_0} \int_0^{r_0} rdr \int_0^{2\pi} d\theta \left(E_r B_\theta^* - E_\theta B_r^*\right) \quad (4.62)$$

Using the expressions for the field quantities in Table 4.3, performing the integral in Equation 4.62, and solving for D, Equation 4.61 becomes

$$W_{L,TM} = \frac{2}{r_0}\left(\frac{R_s}{Z_0}\right)\frac{P_{TM}}{\left[1-\left(f_{co}/f\right)^2\right]^{1/2}} \qquad (4.63)$$

From this quantity, which is the loss per unit length along the axis, we can compute the average loss per unit area in the wall:

$$\frac{\langle P_{L,TM}\rangle}{A} = \frac{W_{L,TM}}{2\pi r_0} = \frac{1}{\pi r_0^2}\left(\frac{R_s}{Z_0}\right)\frac{P_{TM}}{\left[1-\left(f_{co}/f\right)^2\right]^{1/2}} \qquad (4.64)$$

Because of the power loss in the wall, the power carried along the waveguide will also decline exponentially with distance along the waveguide:

$$P_{TM}(z) = P_0 e^{-2\beta z} \qquad (4.65)$$

In this expression,

$$\text{TM modes: } \beta_{TM} = \frac{W_{L,TM}}{2P_{TM}} = \frac{1}{r_0}\left(\frac{R_s}{Z_0}\right)\frac{1}{\left[1-\left(f_{co}/f\right)^2\right]^{1/2}} \qquad (4.66)$$

The derivation of the corresponding expressions for TE waves proceeds analogously, taking account of the fact that the wall current flows in response to both tangential magnetic field components, B_θ and B_z. In this case,

$$W_{L,TE} = \frac{2}{r_0}\left(\frac{R_s}{Z_0}\right)\left[\left(\frac{f_{co}}{f}\right)^2 + \frac{n^2}{\mu_{pn}^2 - n^2}\right]\frac{P_{TE}}{\left[1-\left(f_{co}/f\right)^2\right]^{1/2}} \qquad (4.67)$$

$$\frac{\langle P_{L,TE}\rangle}{A} = \frac{W_{L,TE}}{2\pi r_0} = \frac{1}{\pi r_0^2}\left(\frac{R_s}{Z_0}\right)\left[\left(\frac{f_{co}}{f}\right)^2 + \frac{n^2}{\mu_{pn}^2 - n^2}\right]\frac{P_{TE}}{\left[1-\left(f_{co}/f\right)^2\right]^{1/2}} \qquad (4.68)$$

$$\text{TE modes: } \beta_{TE} = \frac{1}{r_0}\left(\frac{R_s}{Z_0}\right)\left[\left(\frac{f_{co}}{f}\right)^2 + \frac{n^2}{\mu_{pn}^2 - n^2}\right]\frac{1}{\left[1-\left(f_{co}/f\right)^2\right]^{1/2}} \qquad (4.69)$$

A plot of β for several modes in a copper waveguide is shown in Figure 4.8. Remember that β gives us the rate at which the *fields* decrease exponentially with distance along a waveguide with small, but nonzero, resistivity in the wall material, while the power drops exponentially at twice this rate, according to Equation 4.65. Note that the quantities we have derived involve dimensionless ratios except for the inverse dependence on the waveguide radius and area, so that the units in these expressions depend only on the units with which we measure the waveguide radius. We see that this quantity diverges near the cutoff frequency. In fact, physically, we expect that as the group velocity drops to zero at the cutoff frequency, the electromagnetic energy quits propagating along the line, so that the wave, in a sense, sits

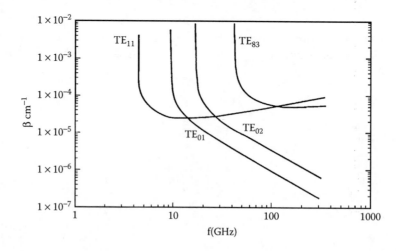

FIGURE 4.8
The attenuation coefficients for the fields and power in a waveguide with lossy walls having the conductivity of copper.

and dissipates its energy. We caution the reader, though, that our expression is not completely accurate in the vicinity of the cutoff frequency, since our underlying approximations begin to break down in this region.

Of course, all of the foregoing relationships between the power and the wall field strengths and between the power and the power dissipated in the walls are single-mode relationships linking the power and electric field for a given mode. As the cross section increases, multiple modes will become available to carry power in the waveguide, and the power density summed across all modes will become important. Regarding power density, a trade-off is many times involved: one would like to limit the number of modes available at the signal frequency in a waveguide or cavity, either to cut off propagation between cavities in a klystron or klystron-like device altogether or to inhibit competition between modes in the generation and transmission of power. As we have shown, in principle, any mode with a cutoff frequency below the frequency of an input signal can carry the wave energy; the exact apportionment of energy among the different modes is determined by the details of the coupling to the waveguide. Because ω_{co} decreases as the cross-sectional dimensions of a waveguide increase, a *single-mode waveguide* must be small enough to allow only one mode to be above cutoff for a given frequency. Achieving high power-handling capacity in HPM systems, however, frequently requires large cross-sectional areas, so that many modes lie above cutoff, particularly at the higher signal frequencies. Such waveguides are said to be *overmoded*, and special care must be taken to prevent energy transfer into unwanted modes. Many of the recent advances in HPM sources have focused on techniques for better control of parasitic modes in over-moded structures: waveguides, cavities, and beam–wave interaction regions.

4.4 Periodic Slow-Wave Structures

Let us return our attention to Figure 4.4 and recall a significant feature of the dispersion curves for both rectangular and circular waveguides: the phase velocity along the waveguide, ω/k_z, of all of the normal modes for rectangular and circular waveguides is greater than the speed of light, c. To be sure, the group velocity for each mode, $\partial\omega/\partial k_z$, lies between –c and c, so that no wave carries energy along the waveguide at a speed greater than the speed of light. Nevertheless, the superluminal phase velocities (i.e., greater than c) present a limitation because, as we shall see, they do not allow interactions with space-charge waves with phase velocities less than c.

The solution to this limitation is to introduce a periodic variation in some property of the waveguide in order to reduce the phase velocities below the speed of light, thus creating normal modes known as *slow waves*. This might involve periodically loading the waveguide with a resonant cavity, or winding a helically wound wire around the inner wall of the waveguide, or periodically varying the radius of the waveguide. In fact, loading a waveguide with resonators spaced azimuthally around the wall is how the interaction region for magnetrons, including relativistic magnetrons, is constructed. The use of slow-wave structures (SWSs) constructed by wrapping a helical winding around the beam transport region, while common in conventional microwave devices operating at subrelativistic voltages (i.e., well below 500 kV), is not seen in relativistic microwave sources, presumably because of concerns about breakdown between turns in the winding at high power and large DC fields. In addition, the low phase velocities that can be achieved with a helix are simply not needed in relativistic devices, as we shall see. On the other hand, periodic variation of the wall radius in relativistic Cerenkov devices (Chapter 9), as well as in Bragg reflectors and mode converters, is a very common configuration.

4.4.1   Axially Varying Slow-Wave Structures

Let us consider this last method of producing a slow-wave structure first. We will only sketch out the process of deriving a dispersion relation, leaving the rather complicated derivation for such devices to the reader who wishes to consult the references.[2-6] The important point is that for slow-wave structures of this kind in relativistic electron beam devices, because the electron velocity is almost c, the waves need not be slowed to any great degree. Therefore, the variation in the wall radius is rather small, so that it can be treated as a small perturbation of a smooth-walled circular waveguide.

For simplicity, we consider variations of the wall radius with distance along the axis of period z_0, assuming no variations with θ,

$$r_w(z) = r_w(z + z_0) \tag{4.70}$$

although we note that structures can vary helically, also. As a consequence of Equation 4.70, the radian frequency of the normal electromagnetic modes of the slow-wave structure ω is a periodic function of the axial wavenumber k_z:

$$\omega(k_z) = \omega(k_z + mh_0) = \omega(k_m) \tag{4.71}$$

where m is any integer, $k_m = k_z + mh_0$, and

$$h_0 \equiv \frac{2\pi}{z_0} \tag{4.72}$$

Consider TM modes and further assume no azimuthal variation (i.e., p = 0); in this case, the only nonzero electromagnetic field components are the radial and axial electric field components, E_r and E_z, and the azimuthal magnetic field component B_θ. The periodicity of the system allows us to employ Floquet's theorem to write E_z as a sum of terms:

$$E_z = \sum_{m=-\infty}^{\infty} E_{zm}(r) e^{i(k_m z - \omega t)} = \left[\sum_{m=-\infty}^{\infty} E_{zm}(r) e^{imh_0 z} \right] e^{i(k_z z - \omega t)} \tag{4.73}$$

Note here that Floquet's theorem essentially amounts to expanding the field in a Fourier series, which is required to match the boundary conditions in Equations 4.5 and 4.6 at the periodically varying boundary of the slow-wave structure. Each of the other field components can be written in a sum of the same form; for each term of the sum at a given value of m, the components are related by relations of the form shown in Table 4.3, with k_z replaced by k_m, and $k_\perp$ replaced by $k_{\perp m}$:

$$k_{\perp m}^2 = \frac{\omega_{co,m}^2}{c^2} = \frac{\omega^2}{c^2} - k_m^2 \tag{4.74}$$

Therefore, as an example,

$$E_{rm} = i \frac{k_m}{k_{\perp m}^2} \frac{\partial E_{zm}}{\partial r} \tag{4.75}$$

In the sum for E_z, each function E_{zm} obeys a differential equation like Equation 4.34, so that, from Table 4.3,

$$E_{zm} = A_m J_0 (k_{\perp m} r) \tag{4.76}$$

At this point, the derivation for TM waves in a periodic slow-wave structure diverges from that of the smooth-walled circular waveguide because, in general, $k_{\perp m} \neq \mu_{0n}/r_0$.* Rather, we need to find a different dispersion relation between ω and k_z. The process by which this is done is recounted in a number of the references. In short, we find the unit tangent and unit normal vectors for the surface of the slow-wave structure; the direction of these unit vectors $\mathbf{n}_t$ and $\mathbf{n}_p$ is straightforwardly determined from the shape of the wall, given by $r_w(z)$, so that $\mathbf{n}_t$ and $\mathbf{n}_p$ will depend on z. We then use these vectors in Equations 4.5 and 4.6, also using the solution for E_{zm} in Equation 4.76 and the expressions for the other field quantities derived from

---

* For clarity, we point out that m is an index on the so-called spatial harmonics in the periodicity relation of Equation 4.70 and the sum in Equation 4.72, while n is the index on the radial ordering of roots of the Bessel functions J_0. By assumption, the azimuthal index p has been chosen to be equal to 0.

E_{zm}, applying them at $r_w(z)$. To avoid having to deal with the z variation of the resulting expressions, we invert the Fourier transform that we employed in Equation 4.74, for example, multiplying the boundary condition for the tangential component of **E** by $\exp(-ilh_0)$ and integrating over z from $-z_0/2$ to $z_0/2$. When we do so, our boundary conditions can be written as a linear algebraic expression of the form

$$\mathbf{D} \cdot \mathbf{A} = 0 \tag{4.77}$$

where **D** is an infinite-dimensional matrix with elements give by

$$D_{lm} = \left(\frac{\omega^2 - k_l k_m c^2}{k_{\perp m}^2 c^2}\right) \int_{-z_0/2}^{z_0/2} e^{i(m-l)h_0 z} J_0 \Big[k_{\perp m} r_w(z) \Big] dz \tag{4.78}$$

and **A** is an infinite-dimensional vector whose elements are the amplitude factors A_m in Equation 4.76. In order for a solution to exist for the vector **A** in Equation 4.77, it must be true that

$$\det [\mathbf{D}] = 0 \tag{4.79}$$

Note that the matrix elements depend on the shape of the wall perturbation through the term in the argument of the Bessel function in the integral. In Figure 4.9, we plot ω vs. k_z for the particular case of a sinusoidally rippled slow-wave structure, for which

$$r_w(z) = r_0 + r_1 \sin (h_0 z) \tag{4.80}$$

with $r_0 = 1.3$ cm, $r_1 = 0.1$ cm, and $z_0 = 1.1$ cm. In the plot, we have let k_z run over the interval from 0 to h_0; the plot is periodic in k_z if we extend the interval. Note that slow waves with $|\omega/k_z| < c$ have been created by the periodicity in Equation 4.71. As we shall discuss, the plot shown there is only approximately correct because of convergence issues with the use of the expansion in Equation 4.73.

The dispersion relation given by Equations 4.78 and 4.79 is clearly more complicated than that for TM waves in circular waveguides with constant radius; however, one can construct an approximate version of the plot shown in Figure 4.9 by properly connecting a set of dispersion diagrams for constant-radius waveguide modes, with wavenumbers properly shifted by multiples of h_0, as shown in Equation 4.71. An example application of this process for the lowest four modes of a sinusoidally rippled slow-wave structure is shown by the solid curves in Figure 4.10.[6] The broken curves in the figure are the dispersion curves of three spatial harmonics of the TM_{0m} modes of a smooth-walled waveguide with constant radius, $r_w = r_0$:

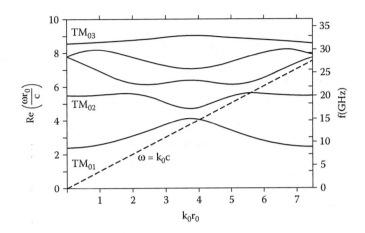

FIGURE 4.9
Dispersion relation computed from Equation 4.79, with $z_0 = 1.1$ cm, $r_0 = 1.3$ cm, and $r_1 = 0.1$ cm in Equation 4.80. (From Swegle, J.A. et al., *Phys. Fluids*, 28, 2882, 1985. With permission.)

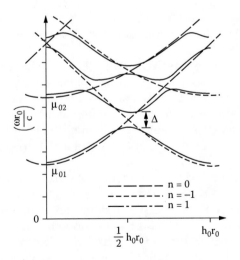

FIGURE 4.10
Construction of the slow-wave-structure dispersion curves from the shifted smooth-walled waveguide curves. These curves are periodic in k_z with period h_0. (From Leifeste, G.T. et al., *J. Appl. Phys.*, 59, 1366, 1986. With permission.)

$$\omega^2 = k_m^2 c^2 + \omega_{co}^2(0,n) \tag{4.81}$$

shifted for three values of m (= −1, 0, 1) in $k_m = k_z + mh_0$. In this expression, since we are considering smooth-walled waveguide modes, the cutoff frequencies are as given in Table 4.3:

$$\omega_{co}(0,n) = \frac{\mu_{on}c}{r_0}$$

When we examine the figure, intuition tells us — and analysis of the dispersion relations shows — that the periodic dispersion curves for the slow-wave structure can be approximated to generally acceptable accuracy by properly connecting the spatial harmonics of the smooth-walled waveguide in Equation 4. 81. Three simple rules govern this process:

1. Curves are constructed from the lowest frequency upward.
2. Every shifted smooth-wall curve is used, and each section of a smooth-wall curve is used only once.
3. SWS dispersion curves for different modes do not intersect. At intersections between the shifted smooth-wall curves, one jumps from one smooth-wall curve to another in tracing the structure curves, leaving a gap between the structure curves that increases in width with r_1/r_0, the relative depth of the wall perturbation.

Note that each curve for the SWS, unlike those for the smooth-wall modes, is limited to a finite range of frequencies, called a *passband*. Passbands for different modes may overlap in frequency. The spatial configurations of the electromagnetic fields associated with a mode will be similar to those of the nearest smooth-wall curve. Thus, a mode might have the characteristics of a TM_{01} mode for one range of k_z, and those of a TM_{02} mode over a different range of k_z (see Problem 7).

Except in the vicinity of the intersection of two smooth-walled curves of the form in Equation 4.81 for different values of axial index m or radial index n [through $\omega_{co}(0, n)$], the dispersion relation for a slow-wave structure is generally well approximated by the nearest smooth-walled waveguide curve of Equation 9.5. Further, the radial variation of the mode pattern for the slow-wave mode is quite similar to that for the closest smooth-walled waveguide mode. Because ω_{co} scales with $1/r_0$ and the periodicity of the dispersion curve scales as $h_0 = 2\pi/z_0$, to lowest order, the dispersion properties of the slow-wave structure are dominated by the average structure radius and its periodicity. The band gaps between modes where two smooth-wall modes intersect and the coupling of the beam and the structure, on the other hand, increases with the ratio r_1/r_0 for a structure such as that of Equation 4.80.

For sinusoidally rippled walls, it has been proven that the series in Equation 4.73, which is based on the so-called *Rayleigh hypothesis*, fails to converge when $h_0 r_1 = 2\pi(r_1/z_0) > 0.448.$[7,8] Comparing the results of calculations using Equation 4.73 with results obtained by numerically solving Maxwell's equations with no assumption of a Floquet expansion, Watanabe et al. found that the dispersion relation in Equation 4.79, with matrix elements given in Equation 4.78, is quite accurate at $h_0 r_1 \sim 5 \times 0.448.$[9] The solutions for the electromagnetic field quantities within the rippled-wall structure derived using a Floquet expansion, however, are noticeably different from the quantities determined by numerical solution of Maxwell's equations. Toward the center of the rippled-wall slow-wave structure, though, the fields computed by the two methods are more similar to one another, so that the effect of the fields on a beam would be approximately the same using either method.

4.4.2 Azimuthally Varying Slow-Wave Structures

In the axially varying slow-wave structures used in Cerenkov devices such as the relativistic backward wave oscillators (BWOs) of Chapter 8, the periodic variation in the wall radius is typically relatively small. In the azimuthally varying slow-wave structures found in magnetrons, on the other hand, the azimuthal variations are substantial. In fact, the slow-wave structure in a magnetron usually consists of N resonators spaced around a cylindrical, or as it is many times known, *reentrant* cavity. A simple example of such a cavity is shown in Figure 4.11. One can see that the volume of a resonator is comparable, although somewhat smaller, than the volume of the central coaxial cavity. Therefore, the technique for deriving a dispersion relation for a magnetron is qualitatively different than that used for an axially varying SWS in a BWO.

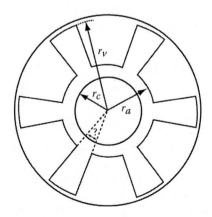

FIGURE 4.11
Cross section of a magnetron, with r_c the cathode radius and r_a the anode radius.

In a magnetron, for simplicity, we neglect axial field variations to lowest order, so that the fields vary only with radius and azimuthal angle, θ. The nonzero field components are E_r, E_θ, and B_z. The two key features of the analytical approach to deriving the dispersion relation for the azimuthally varying slow-wave structure in a magnetron[10] are the following: (1) the electromagnetic fields in the side resonators and in the central coaxial cavity are derived separately, usually under the assumption that the azimuthal electric field E_θ is a constant across the gap where the resonator opens into the central coaxial cavity at $r = r_a$; and (2) the dispersion relation itself results from the resonance that occurs when the admittances of the resonator Y_{res} and the central coaxial cavity $Y_{coaxial}$ add to zero at the gap where the resonator joins the central cavity at $r = r_a$:

$$Y_{res} + Y_{coaxial} = 0 \tag{4.82}$$

where the admittances at the gap are defined by

$$Y = \frac{h \displaystyle\int_{gap} E_\theta H_z d\ell}{\left| \displaystyle\int_{gap} E_\theta d\ell \right|^2} = \frac{h \displaystyle\int_{gap} E_\theta B_z d\theta}{\mu_0 r_a \left| \displaystyle\int_{gap} E_\theta d\theta \right|^2} \tag{4.83}$$

In this expression, the integral is across the gap at $r = r_a$, h is the length of the gap in the axial direction, and $H_z = B_z/\mu_0$. The numerator in the expression is the radial flow of power using the Poynting vector, and the denominator is analogous to a voltage across the gap; since power is proportional to the product of voltage and current, the admittance does indeed have units of resistance^{-1}. According to Kroll,[10] the gap admittance calculated using the fields in a side resonator is given by

$$Y_{res} = i \frac{h}{\psi r_a Z_0} \left[\frac{J_0(kr_a)N_1(kr_v) - J_1(kr_v)N_0(kr_a)}{J_1(kr_a)N_1(kr_v) - J_1(kr_v)N_1(kr_a)} \right] \tag{4.84}$$

where $Z_0 = 377\ \Omega$, ψ is the opening angle of the gap, as shown in Figure 4.11, $k = \omega/c$, and J and N are the Bessel functions of the first and second kind. Note here that Y_{res} is independent of θ, so that it is the same for each of the N resonators in the device. The admittance is a periodic function of k, although the Bessel functions somewhat obscure the real physics involved. The resonator is similar to a rectangular slot resonator with a constant spacing between conducting walls, d, shown in Figure 4.12. The admittance for this rectangular slot resonator is

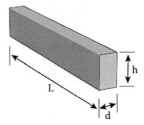

FIGURE 4.12
Dimensions of a rectangular slot resonator.

$$Y_{slot} = -i\frac{h}{dZ_0}cotan(kL)$$ (4.85)

Here, we can clearly see the periodicity, as well as the fact that the admittance has zeroes wherever kL is an odd multiple of $\pi/2$ and poles (where $Y_{slot} \rightarrow \infty$) at multiples of π.

Calculating the admittance of the central coaxial cavity is a more complex process. We see that the coaxial cavity will have a periodic boundary condition on the azimuthal electric field at the anode radius $r = r_a$. We therefore construct our field solutions using a Fourier expansion in θ, writing them in the form, for example,

$$E_\theta(r,\theta,t) = \left[\sum_{s=-\infty}^{\infty} E_{\theta s}(r)e^{is\theta}\right]e^{i(p\theta - \omega t)}$$ (4.86)

Note here that s is the index on the Fourier expansion, analogous to the index m in Equation 4.74 for axially periodic systems, while p is the azimuthal mode number of the solution, analogous to k_0 in Equation 4.74. Solving Maxwell's equations in the coaxial region, with $E_\theta = 0$ at $r = r_c$, yields solutions of the form

$$E_{\theta s}(r) = A_s\left[J_s'(kr) - \frac{J_s'(kr_c)}{N_s'(kr_c)}N_s'(kr)\right]$$ (4.87)

In this expression $J_s' = dJ_s(x)/dx$, and similarly for N_s'. The expression in Equation 4.87 gives one term in the expansion of Equation 4.86. The expressions for E_r and B_z are derived from E_θ in a straightforward fashion. To write an expression for the boundary condition on E_θ at $r = r_a$, let us introduce one more index in order to number the N side resonators from q = 0 to q = N − 1. Now, E_θ vanishes at the conducting wall at $r = r_a$, except in the gaps of the resonators, where it has the fixed magnitude E, but is phase shifted from gap to gap:

$$E_\theta \left(r = r_a, \theta \right) = E e^{i(2\pi p / N)q}, \quad \frac{2\pi q}{N} - \frac{\psi}{2} \le \theta < \frac{2\pi q}{N} + \frac{\psi}{2} \qquad (4.88)$$

It is a straightforward process to compute the constants in the Fourier expansion of Equation 4.86, A_s from Equation 4.87, using the boundary condition at $r = r_a$ and inverting the Fourier transform. The result from Kroll[10] is

$$E_\theta \left(r, \theta \right) = \frac{EN\psi}{\pi} \sum_{m=-\infty}^{\infty} \frac{\sin(s\psi)}{s\psi} \left[\frac{J_s'(kr) N_s'(kr_c) - J_s'(kr_c) N_s'(kr)}{J_s'(kr_a) N_s'(kr_c) - J_s'(kr_c) N_s'(kr_a)} \right] e^{is\theta}, \qquad (4.89)$$

$$s = p + mN$$

Note here that we have suppressed the time dependence for convenience, and only specific values of the index s are allowed, a consequence of the symmetry of the azimuthal boundary condition. Computing H_z from this value of E_θ, one then computes the gap admittance using the coaxial cavity fields, which is shown in the reference to be

$$Y_{coaxial} = i \frac{Nh}{2\pi r_a Z_0} \sum_{m=-\infty}^{\infty} \left[\frac{\sin(s\psi)}{s\psi} \right]^2 \left[\frac{J_s(kr_a) N_s'(kr_c) - J_s'(kr_c) N_s(kr_a)}{J_s'(kr_a) N_s'(kr_c) - J_s'(kr_c) N_s'(kr_a)} \right] \qquad (4.90)$$

In addition to the geometric parameters N, h, ψ, r_c, and r_a, this expression depends on the frequency, through $k = \omega/c$ and the mode number p. One can easily show that in Equation 4.90, the admittance for a given value of p is the same as that for $N - p$.

To find the normal-mode frequencies for a given mode number p, one sets $Y_{coaxial}$ in Equation 4.90 equal to $-Y_{res}$ in Equation 4.84. This equation must be solved numerically, but we can visualize the solutions graphically if we plot the admittances as a function of k. First, Figure 4.13 is a qualitative depiction of the dependence of $Y_{coaxial}$ alone as a function of the normalized variable kr_a for an eight-resonator system (i.e., N = 8). Because $Y_{coaxial}(p) = Y_{coaxial}(N - p)$, only the values for p equal to 0 through 4 are shown. Analysis of Equation 4.90 shows that there are an infinite number of poles and zeroes for the admittance; in the figure, only the poles at k = 0 for p = 0 and at $k \cong 2/(r_a + r_c)$ for p = 1 are shown. Note that this latter pole occurs when $k \sim 2\pi/\lambda \cong 2/(r_a + r_c)$, where λ is the wavelength in the azimuthal direction. Thus, this pole occurs when λ is approximately equal to the circumference of the central coaxial cavity.

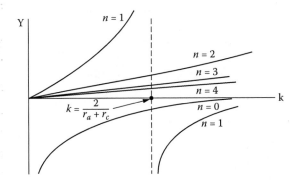

FIGURE 4.13
Qualitative depiction of the dependence of $Y_{coaxial}$ on $k = \omega/c$. (From Kroll, N., *Microwave Magnetrons*, Collins, G.B., Ed., McGraw-Hill, New York, 1948. With permission.)

Figure 4.14 shows simultaneous plots of $Y_{coaxial}$ and $-Y_{res}$. The normal-mode values of $k = \omega/c$ are given by the intersections of the curves for each. The intersections labeled 1 through 4 are the lowest-order modes for the corresponding values of the azimuthal index p. The normal-mode frequency for $p = 0$ is $\omega = 0$. The fact that there is no cutoff frequency above zero is a consequence of the coaxial geometry, in which there is no nonzero cutoff frequency (viewed another way, the modes are like transverse electromagnetic [TEM] modes of a coaxial waveguide that propagate all the way down to $\omega = 0$). We can see that the normal-mode frequencies for the lowest-order modes climb toward a maximum value for $p = N/2$. This frequency must occur at a value of k less than the zero crossing for the curve of $-Y_{res}$. If we approximate Y_{res} by the rectangular slot expression in Equation 4.85, then the frequency for the $p = N/2$ mode must lie below $\omega_{max} \cong \pi c/2L$. The modes labeled 0_1 through 4_1 are the next-higher-order modes, and so on for 0_2 and

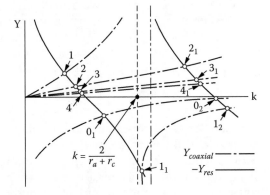

FIGURE 4.14
Qualitative depiction of the dependence of $Y_{coaxial}$ and $-Y_{res}$. The intersections between the two curves give the frequencies of the normal modes for the device. (From Kroll, N., *Microwave Magnetrons*, Collins, G.B., Ed., McGraw-Hill, New York, 1948. With permission.)

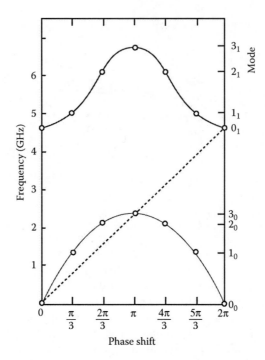

FIGURE 4.15
Dispersion relation for the A6 magnetron, with the phase shift on the horizontal axis $\Delta\phi = 2\pi(p/N)$. (From Palevsky, A. and Bekefi, G., *Phys. Fluids*, 22, 986, 1979. With permission.)

higher. The dispersion relation for the A6 magnetron with $N = 6$ and r_c, r_a, and r_v given by 1.58, 2.11, and 4.11 cm, respectively, is shown in Figure 4.15. Since $L = r_v - r_a = 2$ cm, $f_{max} = \omega_{max}/2\pi = 3.75$ GHz. By comparison, the $p = N/2 = 3$ mode (the so-called π-*mode*) frequency is $f_\pi = 2.34$ GHz, which is considerably lower than f_π. In Chapter 6, we will discuss the dispersion properties and their effect on device performance in greater detail.

4.5 Cavities

An important distinction between waveguides and cavities is that the latter are closed to some degree at the ends, so that electromagnetic radiation is trapped within them. Of course, they are not completely closed, or no radiation would escape, and the ratio of the amount of radiation trapped inside to the amount that escapes is one of the important properties of cavities.

One way to form a simple cavity is to take a section of waveguide and cap both ends with a conducting plate in order to trap the waves inside. Other examples of cavities are shown in Figure 4.16. If the walls and the endplates placed at $z = 0$ and $z = L$ are perfectly conducting, the field

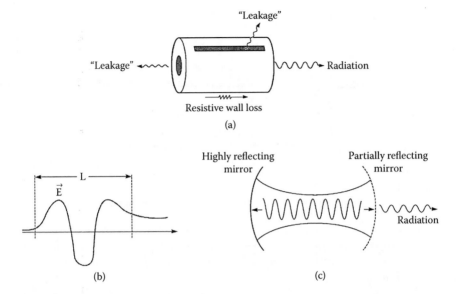

FIGURE 4.16
(a) Cylindrical cavity with partially open ends. Radiation from the right end is usefully extracted, while leakage and absorption in the wall resistance are loss channels. (b) An example of an
electric field envelope in the cavity of (a); note the large field amplitude in the extraction region.
(c) Quasi-optical cavity, with extraction through a partially reflecting mirror on the right.

components for a TM_{0n} wave, for example — E_r, E_θ, and B_z — must vanish
at the ends, so that the electromagnetic waves in the cavity will be reflected
there. Consequently, the axial wavenumber k_z, which was a free parameter
in the infinitely long waveguide, will be restricted to certain discrete values
in the cavity. In the expressions for E_r, E_θ, and B_z, we would replace $\exp(ik_z z)$
by $\sin(k_z z)$ and require $k_z L = q\pi$, with q any integer. This adds a third index
to our designation of the cavity modes; for example, TM_{npq}. The dispersion
curve for ω vs. k_z given in Figure 4.4 is now replaced by a set of points lying
along the former curve, spaced $\Delta k_z = \pi/L$ apart. In actual practice, though,
this ideal, if somewhat artificial, situation is modified, because of losses of
energy from the cavity, either as intended radiation or as parasitic losses to
nonzero wall resistance or to leakage through openings intended* or unintended. Therefore, if we speak of a resonance line for an ideal lossless, closed
cavity with well-defined discrete values of k_z, then an open, lossy cavity has
a nonzero spread in k_z (and, through the dispersion relation, ω) to its line
width. If, for example, a small antenna were inserted into the cavity, the
power coupled into and stored there would have a qualitative dependence
on frequency like that shown in Figure 4.17.

---

* If two different modes are competing within a cavity, one might be suppressed by cutting slots
in the wall at the location of minima in the fields of the desired mode so that the undesired mode
"leaks out" of the cavity, or the slots might be oriented to suppress the flow of the RF wall currents supporting undesired modal field patterns.

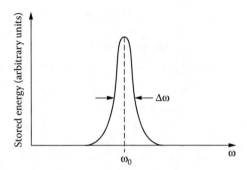

FIGURE 4.17
Stored energy in a cavity with nonzero line width about the line center ω_0.

These notions lead us to the *quality factor*, Q, characterizing cavities. Q is defined very generally, regardless of the cavity shape, as 2π times the ratio of the time-averaged field energy stored in the cavity to the energy lost per cycle:

$$Q = 2\pi \frac{\text{average stored energy}}{\text{energy lost per cycle}} = \omega_0 \frac{\text{average stored energy}}{\text{power loss rate}} \quad (4.91)$$

where ω_0 is the center frequency of the resonant line width of the cavity. This definition holds regardless of the cavity shape. In addition, one can show that Q is related to the resonance line width $\Delta\omega$ of Figure 4.17 by

$$Q = \frac{\omega_0}{2\Delta\omega} \quad (4.92)$$

We have described the power loss from a cavity as arising from two factors: radiated output power and parasitic losses.* Because power losses are additive, we can use the definition in Equation 4.92 to write

$$\frac{1}{Q} = \frac{1}{Q_r} + \frac{1}{Q_p} \quad (4.93)$$

with Q_r and Q_p representing the radiation and parasitic terms, respectively. Both factors will depend, of course, on the cavity configuration and the mode involved. To estimate the radiated contribution, Q_r, return to Equation 4.91

---

* Different terminologies are used for these two contributions to Q. What is called the radiative Q here is sometimes called the diffractive Q — referring in particular to the diffractive output coupling from low group velocity devices such as gyrotrons — or the external Q. Our parasitic Q is also called the resistive Q — apt when the only cavity losses are due to wall losses — or the external Q, which applies when cavity losses are due primarily to radiation of unwanted mode energy through strategically placed wall slits.

to define a radiated energy loss time constant from the cavity, T_r, as the ratio of the stored energy to the power lost from the cavity, in terms of which

$$Q_r = \omega_0 T_r \qquad (4.94)$$

If the cavity has highly reflecting ends with reflection coefficients R_1 and R_2, then

$$T_r \approx \frac{L}{v_g \left(1 - R_1 R_2 \right)} \qquad (4.95)$$

To estimate the group velocity v_g, we write $\omega = \sqrt{\omega_{co}^2 + k_z^2 c^2}$, so that $v_g = k_z c^2 / \omega$. In the limit of large values of k_z, that is, at short wavelengths, $v_g \approx c$. On the other hand, for small k_z or long wavelengths, where v_g is small, k_z is still approximately given by $q\pi/L$, with q the axial mode index. Incorporating these expressions into Equations 4.94 and 4.95, and using the relationship between ω and the free-space wavelength of the radiated microwaves λ, $\omega = 2\pi f = 2\pi/\lambda$, we estimate

$$Q_r \approx 2\pi \frac{L}{\lambda} \frac{1}{\left(1 - R_1 R_2 \right)} \quad \text{(short wavelengths)} \qquad (4.96a)$$

$$\approx 4\pi \frac{L^2}{\lambda^2} \frac{1}{q \left(1 - R_1 R_2 \right)} \quad \text{(long wavelengths)} \qquad (4.96b)$$

As an example, in a low-group-velocity device operated at long wavelengths near cutoff, such as a gyrotron, the long-wavelength limit applies. Such devices are sometimes referred to generally as diffraction generators, since output from the end is largely due to diffraction of the radiation from the (partially) open end. Providing Equation 4.96b holds for such diffraction generators with poorly reflecting end surfaces, the minimum radiative (diffraction-limited) Q of these devices for the lowest axial mode with $q = 1$ is $Q_{r,min} = 4\pi L^2/\lambda^2$. We can see that Q_r decreases with the axial mode number; in other words, radiative losses from a cavity increase with the axial mode number. Overall, the radiative component of Q scales as a power of the length of the cavity measured in free-space wavelengths of the radiation.

Making a general estimate of the parasitic losses in a cavity is a difficult task, particularly when slotted or absorber-loaded cavities are used. An estimate can be made, however, if we limit ourselves to an approximate calculation in a cavity in which the only losses are due to wall resistivity. In this case, we note that the stored energy is proportional to the energy density within the cavity integrated over its volume. The power loss, on the other hand, is proportional to the frequency times the energy density integrated through the volume of the resistive layer roughly skin deep at the cavity wall. In this argument, we cancel out the energy density in the numerator and denominator of Equation 4.91 to obtain

$$Q_p = \frac{\text{cavity volume}}{(\text{surface area}) \times \delta} \times (\text{geometrical factor}) \qquad (4.97)$$

where the geometrical factor is typically of the order unity (see Problem 8).

One might think that maximizing Q is desirable. Certainly, storing energy within the cavity increases the internal feedback, thus lowering the threshold for oscillation and reducing the demand on a power supply to produce the required electron beam current to drive an oscillator. In the world of HPM, however, high currents are relatively easy to produce, so that the challenge is shifted to the tasks of generating and handling high power in the cavity. From each standpoint — generating and handling high power — it is actually desirable to lower Q at high power. In the generation process, we shall see that the driving electron beam power is most effectively utilized when it is several times the threshold value; hence, raising the output power requires raising the threshold level, which translates to lowering Q. In handling microwave power, remember that the output power is proportional to the stored energy divided by Q, so that high Q implies high values of stored energy and larger electric fields in the cavity. This factor raises the wall heating rates and can lead to breakdown.

In closing our discussion of the quality factor, we touch on the issue of *fill time* for cavities, which is the time required for the fields in a cavity to build up to a steady-state level. Reasoning from Equation 4.91, it is apparent that the fill time is proportional to Q. For certain HPM applications involving cavities, fill time is an important additional consideration. In the accelerating cavity of an RF linac, for example, the goal is to transfer energy *to* electrons rather than extract it. Therefore, a trade-off must be made between raising Q to maximize the accelerating field and lowering Q to limit the fill time and resultant requirement it places on RF pulse lengths. As a second example, certain RF pulse compression schemes depend on the dynamic control of a cavity's Q in order to rapidly discharge the HPM energy that was slowly accumulated at lower power. The discharge time is roughly equal to the fill time, and switching to a lower Q permits this rapid discharge.

4.6   Intense Relativistic Electron Beams

High power microwave sources tap the kinetic energy of electrons in an intense relativistic beam or rotating layer to produce the intense microwave fields. To get an idea of the parameter range in which HPM devices operate, remember that the microwave power P is generated at some power efficiency η_P from an electron current I accelerated by a voltage V:

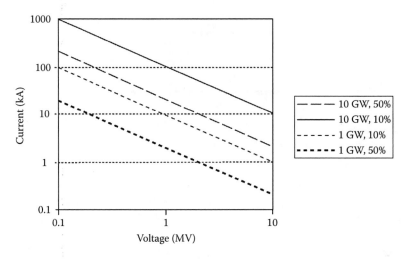

FIGURE 4.18
Beam voltage and current required to produce 1 and 10 GW of microwave output at power efficiencies η_P of 10 and 50%.

$$P = \eta_P V I \qquad (4.98)$$

The microwave power in pulsed HPM sources is of the order of 1 to 10 GW, and efficiencies range from 10 to 50%. In Figure 4.18, we show the required current and voltage to reach 1 and 10 GW at efficiencies of 10 and 50%. One can see that currents of the order of 10 kA, or 10 s, are required for voltages around 1 MV. In most cases, this current is carried in a beam of a specified cross-sectional profile at a high current density (see Problem 9).

We defer the treatment of the technology of producing intense electron beams and layers to the next chapter, where we will discuss cathodes, diodes, and electron guns. In this chapter, though, we discuss the basics of electron flow in diodes — space-charge-limited electron flow in Child–Langmuir diodes (planar geometry) and Langmuir–Blodgett diodes (cylindrical geometry) — as well as the current limit imposed by the magnetic pinching of a high-current beam. Also, we consider electron flow along electron drift tubes and in magnetically insulated electron layers, all as background for our discussion of microwave interactions, which will follow.

4.6.1 Space-Charge-Limited Flow in Diodes

First, consider the one-dimensional planar geometry of Figure 4.19, with a cathode at x = 0 and an anode at x = d. The electric potential at the cathode is $\phi(x = 0) = 0$, and at the anode, $\phi(x = d) = V_0$. An electron current of density $\mathbf{j} = \hat{x}j(x) = -\hat{x}en(x)v_x(x)$ flows from the cathode to the anode, where –e is the

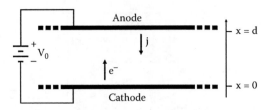

FIGURE 4.19
The ideal Child–Langmuir diode, with infinite parallel-plate electrodes separated by a distance d, with a voltage V_0 across the gap.

charge on an electron, $n(x)$ is the number density of electrons measured in m^{-3}, and $v_x(x)$ is the velocity of an electron. In equilibrium, with no time dependence, ϕ and n are linked by Equation 4.4:

$$\frac{dE_x}{dx} = -\frac{d^2\phi}{dx^2} = -\frac{1}{\varepsilon_0}en \qquad (4.99)$$

where we have replaced the electric field with the gradient of ϕ in this one-dimensional problem. For convenience, let us assume that $V_0 < 500$ kV, so that the problem is nonrelativistic; we therefore relate ϕ and v_x through the conservation of energy, assuming that the electrons leave the cathode with $v_x(0) = 0$:

$$\frac{1}{2}mv_x^2 = e\phi \qquad (4.100)$$

with m the mass of an electron. We close this set of equations by taking the divergence of Ampere's law (Equation 4.1), remembering that $\partial E_x/\partial t = 0$, which yields

$$\frac{dj}{dx} = -e\frac{d(nv_x)}{dx} = 0 \qquad (4.101)$$

This last equation tells us that nv_x is a constant across the gap, so for convenience, we define J, which is the magnitude of the current density across the gap,

$$J = env_x = \text{constant} \qquad (4.102)$$

although we note that we have reversed the sign of this constant from that of the current density. From this set of equations, we can find the following equation governing ϕ between the cathode and anode:

$$\frac{d^2\phi}{dx^2} = \left(\frac{m}{2e}\right)^{1/2} \frac{J}{\varepsilon_0 \phi^{1/2}} \qquad (4.103)$$

In this equation, ϕ vanishes at the cathode. One can also show that as J increases from 0, so that the amount of space-charge in the anode–cathode gap increases, the magnitude of the electric field at the cathode, $d\phi/dx = \phi'$, decreases from V_0/d. The maximum current crossing the gap, which we write as J_{SCL}, is the value at which $\phi'(0) = 0$, which one can show is given in practical units by

$$J_{SCL}\left(\frac{kA}{cm^2}\right) = 2.33\frac{\left[V_0\left(MV\right)\right]^{3/2}}{\left[d\left(cm\right)\right]^2} \qquad (4.104)$$

This is the *Child–Langmuir*,[11] or *space-charge-limited*, current density in a planar electron diode. Now if we can ignore edge effects, so that the current density is uniform across the surface of a cathode with area A, the current I from such a Child–Langmuir diode is $J_{SCL}A$ and the impedance Z for such a diode would be the ratio of the current to the voltage, or

$$Z = \frac{V_0}{I} = 429\frac{\left[d\left(cm\right)\right]^2}{A\left(cm^2\right)}\frac{1}{\left[V_0\left(MV\right)\right]^{1/2}}\ \Omega$$

As an example, a typical value for a diode is $V_0 = 1$ MV, with a gap of d = 1 cm and a cathode diameter of 2 cm. Therefore, $J_{SCL} = 2.33$ kA/cm^2, I = 7.32 kA, and Z = 137 Ω (see Problem 10).

Langmuir and Blodgett solved the same problem in the cylindrical geometry[12] of Figure 4.20. The cathode radius r_c can be either on the inside, so that $r_c = r_i$ and the anode radius $r_a = r_o$, or on the outside, so that $r_c = r_o$. Recasting the expression for $\nabla^2\phi$ in Equation 4.99 and the expression for $\nabla \cdot j$ in the proper form for cylindrical geometry, and retaining Equation 4.100 in the same form, we find that the cylindrical analog of Equation 4.103 is

$$\frac{d^2\phi}{dr^2} + \frac{1}{r}\frac{d\phi}{dr} = \left(\frac{m}{2e}\right)^{1/2}\left(\frac{r_c}{r}\right)\frac{J}{\varepsilon_0\phi^{1/2}} \qquad (4.105)$$

The space-charge-limited (meaning that the electric field at the cathode vanishes) solution to this equation yields an expression for the current density *at the cathode* that is similar in form to Equation 4.104:

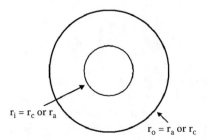

$r_i = r_c$ or r_a

$r_o = r_a$ or r_c

FIGURE 4.20
Cross-sectional view of the cylindrical Langmuir–Blodgett diode.

$$J\left(\frac{kA}{cm^2}\right) = 2.33\frac{\left[V_0\left(MV\right)\right]^{3/2}}{\left[r_a\left(cm\right)r_c\left(cm\right)\right]Y^2} \tag{4.106a}$$

with Y an infinite series solution with more than one form, depending on the ratio r_a/r_c and the desired accuracy of the solution.[12] The form of Y most appropriate for HPM diodes is

$$Y = \left(\frac{r_c}{r_a}\right)^{1/2}\sum_{n=1}^{\infty}B_n\left[\ell n\left(\frac{r_a}{r_c}\right)\right]^n \tag{4.106b}$$

where $B_1 = 1$, $B_2 = 0.1$, $B_3 = 0.01667$, $B_4 = 0.00242$, and so on (14 terms are given in the reference). Remember that Equation 4.106a gives the current density at the cathode of a cylindrical diode; to compute the current, one must multiply by the cathode area, $A_c = 2\pi r_c L$, where L is the length of the emitting area (see Problems 11 and 12).

Our treatments of the flow of electrons in either a planar or cylindrical diode are predicated on the assumption that the diode voltage is nonrelativistic — i.e., $V_0 < 500$ kV — so that the nonrelativistic form in Equation 4.100 applies. In the event that $V_0 > 500$ kV, a relativistic form for the planar Child–Langmuir diode was derived by Jory and Trivelpiece:[13]

$$J\left(\frac{kA}{cm^2}\right) = \frac{2.71}{\left[d\left(cm\right)\right]^2}\left[\left(1+\frac{V_0\left(MV\right)}{0.511}\right)^{1/2} - 0.847\right]^2 \tag{4.107}$$

Note the identical scaling with the anode–cathode gap width in both Equations 4.103 and 4.107. The voltage scaling is quite different, however, with the familiar $V_0^{3/2}$ scaling at low voltages, while at high voltages $J \propto V_0$.

4.6.2 Beam Pinching in High-Current Diodes

In our treatment of the Child–Langmuir diode, we also assumed that the self-magnetic field of the beam current can be neglected. However, at high currents, the strong azimuthal self-magnetic field of the beam causes electron beams to pinch due to the Lorentz force $F_r = v_z \times B_\theta$. The critical beam current above which *beam pinching* occurs is estimated by equating the Larmor radius for an electron at the edge of the beam with the anode–cathode gap spacing:[14]

$$I_{pinch}(kA) = 8.5 \frac{r_c}{d} \left[1 + \left(\frac{V_0(MV)}{0.511} \right) \right] \tag{4.108}$$

where r_c is the cathode electrode radius. Generally speaking, pinching would be a disaster in an HPM source and is to be avoided. This effect is a factor particularly in low-impedance diodes such as those used in virtual cathode oscillators (see Problem 13).

4.6.3 Space-Charge-Limited Electron Beam Flow in a Drift Tube

Once the electron beam has been generated, it will pass through a drift tube on its way to the microwave interaction region. For convenience, let us assume that the drift region is long enough that we can assume it is effectively infinitely long. In addition, we assume that the duration of the beam is long enough that an equilibrium is set up in which we can ignore time-dependent effects. In this case, the space-charge of the beam creates a purely radial electric field between the beam and the drift tube wall, and we can perform a line integral of that field to find the electric potential difference between the beam and the wall. This potential difference reduces the effective accelerating potential of the beam, an effect referred to as *space-charge depression* of the beam energy. Let us calculate the space-charge depression of the beam energy for a thin, annular beam, as shown in Figure 4.21.

For simplicity, we assume that the axial magnetic field is infinitely large and that the beam is infinitely thin (mathematically, the current density could be represented by a delta function in radius), with radius r_b, equal to the mean radius of the thin beam shown in the figure. There is no space-charge inside the drift tube except at $r = r_b$, so inside and outside the beam, the potential is given by

$$\frac{1}{r} \frac{d}{dr} \left(r \frac{d\phi}{dr} \right) = 0 \tag{4.109}$$

The boundary conditions are $d\phi/dx = 0$ at $r = 0$ and $\phi = 0$ at the wall, $r = r_0$. Thus,

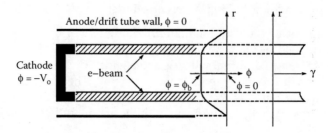

FIGURE 4.21
The potential, ϕ, and relativistic factor, γ, as a function of radius for electrons in an intense annular electron beam. Space-charge in the beam depresses the potential to $-\phi_b$ on the axis. The electron energy varies with radius across the beam (i.e., it is *sheared*), with the highest energy electrons at the outer edge of the beam.

$$\phi(r) = -\phi_b \, , \, 0 \leq r < r_b \tag{4.110}$$

and

$$\phi(r) = A + B\ell n(r) \, , \, r_b < r \leq r_0 \tag{4.111}$$

It remains to find ϕ_b in terms of r_b, r_0, and the space-charge in the beam, which depends on the beam current I_b and the beam velocity

$$v_b = c(1 - 1/\gamma_b^2)^{1/2}$$

In the expression for v_b, γ_b is the effective relativistic factor for the beam electrons; it is given by

$$\gamma_b = 1 + \frac{e}{mc^2}(V_0 - \phi_b) \equiv \gamma_0 - \frac{e\phi_b}{mc^2} \tag{4.112}$$

where $-V_0$ is the potential of the cathode from which the beam electrons were launched and γ_0 is as defined.

Let us eliminate A and B in Equation 4.111. First, ϕ must be continuous across the beam, so that we set $\phi(r_b) = -\phi_b$ in Equation 4.111. Next, we determine the discontinuity in $d\phi/dr$ at r_b by integrating Equation 4.109 across the beam:

$$r_b\left[\frac{d\phi(r = r_b^+)}{dr} - \frac{d\phi(r = r_b^-)}{dr}\right] = \frac{e}{\varepsilon_0}\int_{r_b^-}^{r_b^+} n_b r dr = \frac{1}{2\pi\varepsilon_0 v_b}\int_{r_b^-}^{r_b^+} en_b v_b 2\pi r dr$$

$$= \frac{I_b}{2\pi\varepsilon_0 v_b} \tag{4.113}$$

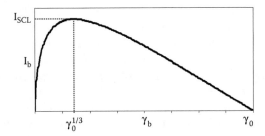

FIGURE 4.22
Plot of the beam current vs. the beam relativistic factor.

In this expression, I_b is the magnitude of the electron beam current. Note that the second term in the brackets on the far left side of this equation vanishes, while the first term is found by differentiating Equation 4.111. Using the continuity of ϕ at r_b, the vanishing of ϕ at r_0, and Equation 4.113, we find

$$I_b = \frac{2\pi\varepsilon_0 v_b}{\ell n\left(r_0 / r_b\right)}\phi_b = \frac{8.5}{\ell n\left(r_0 / r_b\right)}\left(1 - \frac{1}{\gamma_b^2}\right)^{1/2}\left(\gamma_0 - \gamma_b\right) \text{ kA} \qquad (4.114)$$

In this last expression, we have used Equation 4.112 to eliminate the space-charge potential term ϕ_b and the expression for v_b in terms of γ_b to find a relationship between I_b and the actual relativistic factor for the beam electrons, γ_b, which is less than γ_0 — what it would be if there were no space-charge depression.

Figure 4.22 is a plot of I_b vs. γ_b. We see that for each value of I_b, there could be two values of γ_b; in fact, the values of γ_b to the right of the current maximum are the correct values. Differentiating I_b in Equation 4.114 with respect to γ_b, we find that I_b is a maximum for $\gamma_b = \gamma_0^{1/3}$, at which value,

$$I_b\left(\gamma_b = \gamma_0^{1/3}\right) \equiv I_{SCL}\left(\text{annular}\right) = \frac{8.5}{\ell n\left(r_0 / r_b\right)}\left(\gamma_0^{2/3} - 1\right)^{3/2} \qquad (4.115)$$

We call I_{SCL} the *space-charge-limiting current* for the drift tube. Note here that I_{SCL} scales as $V_0^{3/2}$ at low voltages and as V_0 at large voltages, which is a scaling similar to that in the Child–Langmuir diode. In fact, this is the value for a thin, annular beam. One can show that for a beam with a uniform density over a solid cross section of radius r_b,

$$I_b\left(\gamma_b = \gamma_0^{1/3}\right) \equiv I_{SCL}\left(\text{solid}\right) = \frac{8.5}{1 + \ell n\left(r_0 / r_b\right)}\left(\gamma_0^{2/3} - 1\right)^{3/2} \qquad (4.116)$$

(see Problem 14).

One way to neutralize the space-charge fields of the beam, and thereby obviate the space-charge depression and limiting current, is to prefill the drift tube with plasma. As the electron beam passes through the plasma, its space-charge fields will rapidly expel enough plasma electrons that the beam space-charge will be compensated by the remaining plasma ions. An additional salutary effect is that the beam quality is improved by the reduction or elimination of radial variations in the electron velocities. The price one pays, of course, is the added complexity of generating the plasma, and the fact that recombination and further ionization of the plasma by the beam make the self-electric field time dependent.

4.6.4 Beam Rotational Equilibria for Finite Axial Magnetic Fields

When the axial magnetic field is reduced from very large values, at which it can effectively be considered infinite, I_{SCL} is reduced as well. Finite axial magnetic fields also cause the beam to rotate in equilibrium, which serves to balance the forces on the beam so that the net force vanishes. For convenience, consider a nonrelativistic beam. Using a quite general treatment based on the fluid equations, one can show that the azimuthal rotational velocity of electrons in a beam traveling in a cylindrical drift tube is given by[15]

$$v_{eb}^{\pm}(r) = \frac{1}{2}\Omega_{eb}r\left\{1\pm\left[1-4\frac{e^2}{\varepsilon_0 m\Omega_{eb}^2 r^2}\int_0^r dr' r' n_{eb}(r')\right]^{1/2}\right\} \qquad (4.117)$$

where $\Omega_{eb} = eB_z/m$ is the electron cyclotron frequency of the beam electrons in the applied axial magnetic field B_z, with $-e$ and m the charge and mass of an electron, respectively, and n_{eb} the number density of beam electrons. We see that there are two rotational velocities possible for the beam electrons. To see their physical significance, consider the case in which the right-hand term under the square root in Equation 4.117 is much smaller than unity, so that to lowest order in that small quantity

$$v_{eb}^+ \approx \Omega_{eb}r \qquad (4.118a)$$

$$v_{eb}^- \approx \frac{e^2}{\varepsilon_0 m\Omega_{eb}r}\int_0^r dr' r' n_{eb}(r') = -\frac{E_r}{B_z} \qquad (4.118b)$$

In this latter expression, we have used Equation 4.4 to make the replacement

$$E_r = -\frac{e}{\varepsilon_0 r} \int_0^r dr' \, r' \, n_{eb}\left(r'\right)$$

Thus, the two rotational equilibria are designated as follows:

- *Slow rotational mode*, in which the rotational velocity v_{eb}^- of the electron guiding centers has the character of an $\mathbf{E} \times \mathbf{B}$ rotation, although the electrons, on the small scale of a Larmor orbit within the beam, still undergo cyclotron rotation at Ω_{eb} about their guiding centers.
- *Fast rotational mode*, in which the electrons as a whole rotate about the beam axis at a frequency approximately equal to the cyclotron rotational velocity, $\Omega_{eb}r$. Such beams are also called *axis-encircling beams*.

The slow rotational mode is the more commonly seen, although a beam can be induced into the fast rotational mode by passing a slow-rotational-mode beam through a magnetic cusp. Axis-encircling beams created in this way are used in so-called large-orbit gyrotrons (see Problem 15).

4.7 Magnetically Insulated Electron Layers

Not all sources use electron beams. Notably, the magnetron and magnetically insulated line oscillator (MILO) employ a magnetically insulated layer of electrons bounded on one side by an electrode at cathode potential. In this geometry, a magnetic field is applied parallel to the cathode surface, and electrons emitted from the cathode are deflected by the $\mathbf{v} \times \mathbf{B}$ Lorentz force and prevented from reaching the anode. Accordingly, they remain trapped in a layer adjoining the cathode, drifting in the $\mathbf{E} \times \mathbf{B}$ direction, parallel to the cathode. The situation is depicted in Figure 4.23.

The *Hull criterion* for magnetic insulation of the layer — meaning that in equilibrium the electrons cannot cross the gap to reach the anode — is derived from very general arguments based on the conservation of energy and canonical momentum in the direction of net electron drift. Assume that all electrons are emitted from the cathode, that there are no system variations along their direction of drift, and that a time-independent equilibrium is established. In the so-called *smooth-bore magnetron* with an axial magnetic field B_z within a cylindrical geometry with cathode radius $r_c <$ r_a, the anode radius, the electrons drift azimuthally at v_θ with their energy conserved, so that

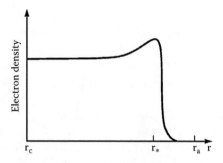

FIGURE 4.23
The variation of the electron density in a layer drifting in crossed electric and magnetic fields. The electrons are emitted from the cathode at r_c, the layer extends outward from the cathode to roughly r_*, and the electrons are insulated by the magnetic field so that they do not cross over to the anode at r_a.

$$(\gamma - 1) mc^2 - e\phi = 0 \tag{4.119}$$

with $\gamma = (1 - v_\theta^2 / c^2)^{-1/2}$. Also, in the absence of azimuthal variations, the canonical momentum in the azimuthal direction is conserved, so that

$$m\gamma v_\theta - \frac{e}{c} A_\theta = 0 \tag{4.120}$$

with A_θ the azimuthal component of the vector potential $\mathbf{A}$, which obeys

$$\mathbf{B} = \nabla \times \mathbf{A} \tag{4.121}$$

The magnetic flux is the surface integral of B_z over the cross-sectional area between the cathode and the anode. Because of the high conductivity of the anode and cathode, we assume that the magnetic flux before the voltage is applied, when B_z is the initial applied magnetic field, is the same as the magnetic flux after the voltage is applied, and the diamagnetic effect of the electron layer modifies the magnetic field in the gap. Then the Hull criterion for magnetic insulation of the gap is as follows:

$$B^* = \frac{mc}{ed_e} \left(\gamma^2 - 1 \right)^{1/2} = \frac{0.17}{d_e \left(cm \right)} \left(\gamma^2 - 1 \right)^{1/2} \ \text{T} \tag{4.122}$$

measured in Tesla, with

$$\gamma = 1 + \frac{eV_0}{mc^2} = 1 + \frac{V_0 \left(kV \right)}{0.511} \tag{4.123}$$

with V_0 the anode–cathode voltage and d_e the effective gap in cylindrical geometry,

$$d_e = \frac{r_a^2 - r_c^2}{2r_a} \tag{4.124}$$

When $B_z > B^*$, the critical magnetic field or Hull field, the electrons drift azimuthally within a bounded electron cloud called the Brillouin layer. As $B_z \to B^*$, the electron cloud extends almost to the anode. As the magnetic field increases, the cloud is confined closer and closer to the cathode. Remarkably, even though Equation 4.122 takes account of collective space-charge and diamagnetic effects from the electron layer, it is exactly the result one would obtain if one set the Larmor radius for a single electron equal to d_e (see Problem 16).

4.8 Microwave-Generating Interactions

We can understand the microwave generation process in HPM sources in terms of resonant interactions between the normal modes of a cavity or waveguide and the natural modes of oscillation in a beam or electron layer. To impose some order on the discussion of microwave sources, we classify them according to various criteria. From the standpoint of the electromagnetic normal modes, sources are said to be either *fast-wave* or *slow-wave* devices. A fast-wave interaction involves a waveguide mode with a phase velocity that is greater than the speed of light, as in the smooth-walled waveguides of Section 4.3, while a slow-wave interaction involves a mode with a phase velocity less than the speed of light, as in the slow-wave structures of Section 4.4.

From the standpoint of the electrons participating in the generation process, we define three classes of devices:

- *O-type*: Devices in which the electrons drift parallel to an externally imposed magnetic field. This field helps to guide the intense beams used in HPM sources and, in some instances, plays an essential role in the microwave generation process.
- *M-type*: Devices in which the electrons drift perpendicularly to crossed electric and magnetic fields.
- *Space-charge*: Devices, possibly involving an applied magnetic field, in which the interaction producing microwaves is intrinsically traceable to an intense space-charge interaction.

The interactions leading to microwave generation in both the O- and M-type devices produce waves that grow up from small perturbations on an initial equilibrium state. The situation is quite different in the space-charge devices. The flow of the electrons in those devices is disturbed in a way that prevents the existence of an equilibrium; however, the resulting oscillatory state has a natural frequency. Thus, there are similarities in the operation of the O- and M-type sources that are not shared with the space-charge devices. We will briefly review the common elements of the first two types of sources before considering each source type individually.

4.8.1   Review of Fundamental Interactions

In O- and M-type devices, microwaves are generated primarily by linear resonances in which the waveguide or cavity normal modes and the electron oscillations must have the same frequency. Further, they both must have the same phase velocity, ω/k_z, so that they advance together along the system axis. Therefore, the values of ω and k_z for the two waves must be equal. In many cases, the properties of each of the electromagnetic and electron waves are largely independent of one another, except at values of ω and k_z in the vicinity of the resonance. Hence, we can estimate the resonance frequency and wavenumber by looking for the intersection in an ω-k_z plot of the dispersion curves for an electromagnetic normal mode (examples of which are shown in Figure 4.4, Figure 4.9, and Figure 4.15) and one of the oscillatory modes of the electrons. However, not all such interactions are truly resonant, permitting the transfer of electron energy to microwaves. The issue of which intersections lead to resonant microwave generation is resolved by deriving and solving the full dispersion relation for a coupled system of electromagnetic waves and electrons. In such a calculation, near microwave-generating resonances ω and k_z are in general complex, with the result that an *instability* exists and the waves exhibit exponential spatial and temporal growth.

More in line with our approach of treating the electromagnetic and particle modes as almost independent, the incremental wave energy of the participating modes, calculable from uncoupled dispersion relations, can be considered to determine if an interaction will be unstable.[16,17] When an intersection in the dispersion diagrams leads to microwave generation, the electron oscillation participating in the resonance is said to be a *negative energy wave* in the following sense. The total energy of the coupled system, including the electromagnetic energy of the normal mode and the summed energy of the electrons, is positive; however, the total energy is higher in the initial equilibrium state with no electron oscillations than it is in the presence of the unstable electron oscillation. The electron oscillation therefore reduces the total energy of the system and has a negative energy in the incremental sense. This energy reduction does not occur for some electron oscillations, which are said to be *positive energy waves*; these will not interact unstably with the electromagnetic normal modes, which always have a positive energy, and they cannot be used to generate microwaves. We will shortly return to this point in the specific case of space-charge waves.

In most sources, wave growth is accompanied by *bunching* of the electrons in one way or another. For example, in some devices, the electrons form spatial bunches, while in others, they bunch in the phase of their rotation about the magnetic field lines. In any event, bunching is an important phenomenon because it enforces the coherence of the electromagnetic waves that are generated.[18] To understand this, think of the generation process as an emission process by the individual electrons. In a uniform beam of electrons, the phases of the waves emitted by the individual electrons are completely random, so that the sum total of the radiation emitted by all of the electrons cancels out. Due to the action of the waves back onto the beam, however, the beam forms into bunches, the phases of the waves from each electron are no longer random, and the net sum of the emitted waves becomes nonzero. Thus, bunching makes the difference between the incoherent spontaneous emission of the initial state and the coherent stimulated emission of the later state. When the beam is strongly bunched,[19] the microwave output power is proportional to the number of bunches, N_b, and the square of the number density of electrons within the bunches, n_b:

$$P \propto N_b n_b^2 \tag{4.125}$$

Of course, the process of wave growth and bunching cannot proceed indefinitely; eventually, if given enough time, the instability will proceed to *saturation* and wave growth ceases, as shown in Figure 4.24. A variety of mechanisms lead to saturation, but two are seen most commonly: either the excess kinetic energy of the electrons is exhausted and the electron oscillation falls out of resonance, or the electron bunches become trapped in the strong potential wells of the electromagnetic waves and begin to reabsorb energy from the wave as they "slosh" around the well.

4.8.2 O-Type Source Interactions

At root, there are only a couple of basic interactions underlying the operation of the O-type devices. One is based on the oscillation of *space-charge waves* on an electron beam. To understand these, imagine a very long, large-diameter beam of electrons drifting along the z axis of a metallic drift tube, confined in the radial direction by a very strong axial magnetic field. If the density of the beam is uniform, axial forces along the beam are negligible, except near the ends. Imagine, though, that the electrons are squeezed together in one area, creating an electron bunch. The effect is like that of squeezing a section of spring together; the space-charge forces between the electrons in the bunch will cause them to repel one another and move apart, overshooting to create higher-density bunches on either side of the original

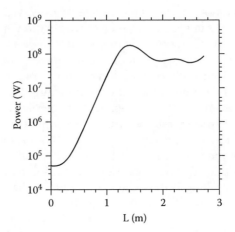

FIGURE 4.24
The growth to saturation of a microwave signal as a function of distance along the interaction region, L.

bunch, which once again repel one another and so move back together, creating an oscillation back and forth. The relation between the oscillation frequency and the axial wavenumber of the oscillations is

$$\omega = k_z v_z \pm \frac{\omega_b}{\gamma_0} \qquad (4.126)$$

where $\omega_b = (n_b e^2 / \varepsilon_0 m \gamma_b)^{1/2}$ is the *beam plasma frequency* for a beam of electron number density n_b and electron velocity v_z, with $\gamma_b = (1 - v_z^2 / c^2)^{-1/2}$ the relativistic factor. The waves with the two frequencies in Equation 4.126 are commonly known as the fast and slow space-charge waves, according to the use of the + or − sign, respectively.

The importance of boundary conditions was stressed in Section 4.2, yet Equation 4.126 contains no geometrical factors and is derived quite simply without regard for the radial boundaries of a real configuration. Analysis shows, though, that the dispersion relation for space-charge waves on a bounded beam in a waveguide is modified in some significant ways when the boundary conditions are taken into account. Figure 4.25 shows the dispersion relation for the space-charge waves in a beam within a waveguide (derived by Brejzman and Ryutov[20]), compared to the infinite cross-section model appropriate to Equation 4.126. The primary difference occurs near $k_z = 0$, where Equation 4.126 gives a wave with a phase velocity greater than the speed of light, whereas the solution to the actual dispersion relation shows that the phase velocity of the space-charge waves is always less than c (note that even the "fast" space-charge wave is a slow wave in this sense). In either case, with or without radial boundary conditions taken into account, the slow space-charge wave is the negative energy wave, while the fast space-

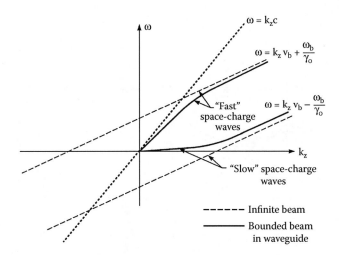

FIGURE 4.25
The space-charge wave dispersion diagram for an unbounded beam model (dashed) and a beam in a waveguide (solid).

charge wave has a positive incremental energy. Microwave generation occurs at a frequency and wavenumber approximately given by the intersection between the dispersion curves for the slow space-charge wave and an electromagnetic normal mode; intersections between curves for the fast space-charge wave curve and a normal mode will not yield a microwave-generating resonance.

The slow space-charge waves, with $\omega/k_z < c$, cannot couple to the normal modes of a smooth-walled waveguide, which have $|\omega/k_z| > c$. Nevertheless, there are ways to achieve the required coupling between the space-charge wave and an electromagnetic normal mode. For example, the phase velocity of the waveguide modes can be reduced below the vacuum speed of light by inserting into the guide a dielectric material with a permittivity, ε, significantly greater than that of vacuum, ε_0 (for example, a rod on axis, a sleeve inside the walls, or even a plasma layer), so that the effective speed of light in the waveguide appearing in Equations 4.20 and 4.21 approaches $c' = c(\varepsilon_0/\varepsilon)^{1/2}$. When c' is made less than v_z, as shown in Figure 4.26a, the resulting resonance between the slow space-charge wave and the electromagnetic mode serves as the basis of the *dielectric Cerenkov maser* (DCM).[21]

The guide modes can also be slowed down by modifying the shape of the walls of the guide to create a *slow-wave structure*, such as those we discussed in Section 4.4. The periodic slow-wave structures of Section 4.4 are the bases of devices like the *backward wave oscillator* (BWO) and *traveling wave tube* (TWT) described in Chapter 8, when the variation in slow-wave-structure radius is along the axis. Distinguishing between the BWO and TWT requires a knowledge of the dispersion curves and the location of the intersection between the slow space-charge wave and the slow-wave-structure curve: BWOs result when the slow-wave-structure dispersion curve has a negative

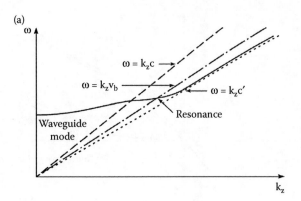

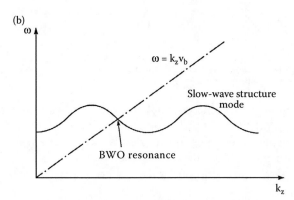

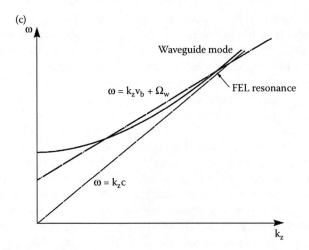

FIGURE 4.26

Uncoupled dispersion relations for systems in which the space-charge waves interact with differing electromagnetic modes: (a) dielectric Cerenkov maser, (b) backward wave oscillator, and (c) free-electron laser.

slope at the point of intersection, so that the group velocity is negative, or backward, while TWTs involve an intersection at a point of positive group velocity. Figure 4.26b illustrates such a resonance for a BWO.

A third means of coupling slow space-charge waves to guide modes involves a modification not of the waveguide dispersion, but of the space-charge wave dispersion. If a periodic transverse magnetic field is applied inside the waveguide, one view of the result is that the dispersion curves for the space-charge waves have been upshifted in frequency by an amount equal to the "wiggle" frequency of the electrons, $\omega_w \approx k_w v_z$, where $k_w = 2\pi/\lambda_w$, with λ_w the periodicity of the magnetic field, so that

$$\omega \approx k_z v_z + \omega_w \tag{4.127}$$

This is the basis for the *free-electron laser* (FEL), or *ubitron*, described in Chapter 10. Hence, in this view of FEL operation, the space-charge waves are made into fast waves capable of interacting with the fast waveguide modes. Alternatively, the radiation process can be viewed as magnetic bremsstrahlung, or coherent synchrotron radiation of the transversely accelerated electrons. Figure 4.26c depicts this resonance. Note that there are two intersections between the upshifted slow space-charge mode and the waveguide dispersion curve. Characteristically, FEL operation is at the upper intersection (see Problem 16).

The *relativistic klystron*, discussed in Chapter 9, makes use of resonant cavities to excite the space-charge waves on the beam. The simplest two-cavity form of the klystron is shown in Figure 4.27. The beam interacts with a normal mode of the first cavity (which is usually driven externally), and the fields associated with this electromagnetic mode modulate the axial drift velocities of the beam electrons and excite space-charge wave growth on the beam. In the drift space between the two cavities, the space-charge wave growth continues. The resulting space-charge bunches then pass into the second cavity, where they excite an electromagnetic cavity mode, the fields of which can be coupled out as radiation. A distinguishing feature of the klystron is that the only communication between the two cavities is through

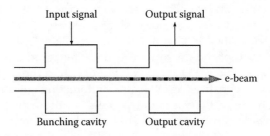

FIGURE 4.27
The simplest two-cavity form of a klystron.

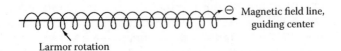

Larmor rotation

FIGURE 4.28
Larmor rotation at the electron cyclotron frequency, ω_c, about the guiding center, which lies along a magnetic field line.

the space-charge waves, since the drift space between the cavities is made sufficiently small that it is cut off to the propagation of the cavity modes between the two (recall that the cutoff frequency $\omega_{co} \propto 1/r_0$). In fact, this requirement becomes quite restrictive as the desired operating frequency increases, since cutting off the lowest frequency waveguide mode, TE_{11}, means that for an operating frequency f, the waveguide radius must be less than a maximum value of $r_{max}(cm) = 8.8/f(GHz)$. One way of getting around this is to cut off all but the few lowest frequency modes, and then design the drift tube between cavities in such a way that the lowest-order modes cannot propagate (e.g., by cutting properly shaped slots in the walls or loading them with a microwave absorber).

An entirely different mechanism for generating microwaves from a magnetized beam of electrons involves electron–cyclotron interactions with the beam electrons. Consider a beam traveling along a reasonably strong guide field, but imagine that the individual electrons have a velocity component perpendicular to $\mathbf{B_0}$, so that they execute small-orbit Larmor rotation about their guiding centers at the cyclotron frequency $\omega_c = eB_0/m\gamma_b$, as shown in Figure 4.28. The individual electrons can then spontaneously emit Doppler-shifted cyclotron radiation of frequency

$$\omega = k_z v_z + s\omega_c \qquad (4.128)$$

with s an integer. Electrons emitting this radiation can interact resonantly with the fast waveguide modes, and because the cyclotron frequency is energy dependent (as $1/\gamma_b$), the rotating electrons will become bunched in their rotational phase (which is entirely different from the spatial bunching of the space-charge waves). This resonance is the basis for a number of *electron cyclotron maser* types, such as the *gyrotron* and *cyclotron autoresonant maser* (CARM), described in Chapter 10 (see Problem 4).

Our interpretation of the microwave-generating resonance in the devices considered so far — the BWO and TWT, the FEL, and the electron cyclotron masers — has been constructed in terms of interacting collective modes. We can alternatively view the radiation process as the action of a coherent ensemble of transversely accelerated electrons that are emitting. In the FEL, where electrons undulate in the wiggler field, and the gyrotron, where the electrons gyrate about field lines, this phenomenon is readily apparent. On the other hand, in the BWO and TWT, the physical picture is modified somewhat. In those devices, the charge density perturbations in the space-charge waves on the beam create image charges in the rippled-wall slow-wave structure, and it is these image charges, moving transversely with the wall ripple, that can be viewed as causing a kind of dipole radiation as they convect along with the beam space-charge waves.

An additional interaction of some interest is that in which two types of beam modes work together parametrically to produce a third wave, an electromagnetic output. To take an example, if a slow space-charge wave becomes large enough in amplitude, it can function as the wiggler to drive an FEL-type interaction, even in the absence of a magneto-static wiggler field. This type of stimulated scattering[22] has been used in two-stage FELs, for example, in which the first stage is either another FEL[23] or a BWO.[24] Such devices are sometimes called *scattrons* in the Soviet and Russian literature.

As a final point in the discussion of O-type devices, we note that the klystron concept — setting some form of bunching in motion in one inter-action region, propagating the beam through a cutoff waveguide section, and then tapping the bunched beam in a separated, downstream interaction region — has quite general applicability and has been employed in the design of several hybrid devices. For example, the *gyroklystron* essentially features gyrotron resonators separated by waveguides in which gyrophase bunching occurs (see Chapter 10). A second example is the *optical klystron* (Chapter 10), which is a variant of the FEL with multiple interaction spaces. In devices of this type, the advantage of the klystron principle is that it makes a pre-bunched beam available to the downstream cavity (or cavities, if the bunching process is enhanced by using more than a single cavity), where the bunching enhances the efficiency.

4.8.3 M-Type Source Interactions

In contrast with O-type devices, electrons in M-type devices drift across a magnetic field in crossed electric and magnetic fields. This type of motion is called an $\mathbf{E} \times \mathbf{B}$ drift. Most commonly, M-type devices involve the inter-action between a slow-wave electron mode and electromagnetic normal modes directed colinearly with the overall drift of the electrons. The primary example of such a device is the *relativistic magnetron*, depicted schematically in Figure 4.29 and discussed in Chapter 7. There, an electron layer drifts azimuthally near the cathode of a cylindrical diode in the radial electric field and applied axial magnetic field. Because the cavity is closed on itself in the azimuthal direction (or reentrant, in the accepted parlance), counterrotating electromagnetic modes can set up a standing wave pattern around the cavity. It has become the convention in the case of magnetrons to characterize the normal cavity modes by the phase shift of the waves between the individual resonators. Electromagnetic modes with phase shifts of π and 2π, known as the π-mode and 2π-mode, are the most commonly seen operating modes in magnetrons. In addition to the magnetron, an amplifying M-type source

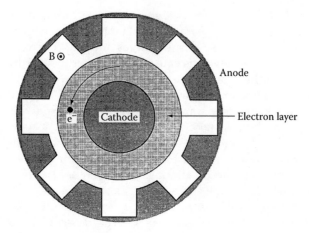

FIGURE 4.29
Schematic depiction of a magnetron cavity, in cross section.

known as the *crossed-field amplifier* has existed in the nonrelativistic regime for some time, and one was briefly explored in the high power regime. More recently, a linear device with high power potential known as a *magnetically insulated line oscillator* (MILO) has been explored.

A fast-wave M-type device has also been examined. In this device, known as a rippled-field magnetron, the slow-wave structure of Figure 4.29 is replaced with a smooth wall into which permanent magnets of alternating polarity have been placed. Thus, the device is like a cylindrical FEL in which the beam has been replaced by a magnetically insulated layer of electrons undergoing $\mathbf{E} \times \mathbf{B}$ drifts.

4.8.4 Space-Charge Devices

Space-charge devices are so named because their operation depends on a strong space-charge-driven phenomenon involving the formation of a virtual cathode. The best-known device of this type is the *virtual cathode oscillator*, or *vircator*. To understand its operation, consider the configuration shown in Figure 4.30. There, an electron beam is generated in a diode with a thin-foil anode that allows the beam to pass through to the drift space beyond. For a cylindrical drift space, the situation is that discussed in Section 4.6.3. If the drift space has grounded walls and a radius r_0, while the beam has radius r_b and was launched from a cathode at voltage $-V_0 < 0$, the space-charge-limited current I_{SCL} for the drift space is given by Equation 4.115 if the beam cross section is annular and Equation 4.116 if the beam cross section is solid. A beam current I_b exceeding the appropriate value of I_{SCL} will cause the formation of a strong potential barrier at the head of the beam, from which a significant number of the beam electrons will be reflected back toward the anode. The reflection point for the electrons is

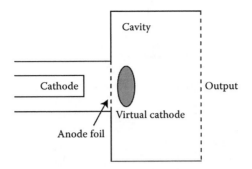

FIGURE 4.30
Schematic depiction of a virtual cathode oscillator, or vircator.

known as the virtual cathode. Essentially, the excess space-charge in the beam is rejected from the drift space. This situation is intrinsically different from the other microwave generation mechanisms we have discussed to this point, because in the earlier cases, we started with a well-defined initial beam state on which perturbations grew up from small amplitudes at early times. In the case of virtual cathode formation, there is a very fundamental breakdown of the beam state with the result that no beam equilibrium is possible. The position of the virtual cathode oscillates axially, shedding charge and behaving like a relaxation oscillator. The motion of the space-charge and potential well of the virtual cathode produces radiation at a frequency that is proportional to the square root of the beam current (for solid beams, it is of the order of the beam plasma frequency, ω_b). The bandwidth of the vircator is quite large, and if the virtual cathode is formed in a resonant cavity, it will interact with the cavity modes to oscillate at a frequency tuned to a cavity mode.

The second type of space-charge device, the *reflex triode*, is configured somewhat similarly to the vircator, as shown in Figure 4.31, except that

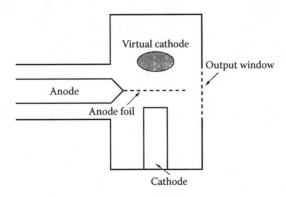

FIGURE 4.31
Schematic depiction of a reflex triode.

TABLE 4.5

Classification of HPM Sources

	Slow Wave	Fast Wave
O-Type	Backward wave oscillator	Free-electron laser (ubitron)
	Traveling wave tube	Optical klystron
	Surface wave oscillator	Gyrotron
	Relativistic diffraction generator	Gyro-BWO
	Orotron	Gyro-TWT
	Flimatron	Cyclotron autoresonance maser
	Multiwave Cerenkov generator	Gyroklystron
	Dielectric Cerenkov maser	
	Plasma Cerenkov maser	
	Relativistic klystron	
M-Type	Relativistic magnetron	Rippled-field magnetron
	Cross-field amplifier	
	Magnetically insulated line oscillator	
Space-Charge	Vircator	
	Reflex triode	

the anode, rather than the cathode, is attached to the center conductor of the transmission line feeding power to the source. If the anode were perfectly transmitting with no absorption of electron energy, an electron from the cathode would drop down to the potential well at the anode, then roll up the other side of the well to the virtual cathode to be reflected back toward the cathode, where it would come to rest after passing back through the anode. Actually, though, some electron energy will be lost in the anode, and electrons will almost, but not quite, reach the emitting cathode, after which they will pass back through the anode, losing a little more energy on each pass, falling short of the original cathode, turning back toward the anode again, and on and on until all of the electron energy is lost. This reflexing back and forth by the electrons provides another source of microwaves for a device known as the reflex triode. A point to note is that in the simple vircator of Figure 4.30, both processes, virtual cathode oscillations and reflexing, can compete with one another unless special care is taken.

As we shall see in Chapter 9, the operation of low-impedance relativistic klystrons has some similarity to the space-charge devices, in that when the beam current exceeds a time-dependent limiting value in the klystron cavity gaps, current bunching takes place almost immediately in the vicinity of the gap.

Table 4.5 summarizes our classification of HPM sources. We can see that there are fast- and slow-wave devices for both the O- and M-type sources. We do not distinguish between fast- and slow-wave types for the space-charge devices. Some source varieties not shown in the table can be found in the later sections of this book, but these represent, in many cases, variations on the basic types discussed here.

4.9 Amplifiers and Oscillators, High- and Low-Current Operating Regimes

The classification system in Table 4.5 is based on the nature of the microwave-generating interactions. We could choose an operational criterion for grouping our sources, however. Let us therefore distinguish between *amplifiers* and *oscillators*. An amplifier produces an output signal that is a larger amplitude version of some input signal; in the absence of an input signal, there is usually no output signal other than low-level amplified noise. An oscillator, on the other hand, requires (1) feedback and (2) sufficient gain to overcome the net losses per cycle, so that an output signal can be generated spontaneously even in the absence of an input signal. Since gain typically increases with beam current, the requirement for sufficient gain is usually translated into a threshold condition on the beam current, called the *start current* for oscillations. Below this value, the system will not oscillate spontaneously, while above it, spontaneous signals will result. The start current is a function of system parameters, including the length of the interaction region.

Amplifiers are characterized by a certain finite bandwidth of input signal frequencies over which useful amplification is possible, while oscillators tend to select a very definite frequency of operation, depending again on the resonant frequency between the beam and the cavity. The output characteristics of an amplifier — frequency, phase, and waveform — can be controlled quite nicely with the provision of a low power master oscillator with the desired features. Designing for high gain and output power while controlling the tendency to break into spontaneous oscillation can be difficult, though. Features of oscillators that are frequently important for applications are *frequency pushing*, the variation of the frequency with source current; *frequency pulling*, the variation of frequency with change of phase of the complex impedance of some load connected to the oscillator cavity; and *phase jitter*, the pulse-to-pulse variation of the initial phase of oscillations. Oscillators enjoy the simplicity attendant to the absence of a master oscillator, although phase and frequency control, to which we will return in the next section, is more complicated than for an amplifier.

Some sources are suited to operation exclusively or primarily as either oscillators or amplifiers. Those usually operated as oscillators are (1) BWOs, because of the feedback inherent in the coupling to a backward wave normal mode; (2) gyrotrons, due to the low-group-velocity, high-Q cavity mode employed; and (3) magnetrons and rippled-field magnetrons, since they use reentrant cavities. Nevertheless, BWOs and gyrotrons can, in principle, function as amplifiers below their start currents. As their name implies, crossed-field amplifiers are designed to be amplifiers by the inclusion of a short absorbing section meant to prevent the complete, 2π circulation of electromagnetic waves around the source azimuth. The other sources of Table 4.5 — cyclotron autoresonance masers (CARMs), FELs, MILOs, TWTs,

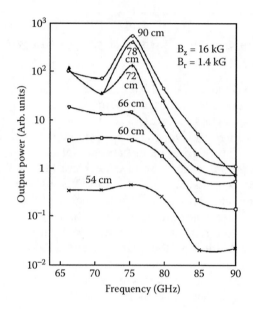

FIGURE 4.32

The output of a super-radiant FEL amplifier. Measurements were made at the frequencies indicated, and points taken for devices of the same length were connected to make the curves shown. (From Gold, S.H. et al., *Phys. Fluids*, 27, 746, 1984. With permission.)

and Cerenkov masers — all have open configurations and forward-directed normal modes, so they all operate primarily as amplifiers. Additionally, they can be run as oscillators if some type of feedback is introduced, such as reflections at the downstream end.

Midway between amplifiers and oscillators, there are the *super-radiant amplifiers*. Although they are not, strictly speaking, a separate device type, these are amplifiers with such a high gain that they will amplify spontaneous noise on the beam to high output power levels. In a sense, the effect is like that of an oscillator, with the difference that the line width of the output (i.e., the spectral width of the output frequency) is broader. Lengthening such a device can narrow the line width about the frequency of the mode with the highest gain. An example of the output from a super-radiant FEL amplifier is shown in Figure 4.32. There, we see the noticeable peaking of output as a function of frequency as the device is lengthened.

A third, very broad means of classifying microwave sources is according to the intensity of the space-charge effects in the device. At low electron densities, typically corresponding to low beam currents, a device is said to be in the *Compton* regime of operation. In this case, collective space-charge effects are negligible and the electrons behave as a coherent ensemble of individual emitters. At high electron densities, with beam currents approaching a significant fraction of the space-charge limit, the collective modes of electron oscillation (e.g., space-charge waves) play the central role in microwave generation. This is called the *Raman* regime of operation. Some authors might distinguish intermediate states between these two extremes, but this simple delineation will serve our purposes.

4.10 Phase and Frequency Control

Phase and frequency control of a microwave signal are desirable in a variety of applications, such as driving radio frequency accelerators, maximizing the spatial power density from a multiple-antenna array, and even electronically steering the output of such an array. As we mentioned in the previous section, perhaps the most straightforward means of controlling the phase of an array of sources is to generate the high power from one or more power amplifiers driven by a relatively lower power master oscillator. This configuration is denoted by the acronym MOPA (master oscillator/power amplifier). The difficulties inherent in the design of high power, high-gain amplifiers and the complexity of MOPA arrays leave room for the use of phase- or frequency-controlled oscillators.

Phase locking and frequency control of oscillators can be achieved in several ways. Perhaps the most common involves the use of the power from a driving master oscillator to determine the phase of a slave oscillator. The important parameter is the injection ratio:

$$\rho = \left(\frac{P_i}{P_0} \right)^{1/2} = \frac{E_i}{E_0} \tag{4.129}$$

where P and E are the power and electric fields in the oscillators, and the subscripts i and 0 denote the injecting (or master) and receiving (or slave) oscillators. Phase locking occurs when the frequency difference between the two oscillators is sufficiently small:

$$\Delta\omega \leq \frac{\omega_0 \rho}{Q} \tag{4.130}$$

with ω_0 the mean frequency of the two oscillators, which are also taken to have the same value of Q. Equation 4.130 is known as *Adler's relation*. The timescale for locking is given approximately by

$$\tau \sim \frac{Q}{2\rho\omega_0} \tag{4.131}$$

For $\rho \sim 1$, the locking range $\Delta\omega$ is maximized and the timescale for locking is minimized. This facilitates locking of oscillators with larger frequency differences and makes arrays tolerant of oscillator variations. The weak coupling regime, $\rho \ll 1$, allows a small oscillator to drive a larger one, but takes a corresponding longer time for locking to occur (see Problem 19). Phase locking at high powers has been studied most thoroughly in relativistic magnetrons (see Chapter 7). A second method for phase locking involves prebunching of a beam before it is injected into a microwave source. This has been done in gyroklystrons. A third method is known as *priming* an oscillator. In this case, a locking signal is injected during the start-up phase of the oscillator, which then locks to the injected signal.

4.11 Summary

Many of the fundamental concepts introduced in this chapter will recur throughout the book. The normal modes of rectangular waveguides are denoted TE_{np} and TM_{np} (and, to avoid inconsistency in notational conventions elsewhere, TE_{pn} and TM_{pn} in circular waveguides), where n and p are indices on the field variations in the transverse plane. A normal mode will not propagate through a waveguide below a characteristic frequency, known as the cutoff frequency. The cutoff frequency scales inversely with the cross-sectional dimensions of the guide, so that single-mode propagation is typically confined to relatively narrow guides. High power transmission through a guide demands a large waveguide cross section and a resulting availability of many modes above cutoff. This is known as overmoded operation.

Slow-wave structures support waves with phase velocities less than the speed of light. The cross-sectional structure of these modes can have the characteristics of the TE and TM modes of a smooth-walled waveguide, although as the frequency and wavenumber vary, the cross-sectional structure of a given slow-wave-structure mode can transition from one TE or TM mode to another.

The normal modes of cavities are characterized by an additional index on the axial variations of the fields within, e.g., TE_{npq} (where the ordering of indices is x-y-z for rectangular coordinates and θ-r-z for cylindrical coordinates, by convention). Cavities are characterized by the *quality factor* Q, which measures energy loss per cycle from the oscillating fields in the cavity. Energy loss, and the contributions to Q, comes primarily from two factors: power coupled out as useful microwave radiation and power lost to parasitics. Low Q operation is in many cases desirable in HPM sources.

Resonant interaction at a common frequency and wavenumber between the waveguide or cavity fields and an oscillatory mode in a body of energetic electrons enables the energy transfer from the electrons that generates HPM. Beam bunching is crucial to enforcing the coherence of the waves that are emitted. The classification scheme used here to organize the devices is based on the character of the interactions underlying the generation mechanism in a device.

Functionally, we can also distinguish between devices on an operational basis, denoting them as amplifiers, oscillators, or super-radiant amplifiers. In some cases, certain devices are better suited to operation in one of these different modes, although others can be operated in each of these three under varying conditions. Oscillators tend to operate at a single frequency, depending on their design features; techniques exist, however, to control the phase and, to an extent, the frequency of the output.

Problems

1. For a length of rectangular waveguide in X-band with dimensions a = 2.286 cm and b = 1.016 cm, (a) What is the WR number? (b) What are the cutoff frequencies for the first four propagating modes? Remember that L(cm) = 2.54 × L (inches).

2. Compute the cutoff frequency for the TE_{11} modes in rectangular and circular waveguides with similar dimensions. For the rectangular waveguide, assume a rectangular cross section with a = b = 2 cm. For the circular waveguide, assume a diameter of $2r_0 = 2$ cm. Repeat the exercise for the TM_{11} mode in both.

3. Gyrotrons operate near the cutoff frequency for the cylindrical interaction cavity. To understand the frequency spacings involved, consider the spacing in cutoff frequencies for a cylindrical waveguide with a radius of 2 cm. (a) First compute the cutoff frequencies for the TE_{11} and TE_{12} modes. What is the spacing in hertz, and how much higher is the second mode than the first, on a percentage basis? (b) Next compute the cutoff frequencies for the TE_{82} and TE_{83} modes. Again, what is the spacing in hertz, and how much higher is the second mode than the first, on a percentage basis?

4. For a maximum allowable field at the wall of 100 kV/cm, (a) What power can a rectangular waveguide propagate in fundamental mode at 3 GHz? (b) What power can a circular waveguide propagate in the lowest-frequency TM mode at 3 GHz? Assume that the waveguide is cut off to propagation for modes above the ones designated here; state the dimensions you use.

5. Consider a circular waveguide carrying 5 GW of power in the TM_{01} mode. How large must the radius be to keep the electric field at the wall at 200 kV/cm? At that radius, how many of the TM_{0n} modes are allowed to propagate in the waveguide?

6. Consulting the formulary, what is the skin depth of aluminum at 10 GHz?

7. Estimate the lower and upper frequencies of the lowest passband for an axially varying slow-wave structure with a sinusoidally rippled wall with a mean radius of 4 cm and a ripple period of 1 cm.

8. Estimate the Q of a copper-lined cavity operating at 35 GHz, with a diameter of 3 cm and a length of 5 cm, with the ends partially capped.

9. Diode impedance is the ratio of voltage over current. What diode impedance is required to produce 5 GW of microwaves from a source with an efficiency of 20% when the diode operates at a voltage of 1.5 MV?

10. For efficient energy transfer, assume you want to match the impedance of a diode to an oil-insulated pulse-forming line with an impedance of 43 Ω. If the voltage on the diode is to be 500 keV, and the gap must be at least 2 cm to avoid gap closure during the pulse due to plasma from the cathode, what cathode diameter do you need?

11. Prove that the expression for the cathode current density in a cylindrical diode (Equation 4.106) yields the expression for the current density in a planar diode (Equation 4.104) in the limit of a near-planar diode, where r_a approaches r_c.

12. Consider a coaxial diode emitting radially from an emitting region $L = 3$ cm long on a cathode at radius $r_c = 1.58$ cm toward an anode at $r_a = 2.1$ cm, with an anode–cathode voltage of 400 kV. These are the radii of the A6 magnetron, but we assume there is no axial magnetic field, so that the current can flow across the gap. Compute the current flowing across the gap. Compare the value of the current to the same current for a planar diode with the same gap and the same emitting area as the cathode in the cylindrical diode.

13. At what voltage will a 40-kA beam from a diode with a 4-cm cathode radius and a 2-cm anode–cathode gap form a pinch?

14. A cylindrical electron beam of uniform current density and radius r_b is emitted from a planar diode with a gap of 2 cm and an accelerating voltage of 1.5 MV. It passes through a planar, thin-foil anode into a cylindrical drift tube 4.5 cm in radius. (a) If we assume that the beam enters the drift space with the same radius as the radius of the circular face of the cathode, which has a radius r_c, what is the minimum value of r_c to get virtual cathode formation? (b) What is the radius of the cathode at which the beam current is twice the space-charge-limiting current in the drift tube?

15. If we inject a 5-kA electron beam with a radius of 2 cm from a 1-MV diode into a drift tube 3 cm in radius, what value of axial magnetic field is required to keep its rotational frequency — the frequency for

rotation of the beam about its axis in the slow rotational mode — below 500 MHz?

16. The basic relativistic magnetron, termed the A6 because it uses six cavities in its resonant structure (see Chapter 7), has a cathode and anode radii of 1.58 and 2.1 cm, respectively, and operates at 4.6 GHz at a voltage of 800 keV. What magnetic field is required to insulate it?

17. A 3-MeV electron beam passes through a wiggler magnet array in which magnets of opposing polarity are spaced 5 cm apart. What is the wiggler frequency in this system?

18. What is the minimum magnetic field required to operate a gyrotron at 100 kV in a 3-cm-radius cavity?

19. Two high power oscillators both have the same 3-GHz frequency and Q = 100, values typical of HPM sources. Their pulse durations are 100 nsec, also typical of HPM sources. The bandwidth of each is 1%. They are driven in parallel by a pulsed power unit, so they radiate simultaneously. They are to be phase locked together to add their powers before being radiated from an antenna. How well coupled do they have to be, i.e., what injection ratio is required if they are to lock within 1/3 of their pulse lengths? If their rise time is Q cycles of the 3-GHz frequency and they are to lock within the rise time, what injection ratio is required?

References

1. Neuber, A.A., Laurent, L., Lau, Y.Y., and Krompholz, H., Windows and RF breakdown, in *High-Power Microwave Sources and Technology*, Barker, R.J. and Schamiloglu, E., Eds., IEEE Press, New York, 2001, p. 346.

2. Kurilko, V.I. et al., Stability of a relativistic electron beam in a periodic cylindrical waveguide, *Sov. Phys. Tech. Phys.*, 24, 1451, 1979 (*Zh. Tekh. Fiz.*, 49, 2569, 1979).

3. Bromborsky, A. and Ruth, B., Calculation of TM_{on} dispersion relations in a corrugated cylindrical waveguide, *IEEE Trans. Microwave Theory Tech.*, MTT-32, 600, 1984.

4. Swegle, J.A., Poukey, J.W., and Leifeste, G.T., Backward wave oscillators with rippled wall resonators: analytic theory and numerical simulation, *Phys. Fluids*, 28, 2882, 1985.

5. Guschina, I.Ya. and Pikunov, V.M., Numerical method of analyzing electromagnetic fields of periodic waveguides, *Sov. J. Commun. Tech. Electron.*, 37, 50, 1992 (*Radiotekh. Elektron.*, 8, 1422, 1992).

6. Leifeste, G.T. et al., Ku-band radiation produced by a relativistic backward wave oscillator, *J. Appl. Phys.*, 59, 1366, 1986.

7. Petit, R. and Cadilhac, M., Sur la diffraction d'une onde plane par un reseau infiniment conducteur, *C. R. Acad. Sci. Paris Ser. B*, 262, 468, 1966.

8. Millar, R.F., On the Rayleigh assumption in scattering by a periodic surface: II, *Proc. Camb. Phil. Soc.*, 69, 217, 1971.

9. Watanabe, T. et al., Range of validity of the Rayleigh hypothesis, *Phys. Rev. E*, 69, 056606-1, 2004.

10. Kroll, N., The unstrapped resonant system, in *Microwave Magnetrons*, Collins, G.B., Ed., McGraw-Hill, New York, 1948, pp. 49–82.

11. Child, C.D., *Phys. Rev.*, 32, 492, 1911; Langmuir, I., The effect of space charge and residual gases on thermionic currents in high vacuum, *Phys. Rev.*, 2, 450, 1913.

12. Langmuir, I. and Blodgett, K.B., Current limited by space charge between coaxial cylinders, *Phys. Rev.*, 22, 347, 1923.

13. Jory, H.R. and Trivelpiece, A.W., Exact relativistic solution for the one-dimensional diode, *J. Appl. Phys.*, 40, 3924, 1969.

14. Friedlander, F., Hechtel, R., Jory, H.R., and Mosher, C., Varian Associates Report DASA-2173, 1965.

15. See, for example, Davidson, R.C., *Theory of Nonneutral Plasmas*, W.A. Benjamin, Reading, MA, 1974.

16. Krall, N.A. and Trivelpiece, A.W., *Principles of Plasma Physics*, McGraw-Hill, New York, 1971, pp. 138–143.

17. Kadomtsev, B.B., Mikhailovskii, A.B., and Timofeev, A.V., Negative energy waves in dispersive media, *Sov. Phys. JETP*, 20, 1517, 1965 (*Zh. Eksp. Tekh. Fiz.*, 47, 2266, 1964).

18. Kuzelev, M.V. and Rukhadze, A.A., Stimulated radiation from high-current relativistic electron beams, *Sov. Phys. Usp.*, 30, 507, 1987 (*Usp. Fiz. Nauk*, 152, 285, 1987).

19. Friedman, M. et al., Relativistic klystron amplifier, *Proc. SPIE*, 873, 92, 1988.

20. Brejzman, B.N. and Ryutov, D.D., Power relativistic electron beams in a plasma and a vacuum (theory), *Nucl. Fusion*, 14, 873, 1974.

21. Walsh, J., Stimulated Cerenkov radiation, in *Novel Sources of Coherent Radiation*, Jacobs, S., Sargent, M., and Scully, M., Eds., Addison-Wesley, Reading, MA, 1978, p. 357.

22. Bratman, V.L. et al., Stimulated scattering of waves on high-current relativistic electron beams: simulation of two-stage free-electron lasers, *Int. J. Electron.*, 59, 247, 1985.

23. Elias, L.R., High-power, CW, efficient, tunable (UV through IR) free electron laser using low-energy electron beams, *Phys. Rev. Lett.*, 42, 977, 1979.

24. Carmel, Y., Granatstein, V.L., and Gover, A., Demonstration of a two-stage backward-wave-oscillation free-electron laser, *Phys. Rev. Lett.*, 51, 566, 1983.

5

Enabling Technologies

5.1 Introduction

The rapid development of high power microwaves (HPM) during the 1980s was due to the availability of a technology base that had been developed for other applications. Figure 5.1 shows the subsystems of an HPM device operating in an HPM facility. Pulsed electrical systems, referred to as *pulsed power*, were developed in the 1960s to enable nuclear weapons effects simulation. Later, inertial confinement fusion became a driver for pulsed power technology. HPM has largely made use of existing pulsed power equipment to perform initial experiments. The intense beams generated by pulsed power devices have been studied in some detail, including beam generation and beam propagation in various media such as vacuum, plasma, and gases. In diagnostics for HPM, the bulk of the devices used are conventional diagnostic techniques with additional attenuation to reduce power levels before the diagnostic data are taken. An exception to this is calorimetry, which must handle high electric fields but little total energy. Some microwave components, such as waveguides, have been borrowed from the conventional microwave technology base. For example, evacuated waveguides of conventional design can support electric fields of 100 kV/cm for short pulses. Similarly, antennas have used direct extrapolation of conventional antenna design with some care taken to avoid air breakdown, with the major exception of the Impulse Radiating Antenna (IRA; see Sections 5.5 and 6.3). With the emergence of facilities specifically designed to study HPM effects and develop HPM technologies, the primary issues are the operational and safety constraints for protection of personnel and ancillary electronic equipment from the effects of high power pulses.

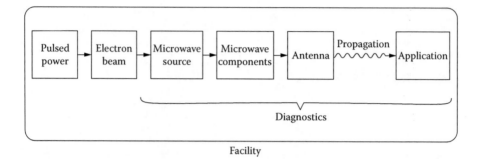

FIGURE 5.1
Flow diagram of HPM system in a facility.

5.2   Pulsed Power

The short, intense pulses of electricity needed to drive HPM loads are obtained by a process of pulse compression. The energy taken from a low-voltage, long-pulse system is compressed in time, increasing both voltage and current at the expense of pulse duration.[1] Electrical requirements of HPM sources encompass a wide range of voltages, currents, and pulse durations.[2] Most initial HPM source physics questions can be adequately addressed in short-pulse experiments (~100 nsec), which are also preferred because they are easy to carry out, relatively inexpensive, and relatively small. There are also difficulties, such as breakdown, in operating high power sources for longer durations. Consequently, the key distinguishing parameters of most devices are the voltage, current, and their ratio, *impedance*. The subset of electrical voltage–impedance parameter space into which HPM source technologies fall is shown in Figure 5.2. With voltages of <100 kV and impedances of >1000 Ω, conventional microwave applications (radar, electronic countermeasures, communications) lie at the low power end (power = V^2/Z). Most, but not all, present-day HPM sources cluster around 10 to 100. Voltages generally span the range of 0.1 to 1 MV, but some sources have been operated up to 4 MV.

The immediate implication of Figure 5.2 is that efficient transfer of electrical energy will require pulsed power technologies that span a substantial impedance range. When circuit elements do not have the same impedance, voltage, current, and energy transferred to the downstream element can be very different (Figure 5.3). The voltage gain V_2/V_1 and energy transfer ratio E_2/E_1 for a transmission line transferring power to a longer output line (so that reflections from the output line can be neglected during the main pulse) are

$$\frac{V_2}{V_1} = \frac{Z_2}{Z_1 + Z_2} \qquad\qquad (5.1)$$

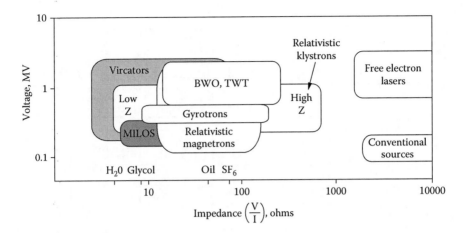

FIGURE 5.2
Voltage and impedance parameter space for HPM and pulsed conventional sources. Optimum impedances for widely used pulse line insulators (water, glycol, oil, SF_6) are shown.

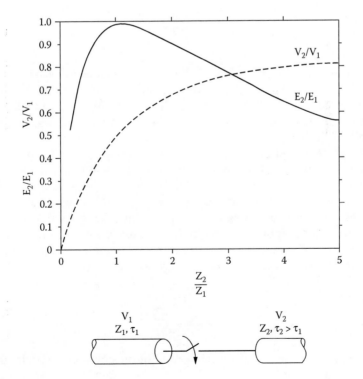

FIGURE 5.3
Energy transfer and voltage ratios between transmission line (1) and load (2) as a function of impedance ratio.

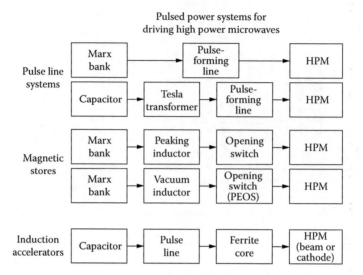

FIGURE 5.4
Types of pulsed power systems for driving HPM loads.

$$\frac{E_2}{E_1} = \frac{4Z_1Z_2}{[Z_1 + Z_2]^2} \tag{5.2}$$

Efficient energy transfer suffers if impedance mismatch exists, especially if the load is of lower impedance than the circuit driving it, so that it looks like a short. Mismatching upward to a higher load impedance looks like an open circuit, but is sometimes used as an inefficient means of increasing load voltage (see Problem 1).

The types of pulsed power systems used in HPM are shown in the block diagrams of Figure 5.4. The most commonly used system is a capacitor bank driving a *pulse-forming line* that, when switched into the load, releases its energy in two transit times at the speed of light in the insulating medium of the line. Coaxial transmission lines span a large range of impedance and voltage depending upon their characteristic breakdown field limits and the insulating dielectric between the electrodes (oil, water, glycol, SF_6, air). The impedance depends upon the dielectric constant, ε:

$$Z = \frac{1}{2\pi}\sqrt{\frac{\mu}{\varepsilon}} \ln b/a \tag{5.3}$$

where b and a are the outer and inner radii of the cylinder, respectively. The energy density,

$$E = 1/2 \ \varepsilon E^2 \tag{5.4}$$

depends linearly upon the dielectric constant, ε. Here E is the electric field at radius r in the dielectric and V the applied voltage:

$$E = \frac{V}{r\varepsilon_r \ln\frac{b}{a}} \qquad (5.5)$$

where ε_r is the relative dielectric constant and b and a the outer and inner radii, respectively. For fixed outer diameter and applied voltage, the electric field at the surface of the inner cylinder is a minimum when $b/a = e = 2.71$. This maximizes energy density while avoiding breakdown. For this condition, the optimum impedance of an oil line is 42; an ethylene glycol line, 9.7; and a distilled water line, 6.7. Optimum coaxial impedances are shown along the horizontal axis of Figure 5.2. On the other hand, liquid energy density is largest for water (0.9 relative to mylar), intermediate for glycol (0.45), and lowest for oil (0.06). The solid dielectric mylar has the highest energy density, but if broken, it will not recover.

Low-impedance microwave loads will therefore be a better match to waterlines, and high-impedance loads will better match oil lines. The duration of the pulse that can be extracted from a line is determined by the electrical length of the line,

$$\tau = \frac{2\varepsilon_r L}{c} \qquad (5.6)$$

where ε_r is the relative dielectric constant and L is the line length. Note that the discharge wave, begun by closing a switch at one end, must make two transits to discharge the line.

Two principal types of pulse-forming line systems are used in HPM. The first is a Marx directly charging a PFL (Figure 5.5). The *Marx generator* is an assembly of capacitors that are charged in parallel and then quickly switched into a series circuit, allowing the original charging voltage to be multiplied by the number of capacitive stages in the Marx. Marx generators are usually insulated with transformer oil to reduce their size. The PFL is switched into the load through a high-voltage closing switch.

The other pulse line scheme uses a capacitor bank at low voltage and a transformer to boost the voltage and charge the pulse line. The advantage of this configuration is in repetitive operation. Marx generators are generally not used at repetition rates above 10 Hz because of losses and fault modes, but transformers are.

For pulse durations of <200 nsec, the PFL is the appropriate pulse compression method, but PFLs become very large for longer pulse lengths. For such lengths, discrete inductive and capacitive components are used to assemble a *pulse-forming network* (PFN), which is designed to generate a Fourier series approximating the desired waveform. An example of a PFN

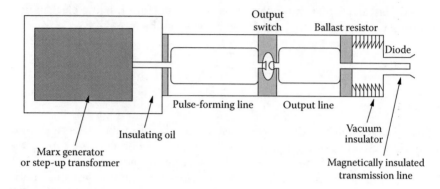

FIGURE 5.5
Schematic of a pulse-forming line system. Many elements are common to other approaches in Figure 5.4.

is shown in Figure 5.6a, and the output pulse from this circuit is shown in Figure 5.6b. In general, PFNs produce a pulse with voltage oscillations, typically an overshoot at the beginning of the pulse and an undershoot at the end. Voltage variations always affect the performance of the source, and one of the key features of matching pulsed power to HPM sources is sensitivity of the load performance to the applied voltage. The concern is usually for voltage flatness during the bulk of the pulse. PFN circuits can be optimized for voltage flatness.

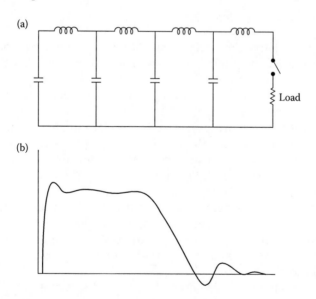

FIGURE 5.6
Pulse-forming networks: (a) typical distributed LC (inductor-capacitor) circuit; (b) output pulse has fast rise, overshoot, and slow fall time with undershoot. Durations are typically from 0.5 to 10 μsec.

Switching of PFLs and PFNs into the load is usually done by a spark gap switch using gas as a dielectric insulating medium. Because most microwave sources require precision in the applied voltage, and spark gaps have substantial variation in pulse-to-pulse voltage output, spark gaps in HPM systems are usually externally triggered. For repetitive operation the insulating gas is circulated to achieve faster recovery, carry away debris evolved from the electrodes, and remove heat dissipated in the gap. The typical timescale for recovery of dielectric strength in an uncirculated gas is ~10 msec; arc switches operating at >1 kHz with gas flow have been reported.[3]

When HPM sources are brought into development, they typically initially operate at powers lower than intended. Therefore, they will draw less current and operate at higher impedance than the experiment has been designed for. A common technique providing a match to a microwave load whose impedance will drop as it is developed is to use a ballast resistor (Figure 5.5), a liquid resistor with adjustable resistance so that the parallel combination with the microwave load provides a match to the pulsed power system.

When the electrical pulse exits the pulse-forming regions (filled with liquid or gas) and enters the vacuum region of the microwave source, it passes through a solid dielectric interface. The typical interface is composed of a stack of insulator rings separated by metal grading rings. The surface of the insulator rings is inclined 45° to the grading rings. The flashover strength of such construction is three to four times that of a straight cylinder. Detailed design is necessary in order to distribute the potential uniformity across the stack, giving roughly equal voltage across each element of the insulator. The typical electric field is 100 kV/cm.

In the vacuum region, insulation is provided by the magnetic field of the pulsed current itself. The typical magnetically insulated transmission line[4] is designed with an impedance about three times that of the load, so that sufficient magnetic field is produced to confine any electrons emitted from the feed.

5.2.1 Magnetic Stores

The second type of pulsed power system for HPM is the *magnetic store* device (Figure 5.4). Interest in magnetic storage is due to the possibility of producing compact accelerators, since the energy density of magnetic storage is one to two orders of magnitude higher than of electrostatic storage systems. Typically, a Marx bank charges a vacuum inductor to peak current, then the key component — the *opening switch* — opens and a large transient voltage appears across the load due to the back-EMF (Electro-Motive Force) of the collapsing magnetic flux in the inductor. Because the peak voltage appears only at the load, and voltage sets the size of components, the storage device can be made compact without danger of internal arcing and shorting. Despite these potential advantages, magnetic storage is not commonly employed for three reasons. First, the L dI/dt voltage across the load has a triangular shape. Only loads not sensitive to applied voltage can be driven efficiently. Second, magnetic storage gives output pulses with impedance lower than that of most sources. Finally, the opening switch is usually destroyed. For a detailed description of opening switch technology, see Guenther et al.[5]

The widely developed opening switches are:

- *Exploding foil or wires*, frequently of copper, immersed in an insulating gas or liquid, open the circuit. They are designed to have high rates of increasing resistance, leading them to vaporize at peak current. Examples of their use are shown below.
- *Plasma erosion opening switch* (PEOS) can in principle be operated repetitively. In the plasma erosion switch, a plasma injected into a vacuum gap between the anode and cathode is removed by conduction of the charge-carriers or magnetic J × B forces associated with the current flow. Removal of the plasma causes the switch to open. Pulse compressions to 50 nsec have been demonstrated on microsecond timescale charging.

5.2.2 Flux Compressors

A very different means of producing electrical pulses is the *magnetic flux compressor*, also termed the *explosive generator* and *magneto-culumative generator*.[6] The very high energy density of explosives, ~1 MJ/m³ compared to capacitors at ~100 J/m³, deforms conductors that enclose trapped magnetic flux. Flux conservation from an initial flux of $I_0 L_o$ gives a later current through the deformed inductor of

$$I(t) = I_0 L_o / L(t) \qquad (5.7)$$

And we can define a final current gain as $G_I = I_f / I_0$. But ideal flux conservation cannot hold true; there are ohmic losses and some flux escapes.[7] A figure-of-merit β for flux compressors is

$$G_I = I_f / I_0 = [L_o / L_f]^\beta \qquad (5.8)$$

An ideal generator has β = 1, but typical values are β ~ 0.6 to 0.8 for the most widely used flux compressor, the helical generator shown in Figure 5.7. Note that the energy gain is

$$G_E = I_f / I_0 = [L_o / L_f]^{2\beta - 1} \qquad (5.9)$$

So no energy gain occurs for β = 0.5, though $G_I > 1$. To maximize both gains, L_o should by large, L_f small, and β ~ 1, or as close as possible.

The *helical generator* operates by a capacitor creating an initial current, therefore a flux in the *stator* winding. Detonation of the high explosive

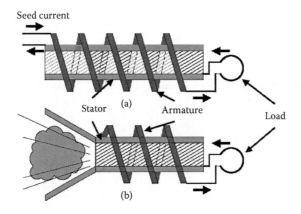

FIGURE 5.7

Helical flux compressor. Detonation of a high explosive converts chemical energy into kinetic energy; flux compression converts it into high current in the winding. Armature expands into a conical form and makes contact with the stator, shorting out turns in the winding as the point of contact progresses down the generator. (From Altgilbers, L., *Magnetocumulative Generators*, Springer-Verlag, New York, 2000. With permission.)

converts chemical energy into kinetic energy; flux compression converts it into electrical energy, a high current in the winding. The *armature* expands into a conical form and makes contact with the stator, shorting out turns in the winding as the point of contact progresses down the generator. The timescale for energy conversion is given roughly by generator length divided by the detonation velocity of the explosive, which is of the order 10 km/sec.

Figure 5.8 shows a helical generator driving a triode vircator with an inductive storage/opening switch pulse compression stage in-between. The current out of the generator (i_z) flows through the storage inductor and fuse, which explodes at peak current. The opening of the circuit creates an over-voltage on the spark gap, which closes, driving the anode to 600 kV positive voltage, and electrons are emitted into the vircator from the cathode (i_4). This all takes about 8 μsec, so microwaves are emitted from the horn before the shock wave from the explosion destroys the device. Microwave output is 200 MW in the S-band (~3 GHz).

Another microwave source that has been powered by flux compressor is the multiwave Cerenkov generator (MWCG; see Chapter 8). Here we see use of transformers, in part to get better rise times. Figure 5.9 shows the circuit: (1) a flux compressor (time-varying inductance and resistance) output is transformed up in voltage to charge an intermediate inductive store (2) with an opening switch (3), a fuse that explodes and produces ~1 MV into a pulse-sharpening output switch (4) into a diode (5), producing an electron beam into the Cerenkov generator. One clever feature is that some current from the transformer is diverted to drive the solenoid around the microwave source. The MWCG in Figure 5.10 produced microwaves similar in power and frequency to those of the above vircator. For a weapon concept that uses the flux compressor, see Section 3.2.2.

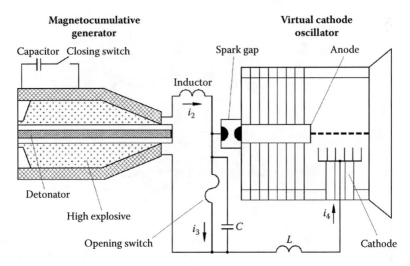

FIGURE 5.8
Helical generator driving a triode vircator through an inductive storage/opening switch pulse compression stage. Current out of the generator flows through the storage inductor and fuse, which explodes at peak current, opening the circuit, creating overvoltage on the spark gap that closes, driving the anode to 600 kV of positive voltage. The cathode emits electrons into the vircator; the microwave output is 200 MW in the S-band (~3 GHz). (From Altgilbers, L., *Magnetocumulative Generators*, Springer-Verlag, New York, 2000. With permission.)

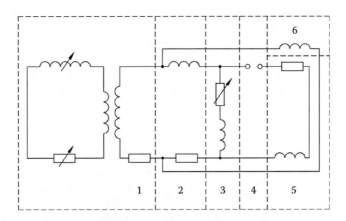

FIGURE 5.9
Circuit of driver for Figure 5.10: (1) flux compressor (time-varying inductance and resistance) output is transformed up in voltage to charge an intermediate inductive store (2) with an opening switch (3), a fuse that explodes and produces ~1 MV into a pulse-sharpening output switch (4) into a diode (5), producing an electron beam into the Cerenkov generator. Note some current from the transformer is diverted to drive the solenoid. (From Altgilbers, L., *Magneto-cumulative Generators*, Springer-Verlag, New York, 2000. With permission.)

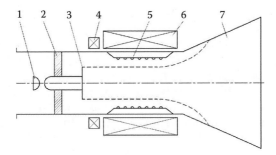

FIGURE 5.10

MWCG microwave source driven by circuit of Figure 5.9: (1) spark gap, (2) insulator, (3) cathode, (4 and 6) electromagnet, (5) slow-wave structure, and (7) antenna. (From Altgilbers, L., *Magnetocumulative Generators*, Springer-Verlag, New York, 2000. With permission.)

The *linear induction accelerator* (LIA) has been developed in the U.S. and U.S.S.R. over the last decade for particle accelerator and radiographic applications.[8] Its principal virtue is that the required accelerating voltage is distributed among a number N of pulse-forming subsystems so that full voltage appears only on the beam or a beam-emitting cathode (Figure 5.11). This technique, essentially a transformer distributed in vacuum, allows substantial downsizing of the pulsed power system. The LIA works as a series of 1:1 pulse transformers, each threaded by an electron beam or a solid cathode called a *voltage adder*. Each transformer section generates an increment of the voltage V on the cathode or beam, so that the peak voltage NV occurs at the

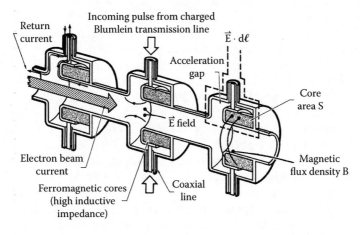

$$\oint \vec{E} \cdot d\vec{\ell} = -\int_s \frac{dB}{dt} \cdot d\vec{S}$$

FIGURE 5.11

Linear induction accelerator, an electron beam accelerated incrementally by induction cavities. Beam can be replaced by a solid cathode, which is then called a voltage adder.

tip of the cathode or in the exiting electron beam. The key component allowing summation of the voltage is a ferrite magnetic core in each of the accelerating cavities. These ferrite magnetic cores provide a high inductive impedance to a pulse arriving from a transmission line, so that the applied pulsed voltage appears as an electric field in the accelerating gap. The leakage current is small until the ferrite magnetic material saturates. At saturation, the cavity becomes a low inductance load and the pulse is shorted. Therefore, the core has a limit on the flux change that it can allow,

$$V\tau = A_c \Delta B \tag{5.10}$$

where τ is the pulse duration, A_c is the magnetic core cross-sectional area, and ΔB is the change of magnetic field in the core, which must be less than the saturation field of the magnetic material. LIAs are useful for short pulses of 100 nsec, but become very large and heavy for long pulses due to the magnetic core size.

A greater output voltage can be obtained by adding more cavities; the system is fundamentally modular. Moreover, because the pulsed power systems that drive the gaps operate at lower voltage, they are smaller and simpler to build. The drawbacks of the induction accelerator are that it is more expensive than pulse line technology for moderate voltages and that it is inefficient for low current. Because the pulse forming is conducted at lower voltage, the induction accelerator is appropriate for repetitive operation. The modular nature of the pulse-forming lines means that cooling of its N switches is easier than in a single accelerator stage, where energy is concentrated in a single output switch. Saturable inductors can also be used to perform several stages of pulse compression, giving an entirely magnetic system. The compression factor available per stage is 3 to 5. Entirely magnetic systems are attractive for long-life, repetitive operation, and thus will be used increasingly in HPM. Induction accelerators have thus far been used to drive free-electron lasers (FELs) and magnetrons (see Chapters 7 and 10). When the beam is replaced by a solid cathode, it is then called a *voltage adder*.

The requirement that HPM systems be operated repetitively, coupled with the desire for a compact, lightweight system, but still operating at both high peak and high average power, puts substantial stress upon pulsed power technologies. Operating repetitively, one has little time to dissipate excess energy due to source inefficiency; therefore, there is a natural coupling of repetitive operation with a preference for efficient sources. The cooling system required to handle dissipated power must be considered a fundamental element in a repetitive system; it may contribute a substantial weight, thus lowering the specific power of the system.

In summary, pulsed power systems used in HPM have been developed for other applications, but have served well. Pulse-forming lines are used for short (<200-nsec) pulse duration. Longer pulses require pulse-forming networks, which have yet to be extensively used in HPM. High-voltage transformers are used for repetitive operation because of their compactness and relative lack of complexity. LIAs (voltage adders or vacuum transformers) are more complex, while offering better average power handling. Magnetic energy storage is an emerging pulsed power technology that may have advantages in compactness, but operates at an impedance substantially lower than that of most loads. The primary challenge in pulsed power for HPM is repetitive operation, which remains relatively undeveloped.

5.3 Electron Beams and Layers

HPM sources tap the kinetic energy of electrons in a beam or layer to produce the intense microwave fields. To estimate the electron currents carried in these beams and layers, note that microwave power P is generated at some power efficiency η_p from an electron current I accelerated by a voltage V: P = η_p VI. The microwave power in pulsed sources is of the order of 1 to 10 GW, efficiencies are 10 to 50%, and voltages typically lie between several hundred kilovolts and several megavolts. Therefore, electron currents of 1 kA to several tens of kiloamps are typically required. In most cases, this current is carried by a beam of a specified cross-sectional profile at a high current density.

5.3.1 Cathode Materials

Production of an intense electron beam or layer begins at the electron source. The most common electron sources used in HPM devices are shown in Table 5.1 and are based on the following emitting phenomena[8] arranged in order of decreasing current density of emission:

- Explosive emission
- Field emission
- Thermionic emission
- Photoemission

In addition to the capacity for emitting at high current density, the attractiveness of each option depends on the requirements for quality of emission, as measured by spreads in the electron velocities and energies, ruggedness, and lifetime. No single source type is ideal for every application; the choice of technology is dependent on the HPM device in which it is to be used and the requirements of that device.

The most commonly used electron sources in present pulsed HPM devices are based on the process of *explosive emission*.[9] When low-impedance pulsed power applies high voltages to a metal cathode, average applied electric fields in excess of roughly 100 kV/cm at the emitting surface are further

TABLE 5.1

Cathode Materials for HPM

Emission Mechanism	Current Density	Current	Lifetime	Repetition Rate	Ancillary Equipment	Comments
Explosive						
Metal	~kA/cm^2	>10 kA	>10^4 shots	>1 kHz	Medium vacuum	Rapid gap closure
Velvet	~kA/cm^2	>10 kA	~10^3 shots	<10 Hz	Medium vacuum	Gas evolution limits repetition rate
Carbon fiber	~kA/cm^2	>10 kA	>10^4 shots	No limit	Medium vacuum	Rapid gap closure
CSi on C fiber	~30 A/cm^2	<1 kA	>10^4 shots	No limit	Medium vacuum	Slow gap closure
Thermionic	Arbitrary	~0.1 kA	>10^3 h	No limit	High vacuum; heater power	Heating to ~1000°C
Photoemission		~0.1 kA	~100 h	No limit	High power laser; very high vacuum	

enhanced by the presence of naturally occurring micropoints.[10] The enhanced fields lead to the flow of large field emission currents through these micropoints, which rapidly heat and explode to form plasma flares within a few nanoseconds. Individual flares then expand and merge in 5 to 20 nsec to form a uniform electron emitter capable of providing current densities up to tens of kA/cm^2. The whole process, while destructive at the microscopic level, is not destructive of the emitting surface as a whole. The microscopic points are regenerated from one shot to the next, so that these sources can be reused many times. Explosive emission sources can be made from a wide variety of materials, including aluminum, graphite, and stainless steel. They produce the highest current densities of all options, are quite rugged, can tolerate poor vacuum conditions (the operationally limiting factor is arcing), and have reasonable lifetimes.

Principal research issues for explosive emission sources are:

- *Gap closure*: Ideally, the emission of electrons from the cathode is uniform and comes from a cool plasma with low closure speed. In reality, the plasma is often, as with a steel or graphite cathode, highly nonuniform with hot spots of rapidly moving plasma (called flares or jets), which rapidly close the diode gap and terminate microwave output. Such plasmas also make diode performance nonreproducible.

- *Uniformity*: Nonuniform or nonreproducible emission causes asymmetries that affect the ability of the beam to generate microwaves. Hot spots mentioned above result when only a few highly localized emission points carry the current. Therefore, solutions lie in producing more emission sites while not exacerbating the closure and outgassing issues.

- *Outgassing*: Outgassing of material during and after the pulse limits both the maximum attainable pulse repetition rate and the maximum burst duration for a given vacuum pumping speed. At any point during a burst, the diode current can be shorted out by an arc through the background neutral gas pressure that builds from pulse to pulse as the outgassed material is released.

To address these problems, use of fibers to produce uniform emission at lower electric fields began with carbon fiber cathodes.[11] Introduction of the CsI-coated cathode made major advances because CsI will light up more evenly due to its photoemissive properties: the copious UV light generated by electronic transitions in the cesium and iodine atoms/ions helps to encourage uniform emission. Many more points photoemit; thus, the current density is lower, ohmic heating is reduced, and the plasma is cooler. The surface then produces only heavy cesium and iodine ions with slow gap closure. Experiments have demonstrated reduction of closure velocity from 1 to 2 cm/μsec with bare carbon fiber to 0.5 cm/μsec when coated with CsI.

Nonexplosive field emission sources use macroscopic points or grooves on a surface to get enhanced electric fields. Arrays of tungsten needles, grooved graphite, and polymer–metallic matrices are examples of emitters of this type[12]; common velvet can also be a field emitter. Microcircuit fabrication techniques have been used to produce experimental arrays of metal-tipped emission points on a SiO_2 substrate.[13] Current densities ranging from several hundred A/cm^2 to several kA/cm^2 are achievable with field emission sources. Their advantages include the ability to operate at field levels below those required for explosive emission and, in this event, to avoid the creation of an expanding plasma at the emitting surface. A disadvantage of these sources relative to explosive emission sources is the lower peak current density achievable; source lifetime, particularly in the cloth emitters, is also a concern.

Thermionic emission is based on the heating of an emitting surface to allow electrons to overcome the work function and escape the surface.[14] No expanding surface plasma is required, and long-pulse to continuous electron beams can be produced. The most common thermionic emitter types are (1) pure materials such as tungsten and LaB_6, (2) oxide cathodes such as BaO, (3) dispenser cathodes, (4) cermet cathodes, and (5) thorium-based cathodes. Tungsten and thorium-based cathodes operate at temperatures too high to make them attractive. LaB_6 and the oxides produce current densities up to several tens of A/cm^2. LaB_6, however, must be polycrystalline at significant emitter sizes, which diminishes emission uniformity due to the differing current densities obtainable from the various crystal planes; further, at the high temperatures required for high-density emission (of the order of 2000 K), the chemical reactivity and mechanical stability of this material pose serious problems. Oxide emitters suffer from ohmic heating in the oxide layer at high current densities, as well as the problem of poisoning, a drastic reduction in emissivity with exposure to air or water vapor, from which they recover slowly and which requires high vacuum for their operation. Dispenser cathodes are used in the wide majority of conventional HPM tubes today. They consist of a porous plug of pressed and sintered tungsten impregnated with a mix of oxide materials. In the type B dispenser cathode, for example, the impregnant is a mixture of $BaO/CaO/Al_2O_3$ in the ratio 5:3:2, while the type M cathode is a dispenser cathode with the additional feature of having its surface coated with about 0.5 μm of a noble metal such as Os, Os-Ru, or Ir. A variant of the M-type dispenser cathode has produced current densities of almost 140 A/cm^2 at an operating temperature of about 1400 K in a vacuum of 10^{-7} torr. Cermet cathodes coated with a layer of Sc_2O_3, known as a scandate cathode, have emitted stable current densities of 100 A/cm^2 at 1225 K for thousands of hours in tests. Uniformity of emission, reproducibility, and a susceptibility to degradation by back ion bombardment are problems to be addressed for the scandates. Overall, the thermionic emitters have the advantage of being able to produce current densities up to the order of 100 A/cm^2 for long or continuous pulses. This current density is lower than those for the explosive and field emitters. Further disadvantages are the need for a cathode heater, which is usually large and heavy, and the tendency for poisoning at pressures above 10^{-7} or so.

In *photoemission* sources, the work function energy for electron emission from materials such as GaAs and Cs_3Sb is supplied by laser illumination.[15] These sources have the useful feature that electron emission can be turned on and off coincident with the laser pulse. Modulating the laser, therefore, has the effect of producing a beam that is effectively bunched. Further, because the electron emission is cold, with very little excess energy over that required for emission, random thermal motions of the electrons imparted in thermionic sources can be eliminated. Current densities as high as 100 to 200 A/cm^2 have been produced. The disadvantages of the photoemitters are the possibility of plasma formation at the surface if the laser intensity is sufficiently high, the complication of adding a laser to the system, and the requirement for vacuums as low as 10^{-10} torr.

The importance of emission quality varies with the type of microwave source. HPM sources with large Doppler upshifts in frequency, such as FELs and cyclotron autoresonant masers (CARMs), require very small spreads in the electron velocities and energies. Gyrotrons and backward wave oscillators (BWOs), for example, tend to be more tolerant of these spreads, and magnetrons and vircators are quite tolerant. In those HPM sources for which emission quality is important, quality is typically specified in terms of fractional velocity and energy spreads, or in terms of the *emittance* and *brightness*.[16] Here, we will define emittance as follows. Imagine that we cut a thin cross-sectional slice in the x-y plane from a group of electrons drifting in the z direction and measure the position and velocity of each electron in the slice. If we then plot the components of each electron in a two-dimensional phase space with axes for the x component of position and $\Theta_x = v_x/v_z$, the emittance ε_x is defined to be the phase space area filled by these points, divided by π; the units of ε_x are $\pi \bullet m \bullet rad$.* A similar definition applies for ε_y. After the electrons leave the source and are accelerated along z, the angle θ_x will decrease as v_z increases. It is therefore common to define a *normalized emittance* by, for example, $\varepsilon_{nx} = \beta\gamma\varepsilon_x$, where $\beta = v_z/c$ and $\gamma = (1 - \beta^2)^{-1/2}$. A third form of the emittance is the *root mean square (rms) emittance*, defined by $\varepsilon_{rms,x} = (<x^2><\theta^2_x> - <x\theta_x>^2)^{1/2}$; ε and ε_{rms} are proportional to one another, with a proportionality constant of the order unity, varying with the exact distribution of electron trajectories. It is important to note that emittance is a measure of both the size and collimation of a beam of electrons. A beam focused to a point might in fact have a relatively large emittance, because of the large value of θ required to bring the beam to a focus. On the other hand, a tightly collimated pencil beam would have a small emittance.

Brightness is defined at each point of a source or beam as the amount of current per unit area per unit solid angle. Rather than deal with this differential quantity, we will deal with the average edge brightness of a source or beam:

$$B = \frac{2I}{\pi^2 \varepsilon_x \varepsilon_y} \qquad (5.11)$$

where I is the total emitted current; the factor π^2 takes account of the factor that appears in both ε_x and ε_y. As with emittance, we also define a normalized brightness, $B_n = B/(\beta\gamma)^2$. A plot of brightness for some different electron sources is shown in Figure 5.12.

5.3.2 Electron Beam Diodes

HPM sources employ electron beams generated from two-electrode diodes. The design of these accelerating structures to produce electron beams of high quality usually involves computer simulation to properly shape the electrodes and magnetic fields (if a field is used). Fundamentally, the basic features of diodes are illustrated by two idealized models, derived in Section 4.6.1:

- The *Child–Langmuir diode*, in practical units

$$J_{SCL}\left(\frac{kA}{cm^2}\right) = 2.33\frac{\left[V_0\left(MV\right)\right]^{3/2}}{\left[d\left(cm\right)\right]^2} \qquad (5.12a)$$

---

* We display the factor π explicitly in the units for emittance because some authors define emittance without division by this factor.

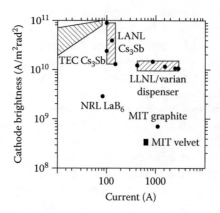

FIGURE 5.12
A comparison of cathode brightness vs. current for several different cathode emitters: field emission (velvet and graphite), thermionic emission (LaB$_6$ and a dispenser cathode), and photoemission (Cs$_3$Sb).

so with d ~ cm and the anode–cathode gap and ~1-MV potential applied from the pulsed power, current densities of kA/cm² and currents tens of kiloamps result. The current and impedance for a cylindrical cathode of radius r$_c$ are

$$I_{SCL}(kA) = J_{SCL}A = 7.35\left[V_0\left(MV\right)\right]^{3/2}\left[\frac{r_c(cm)}{d(cm)}\right]^2 \qquad (5.12b)$$

$$Z_{SCL}(\Omega) = \frac{136}{\sqrt{V_0\left(MV\right)}}\left[\frac{d}{r_c}\right]^2 \qquad (5.12c)$$

Note that impedance is weakly dependent on the voltage, but very dependent on geometry.

- The critical beam current above which *beam pinching* occurs is estimated by equating the Larmor radius for an electron at the edge of the beam with the anode–cathode gap:

$$I_{pinch}(kA) = 8.5\frac{r_c}{d}\left[1+\left(\frac{V_0\left(MV\right)}{0.511}\right)\right] \qquad (5.13)$$

Generally speaking, pinching complicates beam trajectories and is avoided in HPM sources (see Problem 2).

In all high power diodes there is the problem of *gap closure* due to the expansion of the cathode plasma. Gap closure has the effect of making the diode gap a time-dependent quantity: $d = d_0 - v_p t$, where closure velocities v_p are typically 1 to a few cm/μsec. Third, in order to extract the electron beam into a microwave source region, the anode must often be a screen or foil thin enough (i.e., much less than one electron range) that the electrons can pass through without undue angular scattering. The foil becomes a stress point in the design, since it will have a limited lifetime. Also, beam heating of the anode can create an anode plasma, which contributes to gap closure. Finally, the conducting anode will locally short the radial electric field of the beam, which disturbs the beam electron trajectories and introduces further spread in the transverse velocities of the electrons.

5.4 Microwave Pulse Compression

One way of generating high-peak microwave powers is to generate the microwaves in a long pulse at a lower power and then compress the pulse into a shorter duration at a higher peak power. The process can be seen as an extension of direct current (DC) pulse compression, i.e., pulsed power. Microwave pulse compression has been used for many years at low power levels for radar. At higher powers the problem of radio frequency (RF) breakdown limits the number of techniques that can be used. The contemporary work has been driven primarily by the need to obtain higher accelerating gradients in linear colliders, as described in Section 3.4. The practitioners of the art are now developing compressors in X-band and above for particle accelerators such as Compact Linear Collider (CLIC) and for studies of RF breakdown. Compressors have been studied as the basis of HPM weapons, compressing an accelerator-level klystron pulse (see Chapter 9) to ~50 nsec to reach >100 MW in a compact package. However, the Q requirements described below imply very narrowband components, so that such systems are very sensitive to frequency shifts.

The primary quantities of interest in pulse compression are, of course, power multiplication, pulse duration ratio, and overall efficiency. The standard definitions are:

$$M = \frac{P_1}{P_0}, \quad C = \frac{t_0}{t_1}, \quad \eta_c = \frac{M}{C} \tag{5.14}$$

where the initial pulse is characterized by P_0, t_0, and the final pulse is characterized by P_1, t_1. Compression efficiency η_c is generally high, and both M and C are of order 10 from most devices.

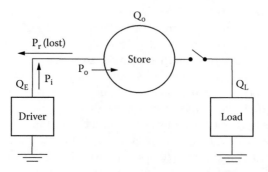

FIGURE 5.13
Microwave driver charges the store; switching into the load generates a high power, short-microwave pulse. Q_E describes re-radiation of power back through the coupling port; Q_L is the extraction of stored energy.

The primary methods used for high power pulse compression are called switched energy storage (SES), SLAC (Stanford Linear Accelerator Center) Energy Development (SLED), and binary power multiplication (BPM). The most widely studied system is SES, which stores energy in a microwave store, then switches it to a load by a large change in the coupling to the load. Before describing specific devices, we first outline the general process of microwave pulse storage and compression.

Energy is stored in a high Q cavity and then released in a low Q external circuit. If a power P_o is injected into a storage cavity with quantity factor Q_o, the stored energy will approach $W_c \simeq P_o Q_o / \omega$. The cavity is then switched into a load for which the load quality factor is Q_L, much smaller than Q_o (Figure 5.13). The power extracted from the cavity will be

$$P_c = \frac{W_c \omega}{Q_L} = \frac{P_o Q_o}{Q_L} \tag{5.15}$$

The power multiplication is

$$M = \frac{P_c}{P_o} = \frac{Q_o}{Q_L} \tag{5.16}$$

If there were no losses, energy would be conserved and power out would increase linearly with the Q-switching ratio, but losses always reduce performance below the above. Storage resonators made of copper have internal Q factors of 10^3 to 10^5. For superconducting resonators, Q values of about 10^9 can be achieved. However, the power multiplication and pulse compression that can be obtained in practice depend upon several other technological factors. Copper cavities are frequently limited by the rise time of the switch element because losses occur during switching. For superconducting cavities, a major limitation is the requirement of very high stability in the resonant frequency of the store and the frequency of the oscillator charging the store. Energy will be reflected from the store if the incoming pulse differs from the resonant frequency of the store. Efficient accumulation of energy in superconducting devices requires extraordinary stability of these two frequencies, of the order $Q_o^{-1} \sim 10^{-9}$.

Some of the injected energy will be reflected from the input port to the store. The external quality factor Q_E describes the re-radiation of power back through the coupling port and is determined by geometry of the port and its field structure. The coupling parameter β is

$$\beta = \frac{Q_o}{Q_E} \tag{5.17}$$

In a steady state the ratio of reflected–to–incident power is

$$\frac{P_r}{P_i} = \frac{(1-\beta)^2}{(1+\beta)^2} \tag{5.18}$$

Therefore if $Q_o = Q_E$, $\beta = 1$ (critical coupling), there is no reflected power. Power is reflected in either the undercoupled ($\beta < 1$) or overcoupled ($\beta > 1$) case. Then the stored energy is reduced and the power released is

$$P_c = 4P_o \left(\frac{\beta}{1+\beta} \right)^2 \tag{5.19}$$

and power multiplication is

$$M = \frac{P_c}{P_o} = \frac{4Q_E}{Q_L} \left[\frac{Q_o}{Q_o + Q_E} \right]^2 \tag{5.20}$$

For good pulse compression, the ordering of Q values is $Q_o \gg Q_E > Q_L$. Since energy is lost to heating the walls, optimization depends on the time taken to fill the store. As an example, consider an S-band conventional klystron charging a copper store cooled to liquid nitrogen temperature. The rectangular waveguide cavity has $Q_o = 4.5 \times 10^4$, so $Q_o/\omega \sim 2.4$ μsec. With a coupling factor $\beta = 3.5$, the 10-MW klystron injects 20 J in 2 μsec, of which 12 J is stored for an efficiency of 60%.

The switched energy storage device most widely studied is shown in Figure 5.14. Energy is slowly stored in the cavity so that RF fields are high. The output branch is an H-plane tee located $\lambda/2$ from the end wall. Therefore, there is a voltage null on the output side branch leading to the load, and little coupling

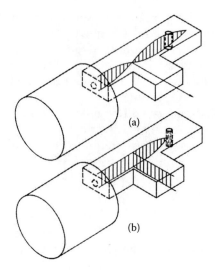

FIGURE 5.14
Switched energy storage system: microwave store and H-plane tee showing electric field envelope in (a) storage state and (b) dump state.

occurs between the cavity and the load. The triggering event is the creation of electrically conducting plasma at $\lambda/4$ from the shorted end of the waveguide, creating a tee null at the switch point and a field maximum at the output aperture. Energy from the cavity then dumps into the load on a short timescale.

The *dump time* from a store depends on the volume, V, of the cavity, the total area, A, of the output ports, and the group velocity, v_g, in the output waveguide. The dump time is crucially dependent upon the geometry of the cavity mode in the store and the precise location of the port in relation to the field geometry. An estimate of the dump time is[17]

$$t_D \equiv t_1 \approx \frac{3V}{Av_g} \left| \frac{B_P}{B_o} \right|^2 \tag{5.21}$$

where B_p is the RF magnetic field at the port and B_o is the maximum magnetic field in the cavity. For example, a spherical cavity of 20 cm radius with four output ports, each with a 15-cm^2 area, gives a dump time of ~100 nsec, about that needed for RF accelerators.

The cavities store an energy density of ~1 kJ/m^3 at electric fields on the order of 100 kV/cm. The switch is the crucial element in this scheme and must support such fields while open during storage time, but switch quickly with low loss when closed at the desired release time. Several switching mechanisms have been studied: spontaneous breakdown of a gas in the waveguide, high-voltage plasma discharge, vacuum arc, and an injected electron beam. The most widely used switch is the plasma discharge tube (triggered or self-breaking, with the former giving better performance).

The original experiments by Birx and coworkers[18] used a superconducting cavity with $Q_o \sim 10^5$ and $Q_L \sim 10^4$, so $M \sim 10$. The Tomsk group used a 3.6-μsec, 1.6-MW S-band klystron. At extraction, pulse durations from 15 to 50 nsec were obtained at powers to 70 MW, giving $M \sim 40$, $C \sim 200$.[19] Extraction efficiency from the resonator is reported as 84%; overall conversion efficiency is about 30%. Alvarez and coworkers[20] report 200-MW, 5-nsec pulses from a 1-μsec, 1-MW drive, so that $M \sim C \sim 200$. However, overall efficiency is reported to be 25%. Detailed work by Bolton and Alvarez has reduced the prepulse in the SES pulse compression scheme. Typically, a prepulse occurs at −40 dB below the main pulse. In their balancing scheme, the decoupling between the resonator and the load was reduced to −70 dB. The Tomsk group[21] used an S-band vircator at 400 MW injected into an SES. Because of the broadband nature of the vircator and the narrow bandwidth of the SES, only 20 MW of the 550-nsec pulse was accepted. The compressed pulse was 11 nsec at 350 MW, giving $M = 17.5$, $C = 45$, and $\eta_c = 38\%$.

We now consider storage systems in which the switching is performed by phase reversal. In the SLED method of pulse compression, two high Q cavities store energy from a klystron. Release of the energy from the store is triggered by a reversal in the phase of the klystron pulse. Switching takes place at low power levels at the amplifier input. Farkas et al.[22] gives an analysis of the energy gain of a SLED system. The maximum obtainable is $M = 9$. However, the shape of the output pulse from SLED is a sharply decaying exponential, which makes the effective power gain substantially less than the theoretical value. SLAC is now operating with 60-MW, 3.5-μsec klystron pulses compressed in SLED to 160 MW, 0.82 μs, giving $M = 2.67$, $C = 4.27$, and an overall compression efficiency of 62%. This attractive efficiency coupled with the relatively simple construction is why the method is currently used worldwide in many electron accelerators.

Because of the limited power gain of SLED, another form that gives a flat output pulse was developed, called the Resonant Line SLED (RELS; also called SLED II). The resonant line shown in Figure 5.15 is simply a transmission line coupled to a source through a network and terminated in a short. The distance between the network and the short must be a multiple of half-guide wavelengths. The final output pulse duration is the round-trip delay between the network and the short. The line is charged and the phase of the input signal is reversed, causing a discharge of the resonant line. In effect, this is very much like the transient charging of a pulse-forming line as described in Section 5.2, except the charging energy is microwave, not DC. Figure 5.16 shows the output field and output power in terms of the round-trip delay time, T_f. The dashed line shows the reversal of phase that triggers the output pulse. Calculations predict that the power gain and compression factors are typically 4, with a compression efficiency of about 80%.

Dissipation loss is minimized by using storage elements consisting of overmoded TE_{01} circular guides. For this mode, loss decreases as the cube of the diameter at fixed frequency. It is possible to stage RELS, limited by line attenuation losses and coupling mismatches. Thus far, a single RELS has been dem-

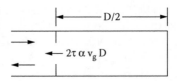

FIGURE 5.15
Schematic of resonant line storage system.

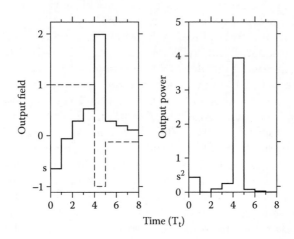

FIGURE 5.16
RELS single-stage output field and output power vs. time in units of T_f. Dashed line shows input phase modulation.

onstrated experimentally at quite low powers. A high power RELS system with a goal of 500 MW was under development at SLAC. Calculations show that for two-stage RELS, power gain could be 10 with an efficiency of about 60%.

The *binary power compressor* (BPC) is a device that multiplies microwave power in binary steps. Each step doubles the input power and halves the pulse length. Compressors of up to three stages have been operated at efficiencies of 45%. The BPC produces a rectangular output pulse. As with RELS, dissipation is reduced with the use of overmoded waveguides. This is an important feature for BPC because waveguides are quite long. As with SLED, the active control element operates only at low power. Passive elements (delay lines and 3-dB couplers) operate at high power. Couplers have been operated at SLAC at power levels in excess of 100 MW.

5.5 Antennas and Propagation

At the interface between the HPM source and free space, the two features of HPM that stress conventional antenna technology are high power and short pulse duration. Consequently, HPM antennas have been direct extrapolations of conventional antenna technology, usually in simple form, with allowance made for high electric field effects and for the shortness of the pulse.

5.5.1 Mode Converters

Between generation and radiation of microwaves, there is an opportunity to convert from one waveguide mode to another. *Mode conversion* goes back to the earliest days of microwaves, when it was called *mode transformation*.[23] This usually means:

- Converting from TE to other TE modes or from TE to TM (because TM, the right mode for interacting with the E_z component of modulations on the electron beam, has an unfortunate null on axis when radiated)

- Converting from circular to rectangular waveguide (so as to connect to antenna hardware)

- Converting directly from a waveguide mode to a radiating mode

The early method was to simply project an electrode from one waveguide into another. For example, the center electrode of a coaxial cable would be inserted into the center of a rectangular waveguide parallel to the short dimension. Such methods can be made efficient, but abrupt transitions will break down at high powers.

To finesse high power breakdown, slow transitions are the method. For example, the TM mode output of the relativistic klystron in Figure 9.20 is converted into the useful rectangular TE_{01} mode using the method shown in Figure 9.26. Radial fins are slowly introduced into the center conductor of the coax at the output end and grow until they reach the outer conductor, where they divide and adiabatically transform into a rectangular cross section. Such a converter has been built for a high-current S-band klystron and has been operated at 20 MW, although it suffered from breakdown due to too-short transition sections.

In gyrotrons, the rotating electrons interact with high-order cavity modes in the millimeter regime.[24] The high frequencies require *quasi-optical* mode converters. They use geometrical optics to ray-trace TM and TE modes, which are essentially decomposed into plane waves, each propagating at angles to the waveguide axis and each undergoing successive reflections from walls and mirrors. Such methods are used extensively for gyrotrons in fusion heating to provide transport of microwaves over long distances with little loss (see Chapters 3 and 10).

Direct conversion from waveguide to radiating waves is exemplified by the Vlasov convertor,[25] shown in Figure 5.23a. The TE mode k-vectors are reflected from a curved conductor and, if dimensions are chosen correctly, produce a Gaussian-shaped beam at an angle to the waveguide axis. For TM modes, Figure 5.23b (see also Figure 7.35) shows a MILO output TM mode converted directly into a Gaussian beam.

5.5.2   Antenna Basics

The *directive gain* of an antenna is its ability to concentrate radiated power in a particular direction. It can be expressed as the ratio of beam area Ω_A to the 4π steradians of all space:

$$G_D = \frac{4\pi}{\Omega_A} = \frac{4\pi}{\Delta\theta\Delta\Phi} \tag{5.22}$$

where the beam widths in the two transverse dimensions are $\Delta\theta$ and $\Delta\phi$. Directivity is frequently defined as the ratio of the power density S in the main beam at a distance R to the power density from an isotopic source radiating total power P:

$$G_D = \frac{4\pi R^2}{P} S \tag{5.23}$$

The *gain*, G, is G_D times the antenna efficiency:

$$G = G_D = \frac{4\pi A\varepsilon}{\lambda^2} \tag{5.24}$$

A is the antenna physical area, or radiating *aperture*, and ε the design-dependent antenna efficiency, due to power fed to the antenna but lost so not radiated, typically 0.5 to 0.8. Here we have introduced directivity in terms of antenna area. The product GP is the *effective radiated power* (ERP). Figure 5.17 graphs power densities for typical HPM cases (see also the nomograph; Figure 3.9). Note that the common way to describe gain is in decibels, G_{dB}. This value corresponds to an actual numerical value of

$$G = 10^{G_{dB}/10} \tag{5.25}$$

so a gain of 100 is called 20 dB (see Problem 3).

The radiation field of an antenna is not a simple spotlight, but has several regions. Moving away from the aperture, the regions are:

- *Reactive near-field region*, where the electromagnetic fields are not yet fully detached from the launching antenna and reactive terms dominate.

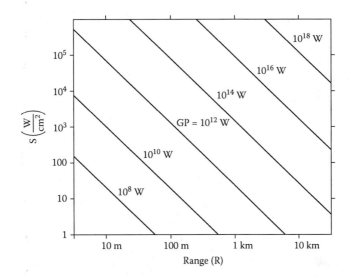

FIGURE 5.17
Power density at range for various effective radiated powers.

- *Radiating near-field region* (also *Fresnel region*) beginning at $0.62\sqrt{\dfrac{D^3}{\lambda}}$,

  where D is the antenna diameter. Radiation fields dominate, but angular field distribution depends on distance from the antenna. The beam is roughly cylindrical with varying diameter and intensity, reaching a peak at about $0.2\,D^2/\lambda$. From this point, the intensity drops monotonically, approaching $1/R^2$ falloff.

- *Radiating far-field region*, where the wave is truly launched, arbitrarily defined as the location at which the phase wave front differs from that of the circular by 5% of a radian:

$$L_{ff} = \frac{2D^2}{\lambda} \tag{5.26}$$

For example, consider the field of a horn antenna, shown in Figure 5.18. Its general features are the same for most antenna types.

The degree of phase uniformity that is acceptable for a test is an important question because it sizes the anechoic room used for testing and limits the power density available from a specific source. Points beyond $0.2L_{ff}$ are used for HPM vulnerability testing. L_{ff} can be located within a specific room by using a lower-gain antenna (smaller antenna or lower frequency), as long as air breakdown is avoided.

One of the most important aspects of antennas is their sidelobes, which are produced by diffraction from edges and feeds. Figure 5.19 shows that the

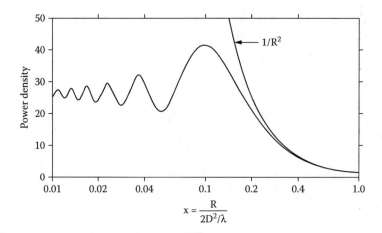

FIGURE 5.18
Horn antenna intensity normalized to unity at $2D^2/\lambda$. Its general features are the same for most antenna types.

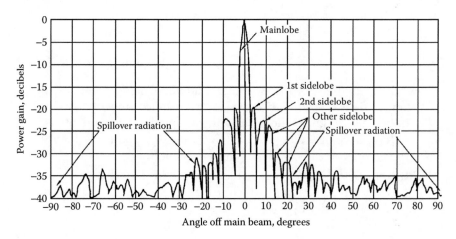

FIGURE 5.19
Example of radiation pattern at high angles to beam.

main lobes are accompanied by lobes at higher angles from diffraction at the antenna edge. There is also a backlobe at 180° to the main beam. In the far field, these effects are small indeed, but near the antenna the fluences can be very large. These elements of the radiation pattern cause concern over potential damage to nearby electronics (fratricide and suicide; see Section 3.2.1).

Once launched into the air, the key issue in HPM antenna design is breakdown. This occurs when the local electric field is sufficient to accelerate an electron to energy high enough to collisionally ionize gas atoms or to cause flashover at the air–waveguide interface. A cascade breakdown ensues, forming a conducting region from which the electromagnetic wave can be reflected or absorbed. The crucial factor is that the electron ionization rate

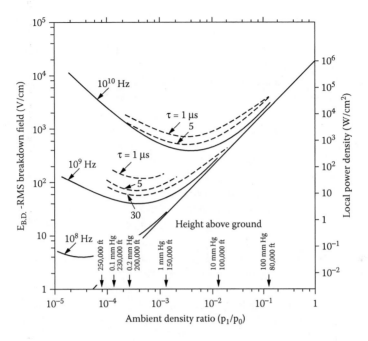

FIGURE 5.20
Microwave breakdown of ambient air at various altitudes, pulse durations, and frequencies.

exceeds the rate electrons are attached to ions. At high pressures, the critical fluence for air breakdown for pulses longer than a microsecond is

$$S\left[\frac{MW}{cm^2}\right] = 1.5\rho^2(atm) \tag{5.27}$$

where ρ is the pressure in units of an atmosphere. This corresponds to a critical field of 24 kV/cm at atmospheric pressure. At high altitudes, only low power densities can propagate. The minimum point is at between 20 and 50 km, depending on frequency (see Figure 3.25). Breakdown thresholds are high at high pressure (because there are many collisions) and at low pressure (because there is no cascade). Lowering the pressure will allow breakdown at lower fluences, as shown in Figure 5.20 [useful formula: E(V/cm) = 19.4 [S(W/cm²)]$^{1/2}$]. The pressure dependence of breakdown is of particular importance for HPM applications that involve radiation from aircraft at high altitudes or from sources beaming either upward from the ground or downward from orbit.

A crucial region is at ~50 km altitude, where a microwave pulse experiences gradual energy loss due to *tail erosion*. The head of the pulse continually ionizes the atmosphere, creating an absorptive plasma through which the tail must propagate. At 50 km, the threshold is about 250 W/cm².[26] For pulses of 100 nsec, a plasma *ionospheric mirror* can form at 50 km that reflects microwaves at low frequencies (~1 GHz) and dissipates by electron attachment and recombination in many milliseconds.

Breakdown is dependent upon pulse duration because it takes time to develop, leading to speculation that pulses shorter than a nanosecond may propagate through the atmosphere at very high fluences (*sneak-through*), i.e., insufficient time is available for breakdown to proceed.

At very high fluences (~10^3 MW/cm^2 at atmospheric pressures), a relativistic regime occurs when electrons achieve velocities approaching light and pondermotive forces come into play, giving a collisionless form of absorption. This regime has not been explored experimentally.

When breakdown occurs, hot spots occur in the near-field region of the antenna, causing reflection of waves, gain falls catastrophically, and the breakdown region spreads. This process is an active research area.

Once launched, the microwave pulse typically propagates in air modified by weather. Absorption in the air by water and oxygen molecules is strongly frequency dependent (Figure 5.21), giving passbands at 35, 94, 140, and 220 GHz. Lower-frequency pulses (<10 GHz) propagate with much greater efficiency over extended distances than the higher bands. Rain and fog increase microwave absorption. At low-microwave frequencies, fog absorptivity is 0.01 to 0.1 dB/km; rain absorptivity is in the same regime for 1 mm/h rain (see Problems 4 and 5).

5.5.3 Narrowband Antennas

Antennas for HPM narrowband have mostly been straightforward extrapolations of conventional antenna types modified to prevent air breakdown. Antenna arrays are just coming into development, principally for use in phase-locked multiple oscillator or amplifier systems. High power antenna subelements, such as phase shifters and splitters, have received little attention to date.

In practice, only a few types of antennas have been used in narrowband HPM. The gains of the principal types are shown in Table 5.2.[27] The most common type is the horn — pyramidal, conical, and transverse electromagnetic (TEM). Its utility follows from its similarity to a waveguide; horns are usually constructed as a flare from a rectangular or circular waveguide extending directly from the source. Conversion to a useful mode is a key issue in all applications. Generally, the fundamental modes (TE$_{10}$ in rectangular guide, TE$_{11}$ in circular) are preferred in part because they radiate onto useful patterns. As long as air breakdown is avoided by having an aperture large enough to reduce the electric fields when the wave enters air, the radiation pattern of horn antennas can be exactly calculated from first principles. Figure 5.22 shows the radiation pattern measured by Sze et al. from a TE$_{10}$ rectangular horn at 2.8 GHz.[28] The pattern is linearly polarized with gradual falloff, making it useful for electronic vulnerability testing (as opposed to TM$_{on}$ patterns, which are polarized with a radial electric field and have a highly nonuniform donut pattern).

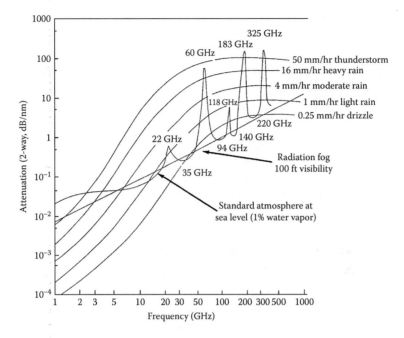

FIGURE 5.21

Atmospheric and rain attenuation for two-way propagation (as in radar). Note vertical scale is for 2-way transmission, as for radar, and nm = nautical mile.

TABLE 5.2

Principal Antenna Types

Antenna	Gain	Comments
Pyramidal and TEM horn	$2\pi ab/\lambda^2$	For standard gain horn; sides a, b; lengths $l_e = b^2/2\lambda$, $l_h = a^2/3\lambda$
Conical horn	$5 D^2/\lambda^2$	$l = D^2/3\lambda$; D = diameter
Parabolic dish	$5.18 D^2/\lambda^2$	
Vlasov	$6.36 [D^2/\lambda^2]/\cos \theta$	Angle of slant-cut θ, $30 < \theta < 60$
Biconic	$120 [\cot \alpha/4]$	α is the bicone opening angle
Helical	$[148 N S/\lambda] D^2/\lambda^2$	N = number of turns; S = spacing between turns
Array of horns	$9.4 AB/\lambda^2$	A, B are sides of the array

Parabolic dish antennas are ubiquitous in conventional microwave applications, but have found little use in narrowband HPM because the centralized feed of the parabolic antenna contains very high fields. However, in ultrawideband, a dish antenna called the Impulse Radiating Antenna (IRA; see Section 6.3) is the leading type. A variant is the Flat Parabolic Surface (FLAPS) antenna discussed in Section 3.2.3.

The Vlasov antenna[29] is well suited to launch narrowband HPM radiation. It has a simple construction with a relatively large feed aperture that helps to suppress or avoid RF breakdown. If its primary aperture is too small to support the microwave fields that it is attempting to launch, it can be easily

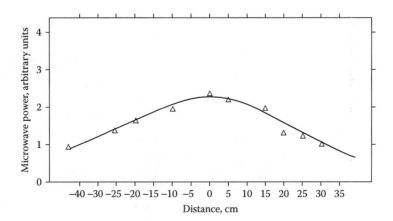

FIGURE 5.22
Horn pattern transverse to axis of propagation.

incorporated into an evacuated bell jar or pressurized gas bag to further extend the air–vacuum interface. Its main attribute is that it takes a TM_{01} feed and produces a directed beam with a nearly Gaussian profile. This makes it a very good match to the cylindrically symmetric microwave sources (MILO, Vircator, Cerenkov, etc.) that generate power in the TM_{0n} modes. The Vlasov antenna takes power from these sources and puts the peak power density at the center of the microwave beam (as opposed to casting a beam with a null at its center, as is usually the case when TM_{0n} modes are directly radiated) without the complication of including a mode converter. In fact, the antenna is the descendant of the mode converter (Figure 5.23a). Figure 5.23b shows a schematic of a Vlasov antenna being fed by a MILO HPM source.

The Vlasov antenna has two main shortcomings. First, the antenna is constructed from a circular waveguide by making a slant cut through the guide's cross section at an angle between 30 and 60°. This does not produce a large aperture (when compared to those characteristic of parabolic reflectors or even pyramidal and conical horns). Therefore, the Vlasov antenna is rarely used in applications requiring high gain (>>20 dB). Second, the propagation angle, θ (the angle between the waveguide axis and the main lobe), is a function of the microwave frequency, f:

$$\theta = 90^\circ - \cos^{-1}\left[\sqrt{1-\left(\frac{f_c}{f}\right)^2}\right] \qquad (5.28)$$

where f_c is the antenna's cutoff frequency. This means that if the frequency chirps within a microwave pulse, then the beam direction will sweep rapidly.

Antenna arrays have been slow to come into use in HPM systems (Figure 5.24). Their utility follows from three factors:

(a)

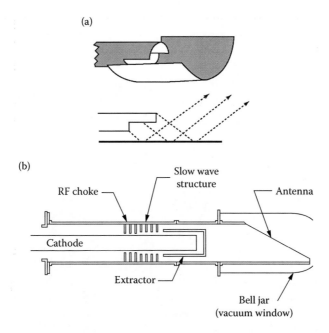

(b)

Slow wave
structure

RF choke

Antenna

Cathode

Extractor

Bell jar
(vacuum window)

FIGURE 5.23
Vlasov antenna. (a) Mode converter, the origin of the antenna. (b) Antenna being fed by a
MILO source.

1. Higher powers will require larger areas to avoid breakdown.
2. Super-high powers (~100 GW) will ultimately be generated from
   phase-locked arrays of sources, which implies multiple-output
   waveguides requiring multiple antennas.
3. Arrays are compatible with rapid electronic tracking and illuminating of targets, with great directivity.

Arrays are termed *broadside* if the direction of maximum radiation is perpendicular to the plane of the array and *end fire* when the radiation maximum is
parallel to the array. End-fire arrays have lower gain, so broadside arrays are
the focus of interest. If array size L is large compared to λ, contributions from
each element will change rapidly as angle is changed from the normal, producing a sharp maximum of beam width ~$2\lambda/L$ with gain L^2/λ^2. The physical
separation of the elements introduces *grating lobes* nearby. If individual aperture separation d is less than λ, grating lobes can be greatly reduced. This
condition is difficult to meet at high power due to air breakdown constraints
on antenna size. Consequently, grating lobes are a problem in practical systems
and must be suppressed by some means, such as power splitting, to increase
the number of elements, or by sidelobe suppression with local absorbers.

The beam can be scanned by controlling the phase of each element separately with phase shifters. For configurations using power amplifiers driven
by a master oscillator (MOPA; see Section 4.10), phase control is done at low

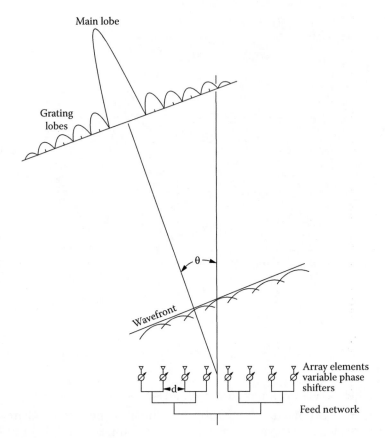

FIGURE 5.24
Antenna array launches wavefront, forming narrow main lobe.

power prior to the amplifier stage, and therefore conventional technology is used. For oscillator arrays, conventional phase shifters (Figure 5.24), even high power ferrite devices, do not approach the required power regime. At present, only electromechanical devices have been developed: waveguide stretchers with piston-driven telescoping U sections and broad-wall waveguide deformation techniques are currently under development. For large arrays with $>10^2$ elements, phase shifter cost will be a limiting factor. In order to achieve dense packing to reduce grating lobes, the cross-sectional dimensions of the shifter should be near those of the waveguide, so that the feeds can be densely packed.

5.5.4   Wideband Antennas

High-gain, low-dispersion wideband antennas present a substantial challenge. When pulse duration becomes of the order of a nanosecond, antenna design becomes much more taxing. The *fill time* of an antenna is several light-transit times across the diameter. If the pulse-length wave entering the antenna is short compared to the fill time, then the full aperture is not used, gain is reduced, and the pulse is dispersed. These effects are important for all applications, but especially for impulse radar (see Chapters 3 and 6). Two approaches are used for impulse radar antennas. The first is to transmit the impulse signal through a low-dispersion antenna with wide bandwidth. The second is to use antenna dispersion to generate an impulse signal from a longer input signal.

The principal criteria for the low-dispersion type of antenna are that it have a wide bandwidth, typically a factor of 2, a high gain, and low sidelobes. These are difficult requirements because:

1. Wavelengths are so large (10 to 100 cm) that high gain is difficult within the antenna sizes imposed by applications.
2. Wide bandwidth makes it difficult to suppress sidelobes.
3. Gain cannot be constant for all wavelengths, so the propagating pulse shape will be distorted.

The transverse electromagnetic (TEM) horn has been used for most impulse radar work to date (Figure 5.25). It is simply two metal strips flared in roughly an exponential shape. Pulses with durations of 1 nsec can be radiated with such horns, and bandwidths are ~1 GHz. Gain tends to be nonuniform across the bandwidth, so that pulse distortion occurs in the far field. A key factor is that the TEM horn transmits a differentiated signal,

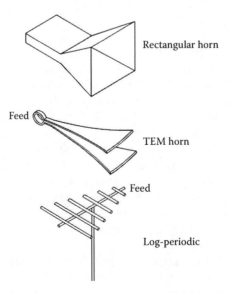

Rectangular horn

Feed

TEM horn

Feed

Log-periodic

FIGURE 5.25
Typical conventional antenna types that are used as wideband HPM antennas.

allowing applied monopulse or unidirectional pulses to be radiated as a single cycle.

The basic approach to producing low dispersion, i.e., little variation of phase velocity with frequency, is to flare the input waveguide into the antenna aperture so that the impedance varies slowly from that of the waveguide to that of free space (377Ω). In the sidelobes, the pulse is stretched to substantially longer than the original duration. Therefore, for impulse radar, the pointing accuracy of both transmitting and receiving antennas is very important to obtaining return signals, i.e., pulses off bore sight can be so long as to be useless.

Rather than minimizing antenna dispersion, dispersion can be used to generate the desired impulse signal by the log-periodic antenna (Figure 5.25). The signal fed to the antenna has strong frequency chirping, i.e., frequency rises during the pulse. The dispersive properties of the antenna, if well matched to the input pulse, can produce the desired compressed output pulse. Designs for high power have not been published. Issues with the log-periodic antenna are large beam width and consequent low gain, and pulse shaping of the feed pulse to give the correct output pulse. Achieving constant gain across the desired bandwidth is also difficult.

However, the above methods are becoming obsolete for many applications due to the introduction of a new type of antenna, the Impulse Radiating Antenna, discussed in Section 6.3.

5.6 Diagnostics

High power devices differ from their conventional relatives in three significant ways. First, of course, "high power" means that high electric fields exist and all diagnostics must avoid breakdown. Second, the pulse durations are short, less than a microsecond and typically less than 100 nsec. This implies fast response on the part of the diagnostics since rise times and critical features can occur on a timescale of a few nanoseconds. Lastly, HPM sources have yet to be operated at high repetition rates or continuously. Therefore, diagnostics are typically required to analyze a single shot or a few shots. Some advantages follow from these features: high power means that it is possible to directly measure the energy of a pulse in a single shot. A low number of shots means that sophisticated data handling is not required, and sampling techniques cannot be used to reduce the data rate.

Historically, HPM diagnostics have borrowed many techniques from conventional microwave diagnostics. They are used to characterize three spatial regions: the source, the radiated field, and the interior of test objects. Here we discuss the diagnosis of the microwave pulse. Diagnostics of plasma processes inside the source are as yet in a primitive state. The sophisticated nonintrusive diagnostics developed in plasma research are beginning to know and understand the plasmas inside HPM devices. To better understand pulse shortening, the HPM community must measure the actual electromagnetic structure and plasma properties of HPM sources. More detailed diagnostics are needed, particularly of microwave field distributions *in situ*, plasma location and motion, and velocity distribution of both electrons and ions. The nonintrusive methods developed by other parts of the plasma physics community, such as that of fusion, should be applied to HPM. Time–frequency analysis of microwave signals and noise spectra can be widely exploited in the analysis of HPM device operation. Future methods will be measurements of time-dependent bandwidth, beam loading (at high currents), and noise spectra.

The principal quantities to be measured can be described by:

$$E(\bar{x},t)=A(t)\sin(\omega t+\phi)S(\bar{x})\qquad(5.29)$$

where E is the RF electric field, A is the peak electric field amplitude, ω is the angular frequency, ϕ is the phase angle, and S is the spatial variation of the electric field. If several frequencies are present, the pulse is defined by a sum of electric fields at various frequencies. The power of the pulse is proportional to the square of the electric field and the energy to its time integral of the power. We will review each of the quantities in Equation 5.29 in turn.

5.6.1 Power

The most common diagnostic method is the use of diodes to rectify microwaves and detect the envelope of a microwave pulse, generating a DC output nonlinear in the instantaneous electric field, giving the power envelope. Typically, probes mounted inside a waveguide feed the detector and an RC low-pass filter passes the lower-frequency power envelope while blocking the microwaves. In addition to rectification, the nonlinear diode can be exploited to produce a power measurement directly. For sufficiently low powers, the diode current is directly proportional to the square of the voltage across the diode, and therefore to the incident microwave power. Square law operation is possible only at power levels less than ~100 μW. Therefore, substantial attenuation is required to stay within the square law domain. If multiple frequencies are present, this technique will still give the total RF power.

Both point contact and Schottky barrier diodes are available in the microwave regime. The principal advantages are that they are relatively inexpensive, have rise times of ~1 nsec, have a reasonably flat frequency response over a limited bandwidth, and are simple to operate. Their disadvantage is that they can handle only a small amount of power, typically 0.1 W for the point contact and about 1 W for the Schottky diodes. Therefore, substantial attenuation of the pulse is necessary before rectification. This is done with directional couplers, which reduce the signal 30 to 70 dB. Minor difficulties with diodes are that they are fragile and must be handled carefully; there is a substantial unit-to-unit variation in their characteristics, and they are very sensitive to x-rays and moderately sensitive to temperature. Because of these latter factors, diodes must be carefully calibrated on a frequent basis.

The key factor in using diodes is accurate attenuation. Commercial components such as directional couplers are sufficient for this task as long as the mode propagating in each component is the one for which the calibration was made. Here sensitivity to calibration error is great. Another method for attenuating the signal is to broadcast the radiation onto a receiving horn. The signal then goes through a directional coupler in the waveguide, and hence to the diode. This method gives the local fluence, and by taking readings in a number of locations, both the pattern and the total radiated power can be deduced.

5.6.2 Frequency

The simplest frequency diagnostic is the direct measurement of E(t) on fast oscilloscopes, but expense is prohibitive. Other methods require interpretation and care in their use (see Problem 6).

5.6.2.1 Bandpass Filters

The basic technique for a coarse survey of frequency is the bandpass filter. They are similar to lumped element electronic filters in performance, but instead use waveguide cutoff to pass a discrete band. In the high power regime, directional couplers are used to reduce power to a level where commercial filters can be used. There are many kinds of microwave filters. They are available in the domain up to 40 GHz and are inexpensive and straightforward in their use. Their disadvantages are that they have a coarse frequency resolution and that they often have periodic attenuation characteristics, allowing frequencies considerably higher than the bandpass to propagate. Figure 5.26a shows the use of a bandpass method in a vircator experiment by Sze et al.[30] The rapid chirp of the vircator is captured by a set of four bandpass filters, and the power envelope in each band is recorded by a crystal detector.

5.6.2.2 Dispersive Lines

This widely employed simple diagnostic for frequency uses the frequency-dependent transit time of a short pulse down a waveguide. Higher-frequency components arrive first. With a sufficient length of waveguide, a measurement of the time delay between the individual peaks in the dispersed signal and the start of the radiation pulse yields the frequency spectrum on a single shot. Dispersive lines come in two types: long evacuated waveguides and solid-state crystals, typically yttrium iron garnet.[31]

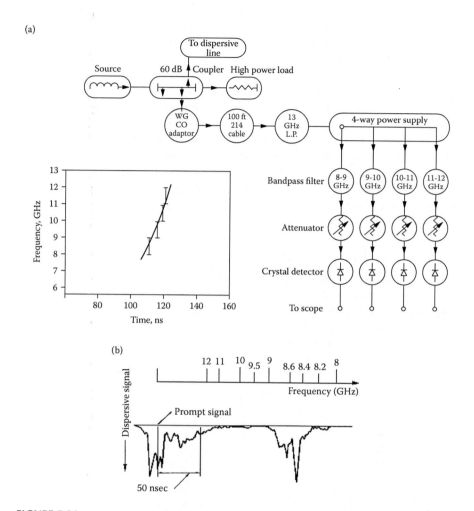

FIGURE 5.26
Frequency measurement methods. (a) Four bandpass filters resolve the vircator source upward chirp (inset). (b) Dispersive line signal for frequency measurement. Timing reference signal ("promp") is superimposed on the dispersed signal.

Waveguide devices operate with waves propagating in the fundamental mode at the frequency-dependent group velocity. The frequency that arrives at a distance L down the waveguide at a transit time t is

$$f = \frac{f_{co}}{\left[1 - \dfrac{L}{ct}\right]^{1/2}} \tag{5.30}$$

For short pulses, the assumption is made that all frequency components are emitted simultaneously. Frequency resolution is directly proportional to pulse length, so only short pulses can be resolved. Resolution is increased if dispersion is enhanced by working near the cutoff frequency. However, waveguide attenuation also increases rapidly near cutoff. An experimental example is shown in Figure 5.26b. An advantage of this method is that only two detectors are required, one at each end of the line.

Waveguides are simple, but long lengths are required for adequate dispersion. They also have high frequency–dependent attenuation. Solid-state lines have higher dispersion and are frequency independent, but require an external magnetic field with a tailored distribution. They also have limited power-handling capability. Both devices have been successfully used, the solid-state devices at frequencies less than 5 GHz and waveguides above 8 GHz. Both methods can achieve dispersions of ~300 nsec/GHz.

5.6.3 Heterodyne Detection

In this method, the microwave signal is mixed with that from a steady oscillator and the beat signal is displayed directly on a fast oscilloscope. Fourier analysis of the heterodyne signal allows the spectrum to be measured. The frequency can be determined by shifting the oscillator frequency over several shots. An example of a Fourier spectrum of a vircator is shown in Figure 5.27.[32] The heterodyne technique has emerged as the most com-

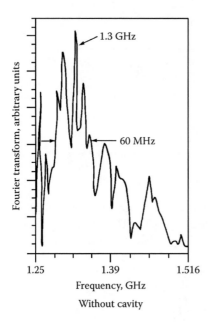
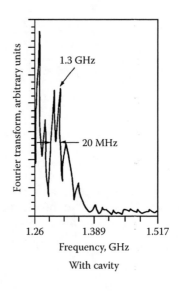

FIGURE 5.27
Fourier spectrum of a vircator. The introduction of a surrounding cavity reduces bandwidth. (See Chapter 10.)

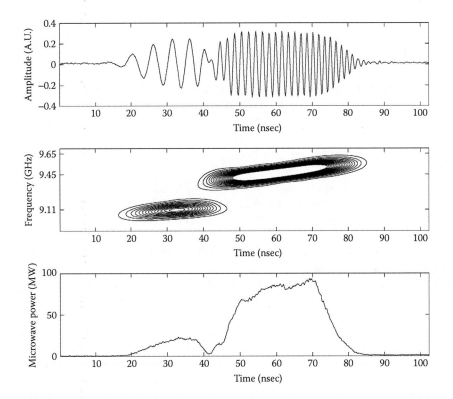

FIGURE 5.28
Mode-shifting data from the University of New Mexico. BWO: upper, heterodyned signal; middle, time–frequency analysis of heterodyne data shows mode (frequency) shift; lower, power radiated. (Reprinted from Barker, R.J. and Schamiloglu, E., Eds., *High-Power Microwave Sources and Technology*, Press Series on RF and Microwave Technology, IEEE Press, New York, 2001, p. 148, Figure 5.17. By permission of IEEE.)

monly used method of measuring frequency because it is an easy-to-use, easy-to-interpret, time-resolved measurement.

5.6.3.1 Time–Frequency Analysis

This powerful diagnostic provides Fourier-analyzed frequency signatures. Figure 5.28 shows *time–frequency analysis* (TFA) of a variety of phenomena inside a BWO.[33] The darker line denotes higher power. Some mechanisms that can be shown in TFA are:

1. *Frequency chirping and detuning* of the resonance condition caused by e-beam voltage fluctuations. Even modest voltage fluctuations can cause large frequency chirping, which causes microwave power fluctuations by detuning from the cavity resonance interaction. For electron beam–driven microwave devices, voltage regulation (flatness) is one of the most crucial parameters.

2. *Mode hopping* between two competing resonance modes is a pulse-shortening mechanism. The conditions under which mode hopping occurs can be critically examined and remedies evaluated with TF analysis.

3. *Mode competition* between two modes that can exist simultaneously and radiate. Mode competition destroys the efficiency of both modes, lowering the total power.

5.6.4 Phase

Phase measurements are required for phase-locking experiments and for measurements of the coherence of a radiation field, i.e., mode competition in the radiated waves. Two methods have been used thus far. Smith et al.[34] used an Anaren phase discriminator, which utilizes a microwave circuit to produce two signals, one proportional to the cosine of the phase difference between the input signals and the other to the sine of the relative phase. The inverse tangent was calculated to give the angle, with care taken to keep track of which quadrant the angle falls into. Rise times of 1 nsec were obtained. This method was used to directly measure the phase difference between resonators in a magnetron and has since been used in multiple magnetron phase-locking experiments (see Chapter 7). The sensitivity is about $10°$. Price et al.[32] used this method to measure relative phase in the radiation pattern of a vircator, showing existence of a single radiated mode.

Friedman et al.[35] used a more sensitive method to measure phase-locking and spectral purity of a relativistic klystron. The method compares the klystron output to the signal injected into the amplifier by a magnetron. Phase variation is measured to be less than $3°$.

5.6.5 Energy

Calorimetry is the fundamental energy diagnostic technique. The power of the pulse can also be determined from the energy measurement by combining it with the pulse shape. Intercepting HPM pulses in a calorimeter can produce electric fields of ~100 kV/cm, so breakdown is an issue. The total energy deposited is typically 1 to 100 J, requiring sensitive measurements of temperature change in the calorimeter. In principle, the calorimeter is quite simple: if a thermally isolated mass m with specific heat c_p absorbs a quantity of energy Q, the temperature rise is

$$T = \frac{Q}{mc_p} \qquad (5.31)$$

In practice, the method is valid only if thermal losses are negligible. Therefore, losses due to convection, conduction, and radiation must be minimized

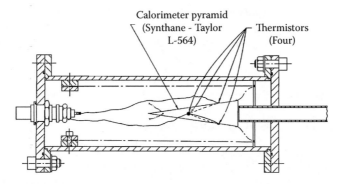

FIGURE 5.29
Construction of a graphite pyramidal microwave calorimeter. Pulses are received from the circular waveguide to the right.

by mounting the absorber material inside an evacuated chamber using thin support members and small transducer leads. In addition, the absorber should represent a matched load to the waveguide it intercepts. Any design is a compromise between large absorber size to better match the waveguide impedance and small size in order to reduce the absorber mass and increase the temperature rise. The absorber material should have a high thermal conductivity to reach thermal equilibrium before losses drain energy away. The typical material is graphite. A common design, shown in Figure 5.29, uses pyramidal graphite absorbers tapered over a distance of 5 to 10 wavelengths. Absorption is typically better than 90%. Orientations both facing and reversed from the Poynting vector have been used, the latter to prevent breakdown.

Temperature rise is usually monitored with thermistors attached to the absorber and connected electrically in series, constituting one arm of an electrical bridge circuit. Temperature changes on the order of a millidegree can be measured. Early et al.[36] have introduced a number of calorimeter designs that use another approach: microwave power is absorbed in a sheet of resistive *space cloth* having the same impedance as the waveguide, and the waveguide is terminated $\lambda/4$ behind the cloth. The cloth heats the intervening layer of air and the pressure is measured. Such methods have the advantage of eliminating electrical connections, and therefore electrical pickup. Infrared measurement of absorber heating has also been used, as have wire calorimeters, which avoid noise problems from x-rays or thermal infrared (IR) signals from the source.

Because source efficiencies are low in many cases, substantial amounts of electrical energy are converted to UV, IR, heated gas, and x-rays. There can also be energy from fields induced from the pulsed power. Use of a calorimeter must be validated by null tests to eliminate these spurious signals. This is frequently the most difficult aspect of experimentation with calorimeters.

Calorimeters are in principle useful over the entire microwave frequency domain. The advantages of the calorimeter are that it operates over a broad band, has relatively simple construction, and can be calibrated without high power sources. Its limitations are that it must match the waveguide imped-ance, which makes low-frequency operation more difficult, and use small absorber mass while meeting all other criteria.

5.7 High Power Microwave Facilities

The primary activities for which HPM facilities are being developed are source research and electronics effects testing. The former is done exclusively indoors and the latter both indoors and outdoors. The requirements of these two applications place many constraints on facilities, the foremost of which are interference-free acquisition of data and safety. Researchers should carefully consider their facility requirements. Operation of an HPM facility is frequently more expensive and time consuming than first anticipated. Experimenters should contemplate using available industrial or government facilities rather than constructing their own, at least in the early stages of their work.

5.7.1 Indoor Facilities

A typical indoor microwave facility does both source development and effects testing, with unified instrumentation and safety systems. The most notable feature of indoor facilities is the anechoic room, i.e., a room in which microwave reflections are minimized by use of absorbers on the wall, as shown in Figure 5.30. The usual method for separating functions is to have the microwave source and its associated pulsed power exterior to the anechoic room with waveguide to transmit the microwaves to an antenna located just inside the room. This location for the antenna maximizes the distance between source and test object, making more efficient use of the anechoic room. The room itself is sized by the requirement that test objects not be too close, i.e., greater than $0.2\ D^2/\lambda$ (see Section 5.5.2). Dimensions of anechoic rooms in use vary from a few meters for high-frequency work to >10 m for large target volumes. Anechoic rooms frequently incorporate a turntable for rotation of test objects and a side room containing ancillary equipment for the test object (pneumatics, control systems, and any diag-nostics for which cable length must be short). Anechoic rooms must be provided with fire safety and audio alarms, and abort switches to prevent firing of the source while humans are present in the room. The absorbing pyramidal anechoic material can provide an attenuation of ~40 dB, produc-ing a *quiet zone* several meters in diameter. The material is hung on the wall of an electromagnetic screen room that provides ~60 dB of attenuation; thus, in a typical experiment, microwave signals cannot be detected outside the anechoic room. This more than fulfills the microwave safety requirements described below. The size and other aspects of the anechoic room usually determine much of the geometry of a facility.

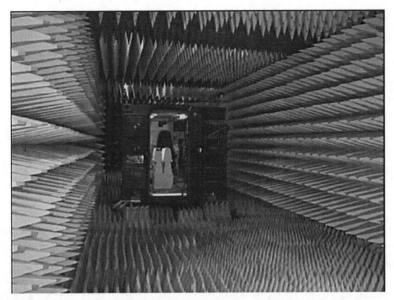

FIGURE 5.30
Anechoic room with absorbing pyramids covering the walls.

Electronic instrumentation for control and diagnostics is carried on con-ductor or (increasingly) fiber-optic cables (most conveniently located over-head in trays) to screen rooms or to control panels. One of the most crucial issues in facilities is proper handling of the ground plane, to avoid ground loops, which cause noise on instrumentation. A minimum requirement is to locate oscilloscopes sufficiently far from microwave sources (which are almost always x-ray sources) to prevent clouding of the screen.

Because of the overlapping requirements of x-ray, microwave, and other safety issues, access to the facility must always be controlled. The basic requirement is to prevent exposure to radiations by creation of *controlled* and *uncontrolled* areas (defined in the U.S. by CFR 49). Uncontrolled areas are those in which people are allowed access because exposure cannot occur. In controlled areas, exposure can occur, and therefore access must be prevented during times of irradiation. The most effective safety procedure is to provide a gate interlock key for each person entering the facility. All areas that can be irradiated should have a personnel access gate at which the interlock key panel is located. A key missing from the interlock panel will disable the pulsed power and prevent radiation. This system requires vigilance in mon-itoring on the part of the facility operator to ensure that personnel carry the keys. All personnel must also carry an irradiation badge at all times in a controlled area. Before firing the machine, the operator must inspect the area to ensure that it is clear of all personnel. The gate must be closed and electrical interlocks set before firing can commence. It is advisable to use rotating beacons and loud distinctive warning sirens strategically located to alert everyone in the area of an impending test. Continuous-wave (CW) anechoic rooms will require a fire safety analysis. Note that anechoic rooms can prevent exterior sirens from being heard inside the chamber. A firing abort switch should be inside the anechoic room.

5.7.2 Outdoor Facilities

Outdoor facilities are fraught with their own issues, especially if they are transportable. First, many experimental subsystems will be weather sensitive, and measures must be taken to ensure against the problems of rain, heat, dust, etc. All radiation leakage must be controlled, usually by local shields or simply by standoff distance. Protection from fratricide damage to nearby electronics is made more difficult because of the absence of anechoic rooms and high-quality shielded rooms for the diagnostics. Site security must be controlled more carefully. Finally, because there are many uses of the microwave spectrum, those contemplating outdoor experiments should consult the *Reference Data for Radio Engineers* and FCC regulations to find if their intended frequency is already in use. Aircraft are the principal concern. Ultimately, no apparatus can ensure the safety of a facility. Eternal vigilance is the price of true site security.

Orion is the state-of-the-art transportable, self-contained HPM test facility first fielded in 1995 and currently in operation (Figure 5.31).[37] The system is housed and transported in five ISO containers, has complete fiber-optic–linked computer controls and data acquisition, and carries its own prime power.[38] At the heart of the system is a suite of four tunable magnetrons (Chapter 7), each capable of delivering 400 to 800 MW. The frequency tuning is continuous, with no gaps in performance over the range from 1.07 to 3.3 GHz. Orion fires 1000 pulses in a burst at repetition rates up to 100 Hz. The thyratron-based modulator that drives the magnetrons has a pulsed power output that can be increased in discrete 50-nsec steps up to a maximum duration of 500 nsec. The microwaves are radiated from a high-efficiency, offset, shaped, parabolic antenna that illuminates a 7×15 m, 3-dB beam spot at a 100-m range.

Orion pulsed power is a two-stage, oil-insulated, thyratron-switched modulator that pulse charges an 11-section ladder network PFN through a step-up transformer and triggered gas output switch.[39] Its main performance features are pulse duration variable, 100 to 500 nsec; voltage variable, 200 to 500 kV; and repetitive up to 100 Hz. Impedance is 50 Ω; rise time, <30 nsec; and flatness, ±8%. Pulse duration is varied by manually removing inductors in 50-nsec increments (Figure 5.32).

The four continuously tunable relativistic magnetrons use stepper-motor tuners and explosive emission cathodes, operate at a vacuum of 10^{-6} to 10^{-7} Torr (cyropumps), use a 10-vane, rising-sun design and two-port extraction, and are surrounded by a conformal Pb x-ray shield. The magnetic field is provided by cyromagnets, up to 10 kG. The cyrofluids can operate for 3 days between refillings.

FIGURE 5.31
The Orion outdoor testing facility. (Reprinted by permission of Smith, I., L-3 Communications, San Leandro, CA.)

FIGURE 5.32
Orion pulsed power. (Reprinted by permission of Smith, I., L-3 Communications, San Leandro, CA.)

After the magnetron, a combiner/attenuator network provides continuous power variation over five orders of magnitude. The waveguide combiner/attenuator consists of a hybrid tee/phase-shifter and power combiner to sum output from two magnetron arms, followed by a hybrid tee/phase-shifter attenuator to vary the radiated power over five orders of magnitude. There is a separate waveguide circuit for the three waveguide bands WR770, -510, and -340. The system is controlled with programmable stepper-motors.

The Orion antenna system, which can be seen in Figure 5.31, includes two off-set, shaped, parabolic reflectors each with two pyramidal feed horns designed to maximize efficiency, reduce sidelobe level, and produce a 7 × 15 m elliptical beam spot at a 100-m range. Gain is 468 (26.7 dB). Modest pointing adjustments of about ±10° are possible with flexible waveguide sections and shims. The antenna and its supports are designed for ease of assembly, storage, and rapid deployment by two workers (see Problem 7).

Orion is a point of departure and reference that other efforts to bring HPM technology out of the laboratory and into the field may draw upon. Features of the Orion facility that should be considered by anyone contemplating an outdoor facility are:

- There are two diesel generators, one for container and ancillary power and one for the HPM system, with a total power of 1.1 MW maximum. Although the maximum microwave average power out of the antenna is only 5 kW, the differences due mostly to losses.

- Separate containers are required because of their very different functions — for control, prime power, and storage (waveguides, antennas, spare parts, tools).

- Many functions must be built in. For transportability, all contents in each container are mounted with shock and vibration isolation provisions. Smoke detectors, fire extinguishers, and audio intercoms are included in each container. Safety interlocks and abort buttons are provided throughout all ISO containers and are incorporated into the Orion safety interlock system.

- The Orion system is completely computer controlled. The computer provides timing, monitoring, and control of the HPM system. The data acquisition system (DAS) computer provides setup, data archiving, and analysis of the DAS digitizers. All links between containers are fiber optic to avoid grounding and shielding problems. The range safety officer is provided external audio communication and video surveillance.

5.7.3 Microwave Safety

Safety issues can usually be addressed in a straightforward way if taken into account in the initial facility design. The hazards of electromagnetic radiation are becoming the focus of intense public scrutiny. Microwave radiation hazards are divided into three groups: biological hazards to personnel; hazards to ordnance, where a potential exists for munitions or electro-explosive devices to be initiated; and hazards to fuel, where there is a potential for spark ignition of volatile combustibles. In the military, these hazards are referred to, respectively, as HERP (personnel), HERO (ordnance), and HERF (fuel).

Absorption of RF energy by the human body is very frequency dependent. In general, the body's absorption is maximized when the long axis of the body is parallel to the incident electric field and its length is 0.4 λ. Therefore, whole-body resonance for humans is maximized at about 70 MHz, where absorption is sevenfold higher than at 2.5 GHz.[40] Since the photon energy of even high-frequency microwaves is two orders of magnitude below the thermal energy of molecules at room temperature, microwaves are definitely not ionizing radiation. Therefore, they are below the threshold at which effects are cumulative, in contrast to x-rays. However, x-rays have become an accepted hazard; public sensitivity to microwave risk is largely due to a lack of familiarity.

The lowest power effect known in humans is *microwave hearing*. A single 10-μsec pulse of microwave energy absorbed in the head at 10 μJ/g results in an acoustic response, i.e., false sound. At 0.1 J/cm^2, nerve activity can be influenced at the cellular level. At higher levels, confusion and a wide variety of mental effects can occur. At the highest energy levels, local heating can elevate the brain temperature. Unfortunately, little is known about short-pulse, high power effects, and it is a subject of considerable experimental study at present.

For purposes of facility operation, some threshold standard must be adopted. The most common approach is to find an exposure level for pulsed HPM where no adverse effects are observed, and then modify it by a safety factor. A commonly used standard is that of the American National Standards Institute (ANSI), based on a specific absorption rate (SAR), the rate of energy absorption per tissue mass. The ANSI standard uses a constant SAR limit of 0.4 W/kg, which is 10% of the value thought to be the threshold for adverse effects. The 1991 ANSI levels as well as HERO limits are shown in Figure 5.33. In the microwave regime, 10^{-2} W/cm^2 should be an adequate level for average powers. In 1992, the IEEE C95.1-1991 safety standard was introduced; it is two-tiered, with a distinction between controlled and uncontrolled environments (the latter for locations where there is exposure of individuals who have no knowledge or control of their exposure, e.g., living quarters or workplaces that are not clearly identified, so not a microwave facility). The maximum permissible exposures for controlled environments are in Table 5.3. The level rises linearly from 0.3 to 3 GHz, then remains constant at 10 mW/cm^2. The time over which variable intensity exposures can be averaged is generally 6 min, but falls off above 15 GHz to 10 sec at 300 GHz. For example, at 10 GHz, 10 mW/cm^2 exposure over 6 min is within the guideline, and so if exposure is to 60 mW/cm^2, only 1 min is within it.

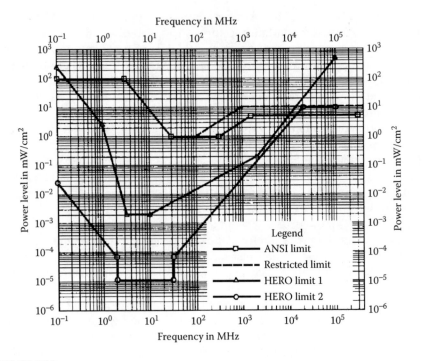

FIGURE 5.33
Safety standards for microwaves.

TABLE 5.3

Safety Guidelines for Microwaves in
Controlled Areas

Frequency	Power Density	Averaging Time
0.3–3 GHz	1–10 mW/cm²	6 min
3–15 GHz	10 mW/cm²	6 min
15–300 GHz	10 mW/cm²	6–0.165 min

For peak power limits, the situation is unsettled. Primates exposed to 5 W/cm² peak experienced no effects. Protecting personnel in a microwave facility at this level may be considered safe. Shielding by wearing conducting suits is very effective; shielding of –50 dB can be obtained. However, such suits are rarely used in HPM. In practice, most facilities use screen/anechoic rooms to attenuate below measurable levels.

The HERO specification in Figure 5.33 is to prevent dudding or premature activation of electro-explosive devices (EEDs), which are alleged to have caused accidents during the Vietnam war. The HERO-2 limit gives the maximum safe fields for bare EEDs; the HERO-1 limit is the safe power level for fully assembled ordnance in normal operations. These levels are experimentally based. Ordnance is more sensitive than humans because it reacts to peak power, whereas human reactions appear to be dominated by average power effects. Of course, ordnance liberates great energy, and therefore the safety requirement is greater. Ordnance is also more easily protected by shielding than humans. The consequences for igniting fuels are large, especially when handling motor vehicle and aviation fuels. Measurements near transmitting antennas have shown that a minimum voltage gradient of 50 kV/cm is required for fuel ignition.

Operating microwave sources in an outdoor site in the U.S. comes under the purview of the Federal Communications Commission and its regulations for radio frequency devices, specifically FCC regulation 18.301, which allocates operating frequencies for industrial, scientific, and medical equipment. These regulations specify radiating bands for which unlimited radiated energy is allowed with the proviso that harmful interference is prohibited, i.e., interfering with radio navigation, safety devices, or radio communication. The experimenter should consult the appropriate federal regulations. As a practical operating point, note that the Trestle EMP simulator in Albuquerque, NM, is located near the city airport and does not operate when field strengths near aircraft will exceed 1 kV/m, which corresponds to a power density of 0.26 W/cm^2. Radar altimeters and other aircraft systems operate at microwave frequencies, so the prudent approach is to find and avoid frequencies and power levels that have been determined to interfere with or damage on-board systems.

5.7.4 X-Ray Safety

X-rays are ionizing radiation produced as a by-product of electron beam microwave sources. The most frequently used unit for radiation is the *rad* (1 rad = 100 erg/g), which is a measure of the radiation energy absorbed in a mass. Exposure to one *roentgen* of radiation results in an absorbed dose of about 1 rad for x-rays. The absorbed dose for biological systems is also frequently expressed in the rem unit (for x-rays, rem = rad). The current standard for an uncontrolled area is 0.5 rem/year.[41] A radiation shield must be designed that, in conjunction with $1/r^2$ falloff, can reduce the fluence in uncontrolled areas to below this level. For safety calculations, one assumes that the full beam energy is deposited in an x-ray converter, i.e., the worst case. For example, a 1-MV, 200-kA, 100-nsec electrical pulse, operating at 10 shots/day, requires about 50 cm of concrete shield. High Z materials such as lead give thinner, heavier shields. Locating this shield close to the source would minimize the volume, in which case lead might be more advantageously used, allowing access to the source. The widely used alternative is to place concrete shields at a distance and use them to define the controlled area. Maze entrances to the radiation area prevent line-of-sight escape of radiation. The interlock safety gate is usually placed at the entrance of the maze to provide the access point to the controlled area. An often overlooked x-ray hazard is *sky shine*: reflection of x-rays from the ceiling or air above the source. Scattering calculations are required to establish the magnitude of the effect, but for the above case, ~2 cm of lead will stop sky shine and other scattered radiation.

In addition to the primary radiation hazards of microwaves and x-rays, most HPM research facilities have other safety issues that must be addressed: (1) high voltage in the pulsed power system; (2) use of insulating sulfur hexafluoride (SF_6) gas, to be used and vented in a controlled manner; and (3) hazardous materials, most especially oil for insulation, which raises the issue of oil fires and oil containment in berms.

Problems

1. The highest energy or power transfer efficiency between pulsed power elements such as pulse lines occurs when the lines are matched in impedance. Mismatching upward, to higher impedance, gives higher voltage but lower current. If that line in turn drives a beam-generating Child–Langmuir diode ($I \sim V^{3/2}$, $Z \sim V^{-1/2}$), will the microwave power be greater than in the matched case?

2. Some types of vircators begin radiating microwaves when the electron beam starts to pinch in the diode, launching electrons on converging trajectories beyond the anode. This increases the density and forms the virtual cathode. This is when the Child–Langmuir current equals the pinch current. For operation at 0.5 MeV, what diode aspect ratio r_c/d is required for starting microwaves? What is the diode impedance?

3. An X-band antenna has a beam width of 18°. If its efficiency is 50%, what is its gain?

4. To produce 1 kW/cm² at a range of 1 km, what effective radiated power (ERP) is required? If the source has 10 GW of peak power, what gain antenna is needed? If the antenna efficiency is 80% and the diameter is 10 m, what is the highest power you can radiate without air breakdown? If it is raining heavily at 16 mm/h, what attenuation will the beam experience if the frequency is 10 GHz?

5. What power density can be produced at a 100-m range from a conical horn operating at 1 GHz? Estimate what power can be radiated at sea level if the only limit is atmospheric breakdown.

6. If you have a source that chirps upward in frequency, increasing 20% during the pulse, which of the frequency diagnostic methods discussed here (bandpass filters, dispersive line, hetrodyne, time–frequency analysis) would you use?

7. The Orion test facility has an effective antenna diameter of 2.5 m. If the target is 100 m away and is 10 m wide, and you are using a

1-GHz beam with 500 MW of power, is the target in the far field of the antenna? What is the power density on the target? If you need to place 90% of the power on target, what could you do? Discuss the practicality of your approach.

Further Reading

Pulsed power topics are described best in the biannual proceedings of the IEEE Pulsed Power Conference. There is also a series of books on special topics edited by Mesyats, Pai, Vitkovitsky, DiCapua, Guenther, and Altgilbers.[1-3,5,6] A comprehensive tome on beams is Humphries,[16] and the best on antennas is Balanis.[27] The best survey of plasma diagnostics is by Hutchinson.[42]

References

1. Mesyats, G.A., *Pulsed Power*, Kluwer Academic/Plenum Publishers, New York, 2005.
2. Pai, S.T. and Zhang, Q., *Introduction to Pulsed Power*, World Scientific, Singapore, 1995.
3. Vitkovitsky, I., *High Power Switching*, Van Nostrand Reinhold Co., New York, 1987.
4. DiCapua, M.S., Magnetically insulated transmission lines, *IEEE Trans. Plasma Sci.*, PS-11, 205, 1983.
5. Guenther, A., Kristiansen, M., and Martin, T., *Opening Switches*, Plenum Press, New York, 1987.
6. Altgilbers, L.L. et al., *Magnetocumulative Generators*, Springer-Verlag, New York, 2000.
7. Neuber, A.A. and Dickens, J.C., Magnetic flux compression generators, *Proc. IEEE*, 92, 1205, 2004.
8. Maenchen, J. et al., Advances in pulsed power-driven radiography systems, *Proc. IEEE*, 92, 1021, 2004.
9. Bugaev, P.S. et al., Explosive electron emission, *Sov. Phys. Usp.*, 18, 51, 1975.
10. Barker, R.J. and Schamiloglu, E., Eds., *High Power Microwave Sources and Technologies*, IEEE/Wiley Press, New York, 2001, chap. 9.
11. Barker, R.J. et al., *Modern Microwave and Millimeter-Wave Power Electronics*, IEEE Press, New York, 2005, chap. 13.
12. Barletta, W.A. et al., Enhancing the brightness of high current electron guns, *Nucl. Instrum. Meth.*, A250, 80, 1986.
13. Barker, R.J. et al., *Modern Microwave and Millimeter-Wave Power Electronics*, IEEE Press, New York, 2005, chap. 8.
14. Thomas, R.E., Thermionic sources for hi-brightness electron guns, in *AIP Conference Proceedings 188*, 1989, p. 191.
15. Oettinger, P.E. et al., Photoelectron sources: selection and analysis, *Nucl. Instrum. Meth.*, A272, 264, 1988.

16. Humphries, S., *Principles of Charged Particle Acceleration*, John Wiley & Sons, New York, 1986.
17. Alvarez, R. et al., Application of microwave energy compression to particle accelerators, *Particle Accelerators*, 11, 125, 1981.
18. Birx, D. et al., Microwave power gain utilizing superconducting resonant energy storage, *Appl. Phys. Lett.*, 32, 68, 1978.
19. Devyatkov, N. et al., Formation of powerful pulses with accumulation of uhf energy in a resonator, *Radiotechnika Elektronika*, 25, 1227, 1980.
20. Alvarez, R.R., Byrne, D., and Johnson, R., Prepulse suppression in microwave pulse-compression cavities, *Rev. Sci. Instrum.*, 57, 2475, 1986.
21. Grigoryev, V., Experimental and theoretical investigations of generation of electromagnetic emission in the vircators, in *BEAMS'90*, Novosibirsk, Russia, 1991, p. 1211.
22. Farkas, Z. et al., Radio frequency pulse compression experiments at SLAC, *SPIE*, 1407, 502, 1991.
23. Montgomery, C.G., Dicke, R.H., and Purcell, E.M., *Principles of Microwave Circuits*, Peter Peregrinus Ltd., London, 1987, chap. 10.
24. Thumm, M.K. and Kasparek, W., Passive high-power microwave components, *IEEE Trans. Plasma Sci.*, 30, 755, 2002.
25. Vlasov, S.N., Zagryadskaya, L.I., and Petelin, M.I., Transformation of a whispering gallery mode, propagating in a circular waveguide, into a beam of waves, *Radio Eng. Elec. Phys.*, 20, 14, 1975.
26. Fenstermacher, D.L. and von Hippel, F., Atmospheric limit on nuclear-powered microwave weapons, *Sci. Global Security*, 2, 301, 1991.
27. Balanis, C.A., *Antenna Theory*, 2nd ed., John Wiley & Sons, New York, 1997.
28. Sze, H. et al., Operating characteristics of a relativistic magnetron with a washer cathode, *IEEE Trans. Plasma Sci.*, PS-15, 327, 1987.
29. Vlasov, S.N. et al., Transformation of an axisymmetric waveguide mode into a linearly polarized Gaussian beam by a smoothly bent elliptic waveguide, *Sov. Tech. Phys. Lett.*, 18, 430, 1992.
30. Sze, H. et al., A radially and axially extracted virtual-cathode oscillator, *IEEE Trans. Plasma Sci.*, PS-13, 492, 1985.
31. Palevsky, A. and Bekefi, G., Microwave emission from pulsed relativistic e-beam diodes. II. The relativistic magnetron, *Phys. Fluids*, 22, 985, 1979.
32. Price, D. et al., Operational features and microwave characteristics of the vircator II experiment, *IEEE Trans. Plasma Sci.*, 16, 177, 1988.
33. Barker, R.J. and Schamiloglu, E., Eds., *High-Power Microwave Sources and Technologies*, IEEE Press, New York, 2001, chap. 5.
34. Smith, R.R. et al., Direct experimental observation of π-mode oscillation in a relativistic magnetron, *IEEE Trans. Plasma Sci.*, 16, 234, 1988.
35. Friedman, M. et al., Externally modulated intense relativistic electron beams, *J. Appl. Phys.*, 64, 3353, 1988.
36. Early, L.M., Ballard, W.P., and Wharton, C.B., Comprehensive approach for diagnosing intense single-pulse microwave pulses, *Rev. Sci. Instrum.*, 57, 2293, 1986.
37. Sabath, F. et al., Overview of four European high-power narrow-band test facilities, *IEEE Trans. Electromag. Compat.*, 46, 329, 2004.
38. Price, D., Levine, J.S., and Benford, J., ORION: A Frequency-Agile HPM Field Test System, paper presented at the 7th National Conference on High Power Microwave Technology, Laurel, MD, 1997.

39. Hammon, J., Lam, S.K., and Pomeroy, S., Transportable 500 kV, High Average Power Modulator with Pulse Length Adjustable from 100 ns to 500 ns, Proceedings of the 22nd IEEE Power Modulator Symp., Boca Raton, FL, 1996.
40. Lerner, E.J., RF radiation: biological effects, *IEEE Spectrum*, 68, 141, 1980.
41. Recommendations of the National Council on Radiation Protection and Measurements, Radiation Protection Design Guidelines for 0.1–100 MeV Particle Accelerator Facilities, NCR Report 51, 1979.
42. Hutchinson, I., *Principles of Plasma Diagnostics*, 2nd ed., Cambridge Univeristy Press, Cambridge, U.K., 2002.

6

Ultrawideband Systems

6.1 Ultrawideband Defined

Ultrawideband (UWB) high power microwave (HPM) sources are quite different from the narrowband sources described in Chapters 7 to 10. The technology grew out of studies of the electromagnetic pulse (EMP) generated by high-altitude nuclear explosions in the 1960s. Technology developments of the last two decades led to the first UWB pulsers capable of transmitting high-peak field signals through a variety of efficient antennas. This chapter briefly reviews the field. Good references are the book by Giri[1] and review articles by Agee et al.[2] and Prather.[3]

What distinguishes UWB? Its spectral density, the power emitted into any narrow-frequency interval, is very low. Figure 6.1 shows typical spectral magnitudes of electric fields, or E-fields, over a wide range of frequency. (This is a *frequency-domain* plot, as distinct from the better-known *time-domain* plot, which shows amplitude vs. time.) It gives perspective on the relation between the regions of HPM: the lower-frequency continuous spectra from natural lightning and high-altitude electromagnetic pulse (HEMP) together with higher-frequency narrowband HPM, including wideband spectra, the so-called ultrawideband.

UWB sources and antennas are of interest for a variety of potential applications that range from jammers to detection of buried objects and space objects, ultrashort radar systems to industrial and law enforcement applications. As a directed energy weapon, UWB offers the prospect of making high electric fields available across a wide range of frequencies to take advantage of coupling mechanisms and internal electronic vulnerabilities at frequencies that are not known *a priori*. The disadvantage of this approach is that in spreading the energy over a wide band of frequencies, the amount of energy actually coupled into a system is reduced by the ratio of the coupling bandwidth divided by the UWB system bandwidth, or a sum of such ratios if multiple coupling bands are available over the full UWB system bandwidth. Many military assets and civilian systems (e.g., nuclear power plants, communications facilities) are suspected of being vulnerable to the effects of

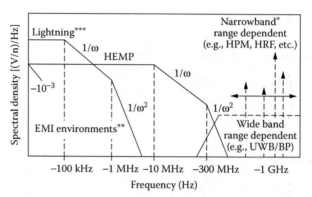

*narrowband extending from −1.5 to 3 GHz
**not necessary HPEM
***significant spectral components up to −20 MHz depending
   on range

FIGURE 6.1
Spectral domains of electromagnetic pulse types. (Reprinted from Giri, D.V. and Tesche, F.M., *IEEE Trans. Electromag. Compat.*, 46, 2, 2004, Figure 1. By permission of IEEE.)

UWB, which would most likely mean interruption and jamming rather than destruction, because the total system energy is spread over such a broad frequency range, while the coupling and vulnerability bandwidths are typically quite narrow.

The types of microwave pulses termed UWB are very different from narrowband microwaves in both the frequency spectrum and the technologies used to generate them. In fact, the UWB terminology has become dated and is being superceded by the more specific terms in Table 6.1, which shows the classification of pulses based on bandwidth.[4] Why are the distinctions between the subbands of UWB important? Because they differ quite a lot in their coupling into electronics, their effects on electronics, and use rather different technologies.

Bandwidth by definition is the half width of power in frequency space: the low- and high-frequency limits are half-power points (3 dB down) from

TABLE 6.1

HPM Band Regions

Bands		Percent Bandwidth, 100 Δf/f	Bandwidth Ratio, 1 + Δf/f	Example
Narrowband		<1%	1.01	Orion (Figures 5.31 and 7.24)
Ultrawideband	Moderate band (mesoband)	1–100%	1.01–3	MATRIX (Figure 6.11)
	Ultramoderate band (subhyperband)	100–163%	3–10	H-Series (Figure 6.16)
	Hyperband	163–200%	>10	Jolt (Figure 6.17)

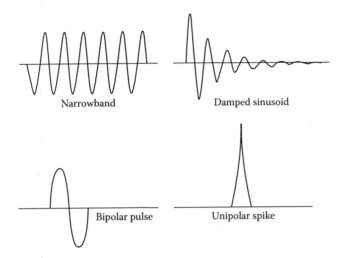

FIGURE 6.2
Electric fields of classes of pulses. As the number of cycles increases, the bandwidth decreases.
(a) Narrowband; (b) damped sinusoid; (c) bipolar pulse; (d) unipolar spike (which does not
radiate as a spike, but as a cycle).

the power peak. (A more general definition is the frequency range that
contains a large fraction of the pulse energy, for example, 90%.) If the frequencies are separated by f, the widely used definitions for bandwidth are
fractional bandwidth $\Delta f/f$, sometimes expressed as percentage bandwidth,
100 $\Delta f/f$, and bandwidth ratio, 1 + $\Delta f/f$ (see Problem 1).

Figure 6.2 shows several types of pulses from narrowband to a simple
voltage spike. The narrow band is a long series of cycles, giving a relatively
narrow bandwidth, usually defined as $\Delta f/f \sim 1\%$. The next most similar waveform is the damped sinusoid, a finite number of cycles. The most common
UWB pulse is bipolar, and the ultimate is a single unipolar spike. The frequency spectrum changes accordingly (Figure 6.3). As the number of cycles
decreases, the bandwidth increases. The spectrum changes from narrow,
when there are many cycles available to determine the frequency, to broad
for shorter pulses. (Mathematically, time and frequency are conjugate variables; the narrower in either time or frequency, the broader you are in the
conjugate variable.) The number of cycles, the bandwidth, BW, and the Q of
the resonant system that produces the pulse (see Chapter 4) are related by

$$N = 1/\text{percentage BW} = Q \qquad (6.1)$$

However, for single cycle pulses, these quantities are not useful (see
Problem 2).

Technologywise, going from narrowband to UWB means a transition from
electron beam devices to devices in which fast switches direct the pulsed
power output directly to an antenna. The wideband spectrum requires a
constant, or nearly constant, gain antenna over a broad frequency range.

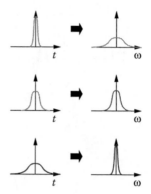

FIGURE 6.3
Time and frequency descriptions of short unipolar pulses. As pulses narrow, they have wider bandwidth.

This is quite difficult to do because antenna gain normally depends strongly upon frequency. The approach now used is to have pulsed power generate a fast rising step in voltage. This is connected to a differentiating antenna (one that produces the time derivative of the incoming pulse), which converts the voltage step into an electric field spike in the radiation field. The antenna of choice in the UWB community is the Impulse Radiating Antenna (IRA).

For narrowband microwave sources, the requirement for a resonance involving a beam of monoenergetic electrons is that the pulsed power must generate a relatively long flat-top voltage pulse. Since the source is narrowband, the antenna can be optimized for the operating frequency.

But for UWB, pulsed power must generate a short voltage spike. The antenna aperture area is typically limited to the area $A < 10\ \text{m}^2$, for example, by the requirement that it can be truck mounted. For such an antenna radiating a peak power P, the average E-field in the aperture of the antenna is

$$E_0 = \sqrt{\frac{PZ_0}{A}} \tag{6.2}$$

where Z_0 is the impedance of free space, 377Ω. But as we shall see, *for UWB the peak-radiated E-field at a distance R depends on the rate of rise of voltage.* The widely used *figure of merit for UWB is the far voltage* (sometimes called *far-field voltage*), the range-normalized peak-radiated E-field (somewhat like the effective radiated power [ERP; see Chapter 5] for narrowband) is

$$V_f = RE_p \tag{6.3}$$

Values are typically of the order 100 kV, but 5 MV has been achieved (see Section 6.4.3).

6.2 Ultrawideband Switching Technologies

The essence of UWB technology is extremely fast switching and nondispersive high-gain antennas. The pulsed power section of a UWB device is a series of electrical pulse compression stages, generating shorter pulses at increasing power as the signal proceeds through the system toward the antenna. The criteria for the switching element are ultrafast closing capability to create very fast rising waveforms, and rapid voltage recovery, so that high repetition rates are achievable.[2] The principal switching technologies used are:

- High-pressure gas or liquid spark gap switches, typically triggered by electrical pulses
- Photoconductive semiconductor switches, which are triggered by lasers

6.2.1 Spark Gap Switches

A fast rising pulse is critical because the rise time determines the high-frequency components of the transmitted spectrum in an antenna such as the IRA, which responds to the derivative of the signal applied to it. To sharpen the rise time on a pulse, a peaking switch may be used. Such switches have been developed since the early 1970s in electromagnetic pulse (EMP) simulators. The essence of the peaking gap is establishment of very high electric fields between very closely spaced electrodes. Contemporary spark channels have gap lengths of 1 mm. Because of the short gap lengths, charge voltages of less than 100 kV can produce spark gap electric fields in excess of MV/cm, resulting in very fast breakdown of the gap.

Spark gaps proceed through several brief intervals: capacitive phase before breakdown and arc formation, resistive phase as the gas channel heats, and an arc phase when the switch is fully closed. The latter is inductive, and very low inductance is essential, less than 1 nH. The rise time is governed by both resistive and inductive times:

$$\tau = \sqrt{\tau_r^2 + \tau_L^2} \qquad (6.4)$$

$$\tau_r = \frac{88ns}{Z^{1/3}E^{4/3}}\rho^{1/2} \qquad (6.5)$$

$$\tau_L = \frac{L_c + L_h}{Z} \qquad (6.6)$$

TABLE 6.2

H-Series of Mesoband Generators

Generator	Operating Voltage, V_0 (kV)	10–90 Rise Time (psec)	dV/dt (kV/sec)	Far Voltage, RE_p (kV), Equation 6.3	Far Voltage/ Source Voltage, RE_p/V_0
H-2	300	250	1.2×10^{15}	350	1.16
H-3	1000	120	8.3×10^{15}	360	0.36
H-5	250	235	1.1×10^{15}	430	1.72

Here Z is the circuit impedance, E the field in the switch, and ρ the gas density in units of atmospheres, and the inductances are those of the arc channel and the housing of the switch. Short gaps mean high electric fields to minimize the resistive phase and, since inductance is proportional to length, the inductive phase as well. The extremely small interelectrode distances do yield high capacitance, even though electrodes are small in area. High-spark-gap capacitance and fast charging times lead to a strong displacement current that produces an undesirable prepulse on the downstream electrode. If not suppressed, it can cause unwanted, lower-frequency components to be radiated before the main pulse.

Both oil and gas are used as insulating media for fast switches. To produce ultrafast switching, the spark gap is dramatically overvolted; that is, the spark gap is charged far in excess of its self-breakdown voltage, which is determined by the gap geometry and pressure. To hold off breakdown, gas UWB gaps operate at pressures in the range of 100 atm. Overvoltages of more than 300% of the self-breakdown voltage are achievable.

High-pressure hydrogen is the gas of choice for high-repetition-rate, peaking UWB switches. Hydrogen has the advantages of supporting higher electric fields and faster recovery time over other atomic gases, in part because it has high ionization energy.

In the 1990s, the H-Series devices developed high-pressure hydrogen switches to produce powerful and very compact mesoband sources. The performance of several H devices, by Air Force Research Laboratory (AFRL) of Albuquerque, can be seen in Table 6.2. Figure 6.4 shows the pulsed power configuration and hydrogen switch in the H-2 system. Pulse compression and voltage multiplication from the initial capacitive store (not shown) occur via the capacitive store/transformer, and the hydrogen switch sends the short pulse onto the antenna. The H-2 generated a 300-kV pulse in a 10-cm-diameter, 40-Ω line with a rise time of about 250 psec and a total pulse length of 1.5 to 2 nsec. This provided an ultrawide spectrum with its center frequency around 150 MHz from a large TEM horn. The H-3 device generated nearly 1 MV into a 40-Ω coaxial line. To achieve the very fast rise time, the H-3 uses a multichannel output switch (Figure 6.5). (Multiple channels lower the switch inductance.) The unique transmission line-to-transmission line-charging technique compresses the pulse and increases the voltage. The first transmission line is charged relatively slowly, then is switched into the

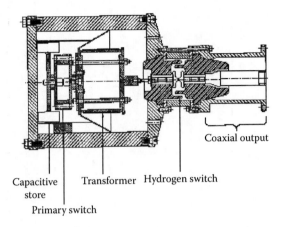

Capacitive store

Primary switch

Transformer   Hydrogen switch

Coaxial output

FIGURE 6.4
H-2 hydrogen switch, about 0.5 m long, 0.3 m in diameter. (Reprinted from Agee, F.J. et al., *IEEE Trans. Plasma Sci.*, 26, 863, 1998, Figure 1. By permission of IEEE.)

second line using a high-pressure hydrogen spark gap. The fast charging of the second transmission line creates a large overvoltage on the high-pressure hydrogen, multichannel, ring-gap switch. The radiated field from H-3 has a rise time of 130 psec and pulse duration of 300 psec, as shown in Figure 6.6. The radiated field was generated using a 0.4-m TEM horn with a peak voltage of 130 kV. The field at 6 m was 60 kV/m. The H-5 is a higher-power version of H-2. Figure 6.7 gives the pulse shape and spectrum. Note the broadband spectrum with substantial variations. The bandwidths of such spectra are difficult to clearly define (see Giri[1] and Section 6.1).

Liquid switches have their own issues. When various types of hydrocarbon oil break down, the streamer in the oil leaves a long-lived, electrically

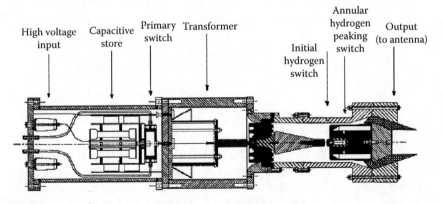

High voltage input   Capacitive store   Primary switch   Transformer   Initial hydrogen switch   Annular hydrogen peaking switch   Output (to antenna)

FIGURE 6.5
Schematic of the H-3 source. Note stages of pulse compression using a sequence of switches to produce faster pulses. (Reprinted from Agee, F.J. et al., *IEEE Trans. Plasma Sci.*, 26, 864, 1998, Figure 3. By permission of IEEE.)

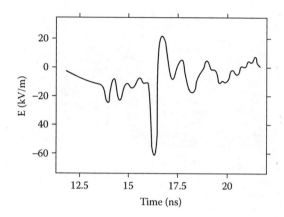

FIGURE 6.6

Field radiated from the H-3 source. (Reprinted from Agee, F.J. et al., *IEEE Trans. Plasma Sci.*, 26, 864, 1998, Figure 4. By permission of IEEE.)

charged streamer containing breakdown by-products that then leave the oil with decreased breakdown strength. The oil recovers more slowly, which limits the possible repetition rate. Titan Pulse Sciences led development of high-repetition-rate oil switches, producing a modulator that contained three oil switches: a transfer switch, a sharpening switch, and a peaking/sharpening switch.[5] The modulator produced pulses with a 10 to 90% rise time of 100 psec, 200-kV pulses at 1500 Hz, as well as 450-kV pulses at 500 Hz.

The Titan Pulse Sciences work investigated the recovery of switches with the direction and rate of oil flow as parameters. If the flow is fast enough to displace a sufficient amount of oil in the gap region, the switch breakdown strength is that of a single pulse, even in repetitive operation. The oil flow rate is a strong function of the switch gap d. The region in the gap of highly stressed oil must be displaced between pulses; the volume to be displaced is proportional to d^3. The displacement distance, and hence the oil velocity, is proportional to d. Therefore, the replacement volumetric rate goes as d^4 (see Problem 3). The flow-rate requirements for the final UWB gap are practically realizable, and the power to drive the fluid is considerable and so may limit the repetition rate.

6.2.2 Solid-State Switches

Photoconductive solid-state (PCSS) switching technology is used for UWB because of very low turn-on jitter, fast rise times, and compact packaging. Such switching has advantages in that it is in principle very efficient (but not in practice), has fewer conversion steps, may allow a degree of agility in frequency and pulse width, and can be used in arrays by phasing of many modules of lower-power units. The principal limitation is power-handling capability. The most likely application is impulse radar because of the low

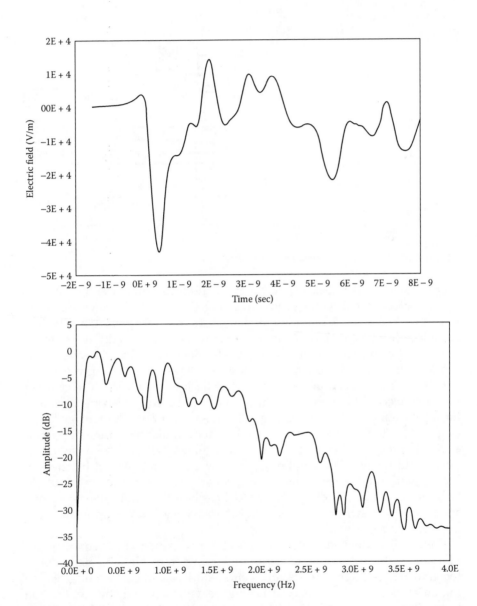

FIGURE 6.7
Radiated field of H-5 at 10 m using a large TEM horn. (Reprinted from Agee, F.J. et al., *IEEE Trans. Plasma Sci.*, 26, 865, 1998, Figure 5. By permission of IEEE.)

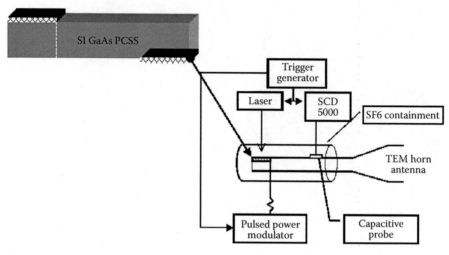

FIGURE 6.8
Schematic of a UWB device driven by a solid-state switch.

energy needed per pulse. The technique is used for lower frequencies ($\lesssim 1$ GHz) to ease laser rise time requirements on the switch. A typical device is shown in Figure 6.8. A pulse line is charged from a voltage source. The line has a transit time equal to the desired microwave half period. The switch is triggered by the laser, which photonically generates charge carriers and connects the center conductor to ground. A voltage wave then travels to the open end of the line, doubles, and reflects back to the switch. The charged line is connected to an antenna and a series of pulses, reflecting back and forth, give a few cycles. This is a mesoband source, of which there are many examples.[6-8] Ideally, the pulses would be rectangular, but parasitic circuit elements produce a rounded profile much like a half cycle of a radio frequency (RF) wave. The amplitude of the wave is determined by the line voltage, and the frequency by the line length. The switch must be able to close in a time that is short compared to the transit time of the line; therefore, subnanosecond switching is essential if frequency content approaching a gigahertz or more is desired.

The photoconductive switch operates by modulation of the conductivity of a semiconductor by photoionization. When a laser photon enters a photoconductor, its energy creates an electron–hole pair in picoseconds, elevating an electron to the valence band and leaving a hole in the conduction band. Reviews of switch development are available.[9,10] Several modes of operation exist:

- In the *linear mode,* one electron–hole pair is produced for each photon absorbed, and therefore, the conductivity is linearly proportional to the total photon flux on the semiconductor material. The typical material is silicon (Si) or doped gallium arsenide (GaAs) or indium phosphide (InP). The carrier lifetime is shorter than that of the optical pulse, so switch conductivity and the output electrical pulse follow the amplitude of the optical pulse. Switching in the linear mode requires ~1 mJ/cm² of optical energy, and so means larger lasers than the other modes.

- In the *lock-on mode*, the laser causes the switch to close, but the switch opens only partway when the laser pulse ceases. The switch (GaAs, InP) recovers when the current through the circuit is shut off by a line reflection or some other circuit mechanism. In the linear mode, carriers live for only a few nanoseconds. In the lock-on mode, the carriers remain for as long as the current flows. The lock-on mode demands less laser power than the linear mode (~1 kW).

- In the *avalanche mode*, the voltage across the switch is higher, so that at a critical electric field, the switch stays closed, even with the laser off. Carrier multiplication occurs at the higher electric field, and the carriers are sustained by the switching processes. The threshold electric field in GaAs is ~8 kV/cm. The energy required to trigger the switch is reduced about two to three orders of magnitude compared to linear switching. Avalanched GaAs has become the PCSS of choice.

The *switching gain* is the ratio of the output power of a photoconducting switch to the laser power necessary to switch it. The gain achievable depends upon the switching medium; gain is greatest in the avalanche mode. For example, Nunnally[10] calculates linear mode gains in GaAs of <2, and avalanche mode gains of 10 to 100; higher values are possible. The lock-on mode has made possible the fast, subnanosecond high-current switching necessary for generation of high power UWB microwave sources. Gallium arsenide is now the most commonly used material, and the principal development issues are lifetime and power handling. The lock-on mode has substantial field in the material, which leads to filamentation of the current and burnout. The lifetime of the switch varies inversely with the power level it passes. Lifetimes are now in excess of 10^9 pulses/switch and decrease with repetition rate, pulse length, and power level. The current developments in photoconductive switching are to improve these performance measures along with switch quality. Although GaAs has become the PCSS of choice, SiC is under study as a possibility.

The most powerful PCSS UWB system was the GEM II, built by Power Spectra, Inc., using GaAs bulk avalanche semiconductor switches (BASSs).[11] The BASS module is very compact, so with a horn output, it can be arrayed to form a planer radiator as in Figure 6.9, a 12 × 12 array of BASS modules that achieved 22.4 kV/m at a 74-m range — a far voltage of almost 1.66 MV — at a repetition rate of 3 kHz. The array measured 1.6 × 1.6 × 0.86 m and weighed 680 kg. It was able to combine the multiple beams and steer the output beam ±30° by switching modules in the correct sequence.

FIGURE 6.9
GEM II, an array of horn outputs driven by bulk avalanche solid-state switches mounted inside a truck. The beam formed by the array can be steered by changing the timing of the individual switches. The output pulse is similar to that in Figure 6.7.

6.3   Ultrawideband Antenna Technologies

Radiating UWB transients requires antennas specifically designed and optimized for this very demanding application. The key consideration for transient antennas is minimizing both frequency and spatial dispersion. Ordinary antennas will generally not transmit fast transients because they have not been corrected for dispersion, which follows from the fact that antenna gain varies as the square of the frequency, while UWB contains a broad range of frequencies. There were two approaches to solving this problem: modification of conventional antennas, usually TEM horns, and invention of a new type, the Impulse Radiating Antenna.

Frequency dispersion can be reduced by modification of some types of conventional antennas. This works for conical TEM horn or a biconic

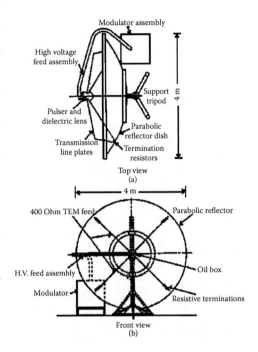

FIGURE 6.10
(a) Configuration of an IRA hyperband system, a pulser driving transmission line feeding a reflector dish. This is the prototype IRA. (b) The later IRA II. (Reprinted from Giri, D.V. et al., *IEEE Trans. Plasma Sci.*, 25, 318, 1997, Figure 1. By permission of IEEE.)

antenna, because they are frequency-independent structures that produce spherical wavefronts. However, because of the finite size of the high-voltage switches that drive them, it is often necessary to add some corrective lenses to the structure to make the wavefront more spherical.

The antenna of choice in the UWB community is the Impulse Radiating Antenna (IRA).[12] The book by Giri[1] is the most complete account, and a useful series of notes appears at Reference 13. The IRA is a means of launching a pulsed hyperband waveform onto a parabolic reflector. A pulser produces the pulse, with the final fast switch located at the focal point, which is also the apex of two transmission lines leading to the dish (Figure 6.10). The driving voltage source is ideally a step function, which in practice is a fast rise impulse with a slower decay. The TEM wave launched from the switch is guided along the conical transmission lines toward the dish. The lines are terminated at the dish, making a load for the pulser electrical circuit that is close to a match.

The key requirement of the geometry is that the dish is illuminated by a spherical TEM wave. If so, the phase center of the wave is fixed, and the IRA is dispersionless, unlike most other antennas. A nearly spherical wave reflects from the dish and propagates as a near-parallel wave into the far field. Since it comes from the focus, a spherical wave illuminating a paraboloidal dish reflects as a plane wave focused at infinity. The IRA concept is complex, and the detailed operation is very complex.[14,15]

The radiated field for a paraboloidal dish is

$$E(R,t) = \frac{dV_p}{dt} \frac{D}{4\pi c f_g R} \tag{6.7}$$

where dV_p/dt is the rate of change of the voltage launched by the pulser (showing the importance of rise time), D is the dish diameter, and f_g is the ratio of the impedance of the transmission lines to that of free space (377Ω). As an example, the prototype IRA system (Giri[1]) has D = 3.66 m, $dV_p/dt \sim$ 1.2×10^{15} volts/sec, and $f_g = 1.06$. Therefore, at a range of 100 m the field is 110 kV/m.

The pulse in the far field has a prepulse, the main pulse, and a following undershoot. The prepulse is negative, with a duration of 2F/c, where F is the focal length of the dish.

The figure of merit adopted for UWB is the far voltage, defined in Equation 6.3. For the prototype IRA, this is 1.1 MV. The best hyperband systems have far-voltage ER products measured in megavolts.

UWB pulses propagate with considerable dispersion off axis, with the fastest pulses on axis, but slower pulse shapes at higher angles. Both rise times and durations of pulses vary with angle. The gain of impulse antennas is an issue in the community because gain in narrowband antennas is defined for single-frequency waveforms. This makes simple forms for gain and beam width difficult to define. A useful quantity somewhat like gain is the figure-of-merit far voltage (Equation 6.3 divided by the voltage on the pulser that produced it):

$$\text{UWB gain} = V_f/V_p = E(R) R/V_p \tag{6.8}$$

where E(R) is the time-domain peak field. For prototype IRA, $V_p \sim$ 115 kV and the gain is 10.

The frequency bandwidth of the radiated field is determined by the rise time of the pulse driving the IRA and the dish diameter. The relation for the upper frequency f_u is

$$f_u t_r = \ln 9/2\pi = 0.35 \tag{6.9}$$

where f_u is the frequency at the upper half-power point in the spectrum and t_r is the 10 to 90 rise time of the pulse injected into the IRA. (This follows from requiring that $\omega t_r = 1$ and that t_r be the 10 to 90 rise time.) The lower limit of frequencies radiated has a half wavelength of about the dish diameter:

$$f_l = c/2D \tag{6.10}$$

For example, the prototype IRA has a 3.66-m dish fed by a pulse with a rise time of 129 psec, so its upper and lower frequencies are 2.7 GHz and 27 MHz, respectively.

When measuring the radiated field, how far away does one need to be to be in the far field? For UWB, the distance corresponding to the $2D^2/\lambda$ distance for narrowband (which comes from requiring that ΔR, the differential travel distance to the observer from the antenna center and edge, be $\lambda/16$) is determined by the condition that the differential travel distance be less than the rise time,

$$R/c < t_r \qquad (6.11)$$

which means

$$R > D^2/2ct_r \qquad (6.12)$$

For prototype IRA, R > 170 m (see Problems 4 and 5).

Building a fast hyperband IRA has complexities:

- At high voltages (>100 kV), the high-field region around the switch is usually insulated by oil and shaped to provide a lens to keep the converging wave spherical. The lens is used at the apex of the antenna to correct the astigmatism introduced by the finite size of the gas switch. With careful design, the wave is corrected by the lens to a near-spherical wavefront, so that, upon reflecting from the parabolic surface, it emerges as a plane wave.

- High-voltage hyperband pulsers producing ~0.1-nsec pulses at ~10^{15} V/sec are difficult to design and build, but the electrical circuits are easy to model with equivalent circuit codes if the components are accurately characterized.

- IRA antennas, on the other hand, are relatively easy to fabricate, but are hard to analyze accurately.

In addition to reflector IRAs, *lens IRAs* consist of a TEM horn with a dielectric lens in the aperture. Focusing lenses at the output of TEM horns correct the nonplanarity of the wavefront and, in creating a plane wave, reduce the spatial dispersion, which increases the rise time. The lens adds substantially to the weight, so lens IRAs are used in applications with small apertures.

6.4 Ultrawideband Systems

We give notable examples of representative UWB devices, by frequency band classes.

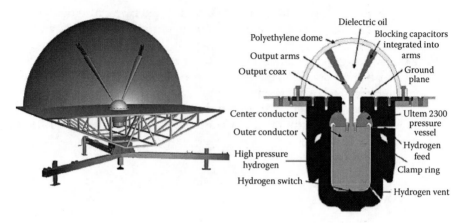

FIGURE 6.11

The MATRIX system (left) is an IRA driven by a coaxial line (right) to produce ringing oscillations. (Reprinted from Prather, W.D. et al., *IEEE Trans. EMC*, 2, 2003, Figure 1 [special issue on IEMI]. By permission of IEEE.)

6.4.1 Mesoband Systems

Mesoband sources have been successfully designed and built to generate HPM energy between 100 and 700 MHz, a range at the lower end of narrowband sources. These sources are typically designed in one of two ways:

- Feed a damped sine waveform into a wideband antenna.
- Switch a wideband transient pulse (from a Marx generator, for example) into a resonant pulse line that is connected to a UWB antenna. Pulse lines of differing lengths can be changed out to radiate a broad range of frequencies.

For the damped-sine/wideband-antenna method, Baum has described concepts that switch a high-voltage transmission line oscillator into a wideband antenna.[16] The oscillator consists of a quarter-wave section of a transmission line charged by a high-voltage source. Frequency and damping constant can be adjustable. This device employs a self-breaking switch at one end of the transmission line and the antenna at the other. When the switch closes, the system generates a damped-sine signal that is fed into a UWB antenna, such as half of an IRA. The half of a reflector over a conducting plane uses the image currents in the plane to complete the reflector. This concept makes the high-voltage problem easier, but the pattern is not the entirely symmetric field pattern of a full parabolic antenna.

For example, MATRIX is a transmission line switched into an IRA.[17] Figure 6.11 shows that it consists of a quarter-wave coaxial transmission line charged to 150 kV connected to a 3.67-m-diameter half IRA. The line, insulated by high-pressure hydrogen, is charged, then switched to ground. An

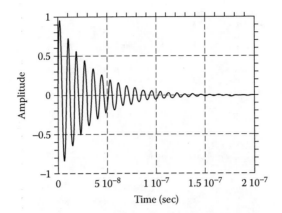

FIGURE 6.12
Calculated electric field from MATRIX at 15 m. (Reprinted from Prather, W.D. et al., *IEEE Trans. EMC*, 2, 2003, Figure 2 [special issue on IEMI]. By permission of IEEE.)

oscillation between the line and the IRA produces a damped sinusoid with frequency of oscillation that can be made adjustable between 180 and 600 MHz. Its radiated waveform is shown in Figure 6.12: a peak electric field of 6 kV/m at 15 m (figure of merit of 90 kV, gain of 0.6 at 150 kV charge) with a percentage bandwidth of about 10% (band ratio of 1.10). With a 300-kV charging supply, this source will radiate energy in the gigawatt range.

For the switched Marx/oscillating antenna method, DIEHL Munitionssysteme of Germany manufactures a number of mesoband sources.[18] One suitcase-sized version, the DS110B (for damped sinusoid), and its electric field are shown in Figure 6.13 and Figure 6.14. The DS110B operates at 750 kV, producing a peak E field of about 70 kV at 2 m (see Problem 6). It is made tunable by changing the length of the loop antenna. The repetition rate is a few hertz operating off a battery, and 100 Hz is available with a more powerful external electrical source.

Mesoband oscillators are also made by BAE Systems in the U.K.[19] These oscillators are nonlinear transmission line solid-state modulators. An example output is shown in Figure 6.15. They generate Q ~ 50 damped sine waves at ~50 kV. Frequencies range from 10 MHz to 2 GHz, with repetition rates in excess of 1 kHz. More advanced systems have the ability to chirp (increase frequency during the pulse) over a 40% range.

6.4.2 Subhyperband Systems

The H-Series systems, which use coaxial pulse-forming lines and high-pressure hydrogen switches as described in Section 6.2, were built at the Air Force Research Laboratory in Albuquerque. One of the most successful designs of the H-Series, the H-2 source, shown in Figure 6.16, is driven by a 0.1m-diameter, 40-Ω coaxial line (Figure 6.4). It generates a fast 0.25-nsec

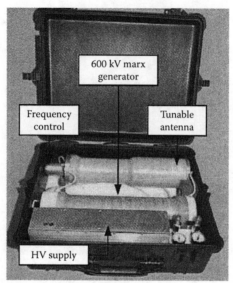

Complete DS110B high-power system

FIGURE 6.13

Suitcase-sized mesoband device, the DS110B (for damped sinusoid). (Reprinted from Prather, W.D. et al., *IEEE Trans. EMC*, 2, 2003, Figure 3 [special issue on IEMI]. By permission of IEEE.)

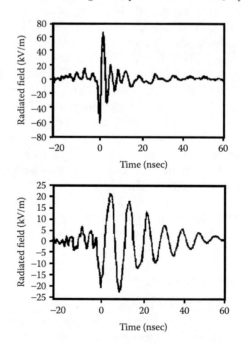

FIGURE 6.14

Tunability of the DS110B. (Reprinted from Prather, W.D. et al., *IEEE Trans. EMC*, 2, 2003, Figure 4 [special issue on IEMI]. By permission of IEEE.)

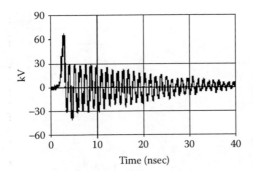

FIGURE 6.15
Voltage output from a nonlinear transmission line. (Reprinted from Prather, W.D. et al., *IEEE Trans. EMC*, 3, 2003, Figure 7 [special issue on IEMI]. By permission of IEEE.)

FIGURE 6.16
H-2 source with large TEM horn. (Reprinted from Prather, W.D. et al., *IEEE Trans. EMC*, 3, 2003, Figure 8 [special issue on IEMI]. By permission of IEEE.)

rise-time 300-kV pulse with a total length of 1.5 to 2 nsec and a considerable amount of late-time, low-frequency ringing. Fired into a TEM horn, the maximum spectral amplitude is at ~150 MHz. Performance is 43 kV/m at 10 m with a rise time of 0.24 nsec, very similar to that shown in Figure 6.7.

6.4.3 Hyperband Systems

The most complex and advanced sources are the hyperband, with band ratio greater than one decade. The state-of-the-art system is *Jolt*, shown in Figure 6.17.[19] The pulsed power (Figure 6.18) is a compact dual-resonant trans-

FIGURE 6.17
The Jolt apparatus. (Reprinted from Prather, W.D. et al., *IEEE Trans. EMC*, 4, 2003, Figure 13 [special issue on IEMI]. By permission of IEEE.)

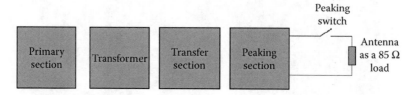

FIGURE 6.18
Jolt circuit elements. (Reprinted by permission of Giri, D.V.)

former that steps 50 kV up to 1.1 MV. A transfer capacitor leads to an oil-insulated peaking switch, which feeds an 85-Ω half IRA. The voltage rise rate is ~5 × 10^{15} V/sec. The ground plane below the half dish serves as the ground reference and an image plane for the IRA. It also provides the lower containment for SF6, which insulates the dome and feed arms. (For half IRAs, Equation 6.7 for the electric field is reduced by 2$^{-3/2}$.)

The half-dish antenna delivers a tightly focused radiated field with a pulse length on the order of 100 psec and a field–range product of approximately 5.3 MV at 200 Hz. This is the highest figure of merit achieved in a hyperband device. The spectrum is broad, with an upper half-power point between 1 and 2 GHz and a lower half-power point around 30 MHz. The radiated waveform in Figure 6.19 has a peak field at 85 m of 62 kV and a rise time of ~0.1 nsec.

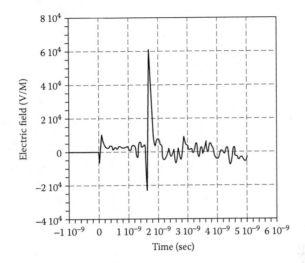

FIGURE 6.19
Measured electric field from Jolt at 85 m. (Reprinted from Prather, W.D. et al., *IEEE Trans. EMC*, 4, 2003, Figure 14 [special issue on IEMI]. By permission of IEEE.)

6.5 Conclusion

The GEM II and Jolt systems will represent the high water mark for UWB systems for some time to come. That would be parallel to the "power derby" in narrowband sources of the 1980s, which ended because of pulse shortening and cost increasing with peak power (see Chapter 1). They could find application as high power jammers for electronic warfare and short-pulse radars. The most likely direction for development of the small mesoband systems is for effects tests and directed energy, because of their compactness and low cost. Commercial versions have already reached the market for effects testing. Current technology mesoband systems operate in the 100- to 500-MHz range. Research efforts are developing devices in the 500-MHz to 1-GHz range.

Problems

1. The fractional bandwidth, percentage bandwidth, and bandwidth ratio are interrelated. Express the bandwidth ratio in terms of the percentage bandwidth. What is the maximum possible value for the percentage bandwidth?

2. How many cycles comprise a mesoband source?

3. How does the power required to drive the oil through the small dimensions of a switch gap depend on the gap size?

4. For comparison, if a narrowband source feeding an IRA operates at the upper frequency f_u, what is the relation between the narrowband and UWB far-field distances?

5. You are given a system requirement for a pulser-fed IRA to produce a figure of merit of 500 kV with an upper frequency of 2 GHz, operating down to 50 MHz. Assuming that the transmission lines are well coupled to free space ($f_g \sim 1$), what size dish antenna and what pulser voltage will be required, and how far away would you need to be to measure the figure of merit in the far field?

6. What are the resonator Q, figure of merit, and gain of the DS110B?

References

1. Giri, D.V., *High-Power Electromagnetic Radiators*, Harvard University Press, Cambridge, MA, 2004, chaps. 3–5.
2. Agee, F.J., Ultra-wideband transmitter research, *IEEE Trans. Plasma Sci.*, 26, 860, 1998.
3. Prather, W.D., Survey of worldwide high power wideband capabilities, *IEEE Trans. Electromag. Compat.*, 46, 335, 2004.
4. Giri, D.V. and Tesche, F.M., Classification of intentional electromagnetic environments (IEME), *IEEE Trans. Electromag. Compat.*, 46, 322, 2004.
5. Champney, P. et al., Development and testing of subnanosecond-rise kilohertz oil switches, in *Proceedings of the IEEE 8th Pulsed Power Conference*, San Diego, CA, 1991, p. 863.
6. Baca, A.G. et al., Photoconductive semiconductor switches, *IEEE Trans. Plasma Sci.*, 25, 124, 1997.
7. Stoudt, D.C. et al., Bistable optically controlled semiconductor switches in a frequency-agile RF source, *IEEE Trans. Plasma Sci.*, 25, 131, 1997.
8. Schoenberg, J.S.H. et al., Ultra-wideband source using gallium arsenide photoconductive semiconductor switches, *IEEE Trans. Plasma Sci.*, 25, 327, 1997.
9. Lee, C.H., Optical control of semiconductor closing and opening switches, *IEEE Trans. Electron Devices*, 37, 2426, 1990.
10. Nunnally, W.C., High-power microwave generation using optically activated semiconductor switches, *IEEE Trans. Electron Devices*, 37, 2426, 1990.
11. Prather, W.D. et al., Ultrawideband sources and antennas: present technology, future challenges, in *Ultra-Wideband, Short Pulse Electromagnetics 3*, Baum, C.E., Carin, L., and Stone, A., Eds., Plenum Press, New York, 1997, p. 381.
12. See papers in Mokole, E.L., Kragalott, M., and Gerlach, K.R., *Ultra-Wideband, Short-Pulse Electromagnetics 6*, Kluwer Academic/Plenum Press, New York, 2003.
13. http://www-e.uni-magdeburg.de/notes.

14. Giri, D.V. et al., Design fabrication and testing of a paraboloidal reflector antenna and pulser system for impulse-like waveforms, *IEEE Trans. Plasma Sci.*, 25, 318, 1997.
15. Farr, E.V., Baum, C.E., and Buchenauer, C.J., Impulse Radiating Antennas Part II, *Ultra-Wideband, Short-Pulse Electromagnetics 2*, Plenum Press, New York, 1995, p. 159.
16. Baum, C.E., 2000 Circuit and Electromagnetic System Design Note 45, Air Force Research Lab, New Mexico (see reference 14).
17. Burger, J. et al., *Design and Development of a High Voltage Coaxial Switch, Ultra-Wideband Short-Pulse Electromagnetics 6*, Plenum Press, New York, 2003, p. 381.
18. Ruffing, K., High Power Microwave and Ultra-Wideband Threats, Aspects and the Concept of the German MOD, AMEREM, 2002, Annapolis, MD.
19. Baum, C.E. et al., JOLT: a highly directive, very intensive, impulse-like radiator, *Proc. IEEE*, 92, 1096, 2004.

7

Relativistic Magnetrons and MILOS

7.1 Introduction

Crossed-field devices generate microwave energy by tapping the kinetic energy of electrons that are drifting perpendicularly to electric and magnetic fields that are oriented at right angles to one another (hence the term *crossed field*). This arrangement prevents breakdown of the anode–cathode gap, confines the electrons within an interaction region, and aligns the electron velocities with the electromagnetic waves. These devices include the *relativistic magnetron*, a high-voltage, high-current version of the well-known conventional magnetron; the *magnetically insulated line oscillator* (MILO), a linear magnetron variant that was invented to take advantage of the very high current capacity of modern pulsed power systems; and the *crossed-field amplifier* (CFA), an amplifying, rather than oscillating, near cousin of the magnetron that has been explored, but not exploited, as a high power source.

The relativistic magnetron, along with the backward wave oscillator and related devices of Chapter 8, is one of the most mature of all high power sources. The operational mechanism is robust; it is a relatively compact source; and it has been operated in the frequency range from just below 1 to about 10 GHz. It is well understood. Over the years, a number of research teams have explored the full range of operational parameters: single-source output in excess of 4 GW; multisource output in a phase-locked configuration of 7 GW total output; repetitive, near-GW peak powers at pulse repetition rates up to 1 kHz; and long-pulse operation with an energy per pulse of about 1200 J. The magnetron is the most tunable high power source, with a 30% tunable range by mechanical means, and can be tuned while firing repetitively. Theoretically, the basic design process has been understood since the 1940s, but modern computer simulation techniques have speeded the process by which major design changes are incorporated, and systems analysis has clarified the major issues involved in understanding the upper limits of achievable repetition rates and energies per pulse. Practically, the major issues of fielding a magnetron-based system have been addressed in the Orion facility, a transportable outdoor microwave effects test system built in 1995.

The MILO has two major attractions. First, its generally low impedance — of the order of 10 Ω, about a tenth of that for a relativistic magnetron — allows for better coupling to compact, lower-voltage power sources such as explosive generators. Second, it uses the current flow in a magnetically insulated transmission line to provide the necessary magnetic field, so that no external magnet is required. Powers of 2 GW with pulse energies of about 1 kJ have been obtained, and repetition rates as high as 100 Hz, with most results obtained in the L-band. The drawbacks of this source are its lack of maturity, particularly relative to the relativistic magnetron; its lower efficiencies* of about 10%, due in part to challenges in extracting the output; and its lack of tunability. Nevertheless, commercial versions have been offered in the U.K.

Relativistic CFAs are the least mature of the high power crossed-field devices. In fact, a single very preliminary exploration of such a device based on a secondary-emission cathode was plagued with problems and terminated without achieving its original output power specification of 300 MW.

The majority of this chapter will be devoted to either the general features of crossed-field devices or the specific features of magnetrons. MILOs and, more briefly, CFAs will be treated in two sections near the end of the chapter.

7.2   History

Arthur Hull invented the magnetron in 1913, and devices built on his original principles in the 1920s and 1930s reached power levels of about 100 W. In 1940, Postumus suggested the use of a solid-copper anode, an innovation independently invented and realized also by Boot and Randall[1,2] and the Soviets Alekseev and Malairov. The initial Boot and Randall magnetron achieved 10 kW and was quickly engineered into an operational device.[3] During World War II, intense technological development of the magnetron occurred at the MIT Radiation Laboratory and other institutions, as described in Collins' classic *Microwave Magnetrons*.[4] Over the course of the war, millions of magnetrons were produced, and magnetron-driven radar probably had a greater impact on the war than any other technology invented in World War II, more so than even the atomic bomb.

Postwar development was equally fast paced into the 1960s, with the development of mode selection techniques such as strapping (shifting the frequency spectrum by connecting alternate resonators, thus preventing

---

* The true comparison of efficiency, however, should take into account the electrical power demand of both the magnetron itself and its field coil, if it is an electromagnet.

mode hopping) and the rising-sun geometry (alternating between two different resonator shapes to separate the modes in frequency). A new generation of higher-power, frequency-agile magnetrons was introduced for electronic warfare applications. The *coaxial magnetron* introduced by Feinstein and Collier allowed stable tuning by use of a cavity surrounding the magnetron resonator.[5] The French introduced the *magnetron injection gun*, now used extensively in gyrotrons, to drive a magnetron by a beam injected from outside the resonator. The *inverted magnetron*, with the cathode on the outside of the device and the anode on the inside, was also introduced. Gradual improvement in the understanding of magnetron operation eventually led to the *crossed-field amplifier*, in which gains of 10 to 20 dB are achieved at power levels of 1 MW or more, with amplification bandwidths of about 10%. Okress documented the state of conventional magnetron development at this time in the two-volume *Cross-Field Microwave Devices*.[6]

The relativistic magnetron was developed as a direct extrapolation of the cavity magnetron, driven at higher currents by pulsed power and cold-cathode technology. The relativistic magnetron is in fact a high-current extension of the conventional magnetron, wherein relativistic voltages are necessary to produce the currents. Bekefi and Orzechowski developed the first such devices in 1976,[7] and the most powerful was the six-vane A6, described extensively by Palevsky and Bekefi.[8] Whereas conventional devices had reached powers of about 10 MW, the first relativistic device produced 900 MW, generating great interest in the U.S. and U.S.S.R. Substantial improvements in the capabilities of magnetrons in the 1980s pushed peak powers up to several gigawatts. Frequencies extended from the original S-band device upward to the X-band and downward to the L-band. The original devices were single shot and uncooled, but cooled devices have been operated repetitively at hundreds of megawatts peak power and average powers well above a kilowatt. Microwave pulse duration is typically 30 nsec, but operation at 600 nsec has been achieved. With the generation of higher radiated energies, extending pulse durations and operating at higher average power have become principal development thrusts. In this regard, one alternative to the maximization of power in single devices has been the phase locking of multiple magnetrons together to build high power arrays of oscillators. One such array of seven magnetrons produced 7 GW of output.

Clark introduced the MILO in 1988.[9] It has the dual attractions of producing high power at low impedance, so that it couples well to lower-voltage, compact power supplies such as low-impedance explosive generators without an externally applied magnetic field. Tapered and double-MILO designs at AEA Technology in the U.K. produced relatively high efficiencies,[10] and versions are commercially available.

Some very preliminary development of high power CFAs was undertaken at the Stanford Linear Accelerator Center (SLAC) in the early 1990s. Although designed to produce 300 MW,[11] several factors limited actual CFA powers to much lower levels.

7.3   Design Principles

Crossed-field devices have two distinguishing features. First, the electrons that supply the energy to generate microwaves are emitted directly from the cathode in the interaction region. No external beam-producing components are needed, so that these devices, specifically magnetrons, can be quite compact. Second, the electrons in the interaction region execute an $E \times B$ drift in applied electric and magnetic fields oriented at right angles to one another, with drift velocity

$$\mathbf{v}_d = \frac{\mathbf{E} \times \mathbf{B}}{|\mathbf{B}|^2} \tag{7.1}$$

Because the space-charge and diamagnetic effects of the electron layer cause E and B to vary with distance from the cathode, $\mathbf{v}_d$ does as well. The resonance requirement in crossed-field devices is that $\mathbf{v}_d$ for some electrons must equal the phase velocity of an electromagnetic wave traveling in the same direction. If z is the direction of propagation, and the wave varies in time t and z as $e^{i(kz - \omega t)}$, then resonance occurs for electrons obeying*

$$|\mathbf{v}_d| = \frac{\omega}{k} \tag{7.2}$$

The fact that the drift and phase velocities are perpendicular to both E and B distinguishes these *M-type devices*, as they are also known, from the parallel-field, or O-type, devices in which the wave propagates parallel to the applied magnetic field (see Chapter 4). Thus, crossed-field devices such as magnetrons and MILOs are M-type Cerenkov sources. Note also that the resonance condition in Equation 7.2 is the same as that for O-type Cerenkov sources in Equation 8.6.

The basic geometry of a cylindrical magnetron is shown in Figure 7.1. The main design features are the following:

- The cathode, which is separated by a gap from the anode structure. The application of a large voltage across the gap causes the explosive emission of electrons from a plasma layer created on the cold cathode (see Chapter 5).

- The interaction region between the cathode and anode, which in operation contains the drifting electrons.

---

* In the cylindrical geometry of magnetrons and CFAs, for a wave with azimuthal and time dependence $\exp[i(n\theta - \omega t)]$, the phase velocity on the right of Equation 8.2 is $\omega r/n$, with r within the interaction region.

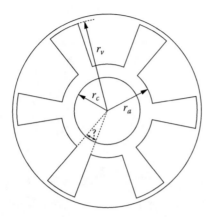

FIGURE 7.1

Basic configuration of a cylindrical magnetron: r_c and r_a are the cathode and anode radii, and r_v is the vane radius.

- The anode and the cavity structure that, together with the dimensions of the interaction region between the anode and cathode, plays the primary role in determining the operating frequencies and output modes. The axial length of the device and any cavities at the end of the device play a secondary role in determining the frequency.

- An applied axial magnetic field B_z, which must be strong enough to prevent electrons from immediately crossing the anode–cathode gap, but not so strong that the azimuthal drift velocity of the electrons, which scales roughly as $1/B_z$, is too slow to allow the resonance described in Equation 7.2.

Microwaves are extracted either *axially*, on the side opposite the pulsed power feed to the device, or *radially* from one or more of the anode cavities. The configuration of a CFA is similar in basic concept to that of a magnetron, as we can see in the schematic of Figure 7.2, which shows a coaxial CFA, with a second cavity located outside of the inner, vane-loaded microwave cavity. In detail, however, there are several additional differences in this amplifying device: microwaves are fed in through one cavity and then circulate around to be extracted at higher amplitude from another cavity; an absorbing sever between the input and output feeds prevents the microwaves from going around a second time; and the number of vanes is larger than that for many relativistic magnetrons. A MILO, depicted schematically in Figure 7.3, differs in two ways. First, it is a linear device with an azimuthal, rather than axial, magnetic field, so that the electrons and microwaves propagate axially rather than azimuthally. Second, there is no external magnetic field; the magnetic field is created by axial current flow in the device.

Electrons emitted from the cathode into the interaction region initially form a layer of electrons that drift azimuthally in the magnetron and CFA and axially in the MILO. To understand the mechanism by which electrons in

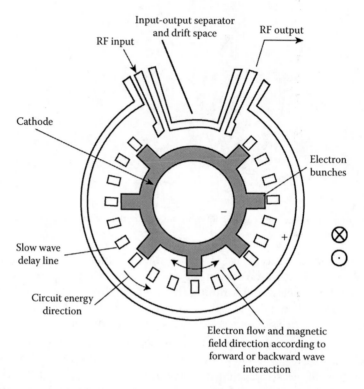

FIGURE 7.2
Schematic of the coaxial CFA investigated at SLAC. (From Eppley, K. and Ko, K., *Proc. SPIE*, 1407, 249, 1991. With permission.)

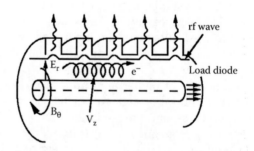

FIGURE 7.3
Schematic of a MILO indicating the drift of a single electron from the cathode sheath and its net axial drift in the radial electric field and azimuthal magnetic field. Radial extraction is shown here, although axial extraction has become more common.

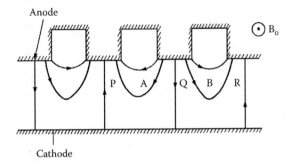

FIGURE 7.4
Behavior of test electrons in the π-mode electric fields of magnetrons and CFAs. (From Lau, Y.Y., *High-Power Microwave Sources*, Artech House, Boston, 1987, p. 309. With permission.)

this layer transfer their energy to the microwaves, consider the two test electrons in Figure 7.4, labeled A and B.[12] For simplicity, the geometry is planar and the microwave fields are those of the π-mode, which reverses phase from one cavity to the next. To be resonant, the electromagnetic wave has a phase velocity that is about the same as the initial drift velocity of the electrons. To start with, both electrons execute $\mathbf{E_0} \times \mathbf{B_0}$ drifts in which the forces due to the applied electric and magnetic fields cancel. The combination of the microwave electric field component, $\mathbf{E_1}$, and the applied magnetic field, $\mathbf{B_0}$, affects each electron in four ways. First, $\mathbf{E_1}$ decelerates A and accelerates B. Second, $\mathbf{E_1} \times \mathbf{B_0}$ pushes A toward the anode and B away from the anode. Third, since the microwave fields are stronger near the slow-wave structure at the anode, A is decelerated even more near the anode, while B is accelerated less as it approaches the smooth cathode. Fourth, as A approaches the anode, it loses potential energy, which is converted to microwave energy, while B gains potential energy, which is supplied by the microwave field energy. Because these effects are stronger for A than for B, more energy is delivered to the microwave fields than is drawn from them, and the fields grow in magnitude.

In concert with the basic gain mechanism, a second phenomenon, *phase focusing*, enhances the gain and leads to the formation of the spokes shown in Figure 7.5, which is taken from computer simulations of the rotating spoke patterns for the A6 magnetron operating in the π- and 2π-modes.[13] To understand phase focusing, return to Figure 7.4 and imagine first that electron A has a drift velocity larger than the phase velocity of the microwave fields. When it advances to location P, where E_{r1} opposes E_{r0}, its drift velocity (E_{r0} + E_{r1})/B_z is reduced, and the drift velocity falls back toward synchronism with the wave. Conversely, if A drifts more slowly with the wave, when it drops back to Q, where the microwave field reinforces the applied field, the drift velocity increases, and A moves back toward synchronism. By this same reasoning, we see that if the drift velocity of electron B is faster than that of the wave, its velocity increases still further when it reaches Q, and if it drifts more slowly than the wave, its velocity tends to decrease even further

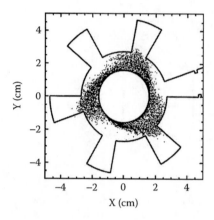
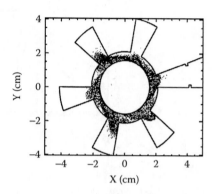

FIGURE 7.5

Spoke patterns from simulations of the A6 magnetron: (a) π-mode operation, with spokes every other cavity (in this snapshot of a rotating pattern); and (b) 2π-mode operation, with spokes at every cavity (again, in a snapshot of a rotating pattern). (From Lemke, R.W. et al., *Phys. Plasmas,* 6, 603, 1999. With permission.)

when it drops back to R. Thus, resonant electrons such as A, which supply energy to the fields, are held in synchronism with the wave, while electrons such as B, which draw energy from the fields, are defocused in phase from the wave.

The spokes carry electrons across the gap, and this radial current flow completes the circuit for an axial current flow through the device. This axial current flow creates an azimuthal magnetic field that, in concert with the radial electric field, creates an axial component of electron velocity. Electrons are thus lost to the anode via the spokes and to the ends of the magnetron via the drift the axial current creates. The locations of loss must be found if x-ray shielding and cooling are required.

In the remainder of this section we address three primary design issues, focusing primarily on the magnetron, although much of the discussion is relevant to its close relative, the CFA:

1. The cold frequency characteristics determined by the configuration of the interaction region and anode cavities (i.e., in the absence of the electrons), which are quite accurate even with the addition of electrons

2. The relationship between the applied voltage and magnetic field, determining where in a parameter space of these two variables is device breakdown prevented and microwave generation allowed

3. The behavior of the device with electrons included, which must largely be done using empirical data and computer simulation because of the highly nonlinear behavior of the operating device

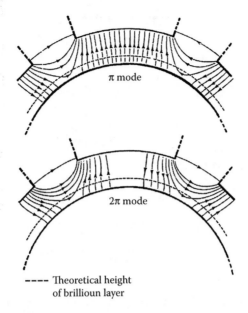

π mode

2π mode

---- Theoretical height
of brillioun layer

FIGURE 7.6
RF electric field patterns for the (a) π- and (b) 2π-modes. (From Palevsky, A. and Bekefi, G., *Phys. Fluids*, 22, 986, 1979. With permission.)

7.3.1 Cold Frequency Characteristics of Magnetrons and CFAs

The coupled resonators of the anode surrounding the annular interaction region between the cathode and anode create a variety of potential operating modes for the magnetron, each with its own characteristic operating frequency and microwave field pattern. Magnetron modes are designated by mode number n for fields that vary azimuthally as $e^{in\theta}$; n is thus the number of times the microwave field pattern repeats during a circuit around the anode. Since the angular spacing between N cavities is $\Delta\theta = 2\pi/N$, the phase shift between adjacent cavity resonators for the nth mode is n $\Delta\theta = 2\pi n/N$ (see Problems 3 to 5). In magnetrons, there are two common operating modes: the *π-mode*, wherein the fields are reversed from one cavity to the next (n = N/2, $\Delta\theta = \pi$), and the *2π-mode*, with fields repeating from cavity to cavity (n = N, $\Delta\theta = 2\pi$). The electric fields for the π- and 2π-modes of the A6 magnetron (N = 6) are shown in Figure 7.6.[8] For further discussion of magnetron mode structure, see Treado.[14]

The dispersion relation for a magnetron relates the frequency for a given mode, ω_n, to either n or the phase shift per cavity, $\Delta\theta$. In the case of a magnetron or CFA, it is usually calculated numerically with an electromagnetic field code or analytically using the technique described in Collins' book.[4] The dispersion relation for magnetrons was discussed in detail in Section 4.4.2. Figure 7.7 shows the dispersion relation for the A6 magnetron,[8] with N = 6 and cathode and anode radii of $r_c = 1.58$ cm and $r_a = 2.11$ cm; the

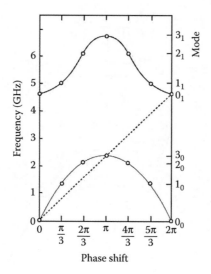

FIGURE 7.7
Dispersion relation, $f_n = \omega_n(\Delta\theta)/2\pi$, for the A6 magnetron. The azimuthal phase velocity of the π- and 2π-modes are the same, although this occurrence is coincidental.

anode cavities have azimuthal opening angles of $\psi = 20°$, and the cavity side walls extend along radial lines to a depth of $r_d = 4.11$ cm, measured from the center of the device. The peak frequency for the modes on the lowest-order curve is the π-mode frequency, about 2.8 GHz. All other modes in that set are doubly degenerate in frequency. The 2π-mode is the lowest-frequency mode in the second-order mode set. The slope of a line drawn from the origin defines the angular phase velocity of a mode. Note that for the A6, the π- and 2π-modes have the same phase velocity; however, they have different Buneman–Hartree conditions (discussed in the next subsection) and field patterns, and therefore are easily separated in practice.

In Collins' book, graphical solution of the dispersion relation reveals that the π-mode frequency lies somewhat below the frequency for which the length of the anode cavities $L_a = r_v - r_a$ is approximately $\lambda/4$, with λ the free-space wavelength of the output radiation. Therefore,

$$f_\pi = \frac{\omega_\pi}{2\pi} = \frac{c}{\lambda} \approx \frac{c}{4L_a} \tag{7.3}$$

In the case of the A6 magnetron, Equation 7.3 yields $f_\pi \approx 3.75$ GHz, which is high by about 60%. This rule of thumb is at least helpful for scoping calculations with somewhat better than factor-of-two accuracy. Further, it points to the relevant scaling parameter for the π-mode frequency.

Magnetrons also have an axial mode structure, particularly when the resonator is shorted at the ends by conducting end caps, or even when it is open on the ends, but its length is not large in comparison to the gap. Under

FIGURE 7.8
L-, S-, and X-band magnetrons, compared to a U.S. penny. (Courtesy of L-3 Communications Pulse Sciences, formerly Physics International.)

these circumstances, the frequency of a given mode is modified by the axial modes according to the relation

$$\left(f_n^g\right)^2 = \left(f_n^0\right)^2 + \left(\frac{gc}{4h}\right)^2 , \; g = 1, 2, 3, \dots \tag{7.4}$$

where h is the resonator length. Clearly, mode competition will be an issue if $h > \lambda$. Typical experimental values for the resonator length are ~0.6 λ, although relativistic magnetrons with $h = 2\lambda$ have been reported in the X-band.[15] Competition among axial modes frequently limits output power in experiments. The most easily observed symptom is beating of the power pulse at the difference frequency, $\Delta f = f_n^{g+1} - f_n^g$.

According to Equations 7.3 and 7.4, the size of a magnetron scales inversely with the operating frequency f, or directly with the free-space wavelength λ. Figure 7.8 allows one to compare the resonator sizes for magnetrons operating in different frequency bands between roughly 1 and 10 GHz. In Table 7.1, for a number of U.S. magnetrons,[14,16,17] we compare the radii of the cathode, anode, and the anode resonator — r_c, r_a, and r_v — as well as the resonator length h, to λ; we also show the π-mode frequency estimated using Equation 7.3 for those devices that operate in that mode (see Problem 6). For the noninverted magnetrons (i.e., those with $r_c < r_a$), we can see that r_c typically ranges between one eighth and one fifth of a wavelength, with the PI L-band magnetron an outlier at $r_c = 0.05\lambda$; r_a ranges from one sixth to about one third λ, with the PI X-band magnetron an outlier on the high end and the PI L-band magnetron an outlier on the low end. The estimate of the π-mode operating frequency in Equation 7.3 is always high, by as much as 60% in a couple of the MIT magnetrons to as little as 6% in the PI X-band device. Resonator lengths are typically a fraction of a wavelength, but the length is almost twice λ in the PI X-band device. We

TABLE 7.1

Dimensions of Example Magnetrons and a Comparison of the Operating Frequency Estimate from Equation 7.3, Valid Only for π-Mode Operation in Noninverted Magnetrons, with the Actual Operating Frequency

	f (GHz)	λ (cm)	r_c (cm)	r_c/λ	r_a (cm)	r_a/λ	r_v (cm)	f_π (GHz) Equation 6.3	f_π/f	h (cm)	h/λ
MIT A6,[a] 2π	4.6	6.5	1.58	0.24	2.1	0.32	4.11	na	—	7.2	1.11
MIT A6,[a] π	2.34	12.8	1.58	0.12	2.1	0.16	4.11	3.73	1.59	7.2	1.11
MIT D6[a]	4.1	7.3	1.88	0.26	2.46	0.34	4.83	na	—	1.88	1.15
MIT K8[a]	2.5	12.0	2.64	0.22	4.02	0.34	6.03	3.73	1.49	2.64	0.60
LLNL,[a] 2π	3.9	7.7	1.58	0.21	2.11	0.27	4.25	na	—	7.2	0.94
PI[b]	1.1	27.0	1.27	0.05	3.18	0.12	8.26	1.47	1.34	20	0.74
PI[c]	2.8	11.0	1.26	0.12	2.1	0.19	4.20	3.57	1.28	7.0	0.64
PI[d]	8.3	3.6	0.67	0.19	1.35	0.39	2.20	8.82	1.06	6.0	1.67
Inverted Magnetrons											
MIT M8[a]	3.7	8.1	4.26	0.526	2.9	0.358	1.28	na	—	5.1	0.59
MIT M10[a]	3.7	8.1	4.64	0.573	3.8	0.469	2.55	na	—	4.6	0.30
NRL[a]	3.2	9.4	21.3	2.27	19.4	2.06		na	—	5.1	0.54
SNL[a]	3.1	9.7	6.79	0.700	5.22	0.538	3.68	na	—	9.2	0.95

Note: Data are from the Massachusetts Institute of Technology (MIT), Lawrence Livermore National Laboratory (LLNL), Physics International (PI, now Titan Pulse Sciences Division), Naval Research Laboratory (NRL), and Sandia National Laboratories (SNL).

[a] Reference 14.
[b] Reference 16.
[c] Reference 17.
[d] Contact James Benford.

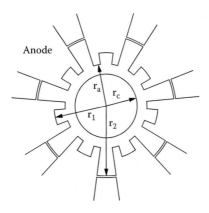

FIGURE 7.9
A rising-sun magnetron geometry with alternating anode cavity lengths. (From Lemke, R.W. et al., *Phys. Plasmas*, 7, 706, 2000. With permission.)

also show dimensions for four inverted magnetrons, for which $r_c > r_a$; we shall consider these later.

Mode competition is a serious issue in all microwave devices because each mode of an oscillator grows from the ambient noise of the Brillouin layer. The key to achieving high power is to avoid mode competition, and so to operate in a single mode. In the A6 magnetron, with only six cavities, the differences in frequency and azimuthal phase velocity, ω_n/n, are sufficiently large that mode competition can be avoided. Under different circumstances, for example, when more resonator cavities are added, more complex configurations, such as the *rising-sun* geometry (Figure 7.9),[18] create greater frequency separation between the π-mode and other modes, preventing mode competition*. We also note, as pointed out in the reference, that any configuration with alternating cavity characteristics constitutes a rising-sun magnetron, including coupling the output from every other cavity, even if the cavities are nominally of equal length.

Figure 7.10 is the dispersion relation for a rising-sun magnetron of the type shown in Figure 7.9.[13] For this device, $r_c = 7.3$ cm, $r_a = 9.7$ cm, $r_1 = 12.5$ cm, and $r_2 = 16.2$ cm; the 14 cavities all have the same angular width of about 10.3°. The dashed line is the dispersion curve for a 14-cavity magnetron with equal cavity depths of $r_d = (r_1 + r_2)/2 = 14.35$ cm. Note the effects of alternating the depths of the cavities. First, the periodicity of the structure is changed, with $N = 7$ identical pairs of cavities, rather than $N = 14$ identical individual cavities. Second, the $n = 7$ mode can now be regarded as either the π-mode of a 14-cavity set (with the fields reversing from one cavity to the next, without regard to its depth) or the 2π-mode of a 7-cavity set (with the fields in phase between cavities of the same depth). Third, the dispersion curve for the $N = 14$ magnetron is now split into two curves for the rising-

---

* Strapping, while a useful practice in conventional magnetrons, may well be impractical at high power due to breakdown problems.

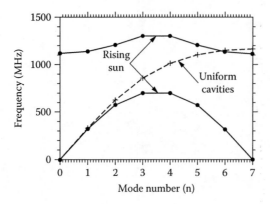

FIGURE 7.10
Dispersion relation for a rising-sun magnetron of the type shown in Figure 7.9, with the dimensions given in the text. (From Lemke, R.W. et al., *Phys. Plasmas*, 7, 706, 2000. With permission.)

sun magnetron; as $r_1 \rightarrow r_2$, the two rising-sun curves coalesce to form the single N = 14 curve. Simulations showed the upper n = 7 mode to provide the most efficient microwave generation. There is mode competition from the upper n = 6 mode; at large values of r_2/r_1, competition between the lower n = 3 and 4 modes becomes an issue also. In addition to this mode competition, there is also a zero-azimuthal-harmonic issue, which we will discuss in Section 7.5, that limits the output power.

7.3.2 Operating Voltage and Magnetic Field

As we stated earlier, for a given voltage, the applied magnetic field must be large enough to form the initial electron sheath around the cathode without allowing breakdown of the gap, but not so large that the rotational velocity of the electrons is too slow to allow resonance with the modes of the resonator formed by the coupled interaction space and anode cavities, as indicated by Equations 7.1 and 7.2.

As we discussed in Section 4.7, the critical magnetic field to prevent breakdown of the anode–cathode gap in the magnetron is the so-called Hull field (see Problem 8), which in the case of a purely axial field is given by[19]

$$B^* = \frac{mc}{ed_e}\left(\gamma^2 - 1\right)^{1/2} \qquad (7.5)$$

where m and –e are the mass and charge of an electron, c is the speed of light,

$$\gamma = 1 + \frac{eV}{mc^2} = 1 + \frac{V(kV)}{0.511} \qquad (7.6)$$

with V the anode–cathode voltage, and d_e the effective gap in cylindrical geometry,

$$d_e = \frac{r_a^2 - r_c^2}{2r_a} \tag{7.7}$$

When $B_z > B^*$, the electrons drift azimuthally within a bounded electron cloud called the Brillouin layer. As $B_z \rightarrow B^*$, the electron cloud extends almost to the anode. As the magnetic field increases, the cloud is confined closer and closer to the cathode. Remarkably, even though Equation 7.5 takes account of collective space-charge and diamagnetic effects from the electron layer, it is exactly the result one would obtain if one set the Larmor radius for a single electron equal to d_e.

In a magnetron, the electric field due to the applied voltage is predominantly radial, E_{r0}. The magnetic field is the sum of contributions from the externally applied axial field, modified by the diamagnetic effects of the azimuthal flow of electrons, and azimuthal components created by axial current flow in the cathode and radial current flow from the spokes. Current flows axially into the magnetron to support both axial current flow off the opposite end of the cathode to the wall at anode potential and radial current flow carried by the spokes in the anode-cathode gap. When flow off the end of the cathode is minimized, and when the azimuthal field from the axial current is relatively small, the magnetic field is predominantly axial. According to the left side of Equation 7.2, the drift velocity of an electron in this case is thus largely azimuthal, approximately given by E_{r0}/B_z, which decreases as B_z increases. Therefore, for a given voltage, there is some maximum B_z above which the electrons will be too slow to achieve resonance with the phase velocity of the electromagnetic wave supported by the anode structure. The relationship between the voltage and magnetic field at this threshold is called the *Buneman–Hartree condition*, which for a relativistic magnetron with an axial magnetic field, ignoring axial geometrical variations, is given by[19]

$$\frac{eV}{mc^2} = \frac{eB_z\omega_n}{mc^2 n} r_a d_e - 1 + \sqrt{1 - \left(\frac{r_a\omega_n}{cn}\right)^2} \tag{7.8}$$

In this expression, ω_n is the frequency in radians per second of the nth azimuthal mode. For a given voltage, this expression gives the maximum B_z for which high power output is expected; alternatively, it is the minimum V, given B_z, for high power operation (see Problems 9 and 10).

When the field from the axial current flow I_z becomes significant, the Buneman–Hartree condition for long cathodes is modified to

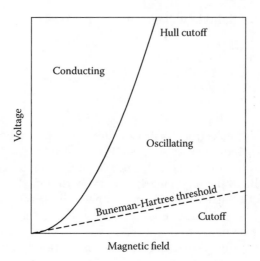

FIGURE 7.11
Magnetron operating domains in B_z-V parameter space.

$$\frac{eV}{mc^2} = \frac{eB_z\omega_n}{mc^2 n} r_a d_e - 1 + \sqrt{\left[1 + b_\phi^2\right]\left[1 - \left(\frac{r_a\omega_n}{cn}\right)^2\right]}$$ (7.9)

where

$$b_\phi = \frac{I_z(kA)}{8.5} \ell n\left(\frac{r_a}{r_c}\right)$$ (7.10)

This expression could also be modified to take account of contributions to b_ϕ in the event that the magnetic field component due to radial current flow became comparable to B_z in magnitude, although that contribution to the field will vary axially.

Figure 7.11 shows the operating domain of the magnetron in B_z-V parameter space. For a given voltage, if the magnetic field is too weak, electrons are allowed to cross the gap and breakdown occurs; if too strong, the oscillations are cut off. The space between the Hull cutoff requirement in Equation 7.5 and the Buneman–Hartree condition of Equation 7.8 or 7.9 is the region of oscillation. Efficient operation usually occurs just to the left of the Buneman–Hartree threshold.

7.3.3 Characteristics of Magnetrons

In operation, the interaction between the electrons and the microwave fields is highly nonlinear. The important parameters, which can be either determined empirically or predicted using computer simulation, are (1) the actual operating frequency; (2) the output power P; (3) the efficiency η, relating the pulsed power electrical input, $P_{PP} = VI$, to the microwave output, P; and (4) the parameter relating the operating current I to the voltage V.

To understand these and other magnetron parameters, consider Table 7.2, with a representative set of parameter values for operating relativistic magnetrons.[14,16,20] Note that most of the magnetrons in the table are truly relativistic, with operating voltages above 500 kV. The operating current spans a broad range, with associated impedances running from just over 10 Ω to several hundred ohms. Similarly, the applied magnetic fields span a broad range, from 0.2 T to about 1 T or slightly higher. Several of the devices generate output power at the level of several gigawatts. Efficiencies of about 30%, or perhaps more, can be reached. Some of the magnetrons are inverted, with $r_a < r_c$. The NRL magnetron is an outlier with respect to size, being much larger than the other devices.

The actual operating frequency will be somewhat lower than the cold-structure frequency discussed in Section 7.3.1. A good example is the MIT A6 magnetron,[8] for which the calculated and measured cold frequencies for the π- and 2π-modes were compared to the hot frequency with electrons, as shown in Table 7.3. The agreement between the calculated and measured cold frequencies is better than 3%, consistent with the fact that the anode length of 7.2 cm is somewhat longer than the free-space wavelength of the 2π-mode, and more than half that of the π-mode, so that the underlying assumption of an infinitely long anode holds.

The microwave output power P is proportional to the electrical power into the device, which is the product of the voltage, V, and the current, I, with the efficiency η the proportionality constant:

$$P = \eta VI \tag{7.11}$$

The impedance Z is the ratio $Z = V/I$, so that one can write

$$P = \eta \frac{V^2}{Z} \tag{7.12}$$

(See Problems 1 and 12.)

Experimental evidence from four different magnetrons operating over different frequency ranges between 1.07 and 3.23 GHz, at voltages between 220 and 500 kV, indicates that P scales with magnetron voltage V as V^m, with m lying between 2.46 ± 0.35 for one device and 2.74 ± 0.41 for another.[21] Thus, for nonrelativistic voltages V < 500 kV, if η is approximately constant, Equation 7.11 indicates that in those magnetrons

$$I \approx KV^{3/2}, V < 500 \text{ kV} \tag{7.13a}$$

TABLE 7.2

Selection of Experimental Results for Magnetrons

	f (GHz)	V (MV)	I (kA)	Z = V/I (Ω)	B (T)	P (GW)	t (nsec)	E (J)	η (%)	N	Mode
MIT A6[a]	4.6	0.8	14	57	1.0	0.5	30	15	4	6	2π
MIT D6[a]	4.1	0.3	25	12	0.49	0.2	30	6	3	6	2p
MIT K8[a]	2.5	1.4	22	64	0.54	0.05	30	1.5	0.2	8	p
LLNL[a]	3.9	0.9	16	56	1.6	4.5	16	72	30	6	2π
SRINP 1[b]	2.4	1.1	4	275	1.2	2.0	50	100	45	6	p
SRINP 2[b]	2.4	0.45	6	75	0.4	0.8	300	240	30	6	p
IAP 1[b]	9.1	0.6	7	86	0.6	0.5	20	10	12	8	p
IAP 2[b]	9.2	0.95	40	24	1.0	4.0	15	60	11	8	p
PI[c]	1.1	0.8	34	24	0.85	3.6	10	36	13	6	p
PI[b]	2.8	0.7	20	35	1.0	3.0	20	60	20	6	p
PI[b]	8.3	1.0	30	33	1.0	0.3	10	3	0.1	6	p
Inverted Magnetrons											
MIT M8[a]	3.7	1.7	4	425	0.46	0.4	30	12	12	8	p
MIT M10[a]	3.7	1.6	9	178	0.7	0.5	30	15	3	10	p
NRL[a]	3.2	0.6	5	120	0.25	0.5	30	15	17	54	p
SNL[a]	3.1	1.0	10	100	0.3	0.25	50	12	3	12	p

Note: Data are from the Massachusetts Institute of Technology (MIT), Lawrence Livermore National Laboratory (LLNL), Scientific-Research Institute of Nuclear Physics in Tomsk, Russia (SRINP), Institute of Applied Physics in Nizhny Novgorod, Russia (IAP), Physics International (PI; now L3 Communications Pulse Sciences), Naval Research Laboratory (NRL), and Sandia National Laboratories (SNL). The parameters in the columns are, from left, the frequency, voltage, current, impedance, magnetic field, output power, pulse length, energy per pulse, efficiency, number of cavities in the anode, and operating mode.

[a] Reference 14.
[b] Reference 20.
[c] Reference 16.

TABLE 7.3

Comparison of the Calculated and Measured Cold
Frequencies for the π- and 2π-Modes of the A6
Magnetron to the Hot Frequency with Electrons[8]

	π-Mode	2π-Mode
Calculated cold frequency (GHz)	2.34	4.60
Measured cold frequency (GHz)	2.41	4.55
Measured hot frequency (GHz)	2.3	4.6

$$P \approx \eta KV^{5/2}, \; V < 500 \text{ kV} \qquad (7.13b)$$

which is the form we would expect for a nonrelativistic Child–Langmuir
diode (see Section 4.6.1), with K known as the perveance. This result is
surprising in that the electron flow in magnetrons is strongly influenced by
the magnetic field, whereas the flow in a Child–Langmuir diode, which has
no axial magnetic field, is assumed to be purely radial. The data suggest that
the radial current transport in a magnetron enabled by the microwave fields
is limited by the space charge in the spokes.

Let us look more closely at the current flow in a magnetron, which occurs
despite the fact that the axial magnetic field lies above the critical Hull field
for insulation given by Equation 7.5. To see this, compare the normalized
current-B_z plot for the smooth-bore magnetron, which has a constant-radius
anode with no cavities, of Figure 7.12[22] with the plots of current and voltage
vs. B_z for the A6 in Figure 7.13.[8] With a smooth anode, the microwave signal
level within the anode–cathode gap is very small and the current, which is
normalized to the Child–Langmuir value for $B_z = 0$, drops to a negligible
level as the magnetic field is increased to B^*. The solid curve in Figure 7.12
is a theoretical value, while the data points are drawn from configurations
with a number of different gap values, $r_a - r_c$. Qualitatively, the drop in the
solid curve and the data as B_z increases is the result of increased travel time
across the gap caused by the curving of the electron trajectories by the
magnetic field. The deviation of the data from the theory is largely the result
of the fact that an axial current flows in the cathode to support the radial
current, and this axial current creates an azimuthal magnetic field component
that modifies the total magnetic field.

The plot of current vs. B_z in Figure 7.13 for an A6 magnetron with anode
cavities shows the effect of the strong microwave fields in the device. As the
magnetic field increases above $B^* = 4.9$ kG (0.49 T) for the given voltage and
gap, the current does not drop to zero, but rather declines continuously with
B_z. This is not to say that the constraint $B_z > B^*$ is irrelevant in a magnetron
or CFA, but rather that, in operation, current is carried by the nonequilibrium
spoke structures.

The plot of power vs. B_z for the A6 shown in Figure 7.14 shows how
sharply the power peaks at the optimum B_z of about 0.76 T, where the output

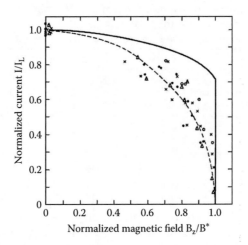

FIGURE 7.12
Current normalized to the Child–Langmuir value for $B_z = 0$, I_L, in a smooth-bore magnetron. The solid curve is a theoretical calculation ignoring the current flow in the cathode, and the data points were taken for a number of different anode–cathode gaps. The dashed curve is a quadratic fit to the data. (From Orzechowski, T.J. and Bekefi, G., *Phys. Fluids*, 22, 978, 1979. With permission.)

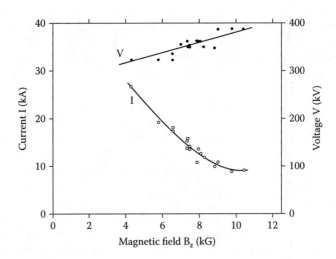

FIGURE 7.13
Current and voltage vs. B_z in the MIT A6 magnetron. (From Palevsky, A. and Bekefi, G., *Phys. Fluids*, 22, 986, 1979. With permission.)

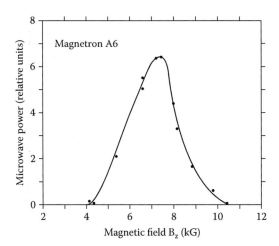

FIGURE 7.14
Output power vs. B_z in the MIT A6 magnetron. (From Palevsky, A. and Bekefi, G., *Phys. Fluids*, 22, 986, 1979. With permission.)

is about 900 MW. The range of operating parameters' regime for the A6 is shown against the Hull and Buneman–Hartree curves in Figure 7.15. The magnetic fields approximately span the region between the Hull cutoff criterion of Equation 7.5 and the Buneman–Hartree condition of Equation 7.8 for the 2π-mode, and the peak power occurs close to the Buneman–Hartree condition. Computer simulations of the A6 indicate that, for the lower-frequency π-mode, as B_z increases, the most efficient operating voltage shifts upward from the Buneman–Hartree voltage.[13] This is because the sheath is too narrow at the Buneman–Hartree voltage and couples poorly to the microwaves; at higher voltages, the sheath becomes thicker and couples more efficiently.

7.3.4 Summary of Magnetron Design Principles

One can scope out many of the basic parameters of a magnetron using the expressions in this chapter. With the dispersion relation derived using the procedure in the reference, we can find the expected operating frequency for a desired operating mode, given the cathode and anode radii and the geometrical parameters of the cavities. With the mode number and frequency in hand, we can find the operating region in V-B_z parameter space using the Hull insulating and Buneman–Hartree criteria. The optimal voltage for a given magnetic field will be close to, but above, the Buneman–Hartree voltage, rising farther above that voltage as the magnetic field increases. If we wish to predict the impedance (or, at lower voltages, perveance), efficiency, and output power in advance of an experiment, we must either work empirically, perhaps making small excursions from known designs and measuring

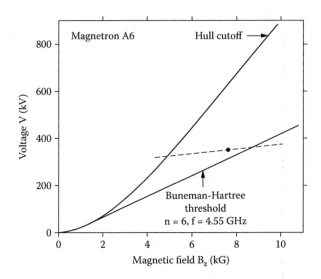

FIGURE 7.15
Buneman–Hartree diagram for the 2π-mode (N = 6) in the MIT A6 magnetron. The solid dot is the peak power operating point, and voltages at which maximum power is achieved at a given magnetic field between about 0.4 and 1.0 T (4 to 10 kG) are shown along the dashed curve. (From Palevsky, A. and Bekefi, G., *Phys. Fluids*, 22, 986, 1979. With permission.)

the resulting currents, or employ a simulation that is normalized to experiment. Nevertheless, we know that the impedance will be within a factor of two of about 100 Ω, and an efficiency of perhaps 20 to 30% can be attained, from which we can predict an expected output power.

7.4 Operational Features

Table 7.4 compares the typical operating parameters for conventional and relativistic magnetrons. Conventional pulsed, high power magnetrons typically operate at power levels up to about 5 MW, although a 60-MW rising-sun magnetron was built with conventional technology.[23] Voltages and currents in 5-MW conventional magnetrons are typically 50 kV or less and about 200 A or slightly more, so that impedance Z ~ 200 to 250 Ω and perveance* K ~ 20 × 10^{-6} A/$V^{3/2}$; the 60-MW magnetron, by comparison, with a secondary-emission dispenser cathode, operated at about 120 kV and 800 A, so that Z $\approx$ 150 Ω and the perveance was about the same as other conventional magnetrons. Referring back to Table 7.2, voltages in relativistic magnetrons are about an order of magnitude higher, and currents are one to two orders

---

* Perveance is many times expressed in units of pervs, with dimensions of A/$V^{3/2}$, or micropervs (= 10^{-6} pervs).

TABLE 7.4

Typical Operating Parameters for Conventional and Relativistic Magnetrons

Parameter	Conventional Magnetrons	Relativistic Magnetrons
Voltage	100 kV	500 kV
Current	~100 A	5–10 kA
Cathode process	Thermionic and secondary emission	Explosive emission
Pulse duration	~1 μsec	~100 nsec
Rise time	200 kV/μsec	~100 kV/nsec
Power	10 MW	~1 GW
Efficiency	~50%	~20–30%

of magnitude higher, with impedance ranging more broadly from 30 Ω upward, which at the lower end matches well with pulsed power machines. Besides the order of magnitude difference in operating voltage, the key difference is the much larger operating currents of relativistic magnetrons, which are enabled by the explosive emission mechanism from cold cathodes. Unfortunately, explosive emission creates a cathode plasma, and the motion of this plasma causes gap closure at velocities typically greater than 1 cm/μsec, limiting pulse lengths to something on the order of 100 nsec in the absence of special measures. In conventional magnetrons, on the other hand, pulse lengths are typically limited by anode heating. Consistent with the higher operating voltages and currents, relativistic magnetrons operate around 1 GW, about 100 times higher than conventional magnetrons, even given their lower efficiencies.

In the previous section, we discussed the theoretical design considerations in choosing, for example, the dimensions of the cathode and anode structure and the operating voltage and magnetic field. Here, we discuss the operational considerations for several situations: single fixed-frequency magnetrons, tunable magnetrons, repetitively operated magnetrons, and, for magnetron-based facilities, notably Orion, a transportable container-based facility for outdoor microwave effects studies.

7.4.1 Fixed-Frequency Magnetrons

The cross section of a typical relativistic magnetron is shown in Figure 7.16.[24] The magnet coils, along with the high-voltage electrical feed from the bottom of the figure, dominate the volume and mass of the magnetron itself. Experimental magnetrons have employed pulsed field coils, but water-cooled or superconducting magnets are needed in repetitively fired magnetrons operating at high average powers. Operation with permanent magnets might be possible, but there are no publications to date on this.

Operationally, the cathode must be built to promote the process of explosive emission, with the rapid and uniform production of the surface plasma that provides the high current densities — of the order of kA/cm^2 — required for space-charge-limited current flow in the radial direction. In addition, the

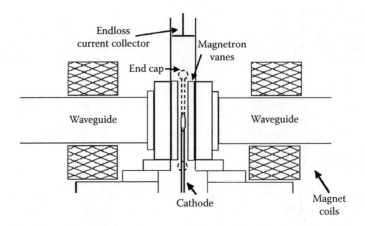

FIGURE 7.16
Cross section of a typical relativistic magnetron. The power feed is at the bottom of the picture.
The cylindrical axis of the magnetron runs along the cathode stalk, and the magnet coils encircle
the axis. (From Lopez, M.R. et al., *IEEE Trans. Plasma Sci.*, 30, 947, 2002. With permission.)

cathode must be designed to maximize the ratio of radial to axial current
flow, since axial electron flow does not contribute to microwave generation.
A number of cathode materials and configurations have been explored; these
materials are placed on the cathode stalk at the slight bulge shown in the
center of the magnetron in Figure 7.16. Placing a washer, or washers, with
its sharp edges in the mid-section of the resonator increased power by as
much as an order of magnitude, as was originally shown by Craig et al.[25]
and quantified by Benford et al.[17] More recently, Saveliev et al. conducted a
systematic investigation of magnetron cathodes comparing the performance
of (1) smooth cylindrical cathodes with a variety of surfaces — velvet covered
by a stainless steel mesh, carbon fabric, a stainless steel mesh, or plain stainless steel; (2) carbon fiber discs; and (3) pins.[26] They found substantial variation in relativistic magnetron performance among the different cathodes. In
terms of peak power and energy per pulse, the cathode material that pro-
moted the most uniform emission — velvet covered by a stainless steel mesh
— was the best performer. Unfortunately, velvet cathodes evidenced considerable damage after 1000 shots, even though performance was still acceptable. The authors judged that stainless steel mesh without the velvet
provided a better compromise between lifetime, power, and energy per pulse.
    Although Phelps eliminated axial end loss in an early experiment by
feeding electrical power to the cathode from both ends,[27] the more common
approach has become the use of the *end caps*, shown in Figure 7.16. Note
from the figure that the end caps are rounded, to reduce the electric fields
at their surfaces. In experiments, the end caps may be coated with glyptal,
a substance that inhibits electron emission; unfortunately, glyptal coatings
are easily damaged and must be replaced rather frequently. Saveliev et al.
showed that end caps almost doubled the efficiency, to 24% in an L-band

rising-sun magnetron operating at over 500 MW.[26] Lopez et al. similarly found that the use of end caps reduced the axial current loss from 80% of the total magnetron current to as little as 12% in a variant of the A6 magnetron[24] with an aluminum cathode coated with glyptal, except at the mid-section of the resonator, where 1-mm carbon fibers were glued to the cathode. Double end caps play additional desirable roles, defining the mode structure, suppressing axial mode competition, and maximizing the electric field in the mid-section of the resonator. As a final point, remember that axial current flow cannot be eliminated, since the axial current flow completes the circuit for the radially directed current in the spokes and since that current flow creates an azimuthal magnetic field and an axial $E_r \times B_\theta$ electron drift; however, axial flow off the end of the cathode, beyond the circuit including radial spoke flow, is a loss and should be suppressed.

Extraction of microwaves from relativistic magnetrons employs two rather different techniques. The most commonly used is *radial extraction* from the ends of one or more of the anode cavities into the fundamental mode of the output waveguide. Sze et al.[28] measured the effect of extraction through multiple ports. In their experiments, frequency did not depend upon the number of extraction ports and mode shifting did not occur. However, there was an optimum number of waveguides. Figure 7.17 shows that the maximum power from the A6 magnetron was obtained most easily with three waveguides, although six gave the same power. Power per waveguide declined steadily as the number of waveguides increased to four, and then was roughly constant. The cavity Q was reduced as more openings were

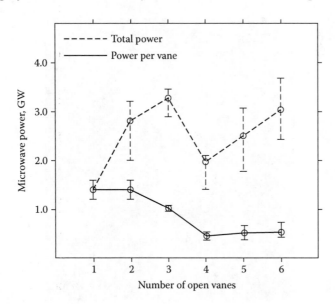

FIGURE 7.17
Power extracted from a six-vane A6 magnetron at Physics International as a function of the number of vanes. (From Sze, H. et al., *IEEE Trans. Plasma Sci.*, PS-15, 327, 1987. With permission.)

made in the resonator. Radio frequency (RF) field declined with Q, and the trade-off between cavity RF field and output power was optimized with three output resonators. We note that extraction from every other vane effectively creates a rising-sun configuration.[18]

Axial extraction of microwave power has also been used with two variations. Ballard et al. extracted from a magnetron with inverted polarity by using a downstream mode converter.[29] Kovalev et al.[30] tapered the magnetron resonator structure in the axial direction to merge with an output waveguide using a method called *diffraction coupling* in Russian works, making a gradual transition from the RF impedance of the magnetron to free-space impedance, as shown in Figure 7.18. In this 9.1-GHz magnetron with an eight-cavity anode, they claimed an output power in excess of 500 MW at 13 to 15% efficiency. Benford and Sze directly compared axial and radial coupling in another X-band magnetron.[31] Their six-vane magnetron was about two vacuum wavelengths long and suffered from axial mode competition. Using radial coupling, they were able to obtain reproducible operation when coupling from only one cavity; with coupling from more cavities, the power level was unstable. With axial coupling using a technique similar to that of Kovalev et al., they were able to raise the efficiency from less than 1% with radial coupling to more than 2%, with stable output power in excess of 350 MW. Figure 7.18 shows the anode of their magnetron, with the vane structure preserved in the flaring axial extraction region.

The radiation patterns for the radial and axial extraction methods differ substantially because each radiates different modes. Radial extraction in

FIGURE 7.18
Axial extraction from a tapered resonator.

rectangular waveguide produces TE_{10} radiation patterns. The power distribution is smooth and linearly polarized. Axial extraction gives TM_{np}, where n is the azimuthal mode number and p is the radial index. Such patterns all have central nulls and therefore are unattractive. Mode converters might be used to convert to a more useful pattern.

Figure 7.16 does not show the diagnostics that would be used in experimental devices and perhaps in some more mature systems. Total current into a magnetron is diagnosed with a Rogowski coil placed in the outer wall encircling the cathode stalk before it enters the magnetron resonator. A second Rogowski coil in the other wall downstream of the cathode end cap would be used to measure axial current loss. Directional couplers are used for power measurements, heterodyne and homodyne systems for frequency and bandwidth, and phase discriminators for phase relations. Close et al. measured the electric field in an evacuated waveguide extracted from the A6 by observing Stark broadening of the hydrogen Balmer line.[32] The result was 100 kV/cm, precisely that deduced from power measurements. Smith et al. measured the phase difference between adjacent resonators directly using a phase discriminator.[33]

7.4.2   Tunable Magnetrons

Mechanically tunable magnetrons require the physical modification of one of the key design parameters. In an initial exploration of mechanical tuning, Spang et al.[34] showed that the variation of the cathode radius of a variant of the A6 design allowed a rather broad frequency range of operation: operating in the 2π-mode, cathode radii varying from about 0.54 to 1.52 cm resulted in frequencies between about 4 and 4.6 GHz, a tunable range of 7% about the 4.3-GHz center frequency. At 4.6 GHz, the peak power was 700 MW with an efficiency of 15%. The π-mode frequency also varied slightly with cathode radius, but over a much reduced range (see Problem 11). Unfortunately, the actual replacement of the cathode was required in these experiments.

The adoption of a rising-sun geometry enabled the construction of a magnetron that can be tuned from shot to shot in a repetitively fired system. Prior to the development of tunable magnetrons, several groups investigated narrowband, high power rising-sun magnetrons. Treado et al. operated a 14-cavity rising-sun magnetron at 60 MW based on conventional technology.[23] When injection locked to a lower-power magnetron to stabilize the operating frequency, the 60-MW magnetron had a bandwidth of about 800 kHz, full width/half maximum (FWHM) about the center frequency of 2.845 GHz. Lemke et al. simulated the operation of 14- and 22-cavity relativistic rising-sun magnetrons.[18]

Levine et al. developed a family of high power, mechanically tunable, 10-resonator rising-sun magnetrons operating in the L- and S-bands.[35] The remarkable anode structure, shown in Figure 7.19, was made from a single piece of stainless steel by electron discharge machining. Without the joints

FIGURE 7.19
Anode structure for the mechanically tunable rising-sun magnetron. (Courtesy of L-3 Communications Pulse Sciences, formerly Physics International.)

and cracks of a device assembled from separate pieces, and using ceramic insulators and cryopumps in the vacuum system, the magnetron can be pumped down to 10^{-8} torr after it has been cleaned and vacuum baked. Note that the vanes are arranged in five closely spaced pairs. Tuning is effected by moving a sliding short in the narrow cavities with parallel walls, varying the depth of those cavities; that variation in depth, which changes the frequency of the device, requires a corresponding adjustment of the magnetic field. In the figure, we can see the circular holes in the outside of the anode block, through which the pins that push the tuning shorts pass. The tuning range for the L-band magnetron is approximately 24% about a center frequency of 1.21 GHz, while that for the S-band magnetron is 33% about a center frequency of 2.82 GHz. One of two extraction ports is seen in the top of the anode block in Figure 7.19. Extraction is thus from cavities with nonparallel walls. Arter and Eastwood's simulations show that the output power remains approximately the same even when additional extraction ports are added.[36]

7.4.3   Repetitive High-Average-Power Magnetrons

The original single-shot experiments with relativistic magnetrons were conducted with the aim of exploring basic physics and engineering issues. In shifting to repetitive operation, one must address several additional engineering and physics issues. On the engineering end, one must provide adequate prime power and the requisite magnetic field on either a continuous or pulsed basis. Further, one must design the pulsed power to provide high voltage and current at the required switching rates. With regard to prime power, we note that the important parameters are the duty factor for the pulsed power, D_{PP}, and the overall system efficiency from prime power to microwave output, η_{SYS}. The duty factor is

$$D_{PP} = R\tau_{PP} \qquad (7.14)$$

where R is the repetition rate and τ_{PP} is the pulse length for the pulsed power on each shot. Note that Equation 7.14 involves τ_{PP}, rather that the microwave pulse length τ, highlighting the waste created by pulse shortening, which reduces the length of the output pulse relative to τ_{PP}. The overall system efficiency is a product of the efficiencies for the prime and pulsed power, microwave generation and extraction, and the antenna. For a radiated power P, the required prime power is

$$P_{PRIME} = \eta_{SYS}D_{PP}P = \eta_{SYS}\ R\tau_{PP}\ P \qquad (7.15)$$

Given that η_{SYS} is the product of several efficiencies, even for magnetron efficiencies of 30 to 40%, the system efficiency could be a few percent.

From the standpoint of magnetron physics, two primary issues for repetitive operation are diode recovery and electrode erosion. The process of explosive emission and the deposition of electron energy in material surfaces release plasma and neutral particles that must clear the system between pulses. Systems must be clean, operating at high vacuum, to limit the surface contaminants that facilitate plasma formation, explaining the use of single-piece anodes, cryopumps (with sufficient pumping speed), and ceramic insulators in the previous subsection.

The second major issue in repetitive operation is electrode erosion. Pin and fiber-enhanced cathodes with longer lifetimes have been developed, but anode erosion is an intrinsically challenging issue in magnetrons, since electron deposition must occur within the cavity (see Problem 7). At high average power levels, water cooling of the anode and the collection surface for axial electron loss is required. To see this, consider the example of anode erosion, since most erosion is observed on the anode. A rough estimate of the magnitude of the erosion problem is that the peak dose for 700-keV electron deposition in most refractory metals is 6 (cal/g)/(cal/cm²). Using S-band magnetron parameters as an example, a single 60-nsec electrical pulse for which 5 kA of electron current strikes 50 cm² of anode produces a surface heat increase of 6 cal/g. Neglecting conduction and radiation, we estimate that the heat of fusion will be exceeded in about 10 shots; ablation will occur thereafter.

Two U.S. industrial firms, GA[37] and AAI,[34] operated magnetrons at 1- to 2-Hz repetition rates for long periods, generating thousands of pulses. The latter used water-cooled vanes, a repetitively fired Marx bank, and pulsed conventional magnets. Intended for operation at a 10-Hz rate, it was limited to 2 Hz by unexpectedly high power demand in the magnets due to eddy current losses in the magnetron structure. In their 4.6-GHz A6 variant operating in the 2π-mode, there was noticeable anode vane erosion at 5000 shots, but no degradation in performance. The metal cathode, however, had to be

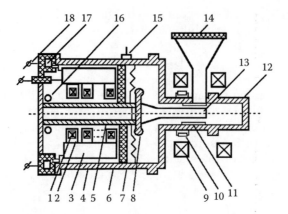

FIGURE 7.20

Schematic of the Tomsk system for driving a magnetron with a linear induction accelerator: (1) ferromagnetic core, (2) magnetizing turn, (3) strip DFL, (4) housing, (5) cathode, (6) helical winding, (7) insulator, (8) protecting shield, (9) magnet system, (10) shunt for operating current of the magnetron, (11) anode unit of the relativistic magnetron, (12) drift tube, (13) cathode, (14) antenna for outlet of microwave radiation, (15) terminal for demagnetizing, (16) total current Rogowski loop, (17) multichannel discharge, and (18) input of high voltage for the DFL charge. (From Vasil'yev, V.V. et al., *Sov. Tech. Phys. Lett.*, 13, 762, 1987. With permission.)

replaced at 1000 to 1500 shots, and the spark gaps in the pulsed power had to be cleaned after 2000 to 5000 shots.

Compact linear induction accelerators developed at the Scientific-Research Institute of Nuclear Physics of the Tomsk Polytechnic Institute in Russia[38] and refined at Physics International (now L-3 Communications Pulse Sciences) in the U.S.[39] provide the necessary power at repetition rates that have been pushed to 1000 Hz in five-pulse bursts (limited by the stored energy available). The system is compact because high voltage is generated by inductively adding relatively lower voltages along a coaxial line; peak voltage, with its volume and mass demands, appears only on the load. Longer lifetime is provided by using magnetic, rather than spark-gap, switches operating at lower voltages. The Tomsk system is shown schematically in Figure 7.20[38]; a photo of the device in Figure 7.21 shows that it is quite compact. The repetition rate is 50 Hz, generating a peak microwave power of 360 MW in the magnetron from a 4-kA, 300-kV, 60-nsec electrical pulse. The ability to operate at higher rates was proven by a three-pulse burst at 160 Hz, achieved by firing during the flat period of the pulsed magnetic field coil. The average power was reported to be about 1 kW.

Five years after the first Tomsk publication, the Physics International group built CLIA (Compact Linear Induction Accelerator) to drive a six-vane, 1.1-GHz magnetron[16]; a 100-nsec version of CLIA without a magnetron is shown in Figure 7.22. The L-band magnetron had produced 3.6 GW in single pulses with a Marx bank/waterline driver, but power was reduced on CLIA due to the mismatch of the 25-Ω magnetron to the 75-Ω output impedance of the driver. The magnetron was modified with the addition of cooling channels

FIGURE 7.21
The Tomsk linear induction accelerator and magnetron system.

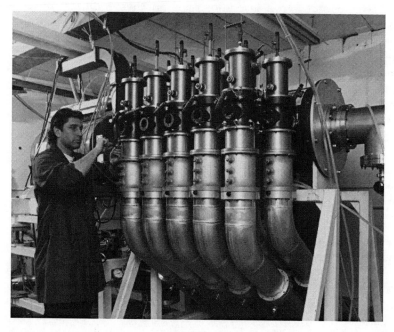

FIGURE 7.22
Inductive adder with 10 pulse-forming lines for the 100-nsec Compact Linear Induction Accelerator. (Courtesy of David Price, L-3 Communications Pulse Sciences.)

TABLE 7.5

Summary of Magnetron Peak and Average Power
Performance on the Compact Linear Induction
Accelerator (CLIA)

Repetition Rate (Hz)	Peak Power (MW)	Average Power (kW)	Number of Shots
100	1000	4.4	50
200	700	6.0	100
250	600	6.3	100
1000[a]	600	25	5

[a] Based on the third shot in a five-shot sequence.

3 mm below the surface and a cryopump to eliminate possible contamination with backstreaming pump oil (base pressure was 4×10^{-6} torr). The performance at four different repetition rates is summarized in Table 7.5. Note that even though the peak power declined with increasing repetition rate, the average power climbed. Note also that even though the 1000-Hz burst was limited to five shots, this repetition rate far exceeded the average-current capability of the system. Pulse-to-pulse reproducibility on this system was remarkable, as was the fact that there was no pulse shortening in this lower-frequency, larger magnetron (r_c and r_a of 1.27 and 3.18 cm vs. 1.3 and 2.1 in the 2.3-GHz π-mode of the A6).

The L-band magnetron also illustrates another empirical observation: anode erosion appears to worsen with increasing frequency, probably due to the reduced anode dimensions and roughly constant electrical current. There was no noticeable erosion on the larger, lower-frequency L-band magnetron after several hundred shots on single-shot generators and several thousand on CLIA; erosion was noticeable on the 4.6-GHz magnetron of Spang et al.[34] after 5000 shots, although performance was unimpaired at that point; and erosion was severe in the X-band magnetrons of Kovalev et al.[15] and Benford and Sze.[31]

7.4.4 Magnetron-Based Testing Systems: MTD-1 and Orion

From the standpoints of size and flexibility, the Mobile Test Device-1 (MTD-1) of AAI[34] and the Orion transportable test facility of Physics International[40] represent the two extremes of magnetron-based testing systems. MTD-1 was designed to minimize size and weight; the core of the system is depicted in Figure 7.23.[34] The pulsed power and 700-MW, A6 variant magnetron fit in a $182 \times 91 \times 182$ cm^3 package, and the auxiliary controls, high-voltage power supplies, and connector panel fit within another $61 \times 71 \times 152$ cm^3 volume. The mass of the system, including heat exchanger and prime power, was 1500 kg, and with a trailer the total was 2200 kg. Power could be radiated from a horn antenna, a lens antenna with a 23-dB gain, or a dielectric rod antenna with a 32-dB gain.

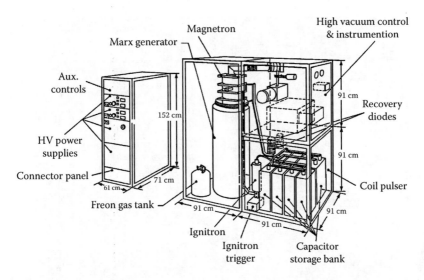

FIGURE 7.23
Diagram of the Mobile Test Device-1 (MTD-1) from AAI Corporation. (From Spang, S.T. et al., *IEEE Trans. Plasma Sci.*, PS-18, 586, 1990. With permission.)

Orion, also described in Section 5.7, was designed to provide a full, trans-portable microwave effects field testing facility in five ISO containers: (1) prime power, including station keeping, and air conditioning; (2) a source container, connected at right angles to the (3) operations container; (4) a control container; and (5) a storage container. The layout of all but the storage con-tainer in a testing configuration is shown in Figure 7.24.[40] The pulse length of the output from the pulsed power system is adjustable in 50-nsec increments between 100 and 500 nsec, with reduced voltage output at the longer pulse length. At the heart of the system, the insertion of one of four mechanically tunable rising-sun magnetrons allows the system to cover the frequency range from 1 to 3 GHz. Each magnetron produces 400 to 800 MW of output power and can be fired repetitively at a rate of 100 Hz in a 1000-pulse burst. Micro-waves are radiated from a parabolic antenna that illuminates a 7×15 m², 3-dB elliptical spot at a 100-m range, and a waveguide combiner/attenuator allows the power to the antenna to be varied over five orders of magnitude.

7.5 Research and Development Issues

The key operational parameters for a relativistic magnetron are:

- Peak power
- Efficiency

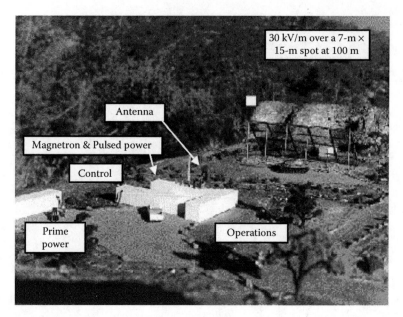

FIGURE 7.24
The Orion system in the field (storage container not shown). The truck in front of the control container provides a sense of scale. (Courtesy of L-3 Communications Pulse Sciences.)

- Pulse duration or energy per pulse
- Pulse repetition rate or average power
- Magnetron lifetime

Within this set, the area of most active research and development is that of pulse duration or, as it is most commonly referred to, *pulse shortening*, stemming from the common observation in all high power microwave (HPM) sources that at the highest power levels, the microwave output pulse is shorter than that of the current and voltage pulses to the magnetron. Further, as the output power increases, the pulse duration decreases.

The issues underlying the *achievable pulse repetition rate* are a combination of physics issues in the source and technological constraints in the prime and pulsed power. In the case of the physics issues, the phenomena are similar to those involved in pulse shortening: plasma is evolved from the electrodes that must then be cleared before the next shot can be fired. On the other hand, an example of a technological constraint is the amount of prime power available, a limitation imposed by either the power feed to a facility or, for a mobile system, volume and mass. A second constraint is the achievable repetition rate of the pulsed power output switch at the required voltage and current.

Maximizing *power* and *efficiency* has also been an active area of research in the past, as has the issue of magnetron *lifetime*. Power, efficiency, and lifetime are clearly interrelated. Maximizing efficiency for a given power helps to lengthen lifetime, since as more energy is extracted from the electrons to make microwaves, less is deposited in the anode, reducing heating and damage. High efficiencies also reduce the demand on the prime power and on high-stress pulsed power components. As we can see in Table 7.2, a number of groups have operated individual relativistic magnetrons at powers of several gigawatts in both the π- and 2π-modes. Efficiencies of some tens of percent in the L- and S-bands have been achieved, with a peak of around 40%. The issues underlying the maximization of peak power and efficiency are a combination of fundamental limitations and open issues that would profit from further research. We defer to Section 7.6 a discussion of the role of fundamentally limiting values for the electric and magnetic fields, which are relatively well understood, as is the scaling with frequency and the need to limit axial current loss. On the other hand, two research issues that we will discuss here are phase locking of multiple sources and extraction.

7.5.1 Pulse Shortening

Unfortunately, progress achieved in radiating higher peak power levels in all classes of HPM sources has been tempered by the pulse-shortening problem.[41] The phenomenon in the case at hand, relativistic magnetrons, is shown in Figure 7.25.[42] The microwave power falls with frequency as $P \sim f^2$ and with pulse length τ as $P \sim \tau^{-5/3}$. This limits radiated energies to a few hundred joules. The implication is clear: to get longer pulses, one must accept less energy per pulse.

Why does pulse shortening matter? Consistent with our earlier notation, let η be the power efficiency, the ratio of instantaneous microwave power P

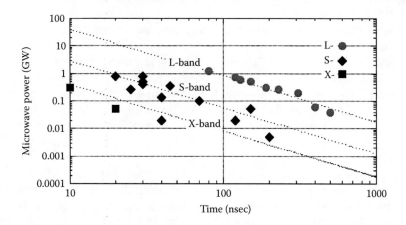

FIGURE 7.25
Pulse length in relativistic magnetrons with explosive emission cathodes decreases as the peak output power increases. Data are from magnetron experiments conducted over the past two decades. All scaling curves are normalized to the L-band, 80 nsec, 1.2-GW point. (From Price, D. et al., *IEEE Trans. Plasma Sci.*, 26, 348, 1998. With permission.)

to electrical pulsed power input $P_{PP} = VI$, and η_E be the energy efficiency, the ratio of microwave energy in the pulse E to the pulsed power electrical energy that drives the beam, E_{PP}. Then, with τ the pulse duration of microwaves and τ_{PP} the pulse duration of the pulsed power voltage and current to the magnetron, the relations are

$$E = P\tau \tag{7.16}$$

$$E_{PP} = P_{PP}\tau_{PP} \tag{7.17}$$

Therefore,

$$\frac{\eta_E}{\eta} = \frac{\tau}{\tau_{PP}} \tag{7.18}$$

where τ/τ_{PP} is the pulse-shortening ratio of the energy efficiency to the power efficiency. Today, high values for the magnetron are $\eta \sim 20$ to 40%, but pulse shortening makes $\eta_E < 10\%$. Returning to a point made in Section 7.4.3, the HPM system designer is concerned with the energy that must be supplied by the prime power and stored in pulsed power because energy drives HPM system volume and mass. Therefore, η_E is frequently more meaningful than η; the distinction between them is the pulse-shortening ratio.

Models to account for pulse shortening in relativistic magnetrons focus on the radial motion of the cathode plasma. The dense, conducting cathode plasma shields the actual cathode and makes the effective cathode radius a time-varying quantity dependent on the radial motion of the plasma surface, $r_c(t)$. Recall that the following all depend on the cathode radius, r_c: the Hull insulation condition (Equation 7.5), the dispersion relation from which we determine the frequencies of the magnetron's different modes of operation (see Section 4.4.2), and the Buneman–Hartree relation (Equation 7.8 or 7.9), which bounds the operating space in V-B_z space and acts as a good approximation for the operating values of these parameters. Gap closure can cause frequency hopping, as the system jumps from one mode to another; the termination of the microwave pulse as the system falls out of resonance; or the remarkable behavior seen in Figure 7.26,[42] in which microwave emission begins when the system dynamically slips into resonance as the effective cathode radius, $r_c(t)$, reaches a value at which the dispersion relation setting the mode frequency and the Buneman–Hartree relation are both satisfied for the given voltage and magnetic field. Note that as the magnetic field is increased in the figure, the microwave pulse turns on later and later in the voltage pulse (and typically terminates before the end of the voltage pulse). While difficult to quantify, we can show that this behavior is consistent with the Buneman–Hartree relation, which we recount from Equation 7.8, making use of Equation 7.7:

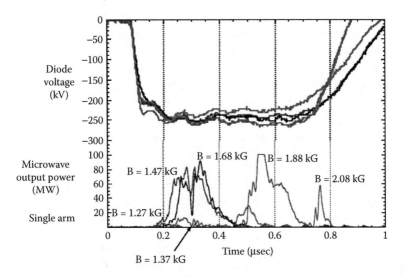

FIGURE 7.26
Observed microwave output pulses for different values of the applied magnetic field (lower curves) and the corresponding voltage pulses on a relativistic magnetron. Modes appearing well below 1.58 kG, at 1.27 and 1.37 kG, are not the fundamental π-mode. (From Price, D. et al., *IEEE Trans. Plasma Sci.*, 26, 348, 1998. With permission.)

$$\frac{eV}{mc^2} = \frac{eB_z\omega_n}{2mc^2n}\left(r_a^2 - r_c^2\right) - 1 + \sqrt{1 - \left(\frac{r_a\omega_n}{cn}\right)^2} \qquad (7.19)$$

In the figure, we focus on the magnetic field values above the two lowest, 1.27 and 1.37 kG, since at those two field values, the magnetron apparently operates in a mode other than the π-mode. At all higher fields, we know that the frequency for the π-mode in which the magnetron operates is quite insensitive to the cathode radius; at cathode radii appropriate to the magnetron in question, the π-mode frequency drops from 1.207 to 1.199 GHz, a difference of only 8 MHz, when the cathode radius increases from 1.778 to 2.078 cm. Therefore, if we assume that $\omega_n = \omega_\pi$ is constant, then the product $B_z(r_a^2 - r_c^2)$ must be constant from one shot to the next as well. Thus, for successively larger values of B_z, Equation 7.19 is satisfied at later and later times, when the expanding cathode plasma makes r_c large enough that the product $B_z(r_a^2 - r_c^2)$ reaches the approximately constant value. Based on the cathode plasma expansion rate inferred from an analysis of the data in the figure, the main plasma ion constituent was inferred to be hydrogen.

The model of expanding cathode plasma suggests three approaches to extending the microwave pulse duration:

- Reduce the plasma temperature by decreasing the cathode current density that heats the plasma. To do this, designers can produce either large devices with large cathode areas that operate at moderate impedance or higher-impedance designs with larger anode–cathode gaps that require high voltage (>1 MV) in order to obtain high power.

- Eliminate the plasma, which requires the development of high-current-density cathodes that do not rely on explosive emission, most likely field emission cathodes without plasma formation. This is an active research topic.

- Increase the plasma ion mass, using heavier cathode materials to increase pulse energy. A magnetron experiment to reduce the pulse-shortening mechanism by heating the resonator to remove low Z contaminants successfully lengthened the pulse from 100 to 175 nsec. A substantial improvement in the Orion magnetrons' long-pulse performance was achieved with carbon–fiber cathode material with cesium iodide surface coating.[43]

Efforts are continuing to eliminate pulse shortening by the above methods.

Just as the changing gap spacing can terminate a pulse, the time-changing voltage provided by pulsed power machines can be problematic. A voltage flatness requirement can be obtained from Figure 7.13, the tuning curve. In scans of the magnetic field at fixed voltage, as is typically done in experiments, a $\Delta B/B$ bandwidth of ±10% or greater is observed. A comparable bandwidth in voltage at a fixed magnetic field is therefore expected. Treado showed that a lack of voltage flatness truncated the microwave pulse in a long-pulse modulator experiment.[14,44] For conventional magnetrons with their much higher impedances, the voltage flatness requirement is a major determinant of the modulator design. The same is likely to occur in relativistic magnetrons as the field matures.

To summarize, evidence indicates that the magnetron pulse duration is determined by gap closure — as confirmed by the remarkable data in Figure 7.26, which shows a magnetron sliding into resonance at different magnetic fields — and the flatness, or lack of flatness, in the voltage pulse. Changes in either the gap or the voltage can take a magnetron out of resonance and terminate a pulse even before the electrical impedance of the device collapses, as has been observed.

7.5.2 Peak Power: Phase-Locking Multiple Sources and Transparent-Cathode Magnetrons

Although the compactness of magnetrons and CFAs is desirable from a packaging standpoint, it becomes a liability as the power, and power density, grows, with gap closure and anode erosion increasingly problematic, particularly at higher frequencies. An obvious solution is to employ multiple sources, either in a master oscillator/power amplifier array with amplifiers such as CFAs or in a phase-locked array of oscillators such as magnetrons. The problems with the former are the lack of a high power CFA and the relatively low gain demonstrated by conventional CFAs, typically of the order of 20 dB, which necessitates the use of a high power master oscillator. For this reason, the focus to date has been phase locking of magnetrons (see Section 4.10), both in the weak-coupling [$\rho = (P_i/P_o)^{1/2} \ll 1$, with P_i and P_o the input and output powers for the magnetron being locked] and strong-coupling ($\rho \sim 1$) regimes.

Chen et al.[45] at MIT and Treado et al.[23] at Varian (now CPI) have explored phase, or injection, locking in the weak-coupling regime. The MIT group drove a 30-MW, 300-nsec relativistic magnetron with a 1-MW conventional device. The locking time was rather long because ρ and locking time are inversely related; however, it does potentially allow one conventional magnetron to drive several relativistic magnetrons, producing a fixed phase relation between groups of high power devices. Chen also studied the effects of nonlinear frequency shift in magnetrons by adding a Duffing term (with a cubic restoring force) to the van der Pol treatment of phase locking.[46] Treado and coworkers locked a 60-MW magnetron based on conventional technology to a 3-MW magnetron.[23]

The Physics International team explored the strong-coupling regime, connecting several magnetrons directly through short waveguides, as shown in Figure 7.27.[47] Operation in this regime minimizes locking time and facilitates phase locking of oscillators with larger natural frequency differences. This latter allows tolerance for oscillator differences in an array.[48] Calculations predict locking in nanoseconds. In experiments with $\rho = 0.6$, phase locking at 1.5 GW occurred in about 10 nsec, as shown in the phase discriminator output in Figure 7.28.[47] Because the phase difference is reproducible from shot to shot, this is an example of phase locking, not merely frequency locking. This is a very practical feature: phase adjustments in the feed of an array to correct for shot-to-shot variations in phase differences would not be possible. The experiments used a second diagnostic to demonstrate locking. Pulses from the two magnetrons were radiated from two antennas onto power density diagnostics in an anechoic room, and a fourfold increase in power density (relative to that of a single magnetron) was observed.

Woo et al.[48] derived a requirement on the length of the waveguide connecting the oscillators:

$$\Delta\omega \leq 2\ \omega\rho\ |\cos \omega\ \tau_c|\ /Q \qquad (7.20)$$

where τ_c is the transit time of the wave in the connecting guide. This relation is an extension of the Adler condition. For $\omega\tau_c = n\pi/2$ with n odd, phase locking cannot occur. Thus, some connecting lengths prevent phase locking. Phase difference can occur in two modes, with a phase difference of either 0 or π between two oscillators. In the experiments of Benford et al.,[47] the latter case is observed (Figure 7.28).

The Physics International group also studied different geometries for phase-locked arrays.[49] They found that the best geometry uses maximum connectivity (Figure 7.29). Calculated locking time for these arrays appears

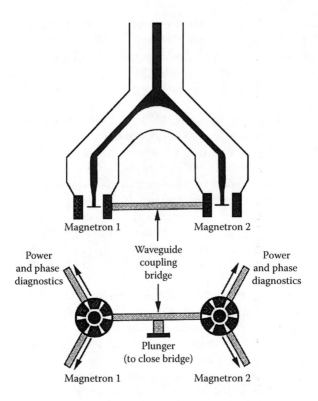

FIGURE 7.27
Two-magnetron phase-locking experiment. (From Benford, J. et al., *Phys. Rev. Lett.*, 62, 969, 1989. With permission.)

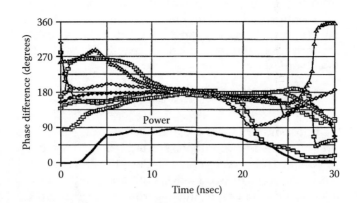

FIGURE 7.28
Phase difference for seven consecutive shots of the two-magnetron experiment. The power pulse shape is shown at the bottom. (From Benford, J. et al., *Phys. Rev. Lett.*, 62, 969, 1989. With permission.)

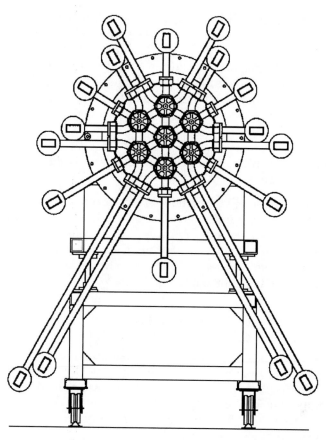

FIGURE 7.29

Seven-magnetron module at Physics International, now L-3 Communications Pulse Sciences. The configuration shown produced the best phase locking among all possibilities considered. (From Levine, J. et al., *Proc. SPIE*, 1226, 60, 1990. With permission.)

to be about that of the two-magnetron experiment. Phase locking was achieved in a seven-magnetron array producing a total output of 3 GW.

Fuks and Schamiloglu have proposed a completely different approach to improving the power and efficiency of magnetrons, the so-called *transparent-cathode magnetron*, in which the solid cathode is replaced by a set of parallel, axial rods, arrayed with equal azimuthal spacing at the radius of a normal cathode.[50] This cathode eliminates the usual boundary condition that the azimuthal component of the RF electric field vanish at the cathode. Thus, the magnitude of the azimuthal electric field component can be larger, which leads to stronger spoke formation, higher power, and higher efficiency. Comparison of simulation results for two versions of the A6 magnetron, one with a solid cathode and the other with a cut cathode consisting of 18 axial rods, shows that while the solid-cathode version produces higher power in the 2π-mode at an operating voltage of 350 kV (0.97 vs. 0.55 GW), the efficiency of the transparent-cathode version is significantly higher (25.5 vs. 15%)

because of the higher current of the solid-cathode version, which has four times the cathode surface area of the new version. Operating at even higher voltages (and an appropriately increased magnetic field), simulations of the transparent-cathode magnetron indicate that it can be operated in the π-mode at 1.5 MV and produce over 9 GW. Because the computer simulations are 2-dimensional, experiments will have to be performed to confirm these encouraging results.

7.5.3 Efficiency: Limiting Axial Current Loss and Radial vs. Axial Extraction

The efficiency we defined in Equation 7.11 in fact combines several efficiencies, all of which are multiplicative in effect: (1) the efficiency of microwave generation, η_μ; (2) the efficiency of power extraction from the device, η_{ex}; (3) the mode conversion efficiency, η_{mc}, if a mode converter is used between the magnetron and the antenna; and (4) the combined efficiency of coupling to, and radiating from, the antenna, η_a. Therefore,

$$\eta = \eta_\mu \eta_{ex} \eta_{mc} \eta_a \qquad (7.21)$$

where we set $\eta_{mc} = 1$ in the absence of a mode converter.

Consider η_μ first. Because microwaves are generated only by the radial component of the current, axial current loss reduces the efficiency of microwave generation; in fact, defining $\eta_e = P/(VI_r)$, where I_r is the radial portion of the total current $I = I_r + I_z$, and I_z is the axial portion of the current, we can easily show that (see Problem 13)

$$\eta_\mu = \eta_e \left(\frac{I_r}{I} \right) \qquad (7.22)$$

Now, axial current flow arises from parasitic losses off the end of the cathode and from axial drifts that flush electrons from the interaction region. Unfortunately, this loss mechanism is self-enhancing, since axial currents increase the azimuthal component of the magnetic field, which in concert with the radial electric field increases the axial $E_r \times B_\theta$ drift of electrons out of the interaction region. The enhanced drop in output power with axial current flow is shown in Figure 7.30.[17] Research into the use of end caps, which in reality are rounded, as shown in Figure 7.16, to reduce the electric fields and corresponding field emission losses there, has minimized parasitic end losses from the end of the cathode. Nevertheless, the intrinsically high operating currents of relativistic magnetrons will always create some B_θ and resulting axial electron loss.

Inverted magnetrons, with the anode on the inside and the cathode on the outside, eliminate the energy loss due to axial current flow because the axially

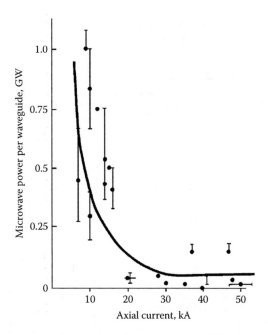

FIGURE 7.30
Reduction in power with increasing axial current loss. (From Benford, J. et al., *IEEE Trans. Plasma Sci.*, PS-13, 538, 1985. With permission.)

flowing electrons return to the cathode. Inverted magnetrons were originally studied at MIT,[32] Stanford,[29] and NRL.[51] In general, all groups reported an overall reduction in output power. The MIT group proposed that their radial extraction method, which required extracted radiation to propagate through the electron space-charge layer, fundamentally limited the efficiency. Ballard proposed that the inverted resonator does not generate RF fields as high as in the normal geometry.[52] The Tomsk group, observing that the length of a voltage pulse in a coaxial diode with inverted polarity is three to four times that of an ordinary diode at equal electric fields, operated an inverted magnetron for almost the entire pulse length of the applied voltage[53] (Figure 7.31). During the 900-nsec pulse, unfortunately the magnetron moved from one mode to another as the voltage changed, in a manner consistent with gap closure. A frequency variation of 30% during the pulse was reported. Consequently, this experiment, although producing the longest pulse yet observed, points to the problematic relationship between impedance and voltage control in high power devices and long-pulse, single-frequency operation.

Efficiency, particularly the efficiency of microwave generation, η_μ and its electronic component η_e, is an important factor in device lifetime. Electron kinetic energy not converted into microwave energy by electrons crossing the gap is deposited in the anode. In contrast to the conventional magnetron, the relativistic magnetron's high-voltage electrons have penetration depths of a few millimeters. Damage occurs in the interior of the vanes and the cathode

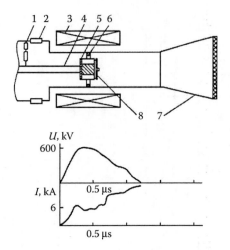

FIGURE 7.31
Long pulse from an inverted magnetron. The elements of the device are the (1) voltage divider, (2) current shunt, (3) solenoid, (4) anode shank, (5) anode, (6) cathode, (7) antenna, and (8) output coupler. (From Vintizenko, I.I. et al., *Sov. Tech. Phys. Lett.*, 13, 256, 1987. With permission.)

and can result in melting and spallation of the surface. Spallation is nonuniform because the rotation of the spokes causes the damage to occur preferentially at one side of the entrance of each resonator, determined by the direction of B_z. Damage is localized on the mid-plane of the resonator. The radial dimensions of a magnetron scale inversely as the frequency f, and the corresponding area over which the energy is deposited scales inversely with f^2, so there is a frequency element to magnetron lifetime as well. Returning to a point made earlier, anode erosion on a large L-band magnetron was not noticeable after thousands of shots, while a C-band magnetron (4.6 GHz) evidenced anode erosion after about 5000 shots, and erosion was severe in X-band magnetrons. To date, specialized materials, such as Elkonite™, tungsten, and other materials that have produced substantial lifetime improvements in high power electrodes in the electrical industry, have not been developed for HPM.

7.6   Fundamental Limitations

In this section, we consider the fundamental limitations of three magnetron parameters: power, efficiency, and frequency.

7.6.1   Power Limits

The output power from a single magnetron is limited fundamentally by a time-dependent analog of the Hull condition, which is defined in the static case in Equation 7.5. Both the RF magnetic field and the external magnetic field are oriented axially, so that insulation fails during a portion of the RF cycle when the peak RF magnetic field cancels the external magnetic field, reducing the total magnetic field to less than B*, the insulating field. In this event, electron loss is enhanced, and the lost electrons strike the anode without producing RF power.

Lemke et al.[18] addressed this issue using computer simulations of rising-sun magnetrons. For a rising-sun magnetron, or for any magnetron with an even number of resonators N, there is a problem with *0-harmonic contamination*, which we explain by noting that the periodicity of the anode structure implies that the electromagnetic fields in the cavity be expanded in a series of the form*

$$B_{zn} = \sum_{m=-\infty}^{\infty} A_m F_m(r) e^{iM(m,n)\theta} e^{-i\omega t}$$ (7.23)

where A_m is the amplitude of the mth space harmonic, $F_m(r)$ is a function describing the radial dependence for the space harmonics, and for a rising-sun magnetron operating in the nth mode,

$$M(m,n) = n + m\left(\frac{N}{2}\right)$$ (7.24)

The expansion in Equation 7.23 is in a series of spatial harmonics, and when $m = -2n/N$, with N even so that m is an integer (and N is always even for a rising-sun configuration), that particular spatial harmonic has no variation in θ, just like the applied magnetic field insulating the gap. Thus, on one half cycle it reinforces the insulation, and on the other it cancels it. In the reference, the authors consider an N = 14 magnetron and show how the strength of the 0-harmonic contribution depends on the ratio of the depths of the alternating resonators. Further, they show how efficiency declines as the strength of the 0-harmonic contribution grows. Figure 7.32 shows the V-B_z parameter space for the magnetron in the simulations; the operating point for peak power is marked by a solid dot, and the bar through it illustrates the amplitude of the 0-harmonic component of the RF magnetic field. Note that this component of the RF field drives the device below the Hull insulation criterion. This fact is reflected in Figure 7.33, which shows the deposited power from electrons striking the anode; note that the spoke power is disproportionately larger on the half cycle where the 0-harmonic

---

* This is the application of Floquet's theorem to a system with azimuthal symmetry, just as the slow-wave structures of the Cerenkov devices in Chapter 8 are described using Floquet's theorem for systems with axial symmetry.

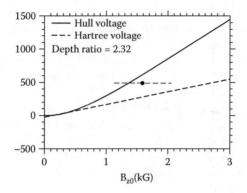

FIGURE 7.32
Operating space for the π-mode (n = 7) in a rising-sun magnetron with N = 14, r_c = 7.3 cm, r_{a1} = 12.5, r_{a2} = 16.2 cm, and cathode length L_c = 20.8 cm, with an operating frequency of 1.052 GHz. (From Lemke, R.W. et al., *Phys. Plasmas*, 7, 706, 2000. With permission.)

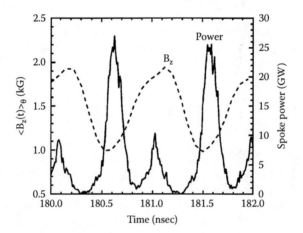

FIGURE 7.33
The spoke power incident on the anode (solid curve) and B_z averaged over θ, which recovers both the static applied field and the time-dependent 0-harmonic component. (From Lemke, R.W. et al., *Phys. Plasmas*, 7, 706, 2000. With permission.)

of the RF magnetic field cancels the applied magnetic field, thus showing the effect of losing insulation for that brief period.

We estimate the limiting condition on the RF magnetic field magnitude B_{RF} as follows:

$$\Delta B_{RF} \leq \frac{1}{2}\left(B^M - B^*\right) \tag{7.25}$$

where B^M is the applied axial magnetic field for maximum power output, as illustrated by the dot in Figure 7.32, slightly less than the Buneman–Hartree field B_{BH}. Since the total power available, proportional to B_{RF}, is proportional to the square of the separation between the Buneman–Hartree and Hull fields, this condition indicates that high power operation occurs at the high-field end of the Buneman–Hartree condition, i.e., high B_z and voltage. Palevsky and Bekefi[8] and Ballard[52] have shown that the A6 magnetron operating in the 2π-mode is limited by this condition to an RF electric field of 750 kV/cm, or an operating power of 3 GW.

Young[28] has derived a non-linear magnetron model that predicts a saturated power level of

$$P_{sat} = \frac{\omega_n}{Q_L} W_{NL} \left(1 - X\right) \qquad (7.26)$$

where W_{NL} is the saturation level of wave energy and $X = \omega_n/2Q_L\Omega$, where Q_L is the loaded Q of the magnetron and Ω is the instability growth rate, taken by Young to be the magnetron instability growth rate:

$$\Omega = \frac{\omega_c}{16} \qquad (7.27)$$

with ω_c the cyclotron frequency. W_{NL} can be expressed in terms of the RF magnetic field energy integrated over the volume of the interaction region, and thus Young's analysis can be compared directly with predictions of Equation 7.25. Sze and coworkers compared the above analysis with A6 performance[28] and found good agreement between Equation 7.26 and experimental results. Note that Equation 7.26 implies an optimum at $Q_L = \omega_n/\Omega$. No experimental work to optimize Q values has been attempted other than Sze's multiple-waveguide extraction experiment. Because $P \sim (B_{RF})^2 V_{res}$, where V_{res} is the resonator volume, $P \sim \lambda^2$, which has been verified experimentally.[16]

Because there is an apparent upper limit on the RF power from a single magnetron, and because pulse duration limits are not yet fully mastered, the total energy presently extracted from single relativistic magnetrons is typically tens of joules, with the exception of the one experiment of the Tomsk group in the Soviet Union,[53] which reported 140 J. Since many applications will require large radiated energies, magnetron research will focus on extension of the pulse duration and on grouping magnetrons together to radiate in phase.

7.6.2 Efficiency Limits

Conventional magnetron efficiency increased steadily from its original development at low efficiencies to contemporary 80 to 90% efficiency in optimized designs. The relativistic magnetron at this phase of its development has achieved values of a few tens of percent. Values as high as 40% have been reported. Because the total extracted power is a product of multiple factors, as shown in Equations 7.21 and 7.22, most notably the electronic efficiency, the effectiveness in minimizing the axial current, and the efficiencies of extraction, mode conversion (if included), and radiation from the antenna, it is difficult to isolate the limiting physical factors.

Ballard has estimated an upper limit on the electronic efficiency η_e.[29] In arriving at the anode, the portion of the DC potential that has not been converted into RF energy appears as kinetic energy deposited in the anode block. Therefore,

$$\eta_e = 1 - \frac{\left(<\gamma_a>-1\right)mc^2}{eV} \tag{7.28}$$

with $<\gamma_a>$ the average relativistic factor for electrons striking the anode. Slater provided a conservative, nonrelativistic estimate of the kinetic energy of the electrons striking the anode by assuming that in the worst case their velocity on impact was the sum of the $E \times B$ drift velocity and a Larmor rotation velocity component aligned with it,[54] arriving at

$$\eta_e^{nr} = 1 - \frac{2V^*}{A^{*2}} \tag{7.29}$$

Ballard generalized that expression to relativistic energies, estimating $<\gamma_a>$ to find

$$\eta_e = 1 - \frac{2V^*}{A^{*2} - V^{*2}} \tag{7.30}$$

where

$$V^* = \frac{eV}{mc^2}, \quad A^* = \frac{eB_z}{mc}d_e \tag{7.31}$$

Comparing Equations 7.29 and 7.30 shows that the maximum possible efficiency is always larger in the nonrelativistic case.

Ballard's expression, Equation 7.30, is compared to the experimental data[14] in Figure 7.34. Clearly, experimental efficiencies are lower than the theoretical maximum, typically by a factor of two or more. Within the magnetron itself, one of the major issues is the axial loss of electrons, as described in Equation 7.22. Loss from the cathode tip can be minimized, but when the axial current in the magnetron is large, there is a substantial $E_r \times B_\theta$ drift in the axial direction that flushes electrons from the resonator before they can produce

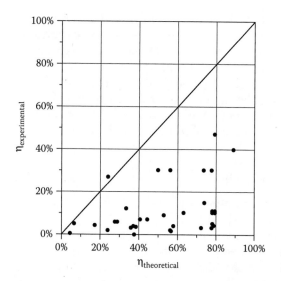

FIGURE 7.34

Comparison of theoretical efficiency, from Equation 7.30, and experimentally observed efficiencies. (From Treado, T., Generation and Diagnosis of Long- and Short-Pulse, High-Power Microwave Radiation from Relativistic Rising-Sun and A6 Magnetrons, Ph.D. thesis, University of North Carolina, Chapel Hill, 1989. With permission.)

high power. The condition that guarantees that electrons have sufficient time to interact with the anode structure before drifting axially out of the anode block is

$$\tau_\theta \leq \tau_z \tag{7.32}$$

with τ_θ the circumferential transit time and τ_z the longitudinal transit time. Since both drifts are due to $E \times B$ forces, this relation becomes, in practical units,

$$I(kA) \leq 0.08 B_z(T) h(cm) \tag{7.33}$$

with h the axial length of the magnetron. For example, in the A6 magnetron, at 0.45 T, the predicted threshold is 24 kA, and the threshold observed by Benford et al.[17] is 15 to 20 kA. The rapid drop in power with increasing axial current at fixed magnetic field shown in Figure 7.30 illustrates the effect. The result is that there is effectively a lower bound on the impedance at which a relativistic magnetron can operate. Combining Equation 7.33 with Equation 7.8 requires $Z > 16\ \Omega$ for an S-band magnetron. Lovelace and Young derived Equation 7.9 to account for axial current flow, and limited experimental data indicate that this condition is followed for high power operation.[17]

Additional explanations for the lower efficiencies seen in Figure 7.34 include two others we have already discussed: possibly suboptimal extraction methods, a topic that has not yet been fully explored, and downstream efficiency issues with antenna coupling and, if included, mode conversion. In addition, we speculate that nonlinear interactions with electrode plasmas dissipate energy, although these interactions are not yet understood. Finally, resonators may not be optimally designed at this stage, just as conventional magnetrons were nonoptimal in their initial embodiments.

7.6.3 Frequency Limits

The two sets of experiments with X-band magnetrons indicate that magnetrons do not scale well to higher frequencies.[30,31] In the experiments, the efficiency was relatively poor, and given the small size of the device, anode damage severely limited the lifetime. Another drawback is the scaling of the magnetic field with wavelength or frequency, which pushes the magnetic field requirement up. To see this, consider the Buneman–Hartree relation, which we rewrite in the form

$$V_{BH} = \frac{\pi B_{BH} r_a^2}{n\lambda}\left(1 - \frac{r_c^2}{r_a^2}\right) - \frac{mc^2}{e}\left(1 - \sqrt{1 - \beta_{ph}^2}\right) \tag{7.34}$$

where $v_{ph} = r_a \omega_n / n$ is the azimuthal rotation velocity of the operating mode of the magnetron and $\beta_{ph} = v_{ph}/c$. Provided the far right term in parentheses is small compared to the first term, and given that π/n is of order unity, we can write

$$V_{BH} \sim \frac{B_{BH} r_a^2}{\lambda}\left(1 - \frac{r_c^2}{r_a^2}\right) \tag{7.35}$$

Therefore, for constant voltage and radial aspect ratio,

$$\frac{B_{BH} r_a^2}{\lambda} \sim \text{constant} \tag{7.36}$$

Now, since $r_a \sim \lambda$, $B_{BH} \propto 1/\lambda$, so for a fixed voltage, higher magnetic fields are required at high frequencies, while the anode block becomes smaller. Gap alignment becomes more difficult, and heating and anode erosion become more substantial. The X-band radiator is already quite small (see Figure 7.8), making extrapolation to shorter wavelengths doubtful.

7.7 MILOs

The *magnetically insulated line oscillator* (MILO) is essentially a linear magnetron that takes advantage of a property of very high power transmission lines: at high voltages, electrons are emitted from the cathode side of the line, but the current carried creates a sufficiently strong magnetic field that these electrons cannot cross the gap. Thus, these lines are self-insulated, and the electrons drift axially in the crossed electric and self-magnetic fields.

In the MILO, a slow-wave structure is added to the anode side of the line, so that the axially drifting electrons in the layer adjacent to the cathode side interact with the axially directed slow electromagnetic wave to generate microwaves. MILOs can have either coaxial or planar geometry. In the cylindrical form, the electric field is radial, the magnetic field is azimuthal, and electron drift is axial. For magnetic insulation to occur, the axial current I_z must exceed the so-called *parapotential current*:[55]

$$I_z > I_p\left(A\right) = 8500\gamma G \ell n\left(\gamma + \sqrt{\gamma^2 - 1}\right) \qquad (7.37)$$

where γ is as defined in Equation 7.6 and G is a geometric factor equal to $[\ell n(b\,/\,a)]^{-1}$ for coaxial cylinders of inner and outer radii b and a, and equal to $W/2\pi d$ for parallel plates of width W separated by gap d (see Problem 14). Because I_p is roughly proportional to γ, the logarithmic term on the right of Equation 7.37 being rather weak, the upper limit on the impedance of a magnetically insulated line is almost independent of voltage: $Z_p = V/I_p \approx 52G^{-1}$ Ω for voltages around 1 MV. A MILO must have a load impedance, such as a diode, in order to establish insulation; if the load current is too small, enough current will flow across the line upstream of the load to reach the value in Equation 7.37. In this regard, Equation 7.37 is the MILO's version of the Hull condition in Equation 7.5 for magnetrons.

Because the magnetic and electric fields both arise from the pulsed power source, their ratio is also a weak function of voltage. The axial drift velocity of electrons is

$$\frac{v_d}{c} = \frac{Z_p}{Z_0} \qquad (7.38)$$

where Z_0 is the transmission line impedance in the absence of magnetically insulated electron flow (see Problems 15 and 16). Because the drift velocity is proportional to Z_p, which is almost independent of voltage, the MILO naturally operates at constant drift velocity, and therefore has none of the tuning requirements of the magnetron. The magnetic field automatically

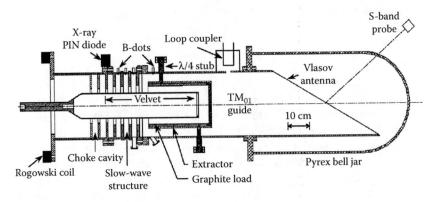

FIGURE 7.35
The cylindrical Hard-Tube MILO of Haworth et al. (From Haworth, M.D. et al., *IEEE Trans. Plasma Sci.*, 26, 312, 1998. With permission.)

adjusts to changing drive voltage, maintaining constant drift velocity.*
Another key practical aspect of the MILO is that the insulating magnetic field is created without the aid of external coils, allowing the elimination of the magnets, their power supply, and any cooling required, and facilitating designs with large cross-sectional areas.

MILO development has proceeded along two major lines. One group is working to optimize a design such as that shown in Figure 7.35,[56] which has been called the *Hard-Tube MILO* (HTMILO). Within this MILO, electrons are emitted from the velvet-covered cathode of 11.43 cm diameter; the velvet speeds the turn-on of the MILO by reducing the emission field at the cathode and making emission more uniform. The outer conductor on the anode side, which comprises the base of the device cavities and the outer wall of the extractor section, has a diameter of 28.575 cm. The role of the two leftmost choke cavities, which are bounded by three vanes of inner diameter 15.24 cm and are thus somewhat deeper than those of the MILO section to their right, is to reflect microwaves back into the MILO section, rather than letting them travel upstream into the pulsed power. Additionally, the impedance mismatch between the choke and MILO sections creates a higher downstream current. The vanes in the four-cavity slow-wave structure have an inner diameter of 17.145 cm, except for the rightmost vane, which functions as a transition into the coaxial extractor section; this vane has an inner diameter of 17.780 cm. The period of all vanes is 3.81 cm. The 28.575-cm outer conductor and a 20.955-cm-diameter inner conductor bound the coaxial extraction section. The velvet-covered cathode and the graphite-coated inner surface of the extractor section form the 500-kV, 60-kA magnetically insulated transmission line (MITL) that carries the insulating current. Microwaves are extracted first into a TM_{01} waveguide and then out through a Vlasov antenna and Pyrex bell–jar radome that supports a base vacuum in

---

* Because of space-charge forces, there is radial variation in v_d, so lack of tuning sensitivity is only true to the first order, but is observed experimentally.

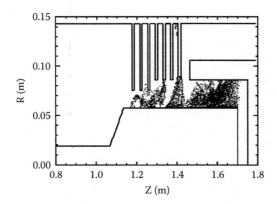

FIGURE 7.36
Electron spoke formation in the HTMILO from simulations with the 2-dimensional particle-in-cell code TWOQUICK. (From Haworth, M.D. et al., *IEEE Trans. Plasma Sci.*, 26, 312, 1998. With permission.)

the device of about 3×10^{-6} torr. The Rogowski coil measures current into the device from the 500-kV, 50-Ω, 300-nsec pulsed power system, the x-ray PIN diodes detect the bremsstrahlung radiation from energetic electrons lost on the vanes, the B-dot probes are used to measure the current at different locations, and the S-band probe is used because its reduced coupling to the 1.2-GHz L-band output is required by the high power levels to which it is exposed. Computer simulation of the HTMILO[56] (Figure 7.36) shows spoke formation characteristic of π-mode operation, which is to be expected in devices that are essentially linear magnetrons.

The process by which this group has improved upon its MILO design illustrates the range of improvements, from obvious to subtle, required to iteratively optimize a device. The device in Figure 7.35 incorporated several design improvements over a 1.5-GW, 70-nsec design described 3 years earlier[57] that were meant to improve both power and pulse length: (1) finger-stock RF joints in the earlier device were replaced with all-stainless-steel brazed construction; (2) the supports of the extractor section were quarter-wavelength stubs, rather than support rods making gravity contact with the extractor; and (3) the operating base pressure was reduced by an order of magnitude. Unfortunately, operation with these design changes initially resulted in no improvement in performance. Combining experimental observations and computer simulation, though, the group made several additional changes. Observation of current loss in the choke section led to the most important change: moving the left end of the cathode taper in Figure 7.35 to the right until it was located under the transition from the choke section to the MILO section. This improvement, in combination with the three others made at the outset, resulted in an increase in power to almost 2 GW and in pulse length to about 175 nsec, which was the length of the available power pulse, after factoring in the turn-on time of the device. However, experiments with the improved device showed that the pulse length and power were not acceptably repeatable, and further computer simulation of the improved design with a shorter cathode showed that the problem was breakdown on the quarter-wavelength supporting stubs for the extractor and loss at the MILO end of the extractor. Further simulations pointed to a new design for the extractor supports and the velvet–metallic interface at the ends of the cathode; a shortening of the length of the cathode within the MITL load region to the minimum necessary to support the magnetic insulation current; a decrease in the inner diameter of the coaxial extractor to 19.30 cm; and an increase in the gap between the last MILO vane and the entrance to the coaxial region. In the absence of experiment, simulations pointed to a further power increase to 3.2 GW.

Subsequent increase of the pulsed power duration from 300 to 600 nsec revealed new issues in the generation of long pulses. Microwave emission would interrupt during the longer pulse, and then turn back on. Using a mix of experimental data and simulation, the problem was traced to a loss of insulation near the transition between the choke and MILO regions, even when the current exceeded the magnetic insulation threshold.[58,59] The magnetic insulation current is defined for unipolar electron flow, however, and simulations showed that the formation of anode plasma allowed the observed loss of insulation. The suggested fix was to shape the end of the cathode near the affected region to reduce the electric fields and electron flow there, thus reducing anode plasma formation.[57]

The other major thrust for MILO development has been motivated by a desire to address two fundamental MILO shortcomings: the energy lost to establishing the insulation current and the difficulty in extracting energy from the device when it operates in the π-mode with an axial group velocity that vanishes. In a first attempt to address this latter problem, a two-section device was designed, with an upstream section running in the π-mode, establishing the frequency and bunching the electrons appropriately, while the downstream section was an amplifier running in the $\pi/2$-mode with a positive group velocity.[60] Unfortunately, the improvements with this design were modest, first because the group velocity even for the $\pi/2$-mode was still rather small, and second because the downstream $\pi/2$-mode section was prone to break into oscillation at the somewhat different frequency of its own π-mode. The proposed improvement was the *tapered MILO* of Figure 7.37.[61]

The tapered MILO in the figure features an upstream three-cavity driver section that runs in its π-mode to bunch the electron flow and establish the operating frequency. In the reference, though, it was emphasized that the driver section is not just a prebuncher for the downstream tapered sections, since there is considerable additional electron emission from these sections. The group velocity in the gentle and sharp taper sections is increasing in order to propagate more microwave energy out of the device, without increasing the group velocity so quickly that the coupling between the drifting electrons and the structure is compromised. Placing the load diode within the device itself makes it possible to extract additional energy from the

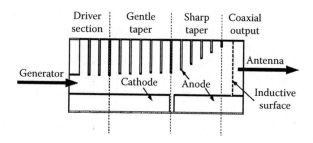

FIGURE 7.37
The tapered MILO. The central anode in this coaxial cylindrical geometry would be attached to the outer conductor by inductive posts. (From Eastwood, J.W. et al., *IEEE Trans. Plasma Sci.*, 26, 698, 1998. With permission.)

electron flow; tapping the energetic electron jets downstream of the gap contributes significantly to the output microwave power. Computer simulations and experiments indicate power levels in excess of 1 GW at efficiencies of 4 to 5%, although simulations also point to possible efficiencies well in excess of 10%.

The advantage of the MILO is its ability to operate at low impedance, so that high power can be achieved at relatively low voltage, particularly in large systems with anode and cathode radii that are close in value (a and b in the geometric factor G of Equation 7.37). MILOs thus couple effectively to low-impedance pulsed power drivers such as waterlines or explosive flux compression generators. Clark and coworkers at Sandia National Laboratories have operated MILOs at 0.72, 1.4, and 2.6 GHz.[62] Power levels in excess of a gigawatt and pulse lengths of hundreds of nanoseconds have been achieved. Breakdown and plasma generation leading to bipolar current loss from the slow-wave structures, as well as cathode jetting, have limited the pulse lengths. System design and construction are simplified by the fact that no external magnetic field coils are required.

The fundamental limitation of MILOs is their low efficiency, which can be traced to two factors. One is intrinsic to the system: the current that provides the insulating magnetic field is generated at high voltages, essentially equal to the operating voltage of the device. Thus, the penalty one pays for design simplicity in not having to provide separate magnetic field coils is the power required to generate the field with a high-voltage, high-current diode, since magnetic field is proportional to current, while power is equal to the product of current and voltage. The second factor contributing to low efficiency is the rather poor extraction efficiency for these devices. In part, this traces to the difficulty of axially extracting power from a device that tends to operate in the π-mode, where the group velocity vanishes. Radial extraction has been tried, but the results have been disappointing to date. The tapered MILO of Eastwood et al. in Figure 7.37 is designed to vary the group velocity at the end of the device in order to improve extraction. The location of the load diode within the MILO region itself is also aimed at recovering some of the energy in the magnetically insulated flow along the line.

Unfortunately, while Eastwood et al. estimate that as much as 40% of the input power is available to be converted to microwaves, and while simulations predict efficiencies in excess of 10%, experimentally measured efficiencies continue to be of the order of a few percent. Nevertheless, if one uses an energy-rich prime and pulsed power system based on explosive flux compression generators (see Sections 2.3 and 5.2.2) — providing adequate voltage and acceptably short rise times — the simplicity vs. efficiency trade may be acceptable. We note that the chemical energy density in high explosives is of the order of 4 MJ/kg.

7.8   Crossed-Field Amplifiers

For completeness, we briefly mention *crossed-field amplifiers* (CFAs). No true HPM version of this device has been successfully built, although one was designed with the intent of operating at 300 MW.[11] This cylindrical device was designed with a coaxial configuration like that originally due to Feinstein and Collier.[5] It was to be quite large, with an anode radius of just over 10.5 cm and 225 anode cavities. The operating mode was to be the $5\pi/6$-mode, and the gain was to be 17 to 20 dB, for a numerical gain of about 50 to 100, so that a master oscillator of 3 to 6 MW was going to be required to reach full power at the operating frequency of 11.4 GHz (see Problem 2).

7.9   Summary

Relativistic magnetrons have produced high power output at moderate efficiency from the L-band to the X-band: power levels of several gigawatts and efficiencies of 20 to 30% are achievable at the lower frequencies, and several hundred megawatts at several percent have been reached at the upper frequency. The technology is relatively mature, and the design of magnetron resonators is well understood: frequencies and modes of the cold resonator can be determined from well-known dispersion relations and fine-tuned for operation using computer simulation, while the experimental values of voltage and applied magnetic field are reasonably predicted by the Hull and Buneman–Hartree resonance conditions. The efficiency of relativistic magnetrons is intrinsically lower than that for nonrelativistic magnetrons; even so, the experimental efficiency of relativistic magnetrons is only about half their theoretical maximum. Consequently, anode erosion limits the lifetime — a problem that worsens as the frequency increases. Plasma evolved from the electrodes has been identified as a primary factor limiting pulse lengths, so that by addressing this issue, progress has been made in increasing the available energy per pulse at high power. In addition, the feasibility of operating relativistic magnetrons at repetition rates of at least a kilohertz has been demonstrated.

The generation of microwaves in magnetrons depends intrinsically on the radial flow of current to the anode, which in turn creates an axial drift of electrons out of the interaction region. Output power is thus fundamentally limited by a lower bound on magnetron impedance of a few tens of ohms. Therefore, further increases in power require either higher-voltage operation of single magnetrons or the phase-locked operation of multiple magnetrons. While this latter approach is by no means optimized, the principle has been demonstrated at several gigawatts.

The MILO promises very high power in a simplified configuration that eliminates the complication of adding separate magnet coils. In principle, one can scale the power up by building large-diameter cathodes, while retaining relatively small anode–cathode spacing. Methodical investigation of the causes of pulse shortening has allowed the extension of pulse lengths to several hundred nanoseconds. Unfortunately, efficiency has been limited to values well below 10%. Extraction is one limiting factor, but, perhaps more importantly, the fact that the insulating magnetic field is generated by current flowing at high voltage — the operating voltage of the device — is a fundamental limitation. Nevertheless, because of its simplicity of construction, robust operating mechanism, and low operating impedance, the MILO may be a good match for systems employing energy-rich, low-impedance explosive flux compression generators for their prime power.

Finally, high power CFAs have been designed for high power, but none have actually been operated.

Future developments in crossed-field devices will probably focus on pushing pulse durations toward a microsecond, mode control over the duration of such long pulses, repetitive operation, and phase locking of multiple units. The serious problem of efficiency in MILOs, as well as the less serious, but nevertheless challenging, problem of efficiency in magnetrons, will also be addressed.

Problems

1. Neglecting gap closure and the time dependence of the microwave source impedance, model a MILO as a constant 10-Ω impedance and a magnetron as a constant 100-Ω impedance. If each were driven by a pulsed power subsystem that could be modeled as a 1-MV voltage source in series with a pulsed power output resistance of Z_{PP}, plot the output power of the microwave source as a function of Z_{PP}.

2. If a CFA operating as an amplifier has a gain of 20 dB, how much input power is required to reach an output power of 1 GW?

3. In cylindrical magnetrons, as well as CFAs, the frequency is a function of the azimuthal mode number n.

   a. For such a magnetron with N anode cavities, what is the value of n if the device is operating in the π-mode?

   b. For a magnetron operating in the 2π-mode, what is the value of n?

4. a. If we design a CFA to operate in the 7π/12-mode, what is the minimum number of vanes, N, that it must have? What is the second-lowest possible number for N?

   b. What value of the mode number n corresponds to the 7π/12-mode for the minimum value of N? What value of n corresponds to the 7π/12-mode for the second-lowest value of N?

5. The frequency spacing between modes, $\Delta\omega_n$, is an important factor in ensuring stable operation in a given mode, designated by the mode number, n.

   a. How does $\Delta\omega_n$ vary with the number of anode vanes, N, for changes in the mode number, Δn, in the vicinity of the π-mode, which operates at frequency ω_π?

   b. How does $\Delta\omega_n$ vary with N for changes Δn in the vicinity of the 3π/2-mode, which operates at frequency $\omega_{3\pi/2}$?

6. Let us estimate the size of magnetrons of different frequencies between the L- and X-bands. Rather than perform the complicated calculation of deriving and solving a dispersion relation (such as that found in Section 4.4.2), proceed empirically with the data of Table 7.1 and Equation 7.3.

   a. Let $r_a = A\lambda$, with A a constant for all frequencies between 1 and 10 GHz. Justify the value of A you choose using the data in Table 7.1. (Note: You will have some latitude in your choice.)

   b. Use Equation 7.3 to estimate r_v (see Figure 7.1) for your values of r_a from part (a) for frequencies between 1 and 10 GHz.

   c. Estimate the length of the magnetron for frequencies between 1 and 10 GHz assuming a length $h = 0.9\lambda$.

   d. Assume that the azimuthal opening angle of the anode cavities is $\psi = 20°$ (see Figure 7.1) for N = 6 anode cavities. Further assume that the outer radius of the anode block is $r_b = r_v + 0.3$ cm. Compute an estimated mass for a stainless steel anode block (use a density for stainless steel of 7.9×10^3 kg/m³).

7. a. If a magnetron of 1-GW output power operates at a power efficiency of 25%, and if all of the electrons emitted from the cathode strike the anode block, how much power is deposited in the block?

   b. If 20% of the electron current in the magnetron escapes axially, how much electrical power is deposited in the anode block?

8. What is the Hull field for magnetic insulation of an S-band magnetron operating at $f = 2.8$ GHz with $r_c = 1.26$ cm and $r_a = 2.1$ cm and an operating voltage of 700 kV? What is the Hull field of the magnetron if the voltage is raised to 1 MV?

9. Plot the Buneman–Hartree curve for an S-band magnetron operating at $f = 2.8$ GHz with $r_c = 1.26$ cm and $r_a = 2.1$ cm for voltages varying between 500 kV and 1 MV. Assume there is no axial current loss.

10. For the MIT A6 magnetron of Table 7.1 and Table 7.2, assuming a magnetic field of 1 T, what is the range of *possible* operating voltages defined by the Hull and Buneman–Hartree conditions? Assume no axial current loss.

11. The magnetron of Spang et al.[34] showed substantial frequency variation with variation of the cathode radius r_c when it operated in the 2π-mode, but relatively little frequency variation as r_c varied when it operated in the π-mode. Explain why, in qualitative terms.

12. An L-band magnetron operating at a 25-Ω impedance, driven by a 25-Ω Marx bank/waterline pulsed power driver, produces 3.6 GW when the open-circuit driving voltage of the pulsed power is 2 MV (so that the voltage across the magnetron is 1 MV). What is the efficiency of the magnetron? If the open-circuit pulsed power voltage is held constant at 2 MV, but a linear induction driver with a 75-Ω impedance is used, instead of the Marx bank/waterline system mentioned earlier, to what level does the output power drop if the magnetron impedance remains constant at 25 Ω? Assume the same efficiency.

13. Prove Equation 7.22.

14. a. For a coaxial MILO with an inner conductor radius of 5.7 cm and an outer conductor radius of 8.7 cm, plot the minimum insulating current, assuming the parapotential flow model of Equation 7.37 for voltages between 500 kV and 2 MV.

    b. For the same MILO, plot the parapotential impedance of the line Z_p over the same voltage range (500 kV to 2 MV).

    c. Estimate the drift velocity of the electrons in the line using Equation 7.38.

15. For a coaxial MILO with an inner conductor radius of 5.7 cm and an outer conductor radius of 8.7 cm, operating at a voltage of 700 kV, using the parapotential flow model of Equation 7.38, compute the drift velocity of the electrons in the line. For a line operating at a π-mode frequency of 2 GHz, estimate the spacing between vanes on the anode.

16. Consider the MILO of the previous problem, operating at a voltage of 700 kV. Assume that the inner line of the MILO is terminated in a flat cathode of circular shape that faces a flat anode. If the full

voltage of 700 kV is imposed across that anode–cathode gap, and if the emission of electron current from the cathode is uniform across the circular face at the space-charge-limited current density, what anode–cathode gap spacing is required in order for the current across the gap to equal the parapotential current for the line, I_p?

References

1. Boot, H.A.H. and Randall, J.T., The cavity magnetron, *IEEE Proc. Radiolocation Conv.*, 928, 1946.
2. Boot, H.A.H. and Randall, J.T., Historical notes on the cavity magnetron, *IEEE Trans. Electron Devices*, ED-23, 724, 1976.
3. Megaw, E.C.S., The high-power pulsed magnetron: a review of early developments, *IEEE Proc. Radiolocation Conv.*, 977, 1946.
4. Collins, J.B., *Microwave Magnetrons*, McGraw-Hill, New York, 1948.
5. Feinstein, J. and Collier, R., A class of waveguide-coupled slow-wave structures, *IRE Trans. Electron Devices*, 6, 9, 1959.
6. Okress, E., *Cross-Field Microwave Devices*, Academic Press, New York, 1961.
7. Bekefi, G. and Orzechowski, T., Giant microwave bursts emitted from a field-emission, relativistic-electron-beam magnetism, *Phys. Rev. Lett.*, 37, 379, 1976.
8. Palevsky, A. and Bekefi, G., Microwave emission from pulsed, relativistic e-beam diodes. II. The multiresonator magnetron, *Phys. Fluids*, 22, 986, 1979.
9. Clark, M.C., Marder, B.M., and Bacon, L.D., Magnetically insulated transmission line oscillator, *Appl. Phys. Lett.*, 52, 78, 1988.
10. Eastwood, J.W., Hawkins, K.C., and Hook, M.P., The tapered MILO, *IEEE Trans. Electron Devices*, 26, 698, 1998.
11. Eppley, K. and Ko, K., Design of a high power cross field amplifier at X band with an internally coupled waveguide, *Proc. SPIE*, 1407, 249, 1991 (available online from the Stanford Linear Accelerator Center SPIRES database: SLAC-PUB-5416).
12. Lau, Y.Y., *High-Power Microwave Sources*, Artech House, Boston, 1987, p. 309.
13. Lemke, R.W., Genoni, T.C., and Spencer, T.A., Three-dimensional particle-in-cell simulation study of a relativistic magnetron, *Phys. Plasmas*, 6, 603, 1999.
14. Treado, T., Generation and Diagnosis of Long- and Short-Pulse, High-Power Microwave Radiation from Relativistic Rising-Sun and A6 Magnetrons, Ph.D. thesis, University of North Carolina, Chapel Hill, 1989.
15. Kovalev, N.F. et al., High-power relativistic 3-cm magnetron, *Sov. Tech. Phys. Lett.*, 6, 197, 1980.
16. Smith, R.R. et al., Development and test of an L-band magnetron, *IEEE Trans. Plasma Sci.*, 19, 628, 1991.
17. Benford, J. et al., Variations on the relativistic magnetron, *IEEE Trans. Plasma Sci.*, PS-13, 538, 1985.
18. Lemke, R.W., Genoni, T.C., and Spencer, T.A., Investigation of rising-sun magnetrons operated at relativistic voltages using three-dimensional particle-in-cell simulation, *Phys. Plasmas*, 7, 706, 2000.
19. For a derivation of the relativistic Hull field and references to earlier derivations, see Lovelace, R.V. and Young, T.F.T., Relativistic Hartree condition for magnetrons: theory and comparison with experiments, *Phys. Fluids*, 28, 2450, 1985.

20. Benford, J., Relativistic magnetrons, in *High-Power Microwave Sources*, Granatstein, V.L. and Alexeff, I., Eds., Artech House, Boston, 1987, p. 351.

21. Price, D. and Benford, J.N., General scaling of pulse shortening in explosive-emission-driven microwave sources, *IEEE Trans. Plasma Sci.*, 26, 256, 1998.

22. Orzechowski, T.J. and Bekefi, G., Microwave emission from pulsed, relativistic e-beam diodes. I. The smooth-bore magnetron, *Phys. Fluids*, 22, 978, 1979.

23. Treado, T.A. et al., High-power, high efficiency, injection-locked, secondary-emission magnetron, *IEEE Trans. Plasma Sci.*, 20, 351, 1992.

24. Lopez, M.R. et al., Cathode effects on a relativistic magnetron driven by a microsecond e-beam accelerator, *IEEE Trans. Plasma Sci.*, 30, 947, 2002.

25. Craig, G., Pettibone, J., and Ensley, D., Symmetrically loaded relativistic magnetron, in *Abstracts for IEEE International Plasma Science Conference Record*, IEEE, Montreal, Canada, 1979.

26. Saveliev, Yu.M. et al., Effect of cathode end caps and a cathode emissive surface on relativistic magnetron operation, *IEEE Trans. Plasma Sci.*, 28, 478, 2000.

27. Phelps, D.A. et al., Rep-rate crossed-field HPM tube development at General Atomics, in *Proceedings of HPM 5*, West Point, NY, 1990, p. 169.

28. Sze, H. et al., Operating characteristics of a relativistic magnetron with a washer cathode, *IEEE Trans. Plasma Sci.*, PS-15, 327, 1987.

29. Ballard, W.P., Self, S.A., and Crawford, F.W., A relativistic magnetron with a thermionic cathode, *J. Appl. Phys.*, 53, 7580, 1982.

30. Kovalev, N.F. et al., Relativistic magnetron with diffraction coupling, *Sov. Tech. Phys. Lett.*, 3, 430, 1977.

31. Benford, J. and Sze, H., Magnetron research at Physics International Company, in *Proceedings of the 3rd National Conference on High-Power Microwave Technology for Defense Applications*, Kirtland AFB, Albuquerque, NM, 1986, p. 44.

32. Close, R.A., Palevsky, A., and Bekefi, G., Radiation measurements from an inverted relativistic magnetron, *J. Appl. Phys.*, 54, 7, 1983.

33. Smith, R.R. et al., Direct experimental observation of π-mode operation in a relativistic magnetron, *IEEE Trans. Plasma Sci.*, PS-16, 234, 1988.

34. Spang, S.T. et al., Relativistic magnetron repetitively pulsed portable HPM transmitter, *IEEE Trans. Plasma Sci.*, PS-18, 586, 1990.

35. Levine, J.S., Harteneck, B.D., and Price, H.D., Frequency agile relativistic magnetrons, *Proc. SPIE*, 2557, 74, 1995.

36. Arter, W. and Eastwood, J.W., Improving the performance of relativistic magnetrons, in *Digest of Technical Papers: International Workshop on High-Power Microwave Generation and Pulse Shortening*, Defence Evaluation and Research Agency, Malvern, U.K., 1997, p. 283.

37. Phelps, D.A. et al., Observations of a repeatable rep-rate IREB-HPM tube, in *Proceedings of the 7th International Conference on High-Power Particle Beams (BEAMS'88)*, 1988, p. 1347.

38. Vasil'yev, V.V. et al., Relativistic magnetron operating in the mode of a train of pulses, *Sov. Tech. Phys. Lett.*, 13, 762, 1987.

39. Ashby, S. et al., High peak and average power with an L-band relativistic magnetron on CLIA, *IEEE Trans. Plasma Sci.*, 20, 344, 1992.

40. Price, D., Levine, J.S., and Benford, J., Orion: A Frequency-Agile Field Test System, paper presented at the 7th National Conference on High-Power Microwave Technology, Laurel, MD, 1997.

41. Benford, J. and Benford, G., Survey of pulse shortening in high power microwave sources, *IEEE Trans. Plasma Sci.*, 25, 311, 1997.

42. Price, D., Levine, J.S., and Benford, J.N., Diode plasma effects on the microwave pulse length from relativistic magnetrons, *IEEE Trans. Plasma Sci.*, 26, 348, 1998.
43. Kerr, B.A. et al., Carbon velvet cathode implementation on the Orion relativistic magnetron, IEEE Pulsed Power Symposium, Basingstoke, UK, 2005, p. 6/1.
44. Treado, T. et al., Temporal study of long-pulse relativistic magnetron operation, *IEEE Trans. Plasma Sci.*, PS-18, 594, 1990.
45. Chen, S.C., Bekefi, G., and Temkin, R., Injection locking of a long-pulse relativistic magnetron, *Proc. SPIE*, 1407, 67, 1991.
46. Chen, S.C., Growth and frequency pushing effects on relativistic magnetron phase-locking, *Proc. SPIE*, 1226, 50, 1990.
47. Benford, J. et al., Phase locking of relativistic magnetrons, *Phys. Rev. Lett.*, 62, 969, 1989.
48. Woo, W. et al., Phase locking of high power microwave oscillators, *J. Appl. Phys.*, 65, 861, 1989.
49. Levine, J., Aiello, N., and Benford, J., Design of a compact phase-locked module of relativistic magnetrons, *Proc. SPIE*, 1226, 60, 1990; Levine, J. et al., Operational characteristics of a phase-locked module of relativistic magnetrons, *Proc. SPIE*, 1407, 74, 1991.
50. Fuks, M. and Schamiloglu, E., Rapid start of oscillations in a magnetron with a "transparent" cathode, *Phys. Rev. Lett.*, 96, 205101, 2005.
51. Black, W.M. et al., A hybrid inverted coaxial magnetron to generate gigawatts of pulsed microwave power, in *Technical Digest, 1979 International Electron Devices Meeting*, Washington, D.C., p. 175; Black, W.M. et al., A high power magnetron for air breakdown studies, in *Technical Digest, 1980 International Electron Devices Meeting*, Washington, D.C., p. 180.
52. Ballard, W.P., Private communication to Benford, J.N.
53. Vintizenko, I.I., Sulakshin, A.S., and Chernogalova, L.F., Generation of microsecond-range microwave pulses in an inverted relativistic magnetron, *Sov. Tech. Phys. Lett.*, 13, 256, 1987.
54. Slater, J., *Microwave Electronics*, Van Nostrand, Princeton, 1950, p. 309.
55. Creedon, J., Relativistic Brillouin flow in the high v/γ limit, *J. Appl. Phys.*, 46, 2946, 1975.
56. Haworth, M.D. et al., Significant pulse-lengthening in a multigigawatt magnetically insulated transmission line oscillator, *IEEE Trans. Plasma Sci.*, 26, 312, 1998.
57. Calico, S.E. et al., Experimental and theoretical investigations of a magnetically insulated line oscillator (MILO), *Proc. SPIE*, 2557, 50, 1995.
58. Haworth, M.D., Luginsland, J.W., and Lemke, R.W., Evidence of a new pulse-shortening mechanism in a load-limited MILO, *IEEE Trans. Plasma Sci.*, 28, 511, 2000.
59. Haworth, M.D. et al., Improved electrostatic design for MILO cathodes, *IEEE Trans. Plasma Sci.*, 30, 992, 2002.
60. An unreferenced design by Ashby mentioned in the following reference.
61. Eastwood, J.W., Hawkins, K.C., and Hook, M.P., The tapered MILO, *IEEE Trans. Plasma Sci.*, 26, 698, 1998.
62. Clark, M.C., Marder, B.M., and Bacon, L.D., Magnetically insulated transmission line oscillator, *Appl. Phys. Lett.*, 52, 78, 1988.

8

BWOs, MWCGs, and O-Type
Cerenkov Devices

8.1 Introduction

The general class of *O-type Cerenkov devices* includes the *relativistic backward wave oscillator* (BWO) and the *relativistic traveling wave tube* (TWT), higher-voltage and -current versions of the conventional BWO and TWT; the *multiwave Cerenkov generator* (MWCG), the highest power producer in this group, a large-cross-section device made possible by operation at relativistic voltages; and the *multiwave diffraction generator* (MWDG) and *relativistic diffraction generator* (RDG), large-cross-section devices capable of gigawatt-level operation at the X-band and above. This class of devices also includes the *dielectric Cerenkov maser* (DCM) and the *plasma Cerenkov maser* (PCM). For their operation, all of these devices depend on some type of *slow-wave structure* (SWS), which reduces the axial phase velocity of the microwaves within the device to a speed slightly below that of the beam electrons and less than the speed of light. Thus, the electrons radiate in these devices in a manner analogous to electrons emitting Cerenkov radiation when they travel through a medium at a speed greater than the local speed of light. They are called *O-type* devices because the electrons travel *along* the axial magnetic field that guides them through the device, rather than across the field, as in M-type devices such as the magnetron. Along with the magnetrons of Chapter 7, relativistic BWOs and MWCGs are the most mature of all high power microwave (HPM) sources, the relativistic BWO being the first true HPM source.

Representative source parameters for BWOs and the multiwave devices are shown in Table 8.1. For reasons that are perhaps more historic than physics based, these devices have been optimized for gigawatt-level power output from the S-band up to 60 GHz, with a peak power of 15 GW exiting the slow-wave structure of an MWCG at a frequency just below 10 GHz. All of the major devices in this class operate at an impedance around 100 Ω, so that voltages approaching, and exceeding, 1 MV are required for a microwave output in the gigawatt range. Output modes for the BWO and MWCG are usually TM_{01}, with an annular intensity pattern, unless a mode converter

TABLE 8.1

Representative Parameters for the O-Type Cerenkov Oscillators

Source Parameters	Sources		
	BWOs	MWCGs	MWDGs, RDGs
Frequency range	3–10 GHz	10–35 GHz	13–60 GHz
Peak power	3 GW	15 GW	4.5 GW
Electronic conversion efficiency	20%	48%	10–20%
Pulse width	~20 nsec	~100 nsec	Up to 700 nsec
Tunable range	300 MHz	150 MHz	NA
Repetition rate	200 Hz	NA	
Output mode	TM_{01}	TM_{01}	Complex
Bandwidth	~1%		
Voltage	0.5–1 MV	1–2 MV	1–2 MV
Electrical impedance	70–120 Ω	100–140 Ω	75–100 Ω
Magnetic field	3–4 T	~2.5 T	~2.8 T

is added downstream. The MWDG and RDG operate in higher-order radial modes with more complicated mode patterns. One of the TWTs with an internal mode converter had a plane-wave-like TE_{11} output.

MWCGs, MWDGs, and RDGs were all developed as large-cross-section devices, with diameters of a few to 10 wavelengths or more. Their strengths are large power production and long-duration/high-energy output pulses, but their large size and the mass and power demand of their magnetic field coils are drawbacks.

More compact BWOs have been optimized to operate in the S- and X-bands at several gigawatts and electrical efficiencies of 10 to 20%. Mechanical tunability in excess of 10% has been demonstrated. They have been taken into the field in a gigawatt-level radar, and they have been operated repetitively in the laboratory at up to 200 Hz with pulse lengths of 45 nsec. At the smaller resonator sizes for these systems, however, peak energy per pulse and average-power-handling capability are issues.

TWT amplifiers have received far less attention, but they have shown promising gains of about 35 to 45 dB, power levels of 400 MW to more than 1 GW, and efficiencies of 25% and more. However, issues of sideband control and uncontrolled oscillation have not yet been fully solved.

8.2 History

Rudolf Kompfner invented the TWT during World War II while working for the British tube establishment,[1] and the BWO followed immediately after the war. Some 25 years later, a BWO utilizing intense relativistic beam technology marked the beginning of the HPM era. It began in 1970 when John Nation of Cornell University used such a beam to generate an estimated 10

MW of microwaves at 0.05% efficiency.[2] Shortly thereafter, a team from the Lebedev Institute in Moscow and the Institute of Applied Physics in Nizhny Novgorod demonstrated the first true HPM source, a relativistic BWO that produced single 10-nsec, 400-MW pulses in the X-band.[3,4] A year after the publication of these results, a similar BWO at Cornell University raised the power to 500 MW,[5] and the race to produce ever-higher power levels using these devices was on. Further development split rather early into two primary lines of research. One led to a new class of devices that used large resonators to produce the highest powers and pulse energies, albeit at the expense of large size and power demand, while the other has continually improved the original BWO in a compact design that is nevertheless limited in pulse energy.

The former line of research involving the Russian Institute of High-Current Electronics (IHCE) in Tomsk, the Institute of Radio Engineering and Electronics (IREE) in Moscow, and Moscow State University pushed forward with a line of sources of increasingly higher power that began with the *surface wave oscillator* (SWO)[6] and *relativistic diffraction generator* (RDG).[7,8] Two modifications led to devices producing much higher output power:

- The use of larger, overmoded devices with diameters substantially greater than a wavelength increased the power-handling capacity of the initial SWO and RDG.
- The slow-wave structure was divided into bunching and output sections separated by a drift space of large enough radius that it was not cut off to the propagation of signal radiation.

The result was the MWCG[9] and the MWDG,[10] as well as two-section RDGs. The combination of these features in MWCGs, RDGs, and MWDGs driven by the powerful single-shot GAMMA accelerator[11] at IHCE — since taken out of service — led to the production of gigawatt power levels[12-14] at frequencies between 9 and 60 GHz. In the highest power experiment, an X-band MWCG coupled 15 GW into its output waveguide.

The second line of research, involving another group at IHCE and the Institute of Applied Physics, and later collaboration with the University of New Mexico and the University of Maryland, focused on improving the original BWO design. Continually improving the BWO, this grouping of institutes and universities has enhanced the efficiency through either axial variations in the slow-wave structure[15] or varying the location of the reflector at the input end of the device[16,17]; reduced the magnetic field at which the device operates[18]; and operated in a repetitive mode.[19] Coupled with one of the SINUS series of repetitive pulsed power machines, this device has powered gigawatt-class radars, including the NAGIRA system.[20,21] This combination could also be the basis for the RANETS-E mobile system that has been advertised as a co-development project for parties interested in a 500-MW microwave weapon.[22] In fact, we used it in the SuperSystem concept of Section 2.5.

Relativistic TWT amplifiers have received less attention than BWOs, in part because the properties of the driving electron beams make it difficult to avoid having the system break into oscillation. Nevertheless, X-band amplifiers of two very different designs have been developed at Cornell University, with the notable achievement of 45% efficiency at hundreds of megawatts output,[23] and at the Institute of Applied Physics (Nizhny Novgorod), with collaboration from the University of New Mexico, with a 45-dB gain at 1.1 GW in a TE_{11} output mode.[24] The master oscillator/power amplifier (MOPA) configuration, with a relativistic BWO driving a TWT amplifier, is more complex, but potentially more powerful in the aggregate. Researchers at Cornell demonstrated a BWO-driven, two-TWT array aimed at producing phase-locked output.[25]

Two types of devices use either a high-dielectric liner (the DCM) or a plasma (the PCM) within a smooth-walled waveguide as their slow-wave structure. From the 1970s through the 1990s, groups at Dartmouth College, the University of California at Irvine, Cornell University, the Tomsk Polytechnic Institute, and the Ecole Polytechnique in Palaiseau, France, produced moderately high powers with the DCM over a wide range of frequencies, notably 200 MW at 3.8 GHz,[26] 580 MW at 8.6 GHz,[27] and 1 MW at 75 GHz.[28] Since the mid-1970s, PCM investigations have largely been confined to the Institute of General Physics in Moscow, with a peak output of 300 MW.[29] In this book, we concentrate far less on DCMs and PCMs because we judge, in the former case, that metal slow-wave structures are more resistant to damage than dielectrics and, in the latter case, that a fixed structure provides greater control of slow-wave-structure properties than plasma.

8.3   Design Principles

The defining feature of Cerenkov devices is the use of a *slow-wave structure*. Consider Figure 8.1. When the phase velocity of the normal mode of the SWS equals that of the drifting electrons, each electron feels an accelerating or decelerating force, depending on the local phase of the wave. As a consequence, the beam bunches, but there is no net exchange of energy, because equal numbers of electrons speed up and slow down. If the electrons are initially somewhat faster than the wave, however, more electrons are decelerated than accelerated, and a resonant transfer of energy to the wave occurs in addition to the bunching. In devices of this type, the exchange of energy causes the wave to grow in amplitude, and the bunching enforces the coherence of the radiation.

Figure 8.2 shows the basic configuration of a Cerenkov device. An electron beam guided by a strong axial magnetic field is injected into a slow-wave structure, where the microwaves are generated; after passing through the

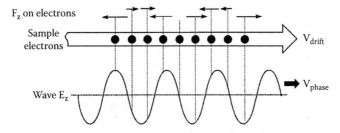

FIGURE 8.1
The bunching effect of the axial electric field of a slow wave on electrons in a beam.

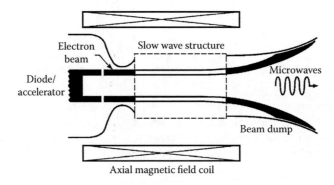

FIGURE 8.2
Basic configuration of the O-type Cerenkov devices. Examples of slow-wave structures are shown in Figure 8.3.

slow-wave structure, the beam flares to the wall in the beam dump. The key elements of these devices are:

- The electron beam, the energy source driving the generation of microwaves, which determines how much power is produced and, as we shall see, the pulse length to a large degree.
- The magnet producing the axial field that guides the beam through the device; this field can affect the output resonance and is a major contributor to system power demands and size and mass.
- The slow-wave structure, which enables the microwave-generating interaction and determines which of the many device types — BWO, MWCG, MWDG, etc. — one is designing, as well as the frequency, output mode structure, and ultimate power- and energy-handling capability. Figure 8.3 shows some examples of slow-wave structures.
- The beam dump, which is a driver for system cooling and x-ray shielding requirements, as described in Chapter 5. Also, if designed improperly, plasma evolved in the beam dump can travel along the axial field lines into the slow-wave structure and limit pulse lengths.

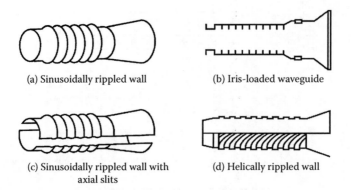

(a) Sinusoidally rippled wall   (b) Iris-loaded waveguide

(c) Sinusoidally rippled wall with   (d) Helically rippled wall
    axial slits

FIGURE 8.3
Examples of slow-wave structures. (From (b) Aleksandrov, A.F. et al., *Sov. Phys. Tech. Phys.*, 26, 997, 1981. With permission. (c) Zaitsev, N.I. et al., *Sov. Tech. Phys. Lett.*, 8, 395, 1982. With permission. (d) Bratman, V.L. et al., *IEEE Trans. Plasma Sci.*, 15, 2, 1987. With permission.)

In this section, we will consider the design relationships between the first three and the output parameters of a device.

The design process is outlined in Figure 8.4, with the shaded boxes at the right representing the main line of process flow:

- The first two steps are interrelated: to acceptable accuracy in these devices, we can treat the SWS and beam separately to estimate the potential operating frequencies and identify the device types (e.g., BWO, TWT, SWO, etc.) associated with each as a function of the transverse dimensions of the SWS and the parameters of the electron beam. Therefore, in Section 8.3.1 we treat the cold SWS in the absence of the beam, and in Section 8.3.2 we add the beam.

- The device types that we identify at the end of the second section fall into two basic classes: *oscillators*, which are by far the more prevalent, and *amplifiers*. For oscillators, there is a minimum beam current required for oscillations to start spontaneously; it is known as the start current, I_{st}. It depends on the mean radius and length of the SWS, the shape and dimensions of the perturbations of the wall, and the reflection coefficients at the ends. Obviously, it also depends on the beam parameters — current, electron energy, and transverse distribution; in some instances, the temporal pulse shape of the beam is also included. Finally, the strength of the magnetic field is significant, although it is usually strong enough that it is ignored unless the frequency of operation lies in a range where a cyclotron resonance competes with the Cerenkov mechanism. The start current is important in two regards. Clearly, the beam current must exceed this threshold for oscillator operation. Of equal importance, though, I_{st} provides a scale for the onset of nonlinear insta-

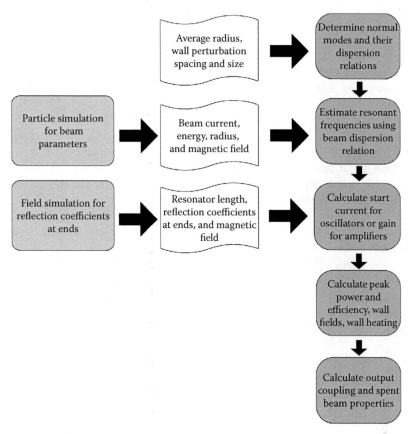

FIGURE 8.4
Basic design flow for O-type Cerenkov devices.

bilities, which occur when the beam current exceeds some multiple of I_{st}. We treat these issues and the subject of gain in amplifiers in Section 8.3.3.

- The output power of oscillators and the linearity and ultimate saturation power of amplifiers are treated at the design stage using computer simulation and at the optimization stage by using computer simulation to augment, and help understand, the experiment. Wall fields and heating can be estimated using analytic estimates based on the understanding of the modes gained earlier. These subjects are all treated in Section 8.3.4.

- Output coupling in these devices is primarily limited to axial extraction of the microwaves through the beam dump. The output modes of BWOs and MWCGs are usually TM_{01}-like. Those for MWDGs and RDGs are more complex. At least one TWT has been designed to provide TE_{11} output.

8.3.1   The Slow-Wave Structure: Dimensions and Frequencies

O-type Cerenkov devices will operate at a resonance frequency and wave-number, ω_0 and k_{z0}, found in the vicinity of the intersection between the dispersion curves for the SWS and those for the so-called *beam line*, $\omega = k_z v_b$. Understanding the dispersion properties of slow-wave structures is thus essential for (1) estimating operating frequencies, (2) calculating the dimensions of the SWS, (3) determining the output mode pattern, and (4) differentiating between the different O-type Cerenkov devices, as we shall see shortly. To a first approximation, the dispersion relationship for a slow-wave structure is actually quite close to that for a cold structure with no beam, so that the SWS and the beam can be treated separately. We therefore start by estimating the dispersion relation for the cold structure.

We discussed the dispersion relation for axially periodic slow-wave structures in Section 4.4.1 as part of our consideration of microwave-generating interactions. These structures are formed by periodically varying the radius of a waveguide, r_w; variations are typically axial, although as shown in Figure 8.3d, they can be helical as well. The two key features playing the dominant role in determining the frequency in O-type Cerenkov devices are the period of the variation in the wall radius, z_0, and the mean radius of the slow-wave structure, r_0. Restricting our attention to purely axial variations, the result of the periodic variation in the wall radius,

$$r_w\left(z\right) = r_w\left(z + z_0\right) \tag{8.1}$$

is that the radian frequency of the normal electromagnetic modes of the slow-wave structure ω is a periodic function of the axial wavenumber k_z:

$$\omega\left(k_z\right) = \omega\left(k_z + nh_0\right) = \omega\left(k_n\right) \tag{8.2}$$

where n is any integer, and

$$h_0 = \frac{2\pi}{z_0} \tag{8.3}$$

For most values of k_z, the detailed shape of the wall plays a secondary part in determining the shape of the curve of $\omega(k_z)$. Except near the points of intersection of the space harmonics of the smooth-walled-waveguide dispersion curves for the transverse magnetic (TM) modes, given, for example, by

$$\omega^2 = k_n^2 c^2 + \omega_{co}^2\left(0,p\right) \tag{8.4}$$

with

$$\omega_{co}(0,p) = \frac{\mu_{op}c}{r_0} \qquad (8.5)$$

the dispersion curves for the slow-wave structure can be approximated by Equation 8.4. In Equation 8.5, μ_{op} is the pth root of the Bessel function J_o. Near the points of intersection for these curves, for different values of the indices n and p, though, the curves are modified in order to produce continuous, periodic curves for ω vs. k_z (see Problems 2 and 3).

To give an example, consider the slow-wave structure of Figure 8.5.[30] The variation of the dispersion relation for the lowest-order mode of the structure with the depth of the wall corrugation is shown in Figure 8.6. This particular curve is important because many, although not all, devices operate in this

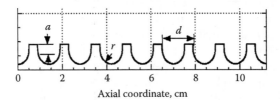

FIGURE 8.5
Wall radius profile for a BWO considered at the University of Maryland, with r the radius of a semicircular curve, a the depth of a rectangular portion of the wall figure, and d the notation for the periodicity of the wall variation in the reference. (From Vlasov, A.N. et al., *IEEE Trans. Plasma Sci.*, 28, 550, 2000. With permission.)

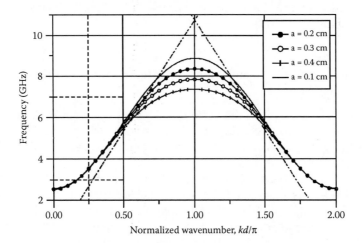

FIGURE 8.6
Calculated dispersion curve for the lowest-order TM_{01}-like mode of a slow-wave structure with a wall radius shaped as shown in Figure 8.5. The four curves correspond to four values of the parameter a in Figure 8.5, while $r = 0.5$ cm, $d = 1.4$ cm, and the distance from the axis to the nearest part of the wall is 4.2 cm. (From Vlasov, A.N. et al., *IEEE Trans. Plasma Sci.*, 28, 550, 2000. With permission.)

lowest-order mode. Note first that the plot is basically independent of the depth of corrugation over about one half of the curve, for k_z in the vicinity of 0 and $2\pi/z_0$. Secondly, we see that over the half of the curve surrounding the π point — where $k_z z_0 = \pi$ — the dispersion curve is increasingly depressed as the corrugation gets deeper. As a check on the estimate in Equation 8.4, consider the lower cutoff frequency for this band, where $k_z = 0$. The mean radius is $r_0 \approx$ 4.5 cm, taking account of the corrugation, so that Equations 8.4 and 8.5 yield $f(k_z = 0) = \omega_{co}/2\pi \approx 2.5$ GHz, which is just what we see in Figure 8.6. Next, consider the upper limit on the frequency for $k_z z_0 = \pi$, which we compute by setting $k_n = \pi/z_0$, to find $f_{max}(k_z = \pi/z_0) \approx 11$ GHz. We see that the wall corrugation drops the frequency considerably below this value, from something less than 9 GHz for a = 0.1 cm to about 7.3 GHz for a = 0.4 GHz. Summarizing, we see that r_0 determines the lower frequency in this passband, while z_0 plays a dominant role in determining the upper frequency. This situation becomes more complex for higher-order modes, though, in which case one must take account of neighboring modes and the exact connection between smooth-walled-waveguide modes in order to understand the mode dispersion curves.

Additional discussion of the dispersion relation for these devices is given in Section 4.4.1. As an additional point here, we remind the reader that the passbands for the different modes can overlap in frequency, although there is no crossing of dispersion curves for the different modes. In addition, we also remind the reader that the transverse variation of the electric and magnetic fields for a given slow-wave-structure mode can vary with k_z; these variations will resemble the smooth-walled-waveguide mode that best approximates the dispersion curve for a selected wavenumber, although the smooth-walled-waveguide mode can vary with k_z along a given slow-wave-structure dispersion curve.

8.3.2 Addition of the Beam: Resonant Interactions for Different Device Types

When we add the beam to the system, it enables the resonant interactions that create the microwave output. In this subsection, we discuss how to estimate the frequencies of these interactions and the types of devices (BWO, TWT, MWCG, etc.) that result.

In the absence of an electron beam, Maxwell's equations are linear, and the dispersion relation for the cold SWS is exact, subject to the approximation that it can be treated as if it were infinitely long. With the addition of the beam, the equations become nonlinear; we can proceed, though, under the assumption that the microwave signal is small enough that it weakly perturbs the beam, which allows us to determine the threshold for operation. The details of the dispersion relation for the coupled SWS–beam system depend on:

- The form of the wall radius axial variation, $r_w(z)$, which determines the shape of the cold slow-wave-structure dispersion curves of ω vs. k_z

- The value of the axial guiding magnetic field, which determines the degree to which electron cyclotron waves affect the beam–wave interaction
- The properties of the electron beam: its cross-sectional dimensions, current I_b, and electron kinetic energy, $(\gamma_b - 1)mc^2$, which determines the properties of the fast and slow space-charge waves* on the beam, which interact resonantly with the slow-wave-structure modes to generate microwaves through the transfer of electron kinetic energy to the electromagnetic fields

Expressions for the dispersion relation taking account of the slow-wave structure and the beam can be found in References 31 to 34.

The fast and slow space-charge waves on the beam have dispersion curves that are shifted upward and downward from the so-called beam lines given by (see Problems 4 and 5)

$$\omega = k_n v_b = \left(k_z + n' h_0\right) v_b \tag{8.6}$$

The degree to which the space-charge wave dispersion curves of $\omega(k_z)$ are shifted above and below these lines (and the shifting is not symmetric for the two waves on the intense electron beams in HPM devices) depends on the current and voltage of the beam (see Problem 6). In the case of an effectively infinite axial magnetic field, for a thin annular-cross-section electron beam of mean radius r_b, the effect of the beam on the dispersion relation is through a term

$$\alpha \equiv \frac{\pi I_b}{\beta_b \gamma_b^3 I_A} \tag{8.7}$$

where $\beta_b = v_b/c$ is the normalized electron velocity and $I_A = 17.1$ kA. Note that α declines strongly in magnitude as γ_b increases, with a scaling like that for the so-called axial plasma frequency in a solid beam.

With no beam, the dispersion relation for the cold SWS allows only real values of ω when solved as a function of real k_z. When we include the beam, however, complex solutions for ω and k_z become possible. The waves grow in time if ω has a positive imaginary part, and in space if the real and imaginary parts of k_z have opposite signs. In general, the full determination of the complex values of ω and k_z depends on the boundary conditions at the ends of the SWS, but we defer that part of the problem to the next subsection.

---

* The terms *fast* and *slow* are sometimes a source of confusion, since both space-charge waves are slow in the sense of having phase velocities less than c. Better terms might have been *faster* and *slower*, but we bow to convention.

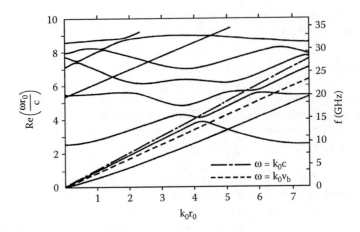

FIGURE 8.7
Dispersion diagram for a thin annular beam with $I_b = 8$ kA, $\gamma_b = 1.91$, and $r_b = 0.5$ cm, within a sinusoidally rippled-wall SWS with $r_0 = 1.3$ cm, $r_1 = 0.1$ cm, and $z_0 = 1.1$ cm. In keeping with the notation of the reference, k_0 corresponds to k_z in our notation. Dimensionless and dimensional values of the frequency are shown on the left and right sides, respectively. (From Swegle, J.A. et al., *Phys. Fluids*, 28, 2882, 1985. With permission.)

As an example of the effect of the beam, Figure 8.7 shows a plot of the solution to the dispersion relation for a slow-wave structure carrying a thin annular beam with $I_b = 8$ kA, $\gamma_b = 1.91$, and $r_b = 0.5$ cm; the SWS wall shape is given by

$$r_w(z) = r_0 + r_1 \sin(h_0 z) \tag{8.8}$$

with $r_0 = 1.3$ cm, $r_1 = 0.1$ cm, and $z_0 = 2\pi/h_0 = 1.1$ cm[33] (see Problem 1). In the plot, k_z is *assumed* to be real (although we emphasize that, in practice, we use the boundary conditions at the ends to complete the set of equations and determine the exact complex values of ω and k_z). Note the beam line in the figure, which actually is continued again at the left side of the plot at the frequency at which it passes off the plot on the right side. The fast and slow space-charge waves straddle the beam line. We see that the fast space-charge wave does not intersect the SWS dispersion curve; rather, one converts to the other as the wavenumber increases. The slow space-charge wave dispersion curve, on the other hand, does intersect the SWS dispersion curve. This is reflective of the fact that an unstable, microwave-generating interaction occurs in the vicinity of the intersection. Further, the curves do not actually intersect, because in Figure 8.7 we are only looking at a plot of the real part of ω vs. the wavenumber. If we examine the complex frequency, we find that it splits into complex conjugates, with opposite signs on the imaginary parts of ω. In Figure 8.8, we show the magnitude of the imaginary part of ω, which is known as the *growth rate* of the instability for real values of the wavenumber. We see that the wave is unstable only for a relatively

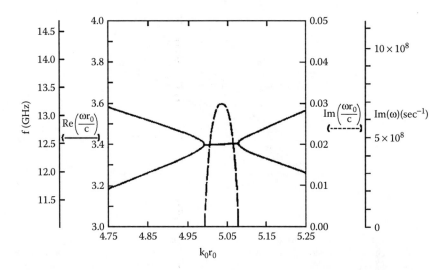

FIGURE 8.8

Detailed view of the real (solid line) and imaginary (dashed line) parts of ω near the lowest-order resonance in Figure 8.7. (From Swegle, J.A. et al., *Phys. Fluids*, 28, 2882, 1985. With permission.)

small range of frequencies in the neighborhood of the intersection between the slow space-charge wave and SWS dispersion curves. Similarly unstable behavior is seen near the intersections at higher frequencies, although the growth rates decline as the real part of the frequency increases.

The operating frequencies, electromagnetic modes, and device types can be determined approximately by finding the intersections between the slow space-charge waves on the beam — with dispersion curves that are usually well approximated by Equation 8.6 — and the cold SWS modes discussed in the previous section. *The different device types — BWOs, TWTs, SWOs, and so on — are distinguished primarily by the slope of the slow-wave-structure dispersion curve at these resonant intersections.* The slope of the SWS curve is the group velocity, $v_g = \partial \omega / \partial k_z$, which is the speed of axial transport of energy by the structure mode. To see how we distinguish the different device types, consider the three example beam lines treated in Figure 8.9, each for a beam of different axial velocity. First, the lowest-velocity beam line in the figure ($v_b = v_{z1}$) intersects the lowest-frequency structure mode at point 1 (BWO), where $v_g < 0$. This is a BWO interaction, and the *backward wave* designation stems from the fact that the structure mode carries energy backward, against the direction of the beam velocity. The same beam line intersects the next higher frequency structure mode at point 2 (TWT), where $v_g > 0$ for the structure. The interaction at this resonance could be exploited to build a TWT, in which the SWS transports electromagnetic energy *forward*.

The dashed beam line in the figure ($v_b = v_{z2}$) intersects the second passband at point 3 (RDG), where $k_z = h_0$. Relativistic diffraction generators are defined

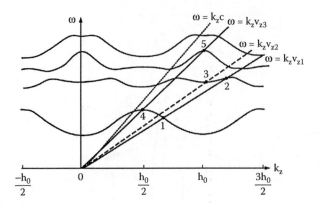

FIGURE 8.9
Example beam lines for three different beams with axial beam velocities $v_{z1} < v_{z2} < v_{z3}$, and their resonances with the slow-wave-structure dispersion curves. Of the intersections (resonances) singled out, 1 is a BWO, 2 is a TWT, 3 is an RDG, 4 is an SWO, and 5 is a Wood's Anomaly (WA). Note that 4 lies below the light line, $\omega = k_z c$.

by these intersections at points where $v_g = 0$ for the SWS mode, except for the special resonance marked point 4 (SWO), to which we return in a moment. The particular RDG resonant chosen in this figure would in fact allow operation in the *2π-mode*, because the wave phase shift from one SWS section to the next, separated by a distance z_0, is $k_z z_0 = 2\pi$.

At higher beam energies, with $v_b = v_{z3}$, the beam line intersects the lowest passband mode at "SWO." This intersection for *π-mode* operation (i.e., $k_z z_0 = \pi$) can occur for both n and n' equal to 0 in Equations 8.4 and 8.6 *only if* the top of the lowest passband lies below the *light line*, $\omega = k_z c$. In this event, the axial electric field will have the character of a *surface wave*, with its amplitude peaking close to the walls of the structure, and evanescently dropping off toward the interior.* Operation at the resonance marked "SWO" thus results in a surface wave oscillator. Qualitatively, to achieve SWO operation, the ripple amplitude must be sufficiently large to depress the peak of the lowest dispersion curve below the light line. The derivations[31,35] for a sinusoidally rippled SWS (Equation 8.8) show that the peak of the lowest dispersion curve lies below the light line in the event that

$$r_1 > 0.3 \frac{z_0^2}{r_0} \tag{8.9}$$

when $r_0^2 \gg 0.6 z_0^2$.

---

* To see this important fact, one must solve Maxwell's equations for the form of the axial field in the SWS, and then the dispersion relation to show that when $\omega < k_z c$, the primary contribution to that field component is described by the I_0 Bessel function, which grows with radius approximately as $\exp(k_\perp r)$.

Finally, the beam line for $v_b = v_{z3}$ intersects the third SWS dispersion curve at point 5 (WA). This is a higher-order 2π-mode that will also support RDG operation, but it differs qualitatively from the 2π-mode at the point marked "RDG" because "WA" is a special point on the dispersion curve referred to in the literature as Wood's anomaly. At this point, the SWS dispersion curve results from the intersection of two smooth-wall TM_{01} dispersion curves given by Equation 8.4 with $n = 0$ and $n = -2$, and $p = 1$ for both. In contrast, the point marked "RDG" results from the $n = -1$ spatial harmonic of the TM_{02} mode near cutoff [i.e., $\omega \approx \omega_{co}(0, 2)$]. Since the SWS curve at "WA" lies below the light line, the mode in question is a surface wave mode, akin to the fields associated with "SWO," rather than a volume wave mode, like the fields associated with "RDG" for $v_0 = v_{z2}$. The efficiency of converting beam energy to diffraction radiation is enhanced at this point,[36] so that it becomes advantageous to operate RDGs.

The multiwave devices, the MWCG and the MWDG, are distinguished from the other devices, first, by the fact that they have overmoded slow-wave structures (i.e., they are several or more wavelengths in diameter) and, second, by the fact that their slow-wave structures are sectioned, with at least two sections separated by a drift space that is *not* cut off to propagation between sections.* In the particular case of the MWCG, the individual sections of the composite SWS have dispersion curves like those for SWOs. In the MWDG, on the other hand, the individual sections are different, so that, for example,[10] the first has the character of a BWO (beam–SWS intersection with $v_g < 0$), while the second has that of an RDG ($v_g = 0$) in the $3\pi/2$-mode.

The sign of v_g at resonance has a fundamental effect on the character of the associated device. When $v_g < 0$, as it is in BWOs, there is an intrinsic feedback mechanism that makes the device act as an oscillator with no driving signal required, provided the beam current exceeds a threshold known as the *start current*, I_{st}, which we will discuss in the next subsection. For those devices where $v_g = 0$ at resonance — SWOs, MWCGs, RDGs, and MWDGs — the SWS effectively becomes a high Q resonator because the electromagnetic energy does not convect through the system (i.e., the situation is akin to the establishment of a standing wave pattern in the SWS, albeit modified by power extraction from the end). Therefore, these devices also operate almost exclusively as oscillators, provided $I_b > I_{st}$.

When $v_g > 0$, energy tends to convect downstream, so that an ideal device with no end reflections will operate as a TWT amplifier for an upstream input signal. Let us emphasize the word *ideal* in connection with amplifiers. In real devices of this type, two nonideal factors modify the situation: (1) given the very high gain associated with the intense electron beam driver, many times noise is spontaneously amplified in a *superradiant* mode of operation; or (2) reflections from the downstream end of the device can provide sufficient feedback that the TWT will break into oscillation above

---

* We emphasize that the intermediate sections are not cut off, so that this arrangement is not confused with the historical practice of *severing* slow-wave structures with cutoff drift sections.

some threshold current. The challenge in designing amplifiers is thus to avoid spontaneous oscillation while obtaining high gain, defined as the ratio of output microwave power, P_{out}, to the input signal power, P_{in}, by

$$P_{out} = GP_{in} \tag{8.10}$$

As we can see, there is a large — in fact, infinite — number of resonances available for microwave generation. A variety of techniques can be applied to make one resonance take precedence over the others in order to generate a clean, single-frequency signal at the maximum possible power. We refer to this as *mode selection*. In many cases, the modes at the lowest frequencies tend to dominate; the first relativistic BWO operated at the lowest, TM_{01}-like resonance,[3] while in the Cornell experiments published a year later,[5] delay-line diagnostics detected power in the two lowest modes on the same pulse. This tendency can be overcome. As we mentioned, when $v_g \approx 0$ at resonance, as it does in SWOs, MWCGs, RDGs, and MWDGs, the slow-wave structure behaves as a high Q resonator, allowing the preferential selection of these resonant modes.[37] Two further examples involve (1) carefully choosing the beam radius and using cyclotron resonance absorption of unwanted mode energy[38] to select TM_{02} operation over TM_{01}, and (2) cutting long slits on either side of a resonator to select modes with azimuthal field variations.[39]

8.3.3   Start Current and Gain

Let us first consider the start current for oscillations, I_{st}. The calculation of I_{st} is quite involved, and the references provide detailed calculations for a number of specific situations:

- BWOs operating at low currents, for which the scaling term from Equation 8.7 is $\alpha^{1/3} \ll 1$, with an infinite magnetic field and with simplified boundary conditions of perfect wave reflection at the beam input end of the SWS and perfect wave transmission at the beam output end[35,40,41]
- BWOs with infinite magnetic fields and general reflection coefficients at the ends[42,43]
- BWOs with finite magnetic fields operating near cyclotron resonance with general reflection coefficients[44]
- SWOs with infinite magnetic fields and general reflection coefficients[37]

We will not detail those calculations, which are beyond the scope of our presentation, but it is valuable to understand the underlying concept. The dispersion relation provides a single relationship between ω and k_z that does not allow us to fully specify them both. In fact, for a device operating at a

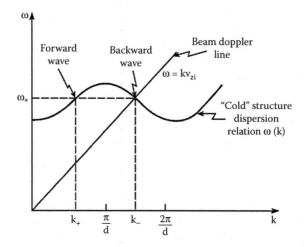

FIGURE 8.10
Schematic depiction of the resonant interaction at $k_z = k_-$ between the lowest-order TM_{01}-like slow-wave-structure mode, a backward wave, and the beam mode, as well as the forward-directed wave at the same frequency as the resonance but with wavenumber k_+. (From Levush, B. et al., *IEEE Trans. Plasma Sci.*, 20, 263, 1992. With permission.)

fixed frequency corresponding to one of the beam–SWS resonances, there are several other nonresonant modes (i.e., away from the intersection of the beam and SWS curves) with the same ω but different values of k_z that will be linearly coupled to the resonant mode by the reflection coefficients at the ends. To take an example, consider Figure 8.10.[42] In the coupled-mode treatment, the resonant BWO mode couples to a forward-directed mode with the same ω but different k_z. By using the boundary conditions at the two ends of the finite-length SWS in conjunction with the dispersion relation for the SWS, one now has three conditions in the three unknowns of ω, k_+, and k_-. In general, all three quantities are complex, and the start current is the minimum value of I_b for which waves grow in time. In other words, when the temporal behavior of quantities is of the form $\exp(i\omega t)$, the imaginary part of ω is positive for $I_b < I_{st}$ and negative (with exponential wave growth) for $I_b > I_{st}$.

As the beam current increases beyond I_{st}, it becomes apparent that there is in fact a hierarchy of start currents.[35] At the lowest value, call it I_{st1}, one axial mode with an axial structure defined by the sum of waves, each with its own value of k_z, oscillates. Above a second, larger value, call it I_{st2}, a second mode with a slightly different frequency and axial mode structure joins the first. A third mode of different frequency and mode structure joins the first two above I_{st3}, and so on as I_b increases. Under the rather restrictive assumptions of Swegle[35] — low current, infinite magnetic field, and idealized end conditions — $I_{st2} = 6.63I_{st1}$ and $I_{st3} = 14.4I_{st1}$.

When one includes general values of the reflection coefficients at the ends of the SWS, the start current is found to be a periodic function of resonator length, L. In particular, when reflections at the output end are added to the calculation, the start current is usually reduced from the idealized value found under the assumption of perfect transmission there, although there are ranges of L for which interference between the coupled waves in the interaction region actually increases I_{st}. Because of the high Q nature of their resonators, start currents are generally lower in SWOs than in BWOs. Overall, start currents are increased when one takes account of the actual time dependence of I_b and γ_b, rather than assuming constant values for each.

For the finite values of the magnetic field, the behavior of I_{st} and device operation overall are modified significantly by the competition between the Cerenkov growth mechanism and absorption by spatial harmonics of the fast cyclotron wave.[44] The effect is most pronounced in the vicinity of the resonant magnetic field value, B_{res}, defined as the field at which the (–1)st spatial harmonic of the fast cyclotron wave resonates with the SWS at the same frequency as the Cerenkov resonance, so that the two compete. Thus, if ω_0 and k_{z0} are the values of frequency and wavenumber at the Cerenkov resonance between the SWS and the beam, then

$$\omega_0 \approx \left(k_{z0} - h_0\right)v_b + \Omega_c = \left(k_{z0} - h_0\right)v_b + \frac{eB_{res}}{m\gamma_b} \tag{8.11}$$

Since the Cerenkov resonance occurs where the beam line of Equation 8.6 is n′ = 0, B_{res} is given by

$$B_{res} \approx \frac{2\pi m\gamma_b v_b}{ez_0} \tag{8.12}$$

The situation for magnetic fields at or near resonance is rather complicated, but generally I_{st} is increased, the stability of the interaction at larger currents is affected, and under some circumstances cyclotron absorption can make device operation impossible. Note also that resonances with higher-order spatial harmonics of the fast cyclotron frequency are possible, so that there are resonant field values at multiples of the value in Equation 8.12.

Although I_b must exceed I_{st} in order for an oscillator to operate, it is possible to have too much of a good thing. Levush et al.,[42] using a formalism and assumptions different from those of Swegle,[35] show that two nonlinear instabilities will modify, and in many cases degrade, the output behavior of a BWO for I_b appreciably above I_{st}. Figure 8.11 is a map of the regions of instability in a two-dimensional parameter space of dimensionless variables; the normalized length on the horizontal axis is the actual SWS length L divided by v_b/ω_0, with ω_0 the operating frequency, while the vertical axis is proportional to the length of the system and the beam current divided by the cube of N = L/z_0.* For the plot, the following parameters are fixed: (1) the

* The normalization parameter for I_b is rather complex, and in the interest of brevity, we refer the reader to Levush et al.[42] for details.

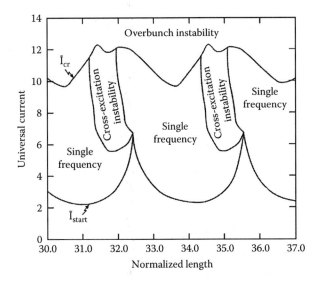

FIGURE 8.11

Regions of BWO operation in a space of normalized device current and normalized device length, assuming a combined reflection coefficient of 0.7 at the two ends of the device. (From Levush, B. et al., *IEEE Trans. Plasma Sci.*, 20, 263, 1992. With permission.)

product of magnitudes of the complex reflection coefficients at the ends, (2) v_b, and (3) the group velocity of the slow-wave-structure wave at resonance. Note that there are regions where one expects (1) single-frequency operation, (2) the *cross-excitation instability*, or (3) the *overbunch instability*.

The cross-excitation instability parameter space extends upward from the cusps in the lower boundary of the single-frequency region for values of I_b > I_{st}. On physical grounds, this happens because those cusps reflect rapid changes in I_{st} as the system shifts from one lowest-order axial mode to the next as the normalized length, $L\omega_0/v_b$, increases. This instability results from a nonlinear reduction in the start current for a competing axial mode in the vicinity of this region where axial modes shift. It manifests itself as shown in Figure 8.12, where the system begins oscillating in a lower-frequency mode, later jumping to a slightly *higher* frequency in a different axial mode; note that in a sense, system performance is improved by the higher-power operation in the latter mode. This instability has been observed experimentally.[45]

The overbunch instability, on the other hand, is a nonlinear effect resulting from the very strong beam bunching that occurs when I_b is substantially larger than I_{st}; this leads to operation in the primary mode as well as a sideband *downshifted* in frequency (similar to behavior observed in klystrons and other O-type sources). This sideband strongly modulates the output signal. Levush et al.[42] note that for much larger currents, with $I_b \sim 30I_{st}$, the microwave fields become so large that electron velocities are actually reversed.

As a rule of thumb, to avoid nonlinear instabilities that create mode hopping and sideband formation, it is desirable to operate an oscillator of this

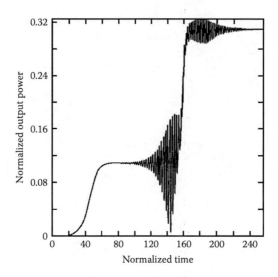

FIGURE 8.12
The behavior of the output power as a function of time when the cross-excitation instability sets in during BWO operation. (From Levush, B. et al., *IEEE Trans. Plasma Sci.*, 20, 263, 1992. With permission.)

type at I_b about three to five times I_{st}. Above about 30 times I_{st}, stochastic operation with broad spectral content results.

As we mentioned in the last section, the issue of start currents does not apply to amplifiers, since it is the goal of amplifier designers to prevent their device from breaking into oscillations. The important parameter for oscillators is the gain, defined as the ratio of output to input microwave power in Equation 8.10. At low-input power levels, G is roughly a constant that can be calculated approximately from the dispersion relation using a real value for ω (so that there is no temporal growth, which would make it an oscillator) to calculate the complex k_z. In the vicinity of a TWT beam–SWS resonance, for fields that vary as $\exp(ik_z z)$, the imaginary part of k_z, $\mathrm{Im}(k_z)$, will be negative, so that the field grows exponentially with length. Because the power is proportional to the square of the electric field, for small values of P_{in} and a SWS of length L,

$$G \approx \exp\left[2\left(\mathrm{Im}\,k_z\right)L\right] \tag{8.13}$$

As P_{in} is increased, P_{out} will begin to saturate, approaching some maximum value for a given beam current I_b, accelerating voltage and device efficiency.

8.3.4 Peak Output Power: The Role of Computer Simulation

There is no reliable analytic theory to compute the output power of an oscillator or the saturation power of an amplifier. To estimate these, and to optimize a device that is being built and tested, we turn to computer simulation. Before we consider the conversion of beam power to output microwave power, let us first consider the beam as the energy source to be tapped to make microwaves. We focus here on the current–voltage relationship for the beam, or more accurately, the relationship between current I_b and electron kinetic energy $(\gamma_b - 1)mc^2$, when it passes through the SWS. Remember from the discussion in Chapter 4 that space-charge effects from the beam will reduce γ_b from its value in the absence of space-charge forces, $\gamma_0 = 1 + eV_0/mc^2$, with V_0 the anode–cathode voltage in the beam diode. For a thin annular beam, γ_b is a function of γ_0, I_b, r_b, and the wall radius, which we approximate by the average radius r_0. Because the kinetic energy decreases as the current increases, the maximum power carried by the beam is a nonlinear function of these parameters; a useful expression for the maximum beam power is given in Vlasov et al.:[30]

$$P_{b,max}\left(GW\right) = \frac{4.354}{\ln\left(r_0 / r_b\right)} W\left(\gamma_0\right)\gamma_0^2 \qquad (8.14)$$

where the nonlinear function W(x) is plotted in Figure 8.13. Note that $P_{b,max}$ depends only on the anode–cathode voltage and the ratio of the wall and beam radii. As the beam approaches the wall, the maximum beam current becomes unlimited, although practically the beam–wall spacing is limited by the fact that with finite magnetic fields, one must allow enough spacing to limit the amount of beam current striking the wall (see Problem 7). As an aside, we note also that as the beam approaches the wall, sources with strong axial fields near the wall — SWOs and MWCGs, as well as RDGs operating near the Wood's anomalies discussed in conjunction with Figure 8.9 — will have stronger beam–field coupling because the axial field in those devices is strongest near the wall.

To understand how one uses Equation 8.14 in the context of a real problem, consider the following example. A reasonable device efficiency is about 20%. Thus, 5 GW of electron beam power is required to produce 1 GW of microwaves. We have a foil-less diode that delivers a beam for which the ratio r_0/r_b is 1.2. Using Equation 8.14 and the plot in Figure 8.13, we see that 5 GW of electron beam power can be delivered at just over 500 kV (or for γ_0, somewhat larger than about 2). If we ignore space-charge depression of the actual beam factor, $\gamma_b \approx \gamma_0$, in order to make a simple estimate, the beam current must be at least 10 kA in order for the beam power to be 5 GW (since $P_b \approx V_0 I_b$), so that the actual diode impedance, $Z = V_0/I_b$, will be approximately 50 Ω. Now remember that this 5 GW is the *maximum beam power* that can be delivered by a diode operating at this voltage. And let us suppose that we wish to design a BWO that operates at a diode impedance of 100 Ω. Then in this case, with $P_b \approx V_0 I_b = V_0^2 / Z$, the beam voltage must be about 707 kV. Both diodes deliver about the same 5 GW of beam power to the BWO. The difference between the approximately 500-kV, 50-Ω beam and the

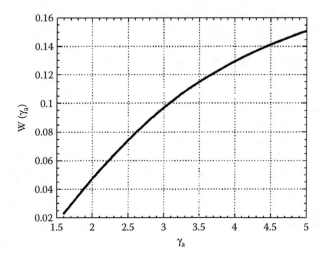

FIGURE 8.13
The function W in Equation 8.14. In the figure, we retain the original notation, with γ_a in the figure corresponding to γ_0 in our notation. (From Vlasov, A.N. et al., *IEEE Trans. Plasma Sci.*, 28, 550, 2000. With permission.)

707-kV, 100-Ω beam is that the former is optimized to deliver the maximum beam power possible at 500 kV, while the 707-kV beam is not optimized, in the sense that a higher beam power could be delivered at a lower impedance. That lower impedance, however, would not necessarily be the optimal impedance for the BWO itself.

Computer simulation with the 2-1/2-D MAGIC particle-in-cell (PIC) code shows the large-signal behavior of an overdriven BWO in Figure 8.14.[33] The output signal from a BWO with a sinusoidally rippled SWS is expressed in terms of a potential obtained by integrating the radial electric field component from the axis to the wall. The initial potential jump is an electrostatic effect due to the injection of the beam with a rise time of 0.2 nsec. The signal then grows exponentially in amplitude with a growth rate of about 1 nsec^{-1}, until growth saturates at about 7.5 nsec. Beyond that point, there is a strong amplitude modulation on the signal indicating the growth of a lower-frequency sideband such as would be created by the overbunch instability. The phase-space plot for the beam electrons given in Figure 8.15, plotted at about 10 nsec, shows the very strong effect of the microwave fields on the axial electron momenta. Near the entrance to the slow-wave structure, the axial momenta are strongly modulated, leading to bunching of the electrons. As the electrons move farther into the structure, the strong potential wells of the microwaves begin to trap some of the electrons, as we see from the large numbers of particles beyond about z = 15 cm that are bunched with small axial momenta. The fraction of the total population of electrons that are trapped trade energy back and forth with the waves with little net exchange, so that they modulate the microwave signal amplitude. Overall, at appreciable distances into the

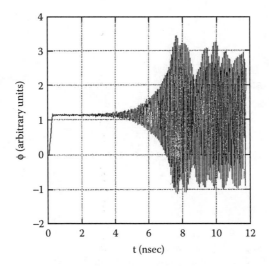

FIGURE 8.14
Time dependence of the signal calculated numerically for a BWO with an infinite axial magnetic field. The vertical axis is calculated by integrating the radial component of the microwave signal. (From Swegle, J.A. et al., *Phys. Fluids*, 28, 2882, 1985. With permission.)

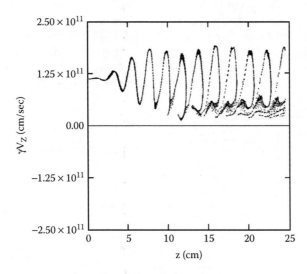

FIGURE 8.15
Phase-space plot of the electron momenta vs. position in a 25-cm-long SWS at t = 10 nsec in the device, with output as shown in Figure 8.14. (From Swegle, J.A. et al., *Phys. Fluids*, 28, 2882, 1985. With permission.)

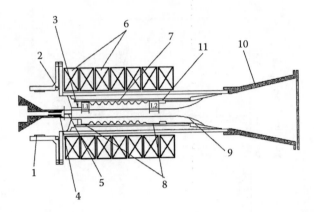

FIGURE 8.16
BWO at the University of New Mexico with both forward and backward shifting. (1) Capacitive voltage divider, (2) Rogowski coil, (3) cutoff neck, (4) cathode, (5) anode–cathode gap, (6) magnetic field coils, (7) slow-wave structure, (8) smooth circular waveguide and shifting lengths L_1 and L_2, (9) electron beam, (10) output horn antenna, and (11) reflection ring. (From Moreland, L.D. et al., *IEEE Trans. Plasma Sci.*, 24, 852, 1996. With permission.)

structure, there are more electrons with momenta less than the injection value at $z = 0$ than there are electrons with momenta over the injection value. Conservation of energy dictates that this energy has gone into the micro-waves. The plots in Figure 8.14 and Figure 8.15 were made under the assumption of an infinite axial magnetic field; in other simulations with finite magnetic fields, the electrons moved radially as well, and some were lost to the walls of the slow-wave structure.

Computer simulations with the TWOQUICK PIC code were also used to investigate the effect of inserting smooth-walled drift sections before and after the SWS in a BWO, as shown in Figure 8.16.[17] When the length of the upstream section L_1 was varied from 0 to 40 mm, Figure 8.17 shows

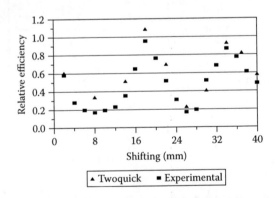

FIGURE 8.17
Variations in BWO power efficiency created by forward shifting with variations in the length L_1 in Figure 8.16. Results of both simulations with the TWOQUICK code and experiments are shown. (From Moreland, L.D. et al., *IEEE Trans. Plasma Sci.*, 24, 852, 1996. With permission.)

that the relative efficiency of microwave production varied by a factor of five in a periodic fashion. Although not shown, the frequency showed a similar variation, from about 9.7 GHz at 1 mm to 9.4 GHz at 6 mm to about 9.63 GHz at 17 mm. The simulations helped elucidate the role of the nonresonant space harmonic of the SWS mode in adjusting the phase between the beam current and the RF voltage to create the effects seen in the figure, effects that had also been considered elsewhere.[16,46] This technique has now become a standard for optimizing BWO output power and efficiency, as well as for providing mechanical tunability. A 670-MW X-band BWO operating at 25% power efficiency was built on the basis of this work, for example.

8.4 Operational Features

In this section, we consider the operational features for three different types of sources: (1) large-diameter multiwave devices, (2) much more compact BWOs, and (3) TWTs. The first two types represent two very different approaches to source construction. The ambitions of the former effort have been directed toward the maximization of peak power and energy per pulse over a very broad frequency range from 10 to 60 GHz, almost without regard to the volume and mass of the system. Much of the most important work was done on a mammoth test bed offering great, but not unlimited, flexibility in pursuing these ambitions. The latter has focused on a very different, but no less ambitious, set of goals: optimization of a multigigawatt BWO-based system in a compact package that operates repetitively and can be taken into the field. We compare some of the major results achieved for each system in Table 8.2. Finally, TWTs remain something of a work in progress, although they offer some unique features, such as an internal mode converter to get near-plane-wave output and the long-term possibility of master oscillator/ power amplifier (MOPA) arrays that could offer high power without stressing any one source beyond its limits.

8.4.1 MWCGs, MWDGs, and RDGs

The peak power and frequency range over which gigawatt-level output power has been produced in the centimeter and millimeter regimes is very impressive: 15 GW at 9.4 GHz in an MWCG[12]; 0.5 to 1 GW at 17 GHz in an MWDG[10]; 3 GW at 31 GHz in an MWCG operating in a superradiant TWT-like mode[47]; and 3.5 GW at 46 GHz and 1 GW at 60 GHz in a two-section RDG.[14] In the series of shots that produced these last results, pulses of up to 700 nsec in duration were produced at somewhat lower power, and pulse energies of up to 350 J were radiated. The efficiencies claimed for these

TABLE 8.2

Comparison of Pulsed Power and Microwave Source Parameters for Representative Experiments Performed with MWCGs and RDGs Using the GAMMA Machine and with High Power BWOs on the SINUS Series of Pulsed Power Generators

E-Beam Generator	GAMMA	GAMMA	GAMMA	SINUS-6 (NAGIRA)	SINUS-6
Voltage	2.1 MV	1.2 MV	1.5 MV	600 kV	1 MV
Current	15 kA	12 kA	20 kA	5 kA	14 kA
Impedance	140 Ω	100 Ω	75 Ω	120 Ω	71 Ω
Pulse length	1 μsec	1 μsec	2.5 μsec	10 nsec (150-Hz repetition rate)	45 nsec
Slow-Wave Structure					
Diameter	14 cm	11.8 cm	11.8 cm		1.5 cm
Period	1.5 cm	4 mm	7 mm		1.61 cm
Total length					9.7 cm ($6z_0$)
Magnetic field	2.45 T		2.8 T		3.7 T
Source	**MWCG**	**MWCG**	**RDG**	**BWO**	**BWO**
Wavelength	3.15 cm	9.72 mm	6.5 mm	3 cm	3.2 cm
Power	15 GW	3 GW	2.8 GW	0.5 GW	3 GW
Duration	60–70 nsec	60–80 nsec	700 nsec	5 nsec	6 nsec
Pulse energy			520 J	2.5 J	20 J
Efficiency	48%	21%	9%	17%	21%

devices are equally impressive, peaking at a quoted value of 50% for the 15-GW MWCG*.

The test bed for many of the major achievements with these devices was GAMMA,[11] a large, Marx-driven, single-shot, pulsed power machine capable of producing 1-μsec voltage pulses when the output was crowbarred, and a pulse of up to 15-μsec duration in a rundown mode when the Marx capacitors just discharged through the diode and any leakage paths. Total electron beam energy ranged up to 140 kJ. It was 6.5 m tall with a footprint of 4.7 × 4.7 m². Figure 8.18 shows the outside of the microwave interaction region of roughly 1 m in length, with magnetic field coils wrapped around the slow-wave structure. The slow-wave structure itself has an inner diameter of about 15 cm, and the output window at the left is 120 cm in diameter. GAMMA has since been shut down, and subsequent experiments with MWCGs use the SINUS-7M machine,[48] a shorter-pulse, somewhat lower-voltage machine.

---

*In a number of instances, there is some ambiguity in the claims for output power; in these instances, the quoted value is for power leaving the source and entering the waveguide for transport to an antenna, while in others, the value is for actual power radiated. Obviously, achieving a high radiated power requires optimization of both source and antenna operation. In some experiments, antenna efficiencies of about 50%, below optimal, reduced the radiated power well below that produced in the source. Here, we do not attempt to resolve this ambiguity for every source, but rather notify the reader and assume that the quoted power levels would in any event be quite close to the value that would be radiated if the antenna performance were optimized.

FIGURE 8.18
The microwave interaction region for HPM experiments conducted on the GAMMA machine at the Institute of High-Current Electronics in Tomsk, Russia. At the center, straps secure the magnetic field coils encircling the slow-wave structure. To the right of the interaction region lies the electron beam diode inside GAMMA. At the left of the picture is the window of the microwave horn radiator, behind which the beam electrons are collected on the walls of a flaring waveguide.

These devices use two (or more) slow-wave sections, separated by a drift space, each with a length roughly comparable to its diameter; in contrast to klystrons, the intervening drift space is not cut off and allows radiation to flow between sections. Examples are shown in Figure 8.19. Basically, the upstream section acts as a buncher, and the downstream section is an extraction section in which the bunched beam radiates; the driving beam current is insufficient to start oscillations in either section alone, but exceeds I_{st} for the aggregate. In the MWCG, each section operates in the π-mode (see Problem 8), like an SWO below its individual start condition, while in the MWDG, the upstream section operates in a backward wave mode and the downstream section operates in a $3\pi/2$ diffraction mode ($v_g \approx 0$). In RDGs, each section operates in the same, typically 2π, diffraction mode. These sources have large cross sections, with diameters D of up to 13 wavelengths in practice. The large diameter reduces the average power intensity and increases the overall power-handling capability of the slow-wave structure. The operating modes of the MWDG and RDG further reduce the fields at the walls because they are volume modes with field maxima closer to the axis (except in the case of those devices operating at the special Wood's anomalies).*

---

* Unfortunately, this point about mode patterns is beyond the scope of our treatment, since it involves an understanding of the details of the relative weighting of each spatial harmonic in the sum making up the expressions for the mode fields, but it is discussed in the MWDG and RDG references.

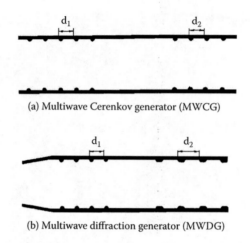

(a) Multiwave Cerenkov generator (MWCG)

(b) Multiwave diffraction generator (MWDG)

FIGURE 8.19
Schematic of the walls of the cylindrical two-section slow-wave structures for an (a) MWCG (with $d_1 = d_2$) and (b) MWDG (or RDG). The electron beam enters from the left. The lengths of the two sections in each structure are, in general, different.

A recent review paper[48] discusses a detailed parameter study of the role of the structure parameters — length of each slow-wave section, length of the intermediate drift tube, and the ripple parameters — as well as the value of the magnetic field and the spacing between the annular beam and the wall, in optimizing MWCG performance on the SINUS-7M. Notably, the use of, and rationale for, three-section structures is discussed, as well as an apparent inverse proportionality between the optimum guiding magnetic field and beam spacing from the wall.

The large ratio of D/λ, however, makes mode control in these devices a challenge. The close spacing of the modes, the shortness of the slow-wave structures relative to their diameter, and the effect of the intense electron beam on the system dispersion characteristics make possible multiwave operation of modes all with the same frequency and similar values of k_z. Nevertheless, the mode control afforded by multiple sections results in a signal with a remarkably narrow spectral width. For example, the spectral width of a 9.4-GHz, 5-GW MWCG[49] with $D/\lambda \approx 5$ was 0.2 to 0.5%; at 35 GHz and 1. 5 GW, with $D/\lambda \approx 13$, the width was still only 0.5%,[47] which in this latter case was on the order of, or less than, the spacing between the normal modes of the slow-wave structure in the absence of the beam.

The experimentally observed increase in both peak power and pulse length with increasing D/λ is also remarkable. In a 9.4-GHz MWCG, when $D/\lambda = 3$, the output power and pulse length[50] were 5 GW and 30 to 50 nsec; increasing D/λ to 5 increased both power and pulse length to 15 GW and 60 to 70 nsec, while increasing efficiency as well. Further increases in D were prevented by the original geometry of the experiment, specifically the diameter of the solenoid providing the axial magnetic field, so subsequent MWCG experiments were performed at 31 and 35 GHz with the same D. At these higher frequencies and smaller values of λ, it was possible to demonstrate GW-level production for $D/\lambda = 13$. With these higher-frequency results in hand as a demonstration of mode control at this large value of D/λ, a series of initial experiments were performed with a larger, 38-cm-diameter structure.[51] At this larger diameter, GAMMA was capable of currents of 60 to 70 kA at voltages of 1.3 to 1.6 MV in 500- to 600-nsec pulses. Significantly, a 4-MJ Marx bank was needed to power the solenoidal field coil providing guide fields of up to 2.25 T. The system was shut down without optimizing the new configuration, however, and preliminary results of 3 GW in the X-band on an MWCG were unremarkable in themselves. We note that the operating beam impedance of almost 25 Ω is much less than the typical 100-Ω value for devices of this type. One other parameter that we shall discuss shortly, the linear beam current density, $I_b/2\pi r_b$, is about twice that for the 15-GW MWCG of Table 8.2, which had an efficiency of 48%, but comparable to that for the 2.8-GW MWCG with an efficiency of 9%.

The higher-frequency devices of Table 8.2 use quite small values of the wall period z_0. In Equation 8.12, we see that the resonant magnetic field for fast-cyclotron-wave effects scales as $1/z_0$. At higher frequencies it thus becomes difficult to operate at magnetic fields well above the resonant value, so that the field becomes a strong determinant of the operating frequency. In the RDGs of Bugaev et al.,[14] the magnetic field was a strong determinant of the frequency. Between 2.3 and 2.4 T, the output frequency was predominately in the 9- to 11.3-mm range, although there was considerable variation in the output frequency over the course of a pulse that extended over several hundred nanoseconds. At an intermediate field of 2.5 T, the output was unstable, with frequency content near 7.2 and 9 to 10 mm. At higher fields, between 2.6 and 2.8 T, the signal was concentrated in the 6.5- to 6.8-mm range. The longest pulse and greatest pulse energy were achieved with a field of 2.8 T, at a voltage of 1.5 MV and a current around 20 kA. At 46 GHz, 2.8 ± 0.8 GW in a 700-nsec pulse was generated, with a pulse energy of 520 J. Over the course of the microwave output pulse, the beam current rose and the voltage dropped continuously; we will return to this subject when we discuss pulse-length issues later in the next section.

The issues of beam control and pulse shortening are very important for these devices. Because these issues are still under investigation, and because they are of general interest, we cover them in Section 8.5.

8.4.2 BWOs

High power BWOs such as those used in the NAGIRA system are more compact and tend to operate for shorter pulses than the MWCG at power levels up to several gigawatts. Historically, the BWOs from the Institute of High-Current Electronics have been driven by electron beams generated with SINUS pulsed power machines, many times in a repetitively fired mode. A summary of SINUS machines is given in Table 8.3, which was compiled from a number of references.[52–58] The largest SINUS machine, the SINUS-7,

TABLE 8.3

Operating Parameters Quoted for High-Current, Pulse-Periodic Nanosecond
Accelerators in the SINUS Family

	Voltage (kV)	Current (kA)	Pulse Length (nsec)	Pulse Energy (J)	PRR (pps)	Average Power (kW)	Reference
SINUS-200	350	3.5	3	3.7	1000	3.7	52
SINUS-M-3	220	0.7	3	0.5	1	0.0005	53
Not named	400	8	25	80	100	8	54
SINUS-4	500	3	30	45	50	2.3	55
SINUS-5	600	10	5	30	100	3	53
SINUS-6	700	7	20	98	100	9.8	56
SINUS-6L	800	8	25	160	1	0.2	53
SINUS-K	450	4.5	25	51	1000	51	52
SINUS-7	1500	50	50	3750	50	188	52
SINUS-700	1000	10	30	300	200	60	52
SINUS-8	200	15	20	60	400	24	53
SINUS-881	610	5.4	25	80	100	8	57
SINUS-13	200	7	7	9.8	100	1	58
SINUS-550-80	550	5	80	220	50	11	A.V. Gunin

is quite large, but the SINUS-500 and SINUS-6 are sized for field deployment
in a complete system the size of a large shipping container. A SINUS-6 at
the University of New Mexico is shown in Figure 8.20. More recently, a
broader range of pulsed power machines has become available to drive these
BWOs: an inductive storage machine from the Institute of Electrophysics in
Ekaterinburg[59]; a modification of the SINUS generator with a spiral-wound
pulse-forming line, which raises the impedance, to better match the impedance of the BWO, and shortens the system for a given pulse length[60]; and
even a one-shot explosively driven generator.[61]

The NAGIRA system discussed in Chapter 2, which also uses a SINUS
machine, has a number of interesting system features. First, the *impedance
transformer* is a tapered transmission line that matches a 120-Ω BWO to the 20-
to 50-Ω output impedance of the resonant Tesla transformer driving it. The
length of the line is related to the length of the output pulse, and it evidently
takes up a considerable fraction of the length of the container. Second, the
BWO uses a superconducting magnet, a feature of systems like this since they
appeared[62] in the late 1970s. As we will discuss in the next section, considerable
effort has been devoted to reducing the magnetic field requirement for systems
of this type. Third, the system uses quasi-optic transmission to the output dish.

BWOs of this type have produced gigawatt-level output in very compact
microwave sources: 3 GW in an X-band device[63] and 5 GW in an S-band
device.[64] Both were driven by a SINUS-7 machine: the X-band BWO with 1-
MV, 14-kA pulses, and the S-band BWO with a 1.2-MV, 15-kA voltage and
current pulse. The X-band BWO was fired repetitively, although it was limited to subhertz firing rates by the magnet of the BWO. The pulse-to-pulse
standard deviation of output power was 3%. The output power was maxi-

FIGURE 8.20
A SINUS-6 accelerator at the University of New Mexico in Albuquerque. The large section labelled "SINUS-6" is the integrated pulse-forming line and Tesla transformer; the cylindrical item on the top is used to trigger the gas switch between the pulse-forming line and the transmission line at the left of the figure.

mized by properly positioning the resonant reflector between the diode and the length of the slow-wave structure, a technique mentioned in Section 8.3.4 that was developed to optimize the phase relationship between the resonant backward space harmonic of the wave and the nonresonant forward space harmonic in the expansion of the fields.[16,17,46] We note that the resonant reflector used in these and other Russian BWO experiments differs from the solid reflector used in a number of other experiments.[17] The resonant reflector consists of a short section of the smooth-walled waveguide between the cathode and slow-wave section that has a significant step increase in radius. The optimized device length was actually quite short: $6z_0 = 9.7$ cm (three times the vacuum output wavelength), and it radiated a TM_{01}-like output mode. The 3-GW, 9.4-GHz pulses lasted 6 nsec, with about 20 J radiated per pulse. A plot of pulse length vs. output power is shown in Figure 8.21. The fit to the data indicates that the energy per pulse is apparently limited by pulse shortening to about 20 J, regardless of the output power. The authors note that at 3 GW, the peak wall fields in a smooth-wall waveguide are about 1.5 MV/cm, and previous work indicates that wall breakdown occurs in about 10 nsec for fields of 1 MV/cm.

The S-band device was referred to as a *resonance BWO*, because its operation depends on properly tailoring the interaction within the device by carefully choosing the length of the slow-wave structure and the length of

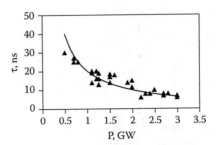

FIGURE 8.21

Microwave pulse duration vs. output power in a single-mode, 3-GW X-band BWO. (From Gunin, A.V. et al., *IEEE Trans. Plasma Sci.*, 26, 326, 1998. With permission.)

the smooth-walled section between the reflector at the beam input end and the slow-wave structure, as well as providing for additional wave reflection at the beam output end of the device. As a consequence, resonance BWOs can be made to be quite short; the SWS was only about 2.5λ, with λ the free-space wavelength of the radiated output, while the smooth-walled insert added about 0.7λ to the length. At peak output power of 5.3 GW, the efficiency of the device was about 30% and the pulse length was 25 nsec, with peak energy per shot around 100 J. Remarkably, the efficiency was rather constant as a function of electron beam power over the range between 5 and 19 GW of electron beam power. The mean diameter of the SWS for this S-band device was about one free-space wavelength.

One compromise route for balancing high output power and size is to increase the diameter of a single-section device, but to operate at a reduced voltage (which would reduce the size of the pulsed power). This approach was considered with a single-section 8.3-GHz SWO with D = 9.8 cm (D/λ ~ 2.7), which has been designed and tested[30] with the express purpose of achieving about 1 GW output at a voltage of 500 kV. Two estimates helped motivate the choice of a larger-diameter SWS:

- Using the expression for the maximum beam power at a given voltage (Equation 8.14), assuming a microwave generation efficiency of 20%, and requiring that the output power be 1 GW and the beam have a 5-mm standoff from the wall leads one to the estimate D/λ ~ 2.

- In a smooth-walled waveguide, the relationship between the power carried in a TM_{mp} mode P and the maximum field at the wall $E_{max,w}$ is (using our expression from Table 4.4)

$$P_{TM} = \frac{\pi r_0^2}{4Z_0}\left(1+\delta_{m,0}\right)\frac{E_{wall,max}^2}{\left[1-\left(\frac{f_{co}}{f}\right)^2\right]^{1/2}} \qquad (8.15)$$

where f_{co} is the cutoff frequency for the given mode and $\delta_{m,0} = 1$ if $m = 0$ and is zero otherwise.

Note that P scales as the area of the guide.

Having chosen D, the authors of Reference 30 considered several wall shapes (none of which was the sinusoidal wall of Equation 8.8), plotted the lowest-frequency dispersion relation for each, and then chose one that gave them an SWO. To enhance power, efficiency, and tunability, they used an upstream reflector with a movable smooth-walled-waveguide section. For details of the design process, refer to the reference. In experiments with this device, a peak output of 500 MW was achieved in a 10.2-J pulse with a 570-kV, 5.8-kA beam. The efficiency was about 15%, and a tunable range of about 120 MHz was achieved about the 8.3-GHz center frequency. Pulse shortening was observed at the highest power level; after several hundred shots, the middle section of the slow-wave structure was covered with damage spots. These spots were concentrated on the semicylindrical ends of the structure ripples, where peak fields exceeded 400 kV/cm.

8.4.3 TWTs

To restate a point made earlier, TWTs have received less attention than the high power Cerenkov oscillators. This may be because of the difficulty of controlling devices that are driven by such a powerful electron beam. As we will see, one of the two major efforts was plagued by sideband generation, while the other had trouble controlling oscillations. Here we concentrate on (1) a Cornell effort using solid "pencil" beams, rather than the more common annular beams, in a two-section, *severed* structure* (unlike the two-section MWDGs, for example, which allow propagation between sections) and (2) an effort involving the Institutes of High-Current Electronics in Tomsk and Applied Physics in Nizhny Novgorod, with collaboration from workers at the University of New Mexico, involving backward wave and traveling wave amplifiers operating in different azimuthal modes and coupled through a mode converter. We will also mention a Cornell effort that used a high power BWO to drive two TWTs in a master oscillator/power amplifier (MOPA) configuration.

The Cornell work using pencil beams in severed structures, which grew out of an effort that began with single-stage TWTs, is summarized in Shiffler et al.[65] In every case, the slow-wave structures had an average radius of 1.32 cm, a ripple depth of 8 mm, and a period of 7 mm, and operation was in the lowest TM_{01}-like mode. At each end, the ripple depth tapered to zero to minimize end reflections and the possibility of breaking into oscillation. Operation was in the X-band at a center frequency of 8.76 GHz. The electron

---

* A *sever* prevents propagation from one section to the next, typically absorbing microwaves as they attempt to pass.

beam had an energy of 850 kV, and the current could be varied from 800 to 1700 A. The single-stage devices had central sections of up to $22z_0$ in length, plus the tapered end regions. Maximum gains of 33 dB at output powers of 110 MW and efficiencies of 11% were achieved. At 70-MW output power, the phase of the output was locked to the phase of a 100-kW input magnetron with an accuracy of ±8°. Above 70 MW, sideband formation occurred, making phase control problematic. Bandwidth was quite narrow: 20 MHz at the 3-dB points on the gain curve.

The two-stage severed amplifier was adopted to prevent self-oscillations at high operating currents. Two amplifiers, each with center sections 22 ripples in length and tapers at the ends, were connected through a 9-mm-radius Poco graphite sever 13.6 cm long that attenuated the microwave signal by 30 dB, but allowed the space-charge waves on the beam to pass unattenuated to couple the two amplifier sections. The total signal output of about 400 MW, averaged over the pulse, was produced at an efficiency of 45%; gain was 37 dB. Bandwidth was considerably larger than that of the single-stage amplifier, of the order of 200 MHz. Unfortunately, sidebands were observed at all signal levels; at peak power, roughly half of the total power was in the sidebands. Computer simulations indicated that the bunching length in this system was quite long, and that the sideband development could be traced at least in part to finite-length effects and end reflections. Remarkably, even at instantaneous power levels up to 500 MW and pulse durations of about 70 nsec, no pulse shortening was observed.

The more complex amplifier from Tomsk, Nizhny Novgorod, and the University of New Mexico[24] produced substantially higher power using an 800-kV, 6-kA beam: up to 1.1 GW with a gain of 47 dB and an efficiency of 23%. The authors of this paper called out two challenging trade-offs in the design of HPM sources generally and HPM amplifiers specifically. First, to prevent breakdown, one must increase the cross-sectional dimensions of the source; however, in doing so, mode selection to produce a coherent output signal becomes more difficult. Second, high gain of the order of 40 to 50 dB is required if one wishes to use a conventional microwave source with power in the 10- to 100-kW range as the master oscillator. High gain, however, increases the chances that the source will break into self-oscillation, particularly under the relatively high noise conditions associated with relativistic electron beams. The authors' answer to these two challenges is the amplifier depicted in block diagram form in Figure 8.22 and schematically in Figure 8.23.

The input signal from a 100-kW, 9.1- to 9.6-GHz tunable magnetron is converted to a TE_{41} mode, which is preamplified in the backward wave amplifier. This amplifier — when run below its start oscillation condition — has a slow-wave structure with a corkscrew ripple like that shown in Figure 8.3d with azimuthal index m′ = 3 on the ripple. Operating in the TE_{41} whispering-gallery mode, with index m = 4, the preamplifier produces a space-charge modulation on the beam with index n = m − m′ = 1, which drives the downstream traveling wave tube in the HE_{11} hybrid mode. This mode is converted to a TE_{11} mode in the downstream waveguide and is radiated as a near-

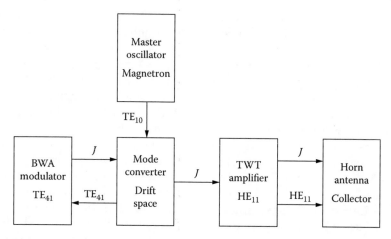

FIGURE 8.22
Block diagram of the 1.1-GW TWT amplifier developed by the Institute of Applied Physics in Nizhny Novgorod and the University of New Mexico. *J* indicates the direction of beam-current flow.

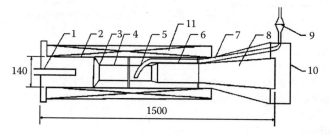

FIGURE 8.23
Schematic of the 1.1-GW TWT amplifier developed by the Institute of Applied Physics in Nizhny Novgorod and the University of New Mexico. (1) Cathode, (2) vacuum chamber wall, (3) matching horn, (4) backward wave amplifier modulator, (5) microwave input and drift space, (6) traveling wave amplifier, (7) input rectangular waveguide, (8) output horn, (9 and 10) input and output microwave windows, and (11) magnetic field-producing solenoid. Scale is in millimeters. (From Abubakirov, E.B. et al., *IEEE Trans. Plasma Sci.*, 30, 1041, 2002. With permission.)

Gaussian beam from a horn antenna. The preamplifier has a mean diameter of 5.6 cm ($D/\lambda \sim 1.8$), while the mean diameter of the traveling wave amplifier is 6.46 cm. Axial magnetic fields up to 5 T could be applied. Gains in excess of 50 dB are possible, but over a narrow range of operating parameters. More desirably, lower gains of 43 to 44 dB can be achieved at comparable efficiencies of 35% or more and over a much wider range of operating parameters.

The system can be run with the backward wave tube either above its oscillation threshold, in which event an output is produced without input from the conventional magnetron, or below it, so that an input is required to realize an output. The parameter governing the transition to self-oscillation in the backward wave tube is the radius of the electron beam; given a particular beam current and voltage, there is a minimum beam radius for oscillation. For beam radii below that value, there are no oscillations; above that value, the backward wave tube goes into oscillation because the strength of the coupling between the beam and the structure reaches a requisite magnitude. A complication in the system is that the beam radius expands with a radial velocity of 1 to 2 cm/μsec over the course of the 500-nsec electron beam pulse. Thus, even when the system begins operation as an amplifier, it will transition to an oscillator when the beam radius expands outward through the threshold radius. When the backward wave tube is allowed to oscillate, the usual cyclotron resonance dip for these circularly polarized systems is seen in the vicinity of about 1.2 T. Higher-output powers, about 1.3 GW, were produced at fields above the dip, although power levels of about 1 GW were seen over a narrow range of fields below the dip. Given the polarization of the system, the polarity as well as the strength of the magnetic field affected performance. (Note: The estimate of the field at which the dip occurs, given in Equation 8.12, does not hold for these polarizations because the estimate does not take account of helically polarized waves.)

When run as an amplifier, the system achieves gains of 47 dB in a bandwidth of about 1% near the resonant frequency of the backward wave tube. Up to a 5% bandwidth could be reached through a coordinated tuning of the magnetic field and the beam radius. Pulse shortening is observed — the pulse length is 250 nsec at 300 MW (75 J, beam radius of 2 cm) and 70 nsec at 1.1 GW (77 J, beam radius of 2.1 cm) — but the authors offer that the benefits of increasing the device diameter have yet to be fully exploited.

Finally, a series of Cornell experiments examined the prototypical master oscillator/power amplifier (MOPA) configuration shown in Figure 8.24[25]

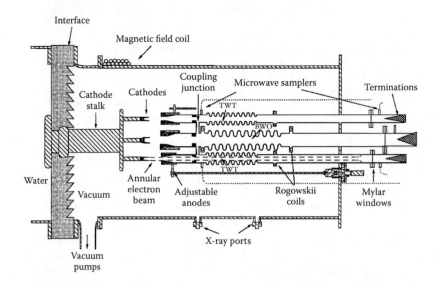

FIGURE 8.24
Layout of the Cornell master oscillator/power amplifier experiment. (From Butler, J.M. and Wharton, C.B., *Proc. SPIE*, 1226, 23, 1990. With permission.)

involving a relativistic BWO driving two relativistic TWTs. Unfortunately, a number of technical problems prevented the system from successful realization, but we discuss the configuration to illustrate the concept. Three cathodes energized by a single pulsed power unit were to provide three beams for a BWO and two TWTs; the BWO was to be the master oscillator for the two TWT amplifiers. The important feature of this arrangement was that it was to take advantage of the fact that pulsed power systems are capable of high-voltage operation at much lower impedances than those of individual BWOs and TWTs, which are typically about 100 Ω (although we note that the Cornell pencil-beam TWT was an order of magnitude higher in impedance). Two TWTs were coupled to a BWO at the upstream coupling junctions with rectangular waveguide sections. The TM_{01}-like mode of the BWO was converted to a rectangular-guide TE_{10} mode, and then reconverted to a TM_{01}-like mode of the TWTs.

8.5   Research and Development Issues

This is a relatively mature class of devices, but there are three major areas of research and development: (1) pulse shortening, which is a affliction for all HPM sources to one degree or another; (2) operation at lower magnetic fields, below the cyclotron-resonant field value, with a long-term goal of using lower-power direct current (DC) electromagnets or even permanent magnets; and (3) axially varying the SWS in order to raise device efficiency.

Beyond those three topics, there is the additional issue of the less mature sources in this class that deserve at least mention: the DCM, PCM, and even plasma-filled BWOs. We will mention these at the end. TWTs are also a less mature source, and in that regard, they qualify as an R&D issue, but we have already discussed the major research in this area.

8.5.1   Pulse Shortening

Pulse shortening involves a suite of factors: (1) breakdown in the power feed to the electron beam diode; (2) plasma motion in the diode that either closes the anode–cathode gap or causes the electron beam radius to increase with time; (3) breakdown processes in the SWS; and (4) beam stability issues within the SWS that cause anomalous electron beam loss. Three issues we will not consider here are plasma formation processes in the beam collector that leak plasma into the SWS; breakdown in the waveguide carrying microwaves away from the source; and breakdown in the antenna under the stress of extremely high power.

Breakdown in power feeds is an occupational hazard in the pulsed power business. One more subtle form of breakdown that caused early problems in some experiments involved parasitic current flow from the cathode, or cathode feed, of the foil-less diode that creates the beam to the opposite wall of the transmission line, which sits at the anode potential.[66] This problem was solved rather straightforwardly by shaping the magnetic field near the electron-emitting surface so that the field lines intercepted a shield on the cathode side designed to catch electrons before they could leak across to the anode side of the feed or transmission line.

Plasma motion in the diode and the effect that it has on gap closure and beam radius growth are more difficult. The electron beam is generated by a process of explosive emission that creates a plasma at the emitting surface from which the beam is drawn. This plasma is confined by an axial magnetic field, and it can expand across the magnetic field, causing beam radius growth, or forward along the magnetic field toward the anode, causing gap closure. When the beam radius grows and the beam approaches the wall, the beam–SWS coupling changes. In the case of the TWT of Abubakirov et al.,[24] the increased coupling raised the gain to the point that the system broke into oscillation and could no longer effectively amplify the input signal. In oscillators, the increased coupling can reduce I_{st}, so that the ratio I_b/I_{st} grows to the point that one of the nonlinear modulational instabilities — the cross-excitation or overbunch instability — becomes a problem. Finally, the increase in r_b can cause the beam to intercept the input iris to the slow-wave structure, exacerbating gap closure problems and causing beam termination. Experiments have shown that cross-field expansion is exacerbated by the penetration of microwave fields into the anode–cathode region.[67]

Two basic approaches have been taken to reduce cross-field expansion. One method was to increase the magnetic field strength as a function of time, in order to hold the plasma in place as it diffused across the field; in effect, these researchers "brought the field to the plasma."[67] Another approach has been to reduce the density of the plasma at the cathode. In this regard, a multipoint cathode consisting of an array of resistor-ballasted carbon fibers was used to reduce the overall plasma density well below that created at the surface of standard annular "cookie cutter" cathodes.[68] Consequently, the cathode plasma density was reduced by two to three orders of magnitude, and an 8-µsec pulse was produced, albeit at the low power of 10 MW. Another means of reducing the plasma density is the use of disk, rather than cookie cutter, cathodes.[69]

Breakdown in the slow-wave structure is evidenced by damage marks on the SWS, and it has received a great deal of attention. In an extensive review of the problem,[70] the emission of plasmas from the wall, and the ambipolar diffusion driven by ions in that plasma, were identified as a primary cause of pulse shortening. Explosive emission is at least one source of the wall plasma, as indicated by the observation of pits on the wall suggestive of this process. Another is adsorbed gases on the slow-wave-structure wall. To reduce this latter effect, the slow-wave structure of a 3-GW, X-band BWO was electrochemically polished and liquid nitrogen traps were added to the vacuum system. As a result, the microwave pulse length was increased from 6 to 26 nsec, and the energy per pulse grew from 20 to about 80 J.

In computer simulations, electrons alone reduced the output power, but did not terminate the pulse. When ions were emitted, however, the output power became strongly dependent on the current density of electron emission from the walls, and above a certain electron current density, microwave emission ceased. The location of the plasma emission was significant: the closer the release of the plasma to the input end of a BWO, the more it affected the microwave output. Plasma at the input end, perhaps due in part to beam interception by the input iris, was most damaging. Thus, BWOs in which the efficiency and power are optimized by adjusting the location of the input-end reflector are most susceptible to this effect.

Additional confirmation for the processes described in Korovin et al.[70] might be drawn from the RDG work of Bugaev et al.[14] There, diagnostics showed that the current received at the output end of these devices grew with time. Further, x-ray diagnostics revealed that the additional current, beyond that of the electron beam, consisted of low-energy electrons carrying as much or more current than the beam by the end of the beam pulse.

During the production of very high power microwaves in the MWCG experiments producing X-band radiation at the level of 10 GW or more, beam transport during the microwave pulse was clearly seen to terminate.[71] This termination occurred despite the fact that a beam of the same current and voltage propagated through a smooth-wall structure of the same radius with the same guide field when no microwaves were generated. Striated damage patterns were observed on the slow-wave-structure walls. It was postulated that the beam broke into filaments and that $\mathbf{E} \times \mathbf{B}$ drifts involving the microwave fields drove these filaments to the wall; however, this phenomenon is still not clearly understood.

Increasing the diameter of the slow-wave structure clearly helps in one regard: Equation 8.15 shows that the maximum electric field at the wall for fixed microwave power drops as $(D/\lambda)^{-2}$. With regard to that field, we note that the maximum fields at the walls of the 3-GW BWO were approximately 1.5 MV/cm. There was also an indication, on the basis of rather limited data from the MWCG experiments, that lower values of the linear beam current density, $I_b/\pi D$, are more desirable than higher values, and there may even be an optimal value.[13] The value in the 15-GW experiment was about 0.35 kA/cm. This is an area where more research is needed.

8.5.2 BWO Operation at Lower Magnetic Fields

A plot of output power and pulse length as a function of magnetic field in the 3-GW BWO is shown in Figure 8.25.[63] The drop in power around 2 T results from the cyclotron resonance dip at the field given in Equation 8.12. Output power peaks around 3.7 T, and from the standpoint of output power or pulse length, there is no advantage to increasing the magnetic field above

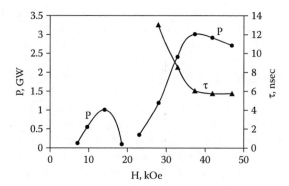

FIGURE 8.25
Microwave power and pulse duration for a single-mode, 3-GW X-band BWO. (From Gunin, A.V. et al., *IEEE Trans. Plasma Sci.*, 26, 326, 1998. With permission.)

that value. Below the cyclotron resonance, we see a local maximum in output power around 1.4 T, although the output power is lower there than it is at 3.7 T. Experiments with a larger-diameter SWS and beam, but a less powerful electron beam, produced 800 MW of output at 24% efficiency; this was intended to reduce the wall fields and show the way toward higher power and energy per pulse. In that larger-diameter device, though, we also note that comparable power was produced at magnetic fields below the resonance.

The use of a larger-diameter SWS and beam is important because beam quality and confinement are also issues below the cyclotron resonance. Preliminary experiments with a 10-GHz BWO showed that 500 MW of output could be produced in a device with a field of 0.7 T near the beam diode and 0.5 T inside the SWS. The DC field coil producing this field consumed 20 kW. At the larger radius, this device required more care in the selection of a TM_{01}-like operating mode.[72]

8.5.3 Axially Varying Slow-Wave Structures to Enhance Efficiency

Axially varying the interaction between the beam and the field as the beam loses energy to the field is a technique applied to increase efficiency in a number of microwave source types. Axial variations can involve modifying the coupling[73] or the phase velocity of the wave.[15] In the former case, a sinusoidally rippled slow-wave structure of total length L had relatively shallow ripples over a length L_{min}, followed by deeper ripples over the remainder. In the portion of the structure with deeper ripples, the magnetic field was decreased to allow the beam to approach the wall. By optimizing this configuration, the microwave generation efficiency was almost tripled, from 11% without the variations to 29% with them, resulting in an output power of 320 MW. In the latter case of varying the phase velocity, a 500-MW pulse was generated at 45% efficiency.

8.5.4 Other O-Type Sources: DCMs, PCMs, and Plasma-Filled BWOs

Dielectric Cerenkov masers (DCMs) and plasma Cerenkov masers (PCMs) operate in smooth-walled waveguides and use different slow-wave structures. In the DCM, slow waves are realized by using either a dielectric liner or rod within a smooth-walled waveguide. The material of the rod or liner is chosen for its high permittivity (or index of refraction), and the slower light speed in the material brings the dispersion curve for this slow-wave structure down below the speed of light; Equation 8.4 is modified to

$$\omega^2 = k_n^2 \bar{c}^2 + \bar{\omega}_{co}^2 (0, p) \qquad (8.16)$$

In this equation, $\bar{c} < c$ because of the presence of the dielectric, and the cutoff frequency in Equation 8.5 has been modified to a value dependent on the modified speed of light. Beam–SWS resonances are possible, typically at high frequencies and wavenumbers (see Problem 10). Power levels in excess of 500 MW have been produced, as discussed in Section 8.2.

Plasma Cerenkov masers use plasma layers and the dispersion properties of the plasma to create slow-wave structures. Once again, a smooth-walled waveguide is used, and a low-energy electron beam (low energy by the standards of pulsed power) is used to create an annular plasma surrounding a high-energy electron beam, or perhaps a thin plasma rod on axis. Power levels of several hundred megawatts have been produced in PCMs, as indicated in Section 8.2.

The filling of BWOs with plasma to increase the allowable electron beam current within the SWS is an approach that traces back almost to the genesis of HPM.[74] The motivation for doing so is that the plasma provides space-charge neutralization of the beam, permitting one to use currents above the space-charge-limiting value for the vacuum SWS. Further, it is hoped that the neutralized beam will have higher quality than the beam in the vacuum SWS. In the initial high-current device, 600 MW of microwaves was generated, but only at 6 to 7% efficiency. A much higher efficiency of 40% was achieved in later work at 400 MW.[75] In the *pasotron* concept (plasma-assisted slow-wave oscillator, which was proven out at only tens of megawatts), the axial guide field was eliminated on the grounds that space-charge neutralization by the plasma would prevent radial expansion of the electron beam.[76]

8.6 Fundamental Limitations

Over the years, a number of groups have developed a substantial understanding of sources of this type. The values of output power, energy per pulse, and efficiency achieved within this group rank among the highest for all HPM sources, and they have produced gigawatt power levels across a remarkable range in frequency, from 3 to 60 GHz. That range could be extended downward to begin to compete with magnetrons, which dominate in the L- and S-bands, although size could become an issue as the operating frequency is reduced. In magnetrons, the lowest passband extends upward in frequency from zero to the π-mode frequency. In the O-type Cerenkov devices, from Figure 8.6 and Equations 8.4 and 8.5, we see that the lowest passband extends upward from a minimum frequency $f \approx 11.5/r_0$, with f and r_0 measured in gigahertz and centimeters, respectively. We also know that the upper end of the passband for the lowest-order mode depends on the periodicity of the slow-wave structure, and that increasing z_0 lowers the frequency of the top of the passband. Thus, as the frequency goes down, device size grows both radially and axially.

The limitations in power, energy per pulse, and efficiency are actually of two classes: *absolute* and *system constrained*. Absolute limitations are the intrinsic limits set by the physics or the state of the technology in the absence of constraints on, for example, size or volume or electrical power consumption. System-constrained limitations, on the other hand, are those imposed by specific mission requirements, such as the demand for a mobile system or for a system with limited electrical demand. In the case of the absolute limits for these devices, we outline the pros and cons of the two major development thrusts in Table 8.4: (1) increasing D/λ to raise power, energy per pulse, and beam-to-microwave conversion efficiency and (2) operating at lower magnetic fields to reduce size, mass, electrical power demands, and temporal operating constraints imposed by pulsed, high-field magnets.

Increasing D/λ should improve the performance of the smaller gigawatt-level BWOs. Increasing the performance of the larger multiwave devices, however, begins to push the limit of ultimate device performance. Experiments on the GAMMA machine demonstrated the ability to operate at values as high as $D/\lambda = 13$. An X-band MWCG at this parameter value would have a slow-wave-structure diameter of about 40 cm. Adding magnetic field coils pushes the system diameter toward a meter. The mass of this magnet will become a major contribution to overall system mass, and the energy storage requirement for the magnet's power supply will grow as D^2, provided the length can be kept constant. These factors are examples of limitations potentially imposed by system constraints. If indeed the empirical observation of an optimal linear current density, $I_b/2\pi r_b$, in the range of 0.3 to 0.5 kA/cm holds — and the data are sufficiently limited that this is considered an issue to be resolved — then the optimum operating current might increase with D, and the operating impedance would drop as $1/D$, if the voltage could be held constant, another uncertainty. Lower impedance either raises overall system efficiency as the source is brought into a better impedance match with the pulsed power or decreases system length if a tapered-line transformer matching the pulsed power and source impedances can be eliminated (see Problem 9). Increasing the operating current also increases the problem associated with the nonlinear cross-excitation and overbunching instabilities, which scale with

TABLE 8.4

Pros and Cons of the Two Main Approaches to Dealing with Device Limitations

Development Area	Pros	Cons
Increase D/λ	Reduce wall fields and lessen the probability of breakdown	Mode control
	Lower impedance, both to reduce voltage required to get requisite beam power and to better match source impedance to pulsed power	Existence of optimum linear beam current density, $I_b/2\pi r_b$, not fully explored
	Lower I_b relative to space-charge-limiting value and improve beam quality and confinement	Beam control issues not fully understood
		Start current must be pushed upward to avoid nonlinear instabilities at high current
Reduce B	Reduce electrical demand for electromagnets	Beam control
	Reduce mass and volume of magnets, including structural support	Output power is optimum only over a narrow range in B
	Make use of nonsuperconducting DC magnets possible	
	Eliminate temporal constraints mandated by the use of pulsed, high-field magnets	

the ratio I_b/I_{st}, and it will be necessary to design devices with higher start currents, which involves such measures as shortening the slow-wave structure or splitting it into multiple segments, to avoid those instabilities.

Decreasing the operating magnetic field addresses a number of system constraints, but at the expense of raising additional issues in beam control. Because many systems use either a superconducting magnet for continuous operation — a system complication — or a long-pulse conventional magnet that limits the number of pulses that can be fired before the magnet pulse terminates, cutting the magnetic field is very desirable.

The role of the beam in pulse shortening is still not fully understood. The use of explosive emission cathodes is a foundation technology for these devices, since these cathodes are necessary to provide the multikiloampere currents that are required. The motion of the cathode plasma is a serious problem, however. Diode shorting and the growth of the beam radius may provide intrinsic limitations on pulse length and, as we saw, the ability to make operational TWTs with substantial energies per pulse. Once the beam enters the slow-wave structure, the underlying causes for the loss of the beam to the wall still need to be better understood before this limitation can be fully addressed.

Ultimately, device powers in unconstrained systems could be pushed into the tens of gigawatts if there is a demand for a system with the associated size, mass, and electrical power demand. Alternatively, one could build compact, more highly optimized BWO-based systems, with longer pulses at current output power levels of several, to perhaps 10 or so, gigawatts, or smaller MWCG-based systems offering similar performance. The limited work in TWTs points to some possible advantages in pulse length, given the lower internal fields in amplifiers, as well as provides a more desirable output mode, but their lower level of development presents more uncertainties.

8.7   Summary

O-type Cerenkov devices are among the most versatile of HPM sources, with gigawatt power levels demonstrated between 3 and 60 GHz, and possibly down into the L-band if that were desired. Further, they have been operated repetitively when driven by the SINUS series of pulsed power machines, which can provide pulse rates up to 1000 Hz. In this repetitive mode of operation, a major consideration is the coil producing the guide magnetic field. Fields are sufficiently high that superconducting coils are used for operation over any appreciable time period.

The pursuit of high power operation has taken two directions, one involving large-diameter multisection slow-wave structures that operate in a multiwave mode (MWCGs, MWDGs, and RDGs) and another involving highly optimized BWOs of much smaller size. Overall, it is recognized that larger-diameter systems are required to achieve both high power and high energies per pulse. This places a burden on the designer to control the output mode of the device. Multiple paths, including multiwave operation, Bragg reflectors, and multisection devices with internal mode converters, are all being pursued in this regard. Ultimately, the power of these devices is limited more by overall system constraints on size and mass than by the operational physics. Devices with power approaching 100 GW would be conceivable if there were any applications that made the associated size, mass, and electrical requirements tolerable.

The bulk of the work in this area has concentrated on oscillators. Prevention of oscillations is very difficult in such high power devices with the transients that accompany the pulsed power producing the beams. Nevertheless, high power TWT amplifiers have been explored, and power levels as high as 1 GW have been reached, although this latter system was plagued by a tendency to go into oscillation as the beam radius expanded over the course of a pulse.

Problems

1. Some slow-wave structures have a sinusoidally rippled wall with a wall radius described by Equation 8.8:

$$r_w(z) = r_0 + r_1 \sin(h_0 z)$$

with $h_0 = 2\pi/z_0$. Plot $r_w(z)$ over the range $0 \le z \le 3z_0$ for $r_0 = 1.5$ cm, $r_1 = 0.15$ cm, and $z_0 = 1.1$ cm.

2. We can approximate the dispersion curve (ω vs. k_z) for a slow-wave structure by using the dispersion curves for the TM_{0p} modes of a smooth-walled waveguide over the appropriate values of k_z. Before we turn to the slow-wave-structure dispersion curves, consider the dispersion curves for the TM_{0p} modes of a cylindrical waveguide of constant radius r_0. The dispersion curves for the smooth-walled waveguide are given by Equation 8.4:

$$\omega^2 = k_n^2 c^2 + \omega_{co}^2(0,p)$$

where, according to Equation 8.5,

$$\omega_{co}(0,p) = \frac{\mu_{op} c}{r_0}$$

with the μ_{0p} solutions to the equation

$$J_0(\mu_{0p}) = 0$$

and

$$k_n = k_z + n h_0$$

a. What are the four smallest values of μ_{0p}?

b. For a waveguide radius of $r_0 = 1.5$ cm, what are the four lowest cutoff frequencies, $f_{co} = \omega_{co}/2\pi$, for the four lowest-frequency TM_{0p} modes of the waveguide?

c. Plot the dispersion relation of Equation 8.4 for the four lowest-frequency TM_{0p} modes for k_z in the range

$$-\frac{3\omega_{co}(0,1)}{c} \le k_z \le \frac{3\omega_{co}(0,1)}{c}$$

d. Just so we recognize the TE_{0p} smooth-walled-waveguide modes if we ever come across them in an experiment, we note that their dispersion relation has the same form as the TM_{0p} modes, except that the cutoff frequencies for these modes are given by the following, with μ_{0p} replaced by v_{0p}, which are the solutions to

$$J_0'(v_{0p}) = 0$$

For our 1.5-cm-radius smooth-walled waveguide, what are the four lowest cutoff frequencies, $f_{co} = \omega_{co}/2\pi$, for the four lowest-frequency TE_{0p} modes of the waveguide?

e. Return to the TM_{0p} modes again. What is the frequency in GHz, $f = \omega/2\pi$, of the TM_{01} mode for $k_z = \omega_{co}/c$? What is the wavelength in centimeters in free space of the wave at that frequency? Within the waveguide, what is the wavelength along the axis for that same wave?

3. Consider a slow-wave structure with a sinusoidally rippled wall described by Equation 8.8, with $r_0 = 1.75$ cm, $r_1 = 0.1$ cm, and $z_0 = 1.2$ cm. Sketch, or plot using your computer, the three space harmonics, for n = −1, 0, and 1 in the expression for k_n, for the lowest three TM_{0p} modes of a smooth-walled waveguide with radius r_0 for k_z between $-3h_0/2$ and $3h_0/2$. Now sketch in the form of the slow-wave-structure dispersion curves for that range of k_z using the procedure outlined in Section 4.4.1.

Each mode of the slow-wave structure has its own continuous dispersion curve. Because the lowest-frequency mode of the SWS is well approximated by space harmonics of the smooth-wall-waveguide TM_{01} mode, we sometimes call it the TM_{01}-like mode, or even the TM_{01} mode, although we have to be careful what we mean when we use the latter shorthand. The dispersion curve for each mode extends over a range of frequencies known as its passband. Sometimes passbands overlap, even though the SWS dispersion curves do not intersect or cross.

a. What is the approximate value of the lowest frequency in the passband of the lowest-order, TM_{01}-like mode of the slow-wave structure? What are the wavenumbers for which the frequency has this value?

b. What is the approximate value of the highest frequency in the passband of the lowest-order, TM_{01}-like mode of the slow-wave structure? What are the wavenumbers for which the frequency has this value?

c. What is the range of frequencies over which the passband extends for the next-higher-order mode? What are the wavenumbers at the lowest and highest frequencies in the passband?

d. In the second-lowest-order passband of part (c), over what range of wavenumbers will the mode have the character of a TM_{01} mode of a smooth-wall waveguide? Over what range of wavenumbers will it have the character of a TM_{02} mode of a smooth-wall waveguide?

e. Would you expect that the TM_{01}-like mode would permit operation as a surface wave oscillator? Check against Equation 8.9.

4. Return to the slow-wave structure of Problem 3 and the dispersion curves we sketched for it. Now we will include a beam in the problem.

a. What velocity would a beam have to have in order for its associated beam line, $\omega = k_z v_b$, to intersect the dispersion curve of the second-lowest-order mode (i.e., the next one up from the TM_{01}-like mode) at the 2π point? The 2π point is where the wavenumber is such that $k_z z_0 = 2\pi$. A mode with this wavenumber goes through a phase shift of 2π for each section of the slow-wave structure.

b. At what values of ω and k_z does a beam line for a beam of the velocity you found in part (a) intersect the lowest, TM_{01}-like mode of the slow-wave structure?

5. The peak of the lowest mode of the slow-wave structure at $k_z = h_0/2$ is located near the intersection of the $n = 0$ and $n = -1$ space harmonics of the smooth-walled-waveguide TM_{01} dispersion curve. However, it is pushed down from that intersection by the formation of a gap between the lowest and the second-lowest dispersion curves for the slow-wave structure. That gap gets larger, and the peak of the dispersion curve is pushed downward, as the ripple, or the perturbation, in the wall of the slow-wave structure increases in size (e.g., as the ratio of r_1/r_0, terms given in Equation 8.9, gets larger).

a. For a slow-wave structure with the dimensions given in Problem 3 ($r_0 = 1.75$ cm, $r_1 = 0.1$ cm, and $z_0 = 1.2$ cm) and a beam with axial velocity $v_b = 2 \times 10^{10}$ cm/sec, how low does the peak frequency for the lowest TM_{01}-like slow-wave-structure mode have to be if the beam line is to intersect that curve at its peak, as in a surface wave oscillator?

b. What is the electron kinetic energy, $(\gamma_b - 1)mc^2$, measured in electron volts?

6. In this chapter, most of our discussion of the addition of an electron beam to the slow-wave structure has used the beam line approximation,

$$\omega = k_z v_b$$

in which we ignore the fact that there are actually two space-charge waves on the beam — which we can see clearly in Figure 8.7 — and ignore the frequency splitting between them. With more accuracy (but less than if we solved the complete dispersion relation in the references), it is customary to approximate the dispersion relation for the slow space-charge wave by

$$\omega = k_z v_b - \omega_q$$

where ω_q is an effective plasma frequency for a beam in a waveguide or SWS. Just as it is a reasonable approximation to ignore the effect of the beam in estimating the dispersion curves for the slow-wave structure, we can estimate the slow space-charge wave dispersion by ignoring the SWS and computing the dispersion curves for a beam in a smooth-walled waveguide.

Find the dispersion relation for the space-charge waves for a thin beam of radius r_b, current I_b, and electron kinetic energy $(\gamma_b - 1)mc^2$ in a waveguide of radius r_0 with perfectly conducting walls. Assume that (1) the beam is guided by a large enough magnetic field that electron motions are only axial along z, (2) all quantities are independent of the azimuthal variable θ and depend on z and t as $\exp[i(k_z z - \omega t)]$, and (3) the electromagnetic fields are transverse magnetic TM_{0p} waves with only E_r, E_z, and B_θ components. Further, assume that beam quantities can be written as a large zero-order piece plus perturbations created by the beam; for example, the electron velocity can be written as $v_b = v_{b0} + v_{b1}$, with $v_{b0} \gg v_{b1}$. Therefore, ignore all products of the small perturbations.

a. First derive the form of the electric and magnetic fields both inside the beam, $0 \le r < r_b$, and outside the beam, $r_b < r < r_0$. Choose the solutions for which the fields are bounded on the axis and obey the boundary condition at the perfectly conducting waveguide wall.

b. Use Newton's law (F = ma) and Maxwell's equations to relate the perturbations to the beam quantities v_{b1}, γ_{b1}, and n_{b1} (density) to the fields of part (a).

c. Match the electric and magnetic fields across the beam, taking account of discontinuities in the azimuthal magnetic field and the radial electric field at the beam. Hint: Because the beam is thin, you only have to integrate the current density across it:

$$\int_{r_b-\varepsilon}^{r_b+\varepsilon} j_b 2\pi r dr = I_b$$

where j_b is the current density in the beam.

    d. The resulting expression is your dispersion relation. It must be solved numerically to find ω vs. k_z, and the solutions will include the space-charge waves as well as electromagnetic waveguide modes. The space-charge modes will have dispersion relations straddling the beam line. The term ω_q in our earlier expression can be found numerically as a function of the beam parameters and k_z.

7. Using Equation 8.14 and the information in Figure 8.13, for a fixed beam–wall standoff of 5 mm (i.e., $r_0 - r_b = 5$ mm), plot the maximum power that an electron beam can provide as a function of wall radius r_0 in a smooth-walled waveguide for voltages V of (a) 500 kV, (b) 1 MV, and (c) 1.5 MV. Also, for each voltage, plot the beam current I_b and the impedance (V/I_b) vs. r_0. Remember that many of these devices operate at an impedance of about 100 Ω.

8. In this problem, we will scope out some of the major parameters of a large-diameter MWCG. We start by fixing the following design parameters:

    • Voltage V = 2 MV

    • Linear beam current density $I_b/2\pi r_b = 0.4$ kA/cm

    • Ratio of diameter D = $2r_0$ to free-space wavelength λ, $D/\lambda = 10$

    • Frequency f = $\omega/2\pi = 10$ GHz

    • Magnetic field B = 2 T

    a. Compute the beam velocity, v_b, neglecting any loss of beam energy to space-charge effects in the slow-wave structure (i.e., simply use the voltage to compute γ_b). We need this for our beam line.

    b. An MWCG operates at the peak of the lowest-frequency, TM_{01}-like passband in the π-mode. Find the value of the slow-wave-structure period z_0 that allows for a resonance with the beam line at the peak. Remember that this is an approximate calculation and that the exact value would depend on a full solution of the dispersion relation. How far in frequency below the intersection of the n = 0 and n = –1 space harmonics of the smooth-walled-waveguide TM_{01} modes does the peak of the SWS curve have to lie?

    c. If the beam spacing from the average wall radius, $r_0 - r_b$, is to be 6 mm, what is the beam current, given the fixed value of linear beam current density? What is the impedance, V/I_b? Is this value of beam current allowed by Equation 8.14?

d. Given the radius of the SWS, assuming that the field coils producing the guide field are placed up against the structure, what is the energy per unit length of SWS stored in the magnetic field, which has energy density $B^2/2\mu_0$?

9. Consider a 100-Ω BWO driven by a pulsed power generator with an output impedance of 50 Ω. The voltage at the BWO is 1 MV, and the BWO generates microwaves with 25% efficiency.

a. What is the output power of this device?

b. What voltage at the pulsed power generator, V_{PP}, is required to provide 1 MV at the BWO?

c. While we make much of source efficiency, the ratio of microwave output power to the electron beam power, system efficiencies are important as well. What is the system efficiency from pulsed power to microwave output? To be specific, what is the ratio of microwave power output to pulsed power input?

10. Dielectric Cerenkov masers (DCMs) use smooth-wall waveguides with dielectric liners to reduce the phase velocity of the normal electromagnetic modes of the structure so they can interact with the beam electrons.

a. Compute the cutoff frequency of the TM_{01} mode, in both radians/sec and Hz, for a cylindrical waveguide with perfectly conducting walls of radius 4 cm.

b. Now, assume a dielectric liner with relative permittivity $\varepsilon = 2$ fills the inside of the waveguide of part (a) between $r = 2$ cm and the wall. Derive a dispersion relation between ω and k_z for this modified waveguide. Show that the form of the dispersion relation will be the same as for the waveguide without the dielectric linear, albeit with a modified cutoff frequency for each mode.

c. The dispersion relation in part (b) will require a numerical solution. Without determining the actual value of the cutoff frequency for the TM_{01} mode, we can denote it ω_{co}. Derive an expression for the resonant frequency at which the beam line for a beam of energy γ_0 intersects the dispersion curve for the dielectric-lined waveguide.

References

1. Kompfner, R., The invention of traveling wave tubes, *IEEE Trans. Electron Dev.*, ED-23, 730, 1976.

2. Nation, J.A., On the coupling of a high-current relativistic electron beam to a slow wave structure, *Appl. Phys. Lett.*, 17, 491, 1970.

3. Kovalev, N.F. et al., Generation of powerful electromagnetic radiation pulses by a beam of relativistic electrons, *JETP Lett.*, 18, 138, 1973 (*ZhETF Pis. Red.*, 18, 232, 1973).

4. Bogdankevich, L.S., Kuzelev, M.V., and Rukhadze, A.A., Plasma microwave electronics, *Sov. Phys. Usp.*, 24, 1, 1981 (*Usp. Fiz. Nauk*, 133, 3, 1981).

5. Carmel, Y. et al., Intense coherent Cerenkov radiation due to the interaction of a relativistic electron beam with a slow-wave structure, *Phys. Rev. Lett.*, 33, 1278, 1974.

6. Aleksandrov, A.F. et al., Excitation of surface waves by a relativistic electron beam in an irised waveguide, *Sov. Phys. Tech. Phys.*, 26, 997, 1981 (*Zh. Tekh. Fiz.*, 51, 1727, 1981).

7. Aleksandrov, A.F. et al., Relativistic source of millimeter-wave diffraction radiation, *Sov. Tech. Phys. Lett.*, 7, 250, 1981 (*Pis'ma Zh. Tekh. Fiz.*, 7, 587, 1981).

8. The *orotron* is similar to an RDG; see Bratman, V.L. et al., The relativistic orotron: a high-power source of coherent millimeter microwaves, *Sov. Tech. Phys. Lett.*, 10, 339, 1984 (*Pis'ma Zh. Tekh. Fiz.*, 10, 807, 1984).

9. Bugaev, S.P. et al., Relativistic multiwave Cerenkov generator, *Sov. Tech. Phys. Lett.*, 9, 596, 1983 (*Pis'ma Zh. Tekh. Fiz.*, 9, 1385, 1983).

10. Bugaev, S.P. et al., Relativistic bulk-wave source with electronic mode selection, *Sov. Tech. Phys. Lett.*, 10, 519, 1984 (*Pis'ma Zh. Tekh. Fiz.*, 10, 1229, 1984).

11. Bastrikov, A.N. et al., "GAMMA" high-current electron accelerator, *Instrum. Exp. Tech.*, 32, 287, 1989 (*Prib. Tekh. Eksp.*, 2, 36, 1989).

12. Bugaev, S.P. et al., Interaction of an electron beam and an electromagnetic field in a multiwave 10^{10} W Cerenkov generator, *Sov. J. Commun. Tech. Electron.*, 32, 79, 1987 (*Radiotekh. Elektron.*, 7, 1488, 1987).

13. Bugaev, S.P. et al., Relativistic multiwave Cerenkov generators, *IEEE Trans. Plasma Sci.*, 18, 525, 1990.

14. Bugaev, S.P. et al., Investigation of a millimeter wavelength range relativistic diffraction generator, *IEEE Trans. Plasma Sci.*, 18, 518, 1990.

15. Korovin, S.D. et al., Relativistic backward wave tube with variable phase velocity, *Sov. Tech. Phys. Lett.*, 18, 265, 1992 (*Pis'ma Zh. Tekh. Fiz.*, 18, 63, 1992).

16. Korovin, S.D. et al., The effect of forward waves in the operation of a uniform relativistic backward wave tube, *Tech. Phys. Lett.*, 20, 5, 1994 (*Pis'ma Zh. Tekh. Fiz.*, 20, 12, 1994).

17. Moreland, L.D. et al., Enhanced frequency agility of high-power relativistic backward wave oscillators, *IEEE Trans. Plasma Sci.*, 24, 852, 1996.

18. Gunin, A.V. et al., Relativistic BWO with electron beam pre-modulation, in *Proceedings of the 12th International Conference on High-Power Particle Beams*, BEAMS'98, Haifa, Israel, 1998, p. 849.

19. Gunin, A.V. et al., Experimental studies of long-lifetime cold cathodes for high-power microwave oscillators, *IEEE Trans. Plasma Sci.*, 28, 537, 2000.

20. Bunkin, B.V. et al., Radar based on a microwave oscillator with a relativistic electron beam, *Sov. Tech. Phys. Lett.*, 18, 299, 1992.

21. Clunie, D. et al., The design, construction and testing of an experimental high power, short-pulse radar, in *Strong Microwaves in Plasmas*, Litvak, A.G., Ed., Russian Academy of Science, Institute of Applied Physics, Nizhny Novgorod, 1997, pp. 886–902.

22. Russia Locks Phasors and Boldly Goes ..., *Asia Times* online, October 21, 2001, http://atimes.com/c-asia/CJ27Ag03.html.

23. Shiffler, D. et al., Gain and efficiency studies of a high power traveling wave tube amplifier, *Proc. SPIE*, 1226, 12, 1990.

24. Abubakirov, E.B. et al., An X-band gigawatt amplifier, *IEEE Trans. Plasma Sci.*, 30, 1041, 2002.

25. Butler, J.M. and Wharton, C.B., Twin traveling-wave tube amplifiers driven by a relativistic backward-wave oscillator, *IEEE Trans. Plasma Sci.*, 24, 884, 1996.

26. Main, W., Cherry, R., and Garate, E., High-power dielectric Cerenkov maser oscillator experiments, *IEEE Trans. Plasma Sci.*, 18, 507, 1990.

27. Didenko, A.N. et al., Cerenkov radiation of high-current relativistic electron beams, *Sov. Tech. Phys. Lett.*, 9, 26, 1983 (*Pis'ma Zh. Tekh. Fiz.*, 9, 60, 1983).

28. Walsh, J., Marshall, T., and Schlesinger, S., Generation of coherent Cerenkov radiation with an intense relativistic electron beam, *Phys. Fluids*, 20, 709, 1977.

29. Kuzelev, M.V. et al., Relativistic high-current plasma microwave electronics: advantages, progress, and outlook, *Sov. J. Plasma Phys.*, 13, 793, 1987 (*Fiz. Plazmy*, 13, 1370, 1987).

30. Vlasov, A.N. et al., Overmoded GW-class surface-wave microwave oscillator, *IEEE Trans. Plasma Sci.*, 28, 550, 2000.

31. Kurilko, V.I. et al., Stability of a relativistic electron beam in a periodic cylindrical waveguide, *Sov. Phys. Tech. Phys.*, 24, 1451, 1979 (*Zh. Tekh. Fiz.*, 49, 2569, 1979).

32. Bromborsky, A. and Ruth, B., Calculation of TM_{on} dispersion relations in a corrugated cylindrical waveguide, *IEEE Trans. Microwave Theory Tech.*, MTT-32, 600, 1984.

33. Swegle, J.A., Poukey, J.W., and Leifeste, G.T., Backward wave oscillators with rippled wall resonators: analytic theory and numerical simulation, *Phys. Fluids*, 28, 2882, 1985.

34. Guschina, I.Ya. and Pikunov, V.M., Numerical method of analyzing electromagnetic fields of periodic waveguides, *Sov. J. Commun. Tech. Electron.*, 37, 50, 1992 (*Radiotekh. Elektron.*, 8, 1422, 1992).

35. Swegle, J.A., Starting conditions for relativistic backward wave oscillators at low currents, *Phys. Fluids*, 30, 1201, 1987.

36. Bugaev, S.P. et al., Features of physical processes in relativistic diffraction oscillators near the 2π oscillation mode, *Sov. J. Commun. Tech. Electron.*, 35, 115, 1990 (*Radiotekh. Elektron.*, 7, 1518, 1990).

37. Miller, S.M. et al., Theory of relativistic backward wave oscillators operating near cutoff, *Phys. Plasmas*, 1, 730, 1994.

38. Abubakirov, E.B. et al., Cyclotron-resonance mode selection in Cerenkov relativistic-electron RF sources, *Sov. Tech. Phys. Lett.*, 9, 230, 1983 (*Pis'ma Zh. Tekh. Fiz.*, 9, 533, 1983).

39. Bratman, V.L. et al., Millimeter-wave HF relativistic electron oscillators, *IEEE Trans. Plasma Sci.*, 15, 2, 1987.

40. Ginzburg, N.S., Kuznetsov, S.P., and Fedoseeva, T.N., Theory of transients in relativistic backward-wave tubes, *Radiophys. Quantum Electron.*, 21, 728, 1978 (*Izv. VUZ Radiofiz.*, 21, 1037, 1978).

41. Belov, N.E. et al., Relativistic carcinotron with a high space charge, *Sov. J. Plasma Phys.*, 9, 454, 1983 (*Fiz. Plazmy*, 9, 785, 1983).

42. Levush, B. et al., Theory of relativistic backward wave oscillators with end reflections, *IEEE Trans. Plasma Sci.*, 20, 263, 1992.

43. Levush, B. et al., Relativistic backward-wave oscillators: theory and experiment, *Phys. Fluids B*, 4, 2293, 1992.

44. Vlasov, A. et al., Relativistic backward-wave oscillators operating near cyclotron resonance, *Phys. Fluids B*, 5, 1625, 1993.

45. Hegeler, F. et al., Studies of relativistic backward-wave oscillator operation in the cross-excitation regime, *IEEE Trans. Plasma Sci.*, 28, 567, 2000.

46. Vlasov, A.N., Nusinovich, G.S., and Levush, B., Effect of the zero spatial harmonic in a slow electromagnetic wave on operation of relativistic backward-wave oscillators, *Phys. Plasmas*, 4, 1402, 1997.

47. Bugaev, S.P. et al., Investigation of a multiwave Cerenkov millimeter-wave oscillator producing gigawatt power levels, *Sov. J. Commun. Tech. Electron.*, 34, 119, 1989 (*Radiotekh. Elektron.*, 2, 400, 1989).

48. Koshelev, V.I. and Deichuly, M.P., Optimization of electron beam-electromagnetic field interaction in multiwave Cerenkov generators, in *High Energy Density Microwaves*, Conference Proceedings 474, Phillips, R.M., Ed., American Institute of Physics, New York, 1999, p. 347.

49. Bugaev, S.P. et al., Atmospheric microwave discharge and study of the coherence of radiation from a multiwave Cerenkov oscillator, *Sov. Phys. Dokl.*, 32, 78, 1988 (*Dokl. Akad. Nauk SSSR*, 298, 92, 1988).

50. Bugaev, S.P. et al., Generation of intense pulses of electromagnetic radiation by relativistic high-current electron beams of microsecond duration, *Sov. Phys. Dokl.*, 29, 471, 1984 (*Dokl. Akad. Nauk SSSR*, 276, 1102, 1984).

51. Bastrikov, A.N. et al., The state of art of investigations of relativistic multiwave microwave generators, in *Proceedings of the 9th International Conference on High-Power Particle Beams, BEAMS'92*, Washington, D.C., 1992, p. 1586.

52. Korovin, S.D. and Rostov, V.V., High-current nanosecond pulse-periodic electron accelerators utilizing a Tesla transformer, *Russian Phys. J.*, 39, 1177, 1996.

53. Data provided during a tour of the Institute of High-Current Electronics, Tomsk.

54. Zagulov, F.Ya. et al., Pulsed high-current nanosecond electron accelerator with pulsing frequency of up to 100 Hz, *Instrum. Exp. Tech.*, 19, 1267, 1976.

55. El'chaninov, A.S. et al., Stability of operation of vacuum diodes of high-current relativistic electron accelerators, *Sov. Phys. Tech. Phys.*, 26, 601, 1981.

56. Bykov, N.M. et al., High-current periodic-pulse electron accelerator with highly stable electron-beam parameters, *Instrum. Exp. Tech.*, 32, 33, 1987.

57. Chen, C. et al., A repetitive X-band relativistic backward-wave oscillator, *IEEE Trans. Plasma Sci.*, 30, 1108, 2002.

58. Bugaev, S.P. et al., Surface sterilization using low-energy nanosecond pulsed electron beams, in *Proceedings of the 10th International Conference on High-Power Particle Beams, BEAMS'94*, San Diego, CA, 1994, p. 817. The accelerator in this paper is referred to as SINUS-13 in Korovin and Rostov.[52]

59. Luybutin, S.K. et al., Nanosecond hybrid modulator for the fast-repetitive driving of X-band, gigawatt-power microwave source, *IEEE Trans. Plasma Sci.*, 33, 1220, 2005.

60. Korovin, S.D. et al., Repetitive nanosecond high-voltage generator based on spiral forming line, 28th International Conference on Plasma Science and 13th IEEE Pulsed Power Conference, *Digest of Papers*, 1249, 2001.

61. Gorbachev, K.V. et al., High-power microwave pulses generated by a resonance relativistic backward wave oscillator with a power supply system based on explosive magnetocumulative generators, *Tech. Phys. Lett.*, 31, 775, 2005 (*Pis'ma Zh. Tekh. Fiz.*, 31, 22, 2005).

62. Belousov, V.I. et al., Intense microwave emission from periodic relativistic electron bunches, *Sov. Tech. Phys. Lett.*, 4, 584, 1978 (*Pis'ma Zh. Tekh. Fiz.*, 4, 1443, 1978).

63. Gunin, A.V. et al., Relativistic X-band BWO with 3-GW output power, *IEEE Trans. Plasma Sci.*, 26, 326, 1998.

64. Kitsanov, S.A. et al., Pulsed 5-GW resonance relativistic BWT for a decimeter wavelength range, *Tech. Phys. Lett.*, 29, 259, 2003 (*Pis'ma Zh. Tekh. Fiz.*, 29, 87, 2003).

65. Shiffler, D. et al., A high-power two stage traveling-wave tube amplifier, *J. Appl. Phys.*, 70, 106, 1991; Shiffler, D. et al., Sideband development in a high-power traveling-wave tube microwave amplifier, *Appl. Phys. Lett.*, 58, 899, 1991.

66. Kovalev, N.F. et al., Parasitic currents in magnetically insulated high-current diodes, *Sov. Tech. Phys. Lett.*, 3, 168, 1977 (*Pis'ma Zh. Tekh. Fiz.*, 3, 413, 1977).

67. Aleksandrov, A.F. et al., Broadening of a relativistic electron beam in Cerenkov-radiation source, *Sov. Tech. Phys. Lett.*, 14, 349, 1988 (*Pis'ma Zh. Tekh. Fiz.*, 14, 783, 1988).

68. Burtsev, V.A. et al., Generation of microsecond microwave pulses by relativistic electron beams, *Sov. Tech. Phys. Lett.*, 9, 617, 1983 (*Pis'ma Zh. Tekh. Fiz.*, 9, 1435, 1983).

69. Hahn, K., Fuks, M.I., and Schamiloglu, E., Initial studies of a long-pulse relativistic backward-wave oscillator using a disk cathode, *IEEE Trans. Plasma Sci.*, 30, 1112, 2002.

70. Korovin, S.D. et al., Pulsewidth limitation in the relativistic backward wave oscillator, *IEEE Trans. Plasma Sci.*, 28, 485, 2000.

71. Bugaev, S.P. et al., Collapse of a relativistic high-current electron beam during generation of high-power electromagnetic radiation pulses, *Radio Eng. Electron. Phys.*, 29, 132, 1984 (*Radiotekh. Elektron.*, 29, 557, 1984).

72. Korovin, S.D., Rostov, V.V., and Totmeninov, E.M., Studies of relativistic backward wave oscillator with low magnetic field, in *Proceedings of the 3rd IEEE International Vacuum Electronics Conference*, Monterey, CA, 2002, p. 53.

73. El'chaninov, A.S. et al., Highly efficient relativistic backward-wave tube, *Sov. Tech. Phys. Lett.*, 6, 191, 1980 (*Pis'ma Zh. Tekh. Fiz.*, 6, 443, 1980).

74. Tkach, Yu.V. et al., Microwave emission in the interaction of a high-current relativistic electron beam with a plasma-filled slow-wave structure, *Sov. J. Plasma Phys.*, 1, 43, 1975 (*Fiz. Plazmy*, 1, 81, 1975).

75. Carmel, Y. et al., Demonstration of efficiency enhancement in a high-power backward-wave oscillator by plasma injection, *Phys. Rev. Lett.*, 62, 2389, 1989.

76. Goebel, D.M., Schumacher, R.W., and Eisenhart, R.L., Performance and pulse shortening effects in a 200-kV PASOTRON HPM source, *IEEE Trans. Plasma Sci.*, 26, 354, 1998.

9

Klystrons and Reltrons

9.1 Introduction

The distinctive feature of *klystrons* and *reltrons* is that the microwave-generating interactions in these devices take place in resonant cavities at discrete locations along the beam. The drift tube connecting the cavities is designed so that electromagnetic wave propagation between the cavities is cut off; without electromagnetic coupling between cavities, they are coupled only by the bunched beam, which drifts from one cavity to the next. To guide the beam along the drift tube, an axial magnetic guide field is applied, so that klystrons and reltrons (in those instances where a magnetic field is applied) are O-type devices. By contrast, in other O-type devices, such as the Cerenkov sources, free-electron lasers, or gyro-devices, microwaves are generated by beam-wave resonances that transfer beam energy to a wave over an extended interaction space or in a series of electromagnetically coupled cavities. The attractions of klystrons are their high power and efficiency, potentially wide bandwidth, and phase and amplitude stability.

All of the truly high power klystrons are relativistic, with beam voltages of the order of 500 kV or higher. As we show in Figure 9.1, though, relativistic klystrons are clearly differentiated into two groups on the basis of the beam impedance and, as we shall see, the bunching mechanisms that accompany low- and high-impedance operation. *High-impedance klystrons*, with impedances of the order of 1 kΩ, operate much the same way as conventional klystrons, albeit at higher voltages. In most cases, high power klystrons in this class have been developed as drivers for the radio frequency linear accelerators (RF linacs) in high-energy electron-positron colliders. There are two subclasses of high-impedance klystrons: those that operate at mildly relativistic voltages near 500 kV and those that operate at fully relativistic voltages of 1 MV or more. *Low-impedance klystrons* with impedances of about 100 Ω or less, on the other hand, feature much higher space-charge forces, which provide for a uniquely powerful bunching mechanism.

The *reltron* is similar to klystrons in that microwave power is extracted from a bunched beam using a set of output cavities; however, it is unique

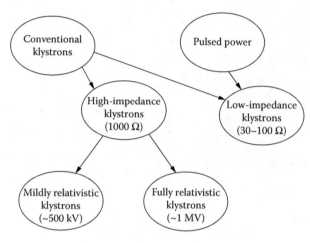

FIGURE 9.1
Classifying the different types of klystrons.

in two respects. First, the bunching mechanism differs from that in a klystron, and, second, the bunched beam is reaccelerated to higher energy to increase the energy withdrawn in the output cavities.

Representative performance parameters for the klystrons and the reltron are shown in Table 9.1. Note several features: (1) these devices are quite efficient by high power microwave (HPM) standards; (2) high power, high-impedance klystrons are optimized for very specialized application to RF linacs, where tunability is of limited interest; and (3) reltrons can be run without a magnetic field in some instances. High-impedance klystrons, par-

TABLE 9.1

Representative Parameters for the Klystrons and Reltron

	Sources			
	High-Impedance Klystrons			
Source Parameters	**Mildly Relativistic Klystrons**	**Fully Relativistic Klystrons**	**Low-Impedance Klystrons**	**Reltrons**
Frequency range	2.856–11.4 GHz	11.4–14 GHz	1–3 GHz	0.7–12 GHz
Peak power	150 MW	100–300 MW	10–15 GW	600 MW
Electronic conversion efficiency	50–60%	35–50%	40%	40%
Pulse width	3 μsec	35 nsec	50–100 nsec	Up to 1 μsec
Tunable range	na	na	—	±15%
Repetition rate	60 Hz	na	na	10 Hz
Output mode	—	—	TM_{01}	TE_{10}
Bandwidth	Limited	Limited	—	0.1%
Voltage	~500 kV	~1 MV	~1 MV	~1–1.5 MV
Electrical impedance	~1 kΩ	200–400 Ω	30 Ω	~700 Ω
Magnetic field	0.2 T	0.5 T	1 T	—

ticularly the mildly relativistic version, and the reltron are the most mature of the devices considered in this chapter. The klystron development efforts at the Stanford Linear Accelerator Center (SLAC) and at the High Energy Accelerator Research Organization (KEK) in Tsukuba, Japan, have created a line of highly optimized klystrons with power output of about 100 MW at frequencies from the S-band to the X-band. Similar technology might be scaled up to 1 GW if proposed multibeam versions were to be developed. Several versions of the reltron producing several hundred megawatts have been developed commercially for microwave effects testing at European laboratories, also over a wide range of frequencies that can be obtained through a mix of mechanical tuning and part change outs. Ultimately, the low-impedance klystron has generated the highest power, with proven performance of over 10 GW at 1.3 GHz. Using annular beams and a bunching mechanism that depends intrinsically on high beam currents, it offers the most potential for operation at gigawatt power levels.

9.2 History

Russell and Sigurd Varian invented the klystron in 1939, replacing the lumped-element L-C resonator of earlier devices with a resonant cavity in which fields interact directly with electron beams.[1] The Varian brothers derived the name *klystron* from the Greek verb *koyzo*, which describes the breaking of waves on a beach. The analogy is to the peaking of the electron density at the point of maximum bunching.

Through the 1940s, magnetrons were superior to klystrons in providing high power sources for radar. In the 1950s, though, klystron development was spurred by a growing understanding of the theory of space-charge waves and by the knowledge gained in the parallel development of traveling wave tubes. The basic configuration was scaled upward, and new design variants were introduced, most notably the *twystron* traveling wave hybrid amplifier and the *extended-interaction klystron*. Both use multiple cavities in the output section, which offers increased efficiency and bandwidth, as well as the added benefit of spreading the final interaction over a longer region to reduce the local fields and help prevent breakdown at high power.[2]

In the 1980s, the path to increased power split between high-impedance klystrons, either mildly or fully relativistic, and low-impedance klystrons. The high-energy physics community has been behind the development of high-impedance klystrons for their use as drivers in electron-positron colliders of increasingly higher energy. One branch of this effort is the so-called *SLAC klystron*, so named because the original S-band klystron of this type was developed at the Stanford Linear Accelerator Center, beginning in 1948, when a 30-MW, 1-μsec S-band klystron was developed for Stanford's Mark III linac.[3] In 1984, the Stanford accelerator was upgraded using 65-MW S-band klystrons,[4] and the current version of this 2.856-GHz klystron produces 3.5-μsec pulses at 65 MW with a repetition rate of 180 Hz; at shorter pulse lengths, 100-MW output levels have been achieved.[5] In the mid-1990s, teams at SLAC and KEK produced a variety of klystrons that generated 150 MW in the S-band,[6] later improved in Germany to 200 MW,[7] and up to 100 MW in the X-band, at four times the original SLAC klystron frequency (11.424 GHz). The desire to reduce the overall electric power demand by a future 1-TeV accelerator that was referred to as the Next Linear Collider (NLC) led to one of the most notable technology advances for this tube, the development of a 75-MW tube with pulsed periodic magnet (PPM) focusing,[8] which eliminated the power demand by the electromagnet used in earlier versions. This line of klystron development stalled in August 2004, however, when the International Technology Recommendation Panel chose a competing technology from DESY (Deutsches Elektronen-Synchrotron) based on super-conducting accelerating structures and lower-frequency, lower-power klystrons over the NLC concept for the future International Linear Collider, which has refocused klystron development in this direction.

Between the mid-1980s and mid-1990s, two groups focused on developing high-impedance klystrons operating at fully relativistic voltages. One group from the Lawrence Berkeley National Laboratory (LBNL) and the Lawrence Livermore National Laboratory (LLNL) explored an 11.424-GHz klystron as an alternative to the SLAC and KEK klystrons. Another group involving Russia's Joint Institute of Nuclear Research (JINR) in Dubna and the Budker Institute of Nuclear Physics (BINP) in Protvino investigated a 14-GHz driver for VLEPP, a Russian electron-positron collider concept.

The low-impedance relativistic klystron development traces to the high-current relativistic klystron amplifier (RKA) of Moshe Friedman, Victor Serlin, and coworkers at the Naval Research Laboratory (NRL) in the 1980s. Working at lower frequencies, in the L- and S-bands, they recorded power in excess of 10 GW in the mid-1990s. Motivated by a desire to push power levels still higher, they designed a triaxial version with an annular electron beam propagating within an annular drift space. The motivation was to increase the beam current while holding the current density relatively constant, and the triaxial configuration was intended to keep the cutoff frequency of the drift tube above the frequency of the microwaves generated. Workers at the Air Force Research Laboratory (AFRL) in Albuquerque and Los Alamos National Laboratory (LANL) took up triaxial RKA research in the late 1990s when the NRL effort halted.

The reltron grew out of early 1990s' investigations of the klystron-like *split-cavity oscillator* (SCO). Reltrons use the low-voltage bunching mechanism of the SCO, but then reaccelerate the bunched beam. This ordering, bunching at low voltage followed by further acceleration, somewhat reverses the order of acceleration and then bunching to take advantage of the greater ease of bunching at lower energies. Energy from the resulting high-energy bunched beam is extracted in klystron-like cavities. The reltron is used extensively for microwave effects studies in Germany, France, and Israel.

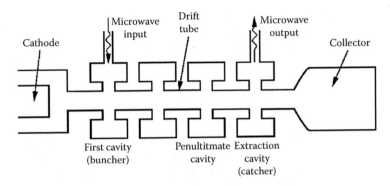

FIGURE 9.2
Basic configuration of a klystron.

9.3   Design Principles

Figure 9.2 shows a klystron in schematic form. The key elements are (1) the electron beam, (2) the axial magnetic field that guides the beam, (3) the interaction circuit, comprised of the cavities and the drift tube for the electron beam, (4) the input and output couplers for the microwaves, and (5) the beam collector. A microwave input signal excites a resonant electromagnetic mode of the first cavity. Each cavity has a gap that opens onto the drift tube, and electrons in the beam passing through the drift tube are accelerated or decelerated by the electric fields from the gap, depending on the phase of the fields as they pass by. Drifting between cavities, the faster electrons overtake the slower electrons, and charge bunches form on the beam. The drift tube separating the cavities is designed so that the electromagnetic fields do not propagate from one cavity to the next. Thus, communication between the cavities is via the bunched beam only. The fields excited in downstream cavities by the bunched beam enhance the bunching, so that the input signal is amplified in downstream cavities until the high power output signal is extracted from the final cavity.

As we shall show, there are some significant differences between the operating mechanisms of high-impedance klystrons, low-impedance klystrons, and reltrons. We therefore describe their operating and design principles in that order. First, though, we discuss the basic considerations in choosing the configuration of these devices.

9.3.1   Voltage, Current, and Magnetic Field

The electron beam is the energy source for the microwaves generated in a klystron. Let I_b be the beam current and V_0 the accelerating voltage for the beam. In *high-impedance klystrons*, such as the SLAC klystron, pencil beams with circular cross sections are employed. The current is typically low enough that the energy of a beam electron, $\gamma_b mc^2$, can be approximated by the expression

$$\gamma_b \approx \gamma_0 \equiv 1 + \frac{eV_0}{mc^2} \tag{9.1}$$

Low-impedance klystrons, on the other hand, use thin, hollow beams with annular cross sections. Because they operate at higher currents, the beam space-charge creates a significant electric field between the beam and the wall. The potential energy in that electrostatic field reduces the beam kinetic energy, so that γ_b is lower than the value in Equation 9.1. In *low-impedance klystrons*, one finds γ_b by solving the equation[9]

$$\gamma_0 = \gamma_b + \frac{I_b}{I_s \beta_b} \equiv \gamma_b + \alpha \gamma_b^3 \tag{9.2}$$

where γ_0 is as defined in Equation 9.1 and $\beta_b = v_b / c = (1 - 1 / \gamma_b^2)^{1/2}$. The factors I_s and α depend on the cross-sectional distribution of beam current; for a thin annular beam of mean radius r_b in a drift tube of radius r_0,

$$I_s = \frac{2\pi\varepsilon_0 mc^3}{e} \frac{1}{\ell n(r_0 / r_b)} = \frac{8.5}{\ell n(r_0 / r_b)} kA \tag{9.3a}$$

$$\alpha = \frac{I_b}{I_s \gamma_b^3 \beta_b} \tag{9.3b}$$

In solving Equation 9.2, one chooses the solution for γ_b that lies in the range $\gamma_0^{1/3} \leq \gamma_b \leq \gamma_0$. Note that as $r_b \rightarrow r_0$, $\alpha \rightarrow 0$ and $\gamma_b \rightarrow \gamma_0$; i.e., as the beam approaches the wall, the potential energy tied up in the electrostatic field in the space between the beam and the wall vanishes. This permits the high operating currents of the low-impedance klystrons, as well as high efficiencies, as long as the beam is kept near the wall (see Problems 1 to 6).

The magnetic field must be large enough to confine the beam within the drift tube and to keep the beam relatively stiff by preventing too much radial motion. For the pencil beams in high-impedance klystrons, the axial magnetic field required to confine the beam current is[10]

$$B_z(T) = \frac{0.34}{r_b(cm)} \left[\frac{I_b(kA)}{8.5 \beta_b \gamma_b} \right]^{1/2} \tag{9.4}$$

As noted in the reference, this is the field required to confine the beam to a radius r_b; however, one must also confine the RF current in the beam, which in a SLAC klystron can be about $4I_b$, so that, as a rule of thumb, the actual magnetic field should be about twice that in Equation 9.4 (see Problem 7).

9.3.2 Drift Tube Radius

The radius of the drift tube is limited by the requirement that the drift tube be cut off to the propagation of electromagnetic modes between cavities. As we showed in Chapter 4, the mode with the lowest cutoff frequency in a cylindrical waveguide of radius r_0 is the TE_{11} mode, for which the cutoff frequency is

$$f_{co}\left(TE_{11}\right) = \frac{\omega_{co}}{2\pi} = \frac{1}{2\pi}\left(\frac{1.84c}{r_0}\right) = \frac{8.79}{r_0\left(cm\right)} \; GHz \tag{9.5}$$

with c the speed of light. Since the operating frequency of the klystron must lie below this value, i.e., $f < f_{11}$, the radius of the drift tube is limited by the relation

$$r_0\left(cm\right) < \frac{8.79}{f\left(GHz\right)} \tag{9.6}$$

Thus, the maximum drift tube radius must decrease as the frequency increases. As an example, at 1 GHz, the upper limit on the drift tube diameter is about 17.6 cm, while at 10 GHz, the limit is 1.76 cm (see Problems 8, 9, and 10).

9.3.3 Klystron Cavities

Figure 9.3 shows examples of cavities used in high power klystrons; the cavity in Figure 9.3a is commonly used in SLAC klystrons, while that in Figure 9.3b is used in the low-impedance klystrons. Klystron cavities operate much like a lumped-element LC oscillator, as indicated by the model in Figure 9.4. Imagine that a pair of highly transmitting meshes with spacing d are placed across the beam tube at the gap and that a time-varying electric field $Ee^{-i\omega t}$ is imposed across the gap. The field induces a time-varying charge on each mesh with amplitude $q = CV = C(Ed)$, where C is the capacitance of the gap between the meshes and V is the voltage, with E assumed constant across the gap. As the charge flows between meshes, a current I flows though the wall of the cavity as shown, creating a magnetic field B. Thus, the gap acts like a capacitance, while the current-carrying walls of the cavity act like an inductor of inductance L, and the series combination of the two, driven by the electron beam, is an LC oscillator.

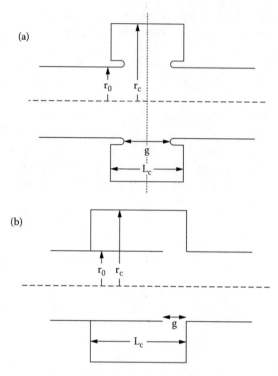

FIGURE 9.3
Cavities commonly used in high power klystrons: (a) a cavity used in a SLAC klystron; (b) a cavity commonly seen in low-impedance klystrons.

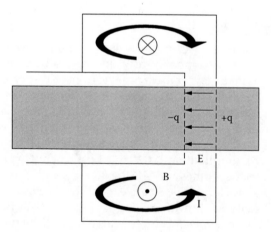

FIGURE 9.4
Model for visualizing a klystron cavity as an LC oscillator.

In terms of the length of a cavity, using the notation of Figure 9.3b, the two extremes are a *pillbox cavity*, with $L_c < (r_c - r_0)$, and a *coaxial cavity*, with $L_c > (r_c - r_0)$. In the latter case, for relatively long cavities with $L_c \gg (r_c - r_0)$, they can be modeled as a transmission line of characteristic impedance $Z_0 = 60 \ln(r_c/r_0)$ Ω, terminated on one end by a short circuit, representing the closed end of the cavity, and at the other end by a capacitance C, representing the cavity gap capacitance. One can show[11] that for a small gap capacitance, with $Z_0 \omega C \ll 1$, the cavity length is approximately given by

$$L_c\left(cm\right) \approx \left(2n+1\right)\frac{\lambda}{4} = \left(2n+1\right)\frac{7.5}{f\left(GHz\right)}, \quad n = 0, 1, 2, \ldots \qquad (9.7)$$

Here, $\lambda = c/f$, and the lowest-order axial mode has $n = 0$ (see Problem 11).

In designing a cavity, one chooses the dimensions and wall materials in order to meet a number of different, and sometimes competing, design criteria. The cavity must be resonant at the operating frequency of choice. Its quality factor, Q, must be low enough that the cavity has the desired bandwidth (since $Q \approx f/\Delta f$, with Δf the bandwidth) and that oscillations can build up on a timescale less than the duration of the electron beam (because, alternatively, one can write $Q \approx \omega E_s/\Delta P$, where E_s is the energy stored in the cavity and ΔP is either the power lost per cycle, which depends on both losses to the wall and the effective loss associated with the buildup of the gap field, or the energy input per cycle required to build up the energy E_s, in which case $Q \approx \omega \tau_B$, with τ_B the time for the buildup of oscillations), but high enough in the case of the input cavity that the input power required to modulate the beam be practical. In addition to Q, one designs the cavity to manage the field magnitudes at the wall and in the gap in order to prevent breakdown.

We compute the desired cavity parameters by solving the electromagnetic field equations. This can be done analytically using some approximations; for example, the fields for the cavity of Figure 9.3a, both inside the cavity and in the drift tube outside, were solved under the approximation that the electric field across the gap be a constant, equal to V_g/g, with V_g the gap voltage.[12] Historically, the characteristics of klystron cavities of the type shown in Figure 9.3b were computed and presented in graphs as functions of the ratios L_c/r_0, g/r_0, and r_c/r_0, with frequencies normalized to c/r_0.[13] Today, however, cavities are typically designed — or cavity designs are refined — using numerical tools such as the code SUPERFISH.[14]

In a multicavity klystron, individual cavities can be separately tuned to maximize gain, efficiency, or bandwith.[15] To maximize gain, each of the cavities is synchronously tuned with the desired signal frequency. To maximize efficiency, bunching is initiated with a set of synchronously tuned cavities, following which one or more of the cavities preceding the final cavity, most notably the penultimate cavity, is tuned upward in frequency so that at the signal frequency ω it appears to be inductive. Consequently,

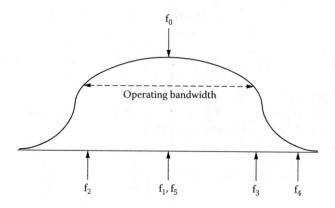

FIGURE 9.5
Qualitative relationship between tuning frequencies for a broadband five-cavity klystron amplifier. Frequencies f_1 and f_5 are for the input and output cavities, respectively, while f_4 is for the penultimate cavity.

the relative phasing of voltage and current is such that electrons in front of a bunch are decelerated, pushing them into the bunch, while electrons trailing a bunch are accelerated, again enhancing the bunching. Thus, the signal in the output cavity is increased in magnitude. Finally, to maximize bandwidth, the intermediate cavities are *stagger tuned* about the central frequency, while the penultimate cavity is tuned upward to enhance bunching, and the output cavity is designed to have wide bandwidth. This tuning scheme is illustrated in Figure 9.5.

In high power klystrons, the electric fields in the output cavity can be many times greater than the breakdown strength, so that the challenge for the designer is to increase the breakdown strength of the output cavity. One design technique that reduces the electric fields in the output section is the use of *extended-interaction* or *traveling wave* output sections, as shown in Figure 9.6. In the former, a small number of cavities with larger gaps are used in place of the single cavities of a conventional klystron. In the latter, the usual bunching cavities are used to generate the signal current on the beam, from which microwave energy is coupled out over a much longer traveling wave section, which is composed of coupled cavities in the figure. Note the dummy load on one end of the traveling wave section to prevent reflections and the buildup of oscillations, while the signal is coupled out on the other end.

9.3.4 Electron Velocity Modulation, Beam Bunching, and Cavity Spacing

The modulation of electron velocities by the electric field from the cavity gaps initiates the process that creates charge bunches on the beam downstream of the cavity. Interestingly, until the gap voltage becomes quite large and the bunching process becomes nonlinear, the evolution of the bunches

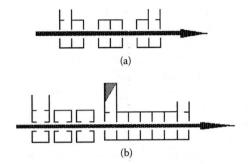

FIGURE 9.6
(a) Extended-interaction and (b) traveling wave structures for klystrons.

is governed by the intrinsic properties of the beam — specifically the electron energy and velocity, and the current density and beam distribution — while the magnitude of the bunching is determined by the intensity of the modulation, which of course is determined by the gap voltage. In this section, we will consider velocity modulation and the bunching process for a simple model relevant to classical, high-impedance klystrons.

To understand electron velocity modulation by the time-varying electric field in the gap of classical klystrons, return to our simplified one-dimensional model of Figure 9.4, ignoring the transverse dimensions and fields in the gap and, to lowest order, the electric field resulting from the beam's space-charge. An electron of charge $-e$ and mass m entering the gap of width g at $t = t_1$ with velocity v_b leaves the gap with velocity $v_b + v_1$, where

$$v_1 = -\frac{eEg}{m\gamma_b^3 v_b} M e^{-i\omega\left(t_1 + \Delta t/2\right)} \tag{9.8}$$

Here, $\gamma_b = [1 - (v_b/c)^2]^{-1/2}$, $\Delta t = g/v_b$ is the travel time across the gap, and

$$M \equiv \frac{\sin\left(\dfrac{\omega\Delta t}{2}\right)}{\left(\dfrac{\omega\Delta t}{2}\right)} \tag{9.9}$$

is the *modulation factor* for a uniform electric field, which is the same for every electron (see Problems 12 and 13). M takes account of the phase average of the time-varying force on an electron as it transits the planar gap, and the form in Equation 9.9 is specific to our assumption that E is constant across the gap. Three points are significant here. First, the magnitude of the modulation is proportional to the gap voltage $V_g = Eg$. Second, the factor γ_b^3 in the denominator of Equation 9.8 rapidly increases the difficulty of perturbing the beam velocity as the beam energy increases. Third, the combination $(g/v_b)M$ is purely sinusoidal in the gap travel time, Δt, with zeroes at $\omega\Delta t = 2n\pi$ $(n \geq 1)$, implying that the velocity of an electron is unchanged as it exits the gap if its travel time is one period of the oscillating field (see Problem 14 to consider alternatives to this modulation factor).

The formation of charge bunches on a velocity-modulated beam is a periodic process. Leaving the gap, the faster electrons overtake the slower electrons and bunches begin to form. Bunch formation is countered by the buildup of space-charge forces as the charge density in a bunch increases, however, and space-charge repulsion eventually causes debunching. The process is like that of compressing and later releasing a spring. The time and distance scale for bunch formation is determined by the interaction of the fast and slow space-charge waves on the beam, which in our simple one-dimensional model have the dispersion relation

$$\omega = k_z v_b \pm \frac{\omega_b}{\gamma_b} \tag{9.10}$$

where the $+$ sign goes with the fast space-charge wave and the $-$ sign with the slow. In this expression, $\omega_b = (\rho_b e/\varepsilon_0 m \gamma_b)^{1/2}$ is the relativistic beam plasma frequency (with ε_0 the permittivity of free space and ρ_b the magnitude of the charge density on the beam), which depends, as we mentioned earlier, only on the properties of the beam, and not on the strength of the modulation. In the small-signal, linear regime of operation, the interaction of these two waves sets the distance from the cavity at which bunching reaches a maximum:

$$z_b = \frac{\pi \gamma_b v_b}{2\omega_b} \equiv \frac{\lambda_b \gamma_b}{4} \propto \frac{\gamma_b^{3/2} v_b^{3/2}}{J_b^{1/2}} \tag{9.11}$$

where $\lambda_b = 2\pi v_b/\omega_b$ and J_b is the beam current density (see Problems 15 and 16). Beyond z_b, the beam debunches, and at $z = 2z_b$, the bunch disappears. The *Applegate diagram* of Figure 9.7, which takes account of space-charge repulsion,[16] shows the trajectories of individual electrons to illustrate this phenomenon. Equation 9.11 shows us that there is a strong increase in bunching distance with the beam energy, as well as a decrease as the beam current density increases. When the beam is generated in a Child–Langmuir diode at voltage V_0 (see Section 4.6.1), we can simplify the scaling still further in the nonrelativistic limit, where $\gamma_b \sim 1$, $v_b \propto V_0^{1/2}$, and $J_b \propto V_0^{3/2}$, to see that z_b is approximately independent of V_0 at low voltage; in the ultrarelativistic limit, however, $\gamma_b \propto V_0$, $v_b \sim c$ and $J \propto V_0$, so that $z_b \propto V_0$.

The key, of course, is to locate each successive cavity near a maximum for bunching from the previous cavity. Because of the importance of this fact, consider several real-world modifications of our simple explanation of mod-

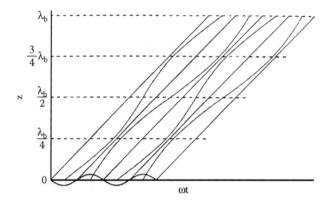

FIGURE 9.7
Applegate diagram taking account of space-charge repulsion within the bunches. The plot is for a nonrelativistic system with λ_b set to unity. (From Gilmour, A.S., Jr., *Microwave Tubes*, Artech House, Dedham, MA, 1986, p. 229. With permission.)

ulation and bunching in classical, high-impedance klystrons (in the next section, we will consider the unique features of low-impedance, relativistic klystrons). For example, real klystron accelerating gaps are not bounded by grids, and real electron beams have finite radial extent. Thus, the effective axial length of the gap is lengthened a bit because the gap equipotential lines bulge near the axis of the drift tube. Without the grids, which short out radial electric fields, and considering the finite radial extent of the beam, the modulation factor of Equation 9.9 must be replaced by the product of axial and radial modulation factors: $M \rightarrow M_a M_r$. The exact form of M_a and M_r depends on the configuration details.[17] In the particular case of a hollow electron beam of radius r_b, for an accelerating cavity like that shown in Figure 9.3a,

$$M_r = J_0\left(k_0 r_0\right) \frac{I_0\left(\theta_b\right)}{I_0\left(\theta_0\right)} \tag{9.12}$$

with $k_0 = 2\pi f/c$, $\theta_b = k_0 r_b/(\gamma_b \beta_b)$, and $\theta_0 = k_0 r_0/(\gamma_b \beta_b)$.[12]

In considering space-charge effects, we note that image charges on the walls of the drift tube reduce the effective value of the beam plasma frequency to a value $\omega_q < \omega_b$. A number of publications present computed values for the reduced plasma frequency.[18,19] The effect of the reduced plasma frequency is to increase z_b in Equation 9.11, and thus the spacing between klystron cavities.

Countering the tendency of the reduced plasma frequency to extend the distance over which bunching peaks, nonlinear effects tend to decrease that distance, as indicated in Figure 9.8. There we can see that the signal current induced on the beam by a first cavity at $z = 0$ peaks at a distance given by Equation 9.11, modified by replacing ω_b with ω_q. Placing a second cavity at z_b creates a stronger modulation that causes the signal current to peak over

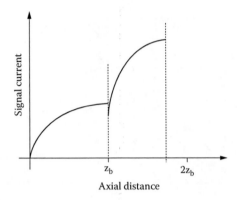

FIGURE 9.8
Nonlinear bunching effects downstream of a second cavity, with the first and second cavities located at $z = 0$ and z_b.

a shorter distance. The reason is that nonlinearities in the modulation process at higher gap voltages create high-energy electrons that overcome the space-charge repulsion force to maximize the signal current component over a shorter distance than the small-signal theory would indicate.[20]

We have also ignored radial motions of the beam to this point, which amounts to an assumption of a strong guiding magnetic field in the axial direction. As the magnetic field is reduced toward the *Brillouin limit*, where the magnetic field is just large enough to counter the space-charge expansion forces, space-charge bunching manifests itself as a modulation in the diameter of the beam also, which tends to reduce the effective beam plasma frequency still further and increase the distance between cavities.[21]

9.3.5   Beam Bunching in Low-Impedance Relativistic Klystrons

The bunching mechanisms in low-impedance relativistic klystrons are particularly intense, so that such devices typically consist of only two bunching cavities plus an extraction cavity. In fact, as we shall see shortly, the bunching mechanisms in the two cavities differ from one another in very significant ways.

The bunching mechanism in the first cavity is similar to that of classical klystrons: velocity modulation in the gap triggers the interaction of the space-charge waves on the beam, which determines the distance over which bunching takes place. A key difference, however, is that the frequencies of the fast and slow space-charge waves on an intense relativistic electron beam in a waveguide are not shifted symmetrically above and below the beam line, $\omega = k_z v_b$, as indicated by the one-dimensional result in Equation 9.10. Rather, as Briggs showed for long wavelengths (i.e., small k_z),[22] the space-charge waves propagating below the cutoff frequency of a waveguide have a dispersion relation given by

$$\omega = k_z v_b \left(\frac{1 \pm \alpha \mu}{1 + \alpha} \right) \tag{9.13}$$

where α is given in Equation 9.3b for a thin hollow beam and

$$\alpha \mu = \frac{1}{\beta_b} \left(\alpha^2 + \frac{\alpha}{\gamma_b^2} \right)^{1/2} \tag{9.14}$$

In solving Equation 9.13, remember that one must also solve Equation 9.2 at the same time to find γ_b and β_b if the known quantities for a given system are the beam current I_b and the anode–cathode voltage in the diode V_0. Using these relations, the bunching maximum downstream of the first cavity is found to be[23]

$$z_b = \frac{\beta_b^2 - \alpha}{4 \left(\alpha^2 + \alpha / \gamma_b^2 \right)^{1/2}} \left(\frac{c}{f} \right) \tag{9.15}$$

The scaling here obviously differs from that in Equation 9.11, derived in one dimension for a solid beam in a classical high-impedance klystron, in that the bunching length depends on the frequency f (see Problem 17). Other, more subtle differences in scaling exist as well, but the most unique feature of the bunching in low-impedance relativistic klystrons is that, as shown in the reference, because of the asymmetric frequency shift of the two space-charge waves, the alternating current (AC) components of the beam current and voltage are partially in phase, whereas in the case considered in the last section, they are out of phase by 90°. As a result, in the case at hand, a portion of the direct current (DC) energy of the beam is converted to AC kinetic energy, which is in turn converted to electromagnetic energy of the micro-waves, a direct DC-to-microwave process that does not occur in traditional high-impedance, low-current, and low-voltage klystrons.

The bunching process in the second cavity is more unique yet. The strength of the fields excited by the bunched beam arriving from the first cavity significantly alters the beam dynamics, creating a gating effect that basically adds an additional bunching effect that occurs right in the cavity. To under-stand this process, we rewrite Equation 9.2 with an additional term on the right to take account of the modification of the electrons' energies by the AC gap voltage V_1:

$$\gamma_0 = \gamma_b + \frac{I_b}{I_s \beta_b} + \frac{|e| V_1}{mc^2} \sin(\omega t) \tag{9.16}$$

One can show that when V_1 exceeds a threshold value,

$$V_1 \sin\left(\omega t\right) > V_{th} \tag{9.17}$$

with

$$V_{th} = \frac{mc^2}{|e|}\left\{\gamma_0 - \left[1 + \left(\frac{I_b}{I_{SCL}}\right)^{2/3}\left(\gamma_0^{2/3} - 1\right)\right]^{3/2}\right\} \tag{9.18}$$

the solution for γ_b and β_b in Equation 9.16 breaks down (see Problem 18). In this equation, I_{SCL} is the space-charge-limiting current discussed in Section 4.6.3, the maximum beam current that can propagate through a drift tube of radius r_0; for a thin annular beam of radius r_b (see Problems 22 and 23),

$$I_{SCL} = \frac{8.5}{\ell n\left(r_0 \, / \, r_b\right)}\left(\gamma_0^{2/3} - 1\right)^{3/2}$$

When the inequality in Equation 9.17 holds, additional current bunching arises at the output of the gap.[24] This additional AC component at the exit of the second gap is estimated to be $I_{1,\text{exit}} = I_b(2 \, / \, \pi)(1 - V_{th}^2 \, / \, V_1^2)^{1/2}$, which is plotted in Figure 9.9. The contribution to the AC current from the second gap rises rapidly as V_1 exceeds V_{th}, asymptotically approaching a maxi-

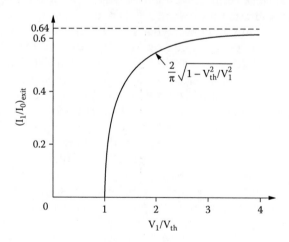

FIGURE 9.9
Current modulation at the exit of the second gap of a low-impedance klystron as a function of the gap voltage. The notation I_0 in the figure, from the reference, corresponds to the notation I_b used in the text. (From Lau, Y.Y. et al., *IEEE Trans. Plasma Sci.*, 18, 553, 1990. With permission.)

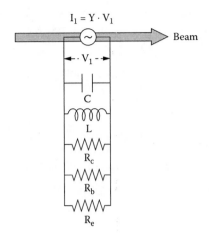

FIGURE 9.10
Lumped-element model of a klystron cavity and its coupling to the beam.

mum of about 0.64. Additional velocity modulation in the gap further bunches the beam downstream.

9.3.6 Circuit Modeling of Klystrons

While the detailed design of klystron cavities and beam tubes involves the use of computer codes that accurately model the electron beam dynamics, many times the properties of the cavities are described in terms of — and the parametric modeling of multicavity klystrons makes use of — lumped-element circuit models. The model for an individual cavity is as shown in Figure 9.10. The inductance and capacitance are arranged in parallel with three resistances: a beam resistance R_b, a cavity resistance R_c, and an external resistance R_e. These resistances represent, respectively, coupling to the beam, to the resistive cavity walls, and out of the cavity through openings or to external absorbers. The AC component of the gap voltage V_1 is related to the AC component of the beam current I_1 by Ohm's law:

$$I_1 = Y V_1 = \frac{1}{Z} V_1 \tag{9.19}$$

where the cavity impedance Z (or admittance Y) is the parallel combination of the shunt impedances in the cavity:

$$Z = \frac{R}{1 + iQ\dfrac{\omega^2 - \omega_0^2}{\omega\omega_0}} = \frac{R}{Q}\frac{1}{\left(\dfrac{1}{Q} + i\dfrac{\omega^2 - \omega_0^2}{\omega\omega_0}\right)} \tag{9.20}$$

with R the parallel combination of R_b, R_c, and R_e. In Equation 9.20, the cavity inductance and capacitance have been replaced by the resonant frequency ω_0 and the cavity quality factor, Q:

$$\omega_0 = \left(LC\right)^{-1/2} \tag{9.21}$$

$$Q \equiv \omega_0 \tau_{RC} = \omega_0 RC \tag{9.22}$$

(see Problems 19 and 20). On the right side of Equation 9.20, we have pulled out the characteristic impedance,

$$Z_c \equiv \frac{R}{Q} \approx \frac{1}{\omega C} \tag{9.23}$$

which is a constant factor for a given cavity geometry in the absence of the beam.[25] The cavity impedance in Equation 9.20 peaks at R when the cavity is resonant, with $\omega = \omega_0$, and drops off away from this frequency with a sharpness dictated by the size of the quality factor, Q.

To sketch out the overall behavior of a multicavity klystron, we generalize the model of Figure 9.10 and model each cavity as shown in Figure 9.11, with an inductance and capacitance, denoted for the mth cavity, L_m and C_m,

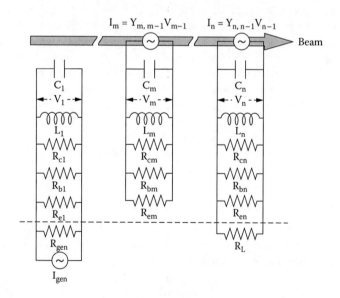

FIGURE 9.11

Lumped-element model for a multicavity klystron, with cavity 1 the input cavity, cavity m representative of an intermediate cavity, and cavity n the output cavity.

plus three parallel resistances, a beam resistance R_{bm}, a cavity resistance R_{cm}, and an external resistance R_{em}. The current source I_{gen} with shunt resistance R_{gen} represents the driving signal source for the klystron. It excites a gap voltage V_1 from the current flow through the cavity impedance. If the cavity is impedance matched to the generator, then $R_{gen} = Z_1$ and

$$V_1 = \frac{1}{2} I_{gen} Z_1$$

and the input power to the first cavity is

$$P_{in,max} = \frac{1}{2} \left| \frac{I_{gen}}{2} \right|^2 R_{gen} = \frac{1}{2} \left| \frac{V_1}{Z_1} \right|^2 R_{gen} \qquad (9.24)$$

This voltage, in combination with time-of-flight effects for the velocity-modulated beam as it travels to the second cavity, creates a signal current in the second cavity,

$$I_2 = Y_{21} V_1 \qquad (9.25)$$

where, in the linear approximation for high-impedance klystrons, the transadmittance from cavity 1 to cavity 2, Y_{21}, is given by[26]

$$Y_{21} = i \frac{4\pi}{Z_0} \left[\frac{I_0(kA)}{17} \frac{1}{(1+s)\beta_b \gamma_b} \right]^{1/2} \sin(k_b d) e^{i\omega d / v_b} \qquad (9.26)$$

with $Z_0 = 377 \ \Omega$ the impedance of free space, $s = 2 \ \ell n(r_0 / r_b)$, and

$$k_b = \frac{2\omega}{3c} \left[(1+s)^{1/2} - s^{1/2} \right] \left[\frac{17}{I_b(kA)} (\beta_b \gamma_b)^5 \right]^{1/2} \qquad (9.27)$$

Now I_2 excites a voltage V_2 in the second cavity via a relation like that in Equation 9.19. One thus proceeds iteratively using Ohm's law in each cavity to get the voltage and Equation 9.26 to get the current in the next cavity until one reaches the last cavity. At that point, one can collect all of the calculations performed previously to express the current in the last, or nth, cavity in terms of the voltage on the first,

$$I_n = Y_{n1} V_1 \qquad (9.28)$$

Assuming that the load resistance R_L is matched to the final cavity, so that $R_L = Z_n$, the output power into R_L can be written

$$P_{out} = \frac{1}{2}\left|\frac{I_n}{2}\right|^2 R_L = \frac{1}{2}\left|\frac{Y_{n1}V_1}{2}\right|^2 R_L \tag{9.29}$$

Combining Equations 9.24 and 9.29, we thus find the gain G for the klystron in the linear regime of operation to be

$$G = \frac{P_{out}}{P_{in,max}} = \frac{|Y_{n1}|^2 |Z_1|^2 R_L}{4R_{gen}} \tag{9.30}$$

9.3.7 Reltron Design Features

The reltron concept is as follows:[27] create a bunched beam at low voltage and then accelerate the electron bunches to relativistic velocities. As we see in Equations 9.11 and 9.15, this order of events makes the bunching easier. Further, the postaccelerated electrons will all be traveling at the same velocity, approximately the speed of light, so that velocity differences within the bunch become insignificant; the bunching created at low voltage in the first cavity is thus "frozen in." Because of the stiffness of the relativistic beam to deceleration, energy can be extracted over several cavities, which reduces the gap voltage per cavity and the chances of breakdown.

In addition to the postacceleration of the bunched beam, the reltron is also distinguished by a special bunching mechanism that, like the low-impedance klystron and the split-cavity oscillator (SCO),[28] is enabled by the use of intense electron beams at currents that lie close to the space-charge-limiting current. The reltron schematic is shown in Figure 9.12, and the bunching

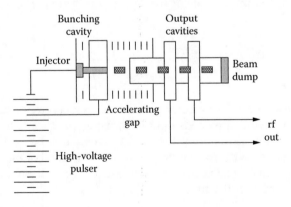

FIGURE 9.12
Schematic of the reltron. (From Miller, R.B. et al., *IEEE Trans. Plasma Sci.*, 20, 332, 1992. With permission.)

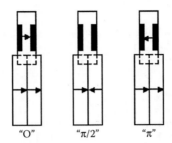

"O"   "π/2"   "π"

FIGURE 9.13
Lowest-order operating modes of the reltron bunching cavity. (From Miller, R.B. et al., *IEEE Trans. Plasma Sci.*, 20, 332, 1992. With permission.)

cavity and its three normal modes — 0, $\pi/2$, and π — are shown in Figure 9.13. The bunching cavity consists of two pillbox cavities coupled through magnetic slots to a third, side-mounted cavity. The electron beam passes only through the two pillbox cavities. The $\pi/2$-mode is unstable, so when the electron beam is injected, the electromagnetic fields in the cavities grow at the expense of the beam energy. When the beam current is an appreciable fraction of the space-charge-limiting current, a phenomenon formally similar to that in low-impedance klystrons comes into play, with the electrons essentially gated by the gap potential.

9.4   Operational Features

In the previous section, we discussed the design principles for high power klystrons and reltrons from a theoretical standpoint. In this section, we consider the operational features of actual devices that have been built and operated. We choose examples drawn from the three different types of devices: high-impedance klystrons employing pencil beams and operating at both near-relativistic and relativistic voltages to produce peak powers of roughly 100 MW; low-impedance klystrons employing intense annular beams to produce peak powers in excess of 1 GW; and reltrons that have produced hundreds of megawatts of power in pulses approaching 1 μsec.

9.4.1   High-Impedance, Near-Relativistic Klystrons

The development of this class of klystrons at SLAC and KEK in support of the NLC concept[29] placed a premium on phase stability, reliability, and cost, including both the capital cost of building the klystrons and the operating cost of the electricity, which in turn is affected by the klystron efficiency and magnet power demand. Although the NLC concept lost out in the competition for the International Linear Collider, the SLAC and KEK klystrons

TABLE 9.2

Key Parameters for the First of Two 150-MW S-band Klystrons
Delivered by SLAC to DESY

Beam voltage	535 kV
Beam current	700 A
Beam microperveance	1.78×10^{-6} A/V$^{3/2}$
RF pulse width at rep rate	3 μsec at 60 Hz
Cathode current density	6 A/cm^2 (max)
Cathode convergence	40:1 (13.3-cm drift-tube diameter)
RF output power	150 MW
Cavity gradients	<360 kV/cm
Saturated gain	~55 dB
Efficiency	40%
Operating frequency	2.998 GHz
Solenoidal magnetic focusing field	0.21 T

remain the highest-power pulsed tubes based on conventional microwave tube technology. To illustrate the features of the devices, we concentrate here on two particular device types: (1) the 150-MW S-band klystrons built by SLAC for DESY, the highest-operating-power klystrons from SLAC, and (2) the 11.424-GHz klystrons developed at SLAC using periodic permanent magnet (PPM) focusing of the electron beam to save on power otherwise expended on an electromagnet.

To give a sense of scale, the 150-MW tubes were about 2.6 m long, with a mass of about 300 kg.[7] Key parameters for the first of the tubes delivered to DESY are shown in Table 9.2.[30] Note first that the voltage is rather high for such a tube, exceeding 500 kV. Second, the cathode current density available from the M-type dispenser cathode originally used for this tube, 6 A/cm^2 or less, forced the designers to compress the solid-beam radius by a 40:1 ratio to reduce it to fit the 13.3-cm diameter drift tube. Lifetime issues later dictated a switch to a scandate cathode. Third, the saturated gain is quite high, about 55 dB. Designers conducted numerical simulations using a 2-dimensional particle-in-cell (PIC) code to fine-tune the cavity placements and frequencies, in the latter instance allowing designers to stagger the first six nonfundamental modes of the cavities to reduce the chances of self-oscillation. Nevertheless, additional work was required to deal with parasitic oscillations in the tube, including the later replacement of a section of the copper drift tube with a threaded and sandblasted stainless steel drift tube to increase signal losses between cavities. Fourth, in order to improve upon the greater than 40% efficiency in this first tube, the second tube featured a two-cell output cavity (an extended-interaction configuration) and an efficiency of about 50%. Finally, the magnetic field of 0.21 T produced by the 15-kW solenoidal field coil surrounding the klystron is about three times the minimum Brillouin field needed to confine and guide the electron beam through the tube.

In later experiments at DESY, the second tube was operated at over 200 MW of peak power in a 1-μsec pulse.[7] This higher power was achieved by

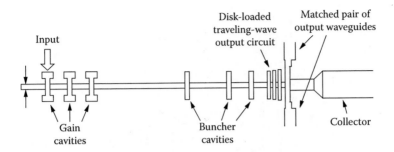

Input

Disk-loaded
traveling-wave
output circuit

Matched pair of
output waveguides

Gain
cavities

Buncher
cavities

Collector

FIGURE 9.14
Configuration for the SLAC 75-MW klystron using a solenoidal magnetic field coil.

raising the operating voltage and current to 610 kV and 780 A, which necessitated an increase in the guiding magnetic field by about 6%.

Concern for the power consumption of solenoidal field magnets in a projected NLC demanding thousands of klystrons first drove SLAC designers, and later designers at KEK, to develop an 11.424-GHz klystron with a PPM magnetic focusing system for the beam. The permanent magnets, which alternate in polarity to reduce their required size and weight,[31] produce a periodically varying magnetic field on axis for which the period must be small compared to the plasma wavelength on the beam for proper focusing (see the discussion surrounding Equation 9.11 and later discussion of the modification of the idealized one-dimensional value by two-dimensional effects in a drift tube). The disadvantage of the magnets is the lack of flexibility that goes with being unable to change the currents in coils to adjust field profiles and strengths without disassembling the tube.

Versions of PPM klystrons rated at 50 and 75 MW have been designed and built. Their configuration is essentially the same as that shown in Figure 9.14, which is that for a 75-MW klystron using a solenoidal field coil.[32] From the left in the figure, the first three cavities are reentrant to enhance the characteristic impedance, R/Q, and are stagger tuned to increase the klystron bandwidth. The next three to the right are pillbox cavities tuned well above the operating frequency to enhance bunching and final output power, as discussed in the previous section. The final cavity is a disk-loaded traveling wave section: four cells for the klystron with the solenoidal coil and five for the PPM klystron. These traveling wave sections spread the power across several cells, reducing the electric fields and preventing breakdown. The added cell in the case of the PPM klystron is a result of the lower perveance of that tube, which reduces the coupling to the cavity. The SLAC PPM klystron itself is shown in Figure 9.15.[33]

Table 9.3 shows parameters for four klystrons from SLAC — one S-band and three X-band, two using electromagnets and two with permanent magnets — and illustrates a number of the design differences within this class of tubes. First, the S-band klystron, with output power higher than the others by a factor of two or three, operates at the highest voltage and current, as

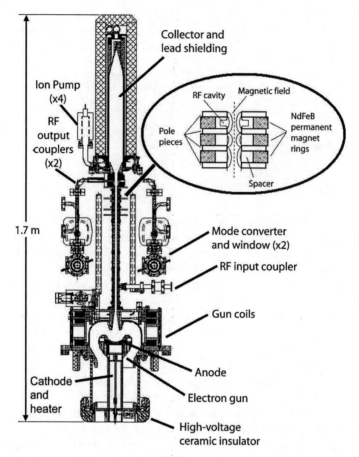

FIGURE 9.15
The 75-MW PP focused klystron (XP3 design).

TABLE 9.3

Comparison of the Second 150-MW S-Band Klystron Built by SLAC for DESY and Three X-Band Klystrons from SLAC, One with a Solenoidal Magnetic Field Coil and Two with PPM Focusing

	DESY, 150 MW	Solenoidal, 75 MW	PPM, 50 MW	PPM, 75 MW
Frequency, GHz	2.998	11.42	11.42	11.42
Beam voltage, kV	525	440	459	490
Beam current, A	704	350	205	274
Beam microperveance, $\mu A/V^{3/2}$	1.85	1.20	0.66	0.80
Cathode current density, A/cm^2	5.04	8.75	7.39	7.71
Cathode convergence	40:1	129:1	144:1	98:1
Axial magnetic field, T	0.20	0.45	0.20	0.17
Saturated gain	~55 dB	~55 dB	~55 dB	~55 dB
Efficiency	~50%	~50%	~60%	~55%

well as the highest perveance. While this makes more beam power available, there is a trade-off: the table shows that even though the gain for each klystron is about the same, the efficiency drops off with increasing perveance. Second, because the drift tube diameter scales approximately inversely with the frequency, the X-band tubes require considerably greater convergence in reducing the electron beam diameter from the cathode to the drift tube. This is the case despite the fact that the cathode current density is significantly higher for these tubes. Third, looking solely at the three X-band klystrons, the magnetic field in the PPM klystrons is half that for the tube with the electromagnet; i.e., the PPM klystrons operate closer to the Brillouin limit. This fact must be taken in context, though: the axial field for the electromagnet is roughly uniform with radius, while in the case of the permanent magnets, the radial dependence of the magnetic field within the drift tube is proportional to the modified Bessel function, $I_0(2\pi r/L)$, where L is the period of the magnetic field variation in the axial direction. This function increases with r from a value of unity on axis, behaving approximately like $\exp(2\pi r/L)/(4\pi^2 r/L)^{1/2}$ for large values of the argument. This behavior stiffens the beam, with a restoring force against radial displacements that increases with the size of the displacement.[34]

Finally, to put the energy savings using PPM technology into context, we consider the 75-MW X-band klystrons. Both are meant to operate at pulse lengths of $\tau_p = 1.5$ μsec and repetition rates of R = 180 Hz. Thus, the average microwave output is

$$\left\langle P_{\mu w} \right\rangle = P_{\mu w, peak} \tau_p R = 20.25 \text{ kW} \tag{9.31}$$

and the average beam power is

$$\left\langle P_{beam} \right\rangle = P_{beam, peak} \tau_p R = \frac{P_{\mu w, peak}}{\eta} \tau_p R \tag{9.32}$$

with η the beam-to-microwave power efficiency. The average beam power is 40.5 kW for the higher-perveance tube with an electromagnet and 36.8 kW for the PPM tube. However, the electromagnet adds an additional 25 kW,[31] so even ignoring the efficiency of converting prime power to electron beam power, the average power demand for the tube with an electromagnet is 65.5 kW, while that for the PPM tube is only 36.8 kW.

9.4.2 High-Impedance, Relativistic Klystrons

Truly relativistic klystrons with operating voltages well above 500 kV were considered as drivers for *two-beam accelerators* (TBAs),[35] in which microwaves generated in sources powered by a lower-voltage/higher-current beam were to be used to drive a higher-energy/lower-current beam in an RF linac to very high energies. The central idea was to reuse the drive beam, passing it through a set of interaction cavities to generate microwaves, then reaccelerating it back to its original energy and cleaning it up to pass it through additional downstream interaction cavities to extract further microwave output. In the originally proposed concept, each drive beam was to be the source of energy for 150 klystrons, which were to be distributed over a distance of about 300 m, with each klystron producing 360 MW.

Two research teams explored the use of klystrons as the microwave source in a TBA, one from Lawrence Livermore National Laboratory (LLNL) and Lawrence Berkeley National Laboratory (LBNL) in the U.S. (with early participation from SLAC) and the other from the Budker Institute of Nuclear Physics (BINP) in Protvino and the Joint Institute of Nuclear Research (JINR) in Dubna, Russia. Both teams' programs were terminated before they were completed, but their work is the source of a great deal of useful research.

The U.S. team's work divides rather neatly into two phases. In the first, experimenters used standard velocity modulation techniques to initiate the bunching process in a single klystron, while in the second, they used a beam-chopping technique in which the beam was electromagnetically displaced back and forth across an aperture by time-varying fields to break the beam into bunches for injection into a series of extraction and reacceleration sections. Four basic device types were explored in the first phase of the work:[36-38]

- SL-3, a multicavity klystron using a conventional electron gun and operating at 8.6 GHz, three times the basic SLAC S-band klystron frequency
- SHARK (for subharmonic-drive klystron), a two-cavity klystron with a 5.7-GHz subharmonic drive and 11.4-GHz output
- SL-4, a high-gain relativistic klystron that produced the highest output power of about 300 MW
- SHARK-2, a three-cavity version of SHARK

As an example, consider the SL-4, shown in Figure 9.16. This configuration was used in the MOK-2 experiments (MOK standing for multi-output klystron), with output power tapped from both the penultimate and final cavities. The cavity parameters for MOK-2 are given in Table 9.4.[37] The penultimate cavity, designed both for power extraction and to further bunch the beam for the final cavity, is a standing wave cavity with a drift tube diameter of 11.4 mm; it is located 21 cm downstream of the final gain cavity. As shown in the table, its characteristic impedance, R/Q, is 175 Ω. To avoid breakdown in this cavity, its output was held below 100 MW. The final cavity is a traveling wave structure with a beam aperture diameter of 14 mm. It consists of six $2\pi/3$-mode cells, has an RF filling time* of about 1 nsec, and the group velocity through it tapers from about 94% of the speed of light at

---

* The filling time is the length of the structure divided by the group velocity of waves through the structure.

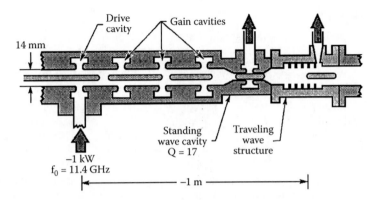

FIGURE 9.16

The SL-4 relativistic klystron used in the MOK-2 (Multi-Output Klystron) experiments. (From Allen, M.A. et al., *Part. Accel.*, 30, 189, 1990. With permission.)

TABLE 9.4

Cavity Parameters for MOK-2

Cavity Number	Cavity Type	Axial Location (cm)	$f - f_0$ (MHz)	Q^a	R/Q^b (Ω)
1	Drive	0	4	315	160
2	Gain	28	-31	117	160
3	Gain	42	23	122	160
4	Gain	63	45	119	160
5	Standing wave output	85	106	17	175
6	Traveling wave output	99[c]	0	—	—

Note: $f_0 = 11.424$ GHz.

[a] Beam-off Q.

[b] Calculated.

[c] To center of cell.

Source: From Allen, M.A. et al., Relativistic Klystrons, paper presented at the Particle Accelerator Conference, Chicago, 1989 (SLAC-PUB-4861; Lawrence Livermore National Laboratory, UCRL-100634; Lawrence Berkeley National Laboratory, LBL-27147; all 1989).

entry to about 90% at the output coupler, to raise efficiency by taking account of the loss of beam energy. At the designed output power level of 250 MW, voltage of 1.3 MV, and RF current of 520 A (*not* the DC current), the peak field level in the traveling wave coupler was to be limited to 40 MV/cm. In experiments, the maximum power extracted from the traveling wave structure itself was about 260 MW. The highest aggregate power from the two extraction cavities together was 290 MW, with 60 MW from the standing wave structure and 230 MW from the traveling wave structure. This latter was reached for a beam voltage and current of 1.3 MV and 600 A, respectively. Thus, for nominal input of 1 kW, the gain was about 55 dB and the efficiency was about 37%. Velocity spread in the injector, which became significant above 1.2 MV, was a limiting factor for the output power.

TABLE 9.5

Comparison of Single Resonant Cavity and Traveling Wave Output Structures at 11.424 GHz

Parameter	Single Resonant Cavity		Traveling Wave Structure	
	Scaling	Value	Scaling	Value
Breakdown power	$1/g^2$	80 MW	$4/L^2$	460 MW[b]
Surface E-field[a]	$1/g$	190 MV/m	$2/L$	80 MV/m
Average E-field[a]	$\Delta V/g$	75 MV/m	$2\,\Delta V/L$	25 MV/m
Filling time	$4Q/\omega$	2 nsec to 95%	L/v_g	1 nsec to 100%
$(R/Q)_\perp$	g	20 Ω	L	100 Ω

Note: g is the gap width of the single resonant cavity, L is the length of the traveling wave structure, ΔV is the beam energy loss, v_g is the group velocity in the traveling wave structure, and $(R/Q)_\perp$ is the transverse mode impedance for the hybrid EM_{11} mode, which leads to breakup of the beam and its loss on the walls.

[a] Field at 80-MW power level.
[b] Extrapolated.

Source: Allen, M.A. et al., Relativistic Klystron Research for Linear Colliders, paper presented at the DPF Summer Study: Snowmass '88, High-Energy Physics in the 1990s, Snowmass, CO, 1988 (SLAC-PUB-4733, 1988).

The use of a traveling wave structure for the output, resulting in a twystron-like device, reduces the risk of breakdown, thereby reducing the peak electric fields by an amount proportional to the length of the structure. Table 9.5 compares output structure parameters and their scaling[36] with the gap width, g, for a single resonant cavity, or the traveling wave structure length, L. Note the favorable reductions of electric fields in the traveling wave structure, as well as the shortened filling time. Unfortunately, the scaling of the transverse mode impedance for the hybrid EH_{11} beam breakup mode is unfavorable, being a factor of five higher for the traveling wave structure. Therefore, additional measures must be taken in the construction of this device to avoid beam breakup problems and the attendant pulse shortening.

Although the second phase of experiments, which were to be performed on the relativistic klystron two-beam accelerator (RTA)[39] shown in Figure 9.17, were never actually conducted, the planned system illustrates the complexity of the relativistic klystrons required for this application. The RTA was laid out over a total distance of 24 m; a 1.2-kA, 1-MV electron beam was to be launched from a dispenser cathode and accelerated to 2.8 MeV before passing into a beam-chopping cavity, which was the same as that used earlier in a device called the Choppertron,[40] shown in Figure 9.18. In this chopper, a beam is deflected back and forth across an aperture by a rotating cavity mode, resulting in a chopped beam with a bunch frequency (11.4 GHz) twice that of the signal in the chopping cavity (5.7 GHz). The decision to use this modulation mechanism was driven by the reduced effect of velocity modulation on relativistic beams, as illustrated in Equation 9.11 or 9.15. To avoid excessive current loss in the chopping cavity, it was necessary that the current bunches

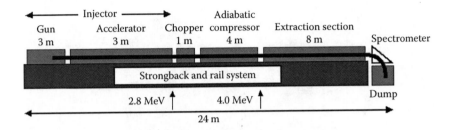

FIGURE 9.17
Layout of the relativistic klystron two-beam accelerator (RTA). (From Houck, T. et al., *IEEE Trans. Plasma Sci.*, 24, 938, 1996. With permission.)

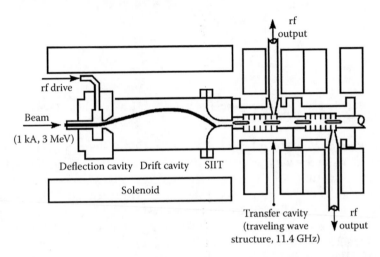

FIGURE 9.18
The Choppertron. The beam deflection cavity at the left was used later in the RTA. (From Allen, M.A. et al., *Part. Accel.*, 30, 189, 1990. With permission.)

created there be longer than desired in the final extraction section. Therefore, the next section of the RTA, the adiabatic compressor section, consisted of a set of additional accelerating cells interspersed with inductively tuned cavities, the latter included for the purpose of shortening the bunches (see Problem 21). The 4-MeV beam exiting the compressor section was to pass into an extraction section consisting of a set of eight klystron microwave extraction cells, each designed to produce 180 MW of 11.424-GHz output, interspersed with reacceleration cells to replace the beam energy lost in an extraction cell.

Ultimately, in a Two-Beam Next Linear Collider (TB-NLC) version (see Section 3.7), beam chopping and microwave extraction were to occur at 2.5 and 10 MeV, respectively. The klystron for that system was to produce 360 MW in the three-cell traveling wave output section shown in Figure 9.19. The beam into that section was to pass through a 16-mm-diameter drift tube with a DC current of 600 A and an RF current of 1150 A. Peak electric fields in the structure were to be 75 MV/m, and output coupling was through a pair of waveguides.

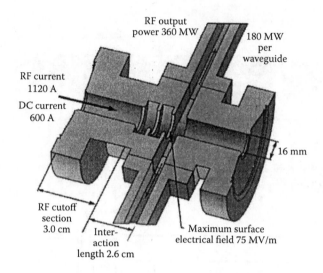

FIGURE 9.19
Planned output section for the proposed Two-Beam Next Linear Collider (TB-NLC). (From Houck, T. et al., *IEEE Trans. Plasma Sci.*, 24, 938, 1996. With permission.)

The key features of the work from BINP and JINR,[41] which has also been terminated, were the very large design gain, over 80 dB, and the large diameter of the drift tube for the beam. The large gain was necessitated by the system design goals of using a solid-state microwave source with an output power of order 1 W as the RF driver for the klystron, and obtaining an output power per klystron of about 100 MW. The design voltage was 1 MV, and the current was 250 A, for a microperveance of 0.25 (i.e., a perveance of 0.25 × 10^{-6} A/V$^{3/2}$), which is lower than that for the SLAC tubes of Table 9.3. To compare with the high-current annular-beam klystrons, this tube had 1 input cavity, 10 gain cavities, and a 22-cell traveling wave output structure; the total length was 70 cm.[42] Within the gain cavity region, PPM focusing of the beam was used, with a maximum field of 0.4 T and a root-mean-square (rms) average field of 0.28 T. The input cavity and each of the gain cavities were two-gap π-mode cavities with a loaded Q of 830. Output power was coupled out on both sides of the traveling wave output section. At a peak output power of 135 MW, the design value for the maximum surface electric field strength in the gain cavities was 300 kV/cm. The design gain was 83 dB and the efficiency 54%. The peak published power for the klystron was 100 MW.

With a relatively large drift tube diameter of 15 mm, the two lowest cutoff frequencies in the cylindrical drift tube between cavities, those for the TE$_{11}$ and TM$_{01}$ modes, are 11.7 and 15.3 GHz, respectively. The lower of these fell below the 14-GHz operating frequency for the klystron. Thus, the designers had to prevent parasitic oscillations in the TE$_{11}$ mode of the beam tube so that it did not propagate through the system. The importance of suppressing this mode was heightened by the fact that its field pattern tends to drive the beam into the wall, terminating operation of the klystron. To accomplish this goal, a lossy waveguide section using a glass–carbon material was inserted between the gain cavities to suppress parasitic oscillations. In particular, between the 9th and 10th cavities, the designers inserted a special filter for the TE_{11} mode as an extra measure to suppress it. Measurements verified that the lossy waveguide sections quenched the TE_{11} parasitic oscillations.

9.4.3   Low-Impedance Klystrons

The intense annular-beam klystron, developed as the RKA at the Naval Research Laboratory,[43] underwent roughly three phases of development: (1) from the late 1980s to the early 1990s, when the original design concept was improved until a maximum power of about 15 GW at 1.3 GHz was achieved; (2) then up to the mid-1990s, with improvements in gap design and output coupling; and (3) from the mid-1990s to the present, with the initial and incomplete development of the triaxial configuration aimed at simultaneously achieving high currents at lower voltages while maintaining the cutoff condition for the beam drift tube (see Problem 25). In addition to the NRL work, this device has been explored by Physics International (now known as L3 Communications Pulse Sciences),[44] Los Alamos National Laboratory,[45] and the Air Force Research Laboratory, where a variant called the relativistic klystron oscillator (RKO) was developed.[46]

In the first stage of development, the RKA produced output powers in excess of 10 GW in the L-band, operating at 1.3 GHz, and in excess of 1 GW in the S-band, operating at 3.5 GHz.[47] Figure 9.20 shows a schematic diagram of the L-band device. The first cavity has a rather large radius to allow microwave signal injection around the foil-less diode into the first cavity; there is one additional bunching cavity, and the beam is collected on the end

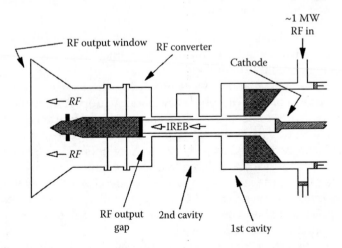

FIGURE 9.20
Schematic diagram of the NRL L-band RKA. (From Serlin, V. and Friedman, M., *IEEE Trans. Plasma Sci.*, 22, 692, 1994. With permission.)

TABLE 9.6

Comparison of Low-Impedance Klystrons from NRL
Operating in the L- and S-Bands[47]

	L-Band (1.3 GHz)	S-Band (3.5 GHz)
Electron Beam		
Beam voltage (kV)	1000.0	~500
Beam current (kA)	35	~5
Beam pulse duration (nsec)	160	120
Beam mean radius (cm)	6.6	2.35
Beam thickness (mm)	4	N/A
Drift tube radius (cm)	7	2.4
First Cavity		
Computed gap voltage (kV)	50	~25
Cavity impedance (Ω)	~50	50
Cavity length (cm)	N/A	19
Distance to current maximum (cm)	39	12
RF current at peak (kA)	1.5	0.3
Second Cavity		
Cavity impedance (Ω)	10	20
Cavity length (cm)	17	11
Distance to current maximum (cm)	39	14
RF current at peak (kA)	17	5
Output power (GW)	15	1.7
Microwave pulse length (nsec)	N/A	40 (triangular pulse)
Power efficiency (%)	~40	~60

Note: N/A = not available.

of the suspended inner conductor of the coaxial output converter, in which a TEM-like mode coupled out through the extraction gap is converted to a TM_{01} output mode. Table 9.6 compares the key design and operating parameters for the L- and S-band devices producing the largest output powers. Note that the radii of the drift tubes vary in rough proportion to the operating frequency, which is to be expected on the basis of Equation 9.6. Similarly, because the cavities have the same impedances in each device [e.g., the driven cavity in each case has an impedance of about $Z_0 = 60 \ln(r_c/r_0) \ \Omega \approx 50 \ \Omega$], the cavity radii scale with the frequency from one device to the other. On the other hand, the larger L-band device operates at higher voltage and current, as well as lower impedance. Further, the RF component of the current is much larger, and the output power is almost an order of magnitude higher than that of the S-band device.

Because of the rather large size of the cavities, they are able to support a number of modes, some lower in frequency than the operating frequency of the device. The designers addressed this issue in several ways. First, the driven and bunching cavities had different impedances and lengths to offset the competing modes. Thus, the two cavities were originally designed to have impedances of 50 Ω for the driven cavity and 20 Ω for the other (although, as we shall discuss, the impedance of the second cavity was later reduced in the L-band klystron). In the S-band klystron, the driven cavity had a length of about $(9/4)\lambda$, while the other had a length of about $(5/4)\lambda$ (see Equation 9.7); both, incidentally, were designed to be relatively narrow in the radial direction compared to their length in order to make their resonant modes more TEM-like. Mode control in the driven cavity of the L-band device was originally found to be dependent on the magnitude of the axial magnetic field, which was attributed to the fact that it was made of a nonmagnetic stainless steel with a weakly field-dependent permeability, μ. Lining the inner wall of the cavity with copper eliminated this problem. As an alternative means of mode control, thin, resistive nichrome wires were strung radially beween the inner and outer walls of the 20-Ω cavity in the S-band klystron at the axial location of the nodes in the field pattern for the mode resonant at the 3.5-GHz operating frequency, which reduced the Q for competing modes while leaving the Q of the resonant mode unaffected.

Beam bunching downstream of the first cavity, and the accompanying generation of an RF current component, proceeds as expected from the linear theory; in the L-band device

$$I_1\left(kA\right) = 1.5\sin\left(0.04z\right) \tag{9.33}$$

An oscilloscope trace of the measured current is shown in Figure 9.21. Earlier published data in a somewhat smaller, lower-power L-band klystron showed that the magnitude of the RF current was proportional to the square root of the input microwave power to the first cavity, which in turn is proportional to the magnitude of the gap voltage in the first cavity,[43] as expected.

As discussed in Section 9.3.5, when the gap voltage in the second cavity exceeds the threshold value given in Equation 9.18, a new mechanism creates

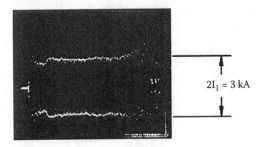

FIGURE 9.21
Oscilloscope trace of the envelope of the RF current I_1 at a distance beyond the first gap for which I_1 is a maximum. (From Serlin, V. and Friedman, M., *IEEE Trans. Plasma Sci.*, 22, 692, 1994. With permission.)

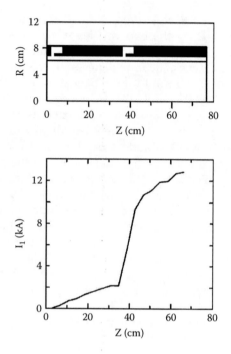

FIGURE 9.22
The geometry employed in computer simulations of an L-band RKA, and the RF current as a function of distance along the simulation geometry. (From Friedman, M. et al., *Rev. Sci. Instrum.*, 61, 171, 1990. With permission.)

a time-periodic gate in the gap, which actually causes the slow space-charge wave on the beam to cease propagating, so that substantial additional bunching occurs in the immediate vicinity of the gap. Computer simulations shown in Figure 9.22 illustrate the process on a 16-kA beam accelerated in a 600-kV diode; the beam had a diameter of 12.6 cm and a thickness of 0.2 cm.[48] In the L-band experiments, the first gap at z = 2.8 cm was driven at 1.3 GHz, and at z = 30 cm, I_1 = 2.6 kA. Just 2 cm past the second gap, however, I_1 = 5.5 kA, and I_1 continues to increase with distance from the gap, reaching 12.8 kA at 34 cm beyond the gap.

In initial experiments with a 20-Ω second cavity, I_1 was lower than expectations, which was blamed on beam loading of the cavity. Lowering the second cavity impedance to 10 Ω, with provision for mechanical tuning of the cavity while in operation, pushed the RF current back up to a value of almost 17 kA, as shown in Figure 9.23 for a 6-GW, L-band RKA. Note that the rise of the current is still rather slow.

The output signal in Figure 9.24 rises more rapidly than I_1 in the previous figure, but it begins to drop off immediately after reaching its peak about 20 nsec into the 80-nsec pulse. Even at this lower power level, the electric fields near the tip of the converter in Figure 9.20 are approximately 500 kV/cm, and open-shutter photographs of the tip of the converter and the radiating

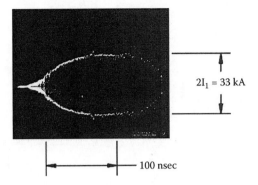

$2I_1 = 33\ \text{kA}$

100 nsec

FIGURE 9.23
Oscilloscope trace of the envelope of the RF current I_1 at a distance beyond the second gap for which I_1 is a maximum. (From Serlin, V. and Friedman, M., *IEEE Trans. Plasma Sci.*, 22, 692, 1994. With permission.)

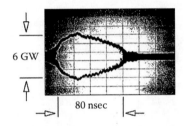

6 GW

80 nsec

FIGURE 9.24
Oscilloscope trace of the RF signal extracted into the atmosphere. (From Serlin, V. and Friedman, M., *IEEE Trans. Plasma Sci.*, 22, 692, 1994. With permission.)

horn showed electron emission from the converter. At power levels near 15 GW, the fields near the converter tip were approximately 800 kV/cm, and the power levels were erratic.

The power output in the S-band device was initially disappointingly low: about 150 MW. The power was increased when a second, tunable cavity was added to share the output gap with the output converter carrying power to the antenna, as indicated in Figure 9.25. When the added cavity was tuned exactly to the output frequency, power was increased to the maximum of 1.7 GW. The stub shown at the left end of the center portion of the extraction section in Figure 9.20 provides a second tunable element in both extraction sections (we note that it was also used on the S-band klystron, although it is not shown in Figure 9.25). The reflection coefficient for radiation between the upstream and downstream sections of the extraction section is a function of the height and location of the tuning stub.[49]

The TM mode output of the design in Figure 9.20 has been converted into the useful rectangular TE_{01} mode using the method shown in Figure 9.26.[50] Radial fins are slowly introduced into the center conductor of the coax at the output end and grow until they reach the outer conductor, where they

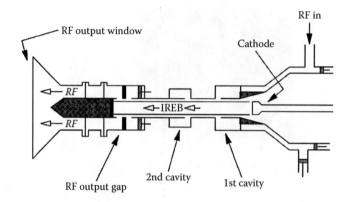

FIGURE 9.25
Schematic diagram of the NRL S-band RKA. (From Serlin, V. and Friedman, M., *IEEE Trans. Plasma Sci.*, 22, 692, 1994. With permission.)

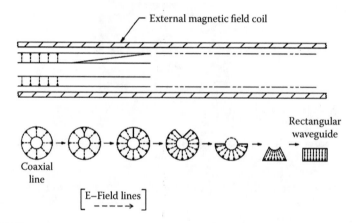

FIGURE 9.26
Mode converter smoothly changes a TM mode to a rectangular TE_{01} mode at the output of the relativistic klystron. (From Serlin, V. et al., *Proc. SPIE*, 1407, 8, 1991. With permission.)

divide and adiabatically transform into a rectangular cross section. Such a converter has been built for a high-current S-band klystron[51] and has been operated at 20 MW, although it suffered from breakdown due to short transition sections.

At the end of this first phase of development, the NRL team identified five problem areas to be addressed in the further development of their low-impedance klystrons:[51] (1) alignment of the drift region, the electron beam, and the axial magnetic field; (2) magnetic field uniformity; (3) beam loading of the cavities; (4) vacuum quality; and (5) x-ray bombardment. In the case of magnetic field uniformity, for example, they derived a requirement on the maximum allowable variation in the axial magnetic field,

$$\Delta B_z = 2B_z \left(\frac{r_0 - r_b}{r_b} \right) \tag{9.34}$$

They noted that while this uniformity is easily achieved with a DC magnetic field, the use of pulsed electromagnetic field coils requires that one take
account of eddy currents, skin depth, and magnetic field penetration through
materials, which typically translates to a requirement on the pulse length
for the magnet. In the case of the vacuum quality, they commented that the
vacuum level of 10^{-5} torr, customary for pulsed power experiments, was
inadequate at the highest power, with the result that breakdown could be
observed in either or both of the cavities, as well as in the converter region,
and that electron beam propagation could become unstable. The x-ray dose
generated in a relativistic electron beam system, which scales roughly
as $V_0^3 I_b$, with V_0 the accelerating voltage, necessitates the use of heavy shielding and creates additional breakdown problems in the klystron due to x-rays
hitting the wall, producing electrons. To reduce the dose, one can reduce the
operating voltage and increase the current proportionately, so that the dose
decreases as V_0^{-2}. To achieve that goal, the triaxial klystron was pursued in
the third stage of development, which we will describe in Section 9.5. Here,
though, we describe the second phase of development, which addressed
beam loading of the cavities as well as the conversion of electrostatic potential energy to beam kinetic energy in order to increase efficiency.

Two major steps were taken during the second phase of RKA development.
First, the gaps in the klystron cavities were opened up considerably, and a
cavity field pattern featuring a reversal of the axial electric field in the gap was
employed, as indicated in Figure 9.27.[52] To prevent the buildup of electrostatic
potential energy in the effectively wider drift tube in the vicinity of the gap,
metallic washers connected by thin conducting rods acting like inductors were
placed in the gap. Therefore, the gap was shorted out for the DC component
of the beam current, while it was open to the RF component of the current.
The wider gap reduced gap capacitance and beam loading; it also reduced
cavity Q, thus increasing the bandwidth of the device. The washer structure
prevented the penetration of radial electric field components, but allowed the
penetration of the azimuthal magnetic field, which had the fortuitous effect of
allowing the generation of an inductive electric field component that heightened the bunching effects in the second RKA cavity. In experiments comparing
wide- and narrow-gap RKAs, the former produced much higher RF current
components for the same beam; results are shown in Table 9.7.

The second step was the tapered reduction of the inner radii of the washers
in the extraction gap to recover a portion of the potential energy stored in
the decreasing gap between the beam and the wall, converting it into kinetic
energy of the beam electrons, which could in turn be converted to microwave
energy and extracted.[53] Remarkably, a device with this feature operated more
efficiently at a low axial magnetic field, with 60% power efficiency and 50%

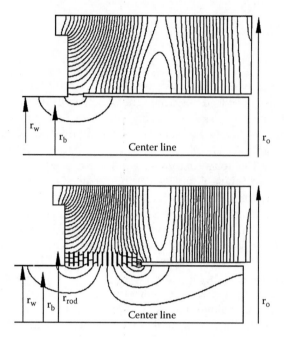

FIGURE 9.27
Electric field line calculations with SUPERFISH for narrow (upper) and wide (lower) gaps. (From Friedman, M. et al., *Phys. Rev. Lett.*, 74, 322, 1995. With permission.)

energy efficiency at 0.3 T; at 0.8 T, the output pulse length was considerably shorter, resulting in a lower energy efficiency. The reason was that at the lower magnetic field, electrons slowed by giving energy up to the microwaves were lost to the washers, while at the stronger magnetic field, those electrons remained in the gap, continuing to absorb microwave energy.

9.4.4 Reltrons

Reltrons are a compact source offering high efficiencies of 30 to 40% and a wide range of operating frequencies, from UHF (700 MHz) to X-band (12 GHz). This broad range was achieved in a system using several interchangeable modulating cavities and output sections at the fundamental frequency and multiples of two and three times the fundamental frequency.[54] An L-band reltron is shown in Figure 9.28. In L-band experiments, a 250-kV, 1.35-kA electron beam was bunched and then postaccelerated an additional 850 kV to produce a 600-MW microwave output at 1 GHz. The efficiency of converting electron beam power to microwave power output was about 40%. Pulse energy in the microwaves was about 200 J. Data from a long-pulse experiment in the L-band with output power of 100 MW over a pulse length of about 1 μsec are shown in Figure 9.29 (see Problem 26). Note that the microwave power begins about 200 nsec after the electron beam is injected. This time delay is the filling time of the modulation cavity, $\tau_f = Q/\omega$.

TABLE 9.7

Comparison of Wide- and Narrow-Gap RKAs

	Wide Gaps	Narrow Gaps
Beam current (kA)	16	16
Diode voltage (kV)	500	500
Beam mean radius (cm)	6.3	6.3
Drift tube radius (cm)	6.7	6.7
Washer inner radius (cm)	6.7	—
Washer outer radius (cm)	9.2	—
Washer thickness (cm)	0.075	—
Number of washers	23	—
First Cavity		
Gap width (cm)	10	~2
Q	~200	1000
RF current (kA)	4	5
Bandwidth	±5 MHz reduces RF current by 20%	<1 MHz reduces RF by 50%
Second Cavity		
Gap width (cm)	10	~2
Q	~450	1000
RF current (kA)	>40	14
Bandwidth	~3 MHz	—

Source: From Friedman, M. et al., Intense electron beam modulation by inductively loaded wide gaps for relativistic klystron amplifiers, *Phys. Rev. Lett.*, 74, 322, 1995.

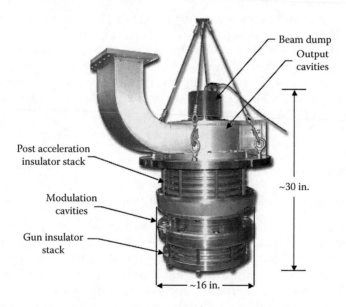

FIGURE 9.28
An L-band reltron. (Photograph provided by L3 Communications Pulse Sciences.)

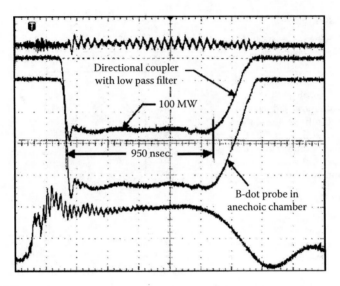

FIGURE 9.29
Long-pulse performance by an L-band reltron. (Courtesy of L3 Communications Pulse Sciences.) From the top down, the traces are for the local oscillator signal, the microwave power, the microwave signal in the far field, and the driving voltage. The horizontal sweep is 200 nsec/division.

In S-band operation, at 3 GHz, a 200-kV, 1-kA beam postaccelerated to 750 kV produced 350 MW of output power in 40-J pulses. Additional experiments demonstrated mechanical tunability of the output frequency by deforming the bunching cavity, giving a tunable range of ±13% and third-harmonic operation by using the 1-GHz bunching cavity with the 3-GHz output cavity. Based on these results, the performance chart shown in Figure 9.30 gives a rough idea of reltron scaling.[55] There, curves of constant output power and efficiency are plotted against the injector and postacceleration voltages. Reltrons continue to develop as a commercial product and are widely used in HPM effects testing facilities.

9.5 Research and Development Issues

In this section we focus on the areas of research and development that have been aimed at revolutionary, as opposed to evolutionary, changes in both the high- and low-impedance klystrons. In the former realm, researchers have proposed multibeam klystrons, in which a number of beams, each of which individually is generated at high impedance, are exposed collectively to the fields from the same cavities. In the latter realm of low-impedance klystrons, the focus has been on a larger-diameter, higher-current triaxial configuration.

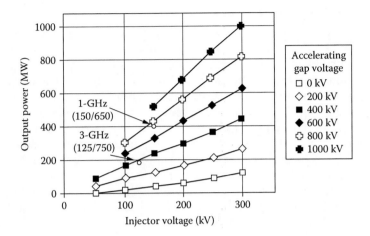

FIGURE 9.30
Reltron output power from calculations and data for different values of injector voltage. (From Miller, R.B. et al., *IEEE Trans. Plasma Sci.*, 20, 332, 1992. With permission.)

9.5.1 High Power Multibeam and Sheet-Beam Klystrons

SLAC researchers have considered, but not yet built, two design alternatives derived from SLAC klystron technology: the gigawatt multibeam klystron (GMBK),[56] aimed at truly high power operation, and the sheet-beam klystron (SBK),[57] conceived as a lower-cost alternative to more traditional pencil-beam klystrons. In the case of the former, multibeam klystrons have been around for some time, and a number of companies sell commercial versions that operate at the multimegawatt level, offering advantages in low-voltage, high-efficiency, wide-bandwidth performance. Figure 9.31 gives a sense of the size of a proposed gigawatt-level GMBK. Ten independent beams are to be launched from thermionic cathodes (necessitating a vacuum of $\sim 10^{-8}$ torr)

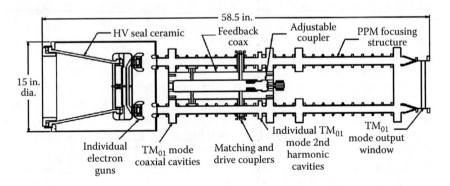

FIGURE 9.31
Proposed design for the gigawatt multibeam klystron (GMBK). (Courtesy of George Caryotakis, SLAC.)

at current densities of about 40 A/cm², which is within the state of the art, with a microperveance of 1.4 (which is comparable to the microperveance of the SLAC tubes shown in Table 9.3). The design voltage is 600 kV, and the total current is 6.7 kA, with a projected pulse length of 1 μsec and repetition rate of 10 Hz. The beams pass through four common stagger-tuned cavities, as well as an individual second-harmonic cavity for each beam to enhance bunching. Internal feedback and a gain of 30 dB would allow the tube to oscillate at a frequency of 1.5 GHz without a separate driving source. PPM focusing is used for each beam line separately. With a design output power of 2 GW, the efficiency is to be 50%; fields of 200 kV/cm in the output cavity are comparable to those in previous SLAC klystrons. Although the project was terminated before the tube was constructed, it was to have a length of about 1.5 m and a mass of about 82 kg, both parameters stated without including the power supply and auxiliary equipment such as cathode heaters and vacuum pumps.

The GMBK is one direction of possible development for conventional (SLAC) klystron development aimed at genuine HPM performance. The sheet-beam klystron is another suggested direction for development aimed at producing a simpler, lower-cost alternative to SLAC klystrons operating around 100 MW. The sheet-beam klystron employs a planar, strip-like electron beam, much wider in one cross-sectional dimension than the other. Although first proposed in the Soviet Union in the late 1930s, and despite the advantages its large lateral dimensions offer in terms of reduced cathode current density and lower power density in the cavities, sheet-beam klystron operation is complicated by the need for large drift tubes and overmoded cavities and by the attendant questions about beam transport. Further, the design of a sheet-beam electron gun promises to be a demanding engineering challenge, and to date, none have been built for this application. Nevertheless, the cost-driven requirement of the Next Linear Collider for a 150-MW peak, 50-kW average, 11.4-GHz klystron was regarded as potentially too demanding for a pencil-beam klystron with a 1-cm bore in the drift tube, which led to the recommendation for the double sheet-beam klystron shown in Figure 9.32.[57] This device features two sheet-beam klystrons operated side by side, each generating 75 MW of X-band radiation for a total of 150 MW, assuming roughly 50% power efficiency. Each klystron in this pair has an input cavity followed by two gain cavities, a penultimate cavity, and an output cavity. In this design concept, the beams from 450-kV, 320-A electron guns with a cathode current density of 5 A/cm² are magnetically compressed to a current density of 50 A/cm². The beam cross-sectional dimensions are 0.8 × 8 cm², and the drift tube has a height of 1.2 cm. The input, gain, and penultimate cavities are so-called triplets, consisting of three closely coupled cavities, as one can see in Figure 9.32, a cavity design that solved the problem of inadequate cavity coupling for a singlet cavity. The output cavity, on the other hand, has five of these cavities, and output from each beam and output cavity is coupled out through its own window. Calculations indicate a cavity isolation of 50 dB over 5 cm, and MAGIC simulations point to a 68-dB overall

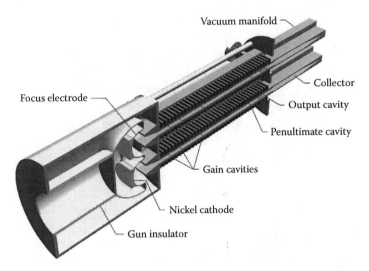

FIGURE 9.32
Cutaway schematic drawing of the double sheet-beam klystron. (Courtesy of George Caryotakis, SLAC.)

tube gain. Although the gun design issues were not yet fully addressed, the suggested performance, the greater simplicity of the device relative to a multibeam klystron, and the technical risk associated with a pencil-beam, 150-MW klystron argued in favor of building a SBK or double SBK for further experimental investigation.

9.5.2 Low-Impedance Annular-Beam Klystrons

Following the first two phases of development for low-impedance, annular-beam klystrons, the NRL group proposed a major design modification, the triaxial klystron with an annular electron beam propagating through a drift space bounded by both inner and outer cylindrical walls (see Problem 24).[47] The schematic layout of an experimental version is shown in Figure 9.33.[58] This configuration has two major advantages over the coaxial version. First, provided the transverse electromagnetic modes are suppressed — and they did not pose a problem in the experiments — this geometry can support large-radius electron beams as long as the distance between the inner and outer conductors is limited to about half of the vacuum wavelength of the microwaves produced. Second, the space-charge-limiting current for this geometry is, in principle, about twice that of the coaxial geometry, which we can see from the expression for the space-charge-limiting current for a thin annular beam, which is of the general form

$$I_{SCL} = I_s \left(\gamma_0^{2/3} - 1 \right)^{3/2} \tag{9.35}$$

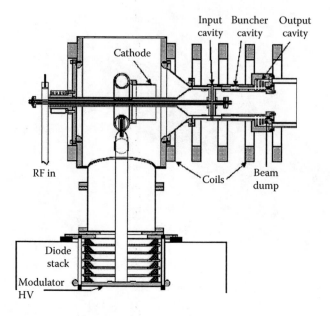

FIGURE 9.33
Layout of an experimental version of the triaxial klystron. (From Pasour, J. et al., *High Energy Density and High Power RF, 6th Workshop*, Berkeley Springs, WV, 2003, p. 141. With permission.)

The expression for I_s is given by Equation 9.3a for a coaxial klystron with beam radius r_b and drift tube radius of r_0; in a triaxial klystron with inner and outer drift tube radii of r_i and r_o, however,

$$I_s = 8.5 \left[\frac{1}{\ell n\left(r_o / r_b\right)} + \frac{1}{\ell n\left(r_b / r_i\right)} \right] \text{ kA} \qquad (9.36)$$

(see Problem 9.27). On a practical basis, though, if beam transport requirements limit the beam–wall standoff to a distance Δ that is approximately independent of beam radius, then

$$I_{SCL} \approx 8.5 \left(\frac{2r_b}{\Delta} \right) \left(\gamma_0^{2/3} - 1 \right)^{3/2} \text{ kA} \qquad (9.37)$$

This expression shows that the freedom to build large-radius triaxial klystrons while maintaining the cutoff condition between cavities provides an open-ended scaling with radius of the current-carrying capacity of the device. Two additional advantages of the triaxial device are (1) that a lower magnetic field can be used for beam confinement, in a geometry that would allow permanent magnets to be used, and (2) the inner conductor helps stabilize streaming and velocity–shear instabilities of the electron beam.

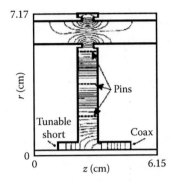

FIGURE 9.34
Input cavity of the experimental triaxial klystron of Figure 9.33. The field pattern was calculated with SUPERFISH. (From Pasour, J. et al., *High Energy Density and High Power RF, 6th Workshop*, Berkeley Springs, WV, 2003, p. 141. With permission.)

Returning to the experimental layout of Figure 9.33, we see first that the negative voltage pulse is fed to a corona ring to which the cylindrical cathode is attached. This configuration allows RF feed to the first cavity, as well as the support for the grounded inner conductor, to pass through the cathode ring and cylinder. The cathode emitter is CsI-coated graphite fiber, which produces a 4-kA beam at 400 kV, for a beam impedance of 100 Ω. The cathode radius is 7.2 cm; a slightly converging magnetic field of 0.2 to 0.3 T reduces the beam radius to $r_b = 6.25$ cm, at which it was optimized for microwave generation in the downstream cavities. The beam dump at the far end of the drift tube is a tapered graphite ring between the inner and outer conductors.

The signal is fed through a coax to a radial input cavity with a radius chosen to be an odd number of quarter wavelengths in order to maximize the axial component of the electric field at the beam radius. The cavity and a SUPERFISH calculation of the field lines are shown in Figure 9.34. For the 9.3-GHz frequency of the experimental klystron, the cavity extended $(7/4)\lambda$ in the radial direction; the field structure was reinforced by placing thin conducting pins at the expected location of the field nulls. These pins also provide additional mechanical support for the inner conductor as well as a return path for the beam current along the inner conductor. The bunching cavity downstream of the input cavity is a five-gap standing wave structure operating in the π-mode with a cavity spacing of 1.3 cm and a depth of a quarter wavelength, measured from the axis of the drift tube to the cavity wall. The field pattern for this structure is shown in Figure 9.35. A four-gap version of the output cavity is shown in Figure 9.36a. The gaps in the outer wall extract microwave power into a coaxial line; the gap spacing is tapered to maintain synchronism with the beam electrons as they lose energy and slow down. Slots cut into the inner wall match the gaps in the outer wall. Conducting pins in the outer gaps reinforce the desired field pattern there and also provide added mechanical strength for the structure. SUPERFISH calculations were used to establish the basic field

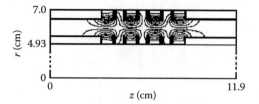

FIGURE 9.35
Five-gap standing wave bunching cavity for the experimental triaxial klystron of Figure 9.33. (From Pasour, J. et al., *High Energy Density and High Power RF, 6th Workshop*, Berkeley Springs, WV, 2003, p. 141. With permission.)

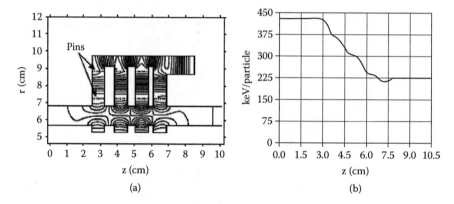

FIGURE 9.36
(a) Geometry of the output cavity used in a MAGIC simulation of the triaxial klystron of Figure 9.33. (b) Energy loss of electrons from a 440-kV, 5-kA beam with 75% current modulation, computed with MAGIC. (From Pasour, J. et al., *High Energy Density and High Power RF, 6th Workshop*, Berkeley Springs, WV, 2003, p. 141. With permission.)

structure in the output cavity, then MAGIC particle-in-cell simulations were employed to optimize the energy extraction efficiency from a prebunched beam assumed to have 75% current modulation, based on simulations of the upstream beam dynamics. MAGIC results indicated 50% energy extraction from a 440-kV, 5-kA beam, for a calculated output of 1.1 GW. The computed loss of electron energy in the output cavity is illustrated in Figure 9.36b, which shows that the bulk of the energy is lost in the first three gaps. The peak electric field of about 400 kV in the output gaps is about half of the Kilpatrick breakdown field.

In actual experiments, a commercial 250-kW pulsed magnetron provided the input signal. Current modulation from the first cavity reached a maximum about 10 cm downstream. An output power of about 300 MW was reached, for a gain of just above 30 dB. For a 420-kV, 3.5-kA electron beam, the efficiency was about 20%. Increasing the beam power also increased the output microwave power, but at the expense of pulse shortening by erosion at the tail end of the microwave pulse. Decreasing the magnetic field somewhat appeared to increase the pulse length, suggesting that multipactoring might be occurring in the output cavity. Multipactoring in the input cavity was a problem that was solved by polishing the cavity surfaces and reducing the sharpness of the cavity edges. In future experiments, multipactoring was expected to be reduced further by the use of polished copper cavity walls, rather than bronze and brass; improved vacuum; and better isolation of the beam dump. It was also expected that a redesign of the structure to allow better field penetration by the pulsed magnetic field, or even the adoption of permanent magnet focusing, would reduce beam interception by the walls, increasing efficiency further.

This set of experiments clearly demonstrated the desirable features of the triaxial klystron for high-frequency operation in the X-band. The triaxial geometry allowed the use of a beam with a radius almost as large as that of the L-band coaxial system (see Table 9.6). The impedance of roughly 100 Ω is comparable to that of the S-band coaxial klystron in Table 9.6, but at three times that of the L-band klystron, the X-band triaxial klystron does not have the beam power to produce multigigawatt output. Nor, in this initial set of experiments, does it have the necessary efficiency. Although the authors were concerned about the magnetic field penetration in the structure and the resulting effect on beam transport through the device, it was also not clear if the multigap bunching cavity produced the necessary total gap voltage to take advantage of the unique bunching mechanism described by Equations 9.17 and 9.18. Nevertheless, multigap cavities reduce the field stresses that can lead to breakdown.

The slight pulse shortening observed in this device is much less pronounced than that seen in the coaxial klystrons. When Friedman's NRL team raised the output power of their coaxial klystron to 15 GW, the output pulse was shortened to about 50 nsec, with dubious reproducibility. Repeatable power levels of about 6 GW in 100-nsec pulses were achieved. This problem of pulse shortening appears to be common to all high power microwave devices. As we have discussed, many of the problems in low-impedance, annular-beam klystrons have been identified. The triaxial configuration, which allows lower-impedance operation, solves some of these problems by reducing the required operating voltage to achieve a given power, thus reducing the voltage stresses that can create breakdown conditions at a number of locations, while also reducing both the generation of x-rays in the beam collector and the linear, circumferential current density of the beam, which eases beam transport problems. The foregoing tend to erode the microwave pulse from the back. Beam loading of the gap, on the other hand, tends to erode the head of the pulse; this problem was addressed with the use of large gaps. Multigap cavities may address the problem as well, although achieving adequate gain with such cavities may require additional cavities. The need for better vacuum has been recognized, although to date a system operating at higher vacuum has not been explored.

9.6 Fundamental Limitations

The maximum power output that can be obtained from a klystron is the smaller of two limits: that available from the beam and that determined by the breakdown strength at a number of points in the device. The former is expressed in terms of the efficiency, η, and the beam power:

$$P_{out} = \eta V_0 I_b \qquad (9.38)$$

Here, we use the anode–cathode voltage in the electron beam diode, V_0, and the electron beam current, I_b. The power efficiency for both high- and low-impedance klystrons has reached about 50%. We can expect that this is practically about the limit for high power klystrons, although, as we discussed in the case of high-impedance klystrons employing PPM focusing, overall system efficiencies can be improved somewhat with the use of permanent magnets, which draw no electrical power, in comparison with pulsed or superconducting electromagnets.

For several reasons, it is more desirable to increase the current rather than the voltage in Equation 9.38. According to the simplified model underlying Equation 9.11, lower voltages correspond to shorter bunching lengths, which reduces device length. We also know that lower voltages reduce both the size of the pulsed power and the chances for breakdown there. Finally, reducing the voltage limits the generation of x-rays in the beam collector, as well as those generated by beam interception within the klystron circuit.

9.6.1 Pencil-Beam Klystrons

The scaling of current with voltage and the terms in which that scaling is expressed vary with both voltage and the preferences of the research communities associated with different device technologies. At nonrelativistic voltages below about 500 kV, and even at voltages somewhat higher in the community associated with high-impedance, pencil-beam klystrons, it is common to express the current in terms of the perveance, K: $I_b = K V_0^{3/2}$. Thus, we can rewrite Equation 9.38:

$$P_{out} = \eta K V_0^{5/2}, \ V_0 < 500 \text{ kV} \qquad (9.39)$$

In the case of the high-impedance and low-perveance pencil-beam klystrons, experimental data from Thomson CSF (now Thales) klystrons shown in Figure 9.37 indicate that the efficiency declines as the perveance grows[59]; however, the reference shows that the efficiency can be recovered by adding more bunching cavities, at the cost of added length and complexity.

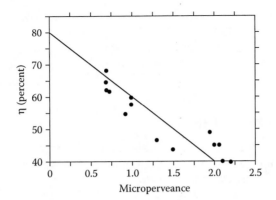

FIGURE 9.37
Reduction of the efficiency with increasing perveance in low-impedance, pencil-beam klystrons.
(From Palmer, R. et al., *Part. Accel.*, 30, 197, 1990. With permission.)

Two approaches to increasing the power of this class of klystron were discussed in Section 9.5.1: multibeam and sheet-beam klystrons. The former currently represents the lower-risk approach, both because there is greater experience with multibeam klystrons at lower voltage and power and because even sheet-beam devices would probably have to use multiple beams to approach gigawatt-level performance. The two left columns of Table 9.8 contain a comparison of the 150-MW, S-band SLAC klystron produced for DESY and the proposed multibeam gigawatt klystron (MBGK). Both operate at comparable voltages, and the perveance per beam in the MBGK is actually somewhat lower than that in the lower-power device. By using 10 individual beams, though, the total current in the MBGK is actually about 10 times higher, and the output power, assuming that 50% efficiency can be achieved, is about 7 times that in the single-beam S-band klystron.

TABLE 9.8

Comparison of SLAC Klystron and NRL/RKA and the Enhanced Current Versions Proposed for Each, the GMBK and Triaxial RKA, Respectively

	SLAC/DESY S-Band	GMBK (proposed)	RKA	Triaxial RKA (proposed)
Frequency, GHz	2.998	1.5	1.3	1.3
Beam voltage, kV	525	600	1,000	700
Beam current, A	704	6700	30,000	100,000
Beam microperveance, $\mu A/V^{3/2}$	1.85	1.4[a]	30	171
Beam impedance, Ω	746	90	33	7
Axial magnetic field, T	0.20		1.0	
Peak output power, GW	0.15	2.0	15.0	
Saturated gain	~55 dB	36 dB		
Efficiency	~50%	50%	~50%	

[a] Per beam, for each of 10 beams.

Obviously, employing multiple beams brings its own complexity to devices of this type, although this technology base will continue to grow at lower power levels, since multibeam klystrons are currently a front-runner for the new S-band superconducting approach for high-energy electron positron colliders.[60]

9.6.2   Annular-Beam Klystrons

In the case of low-impedance, annular-beam klystrons, which operate at voltages of about 500 kV and up, the impedance Z is more appropriate and more commonly used, $I_b = V_0/Z$; now

$$P_{out} = \eta \frac{V_0^2}{Z} \text{ , } V_0 \geq 500 \text{ kV} \qquad (9.40)$$

Once again, to achieve high power while limiting the voltage, one wants to operate at low impedance. The right two columns of Table 9.8 illustrate the parameters for both the 15-GW coaxial klystron and a proposed high power triaxial device. These two klystrons differ from the pencil-beam devices in two important ways: the beams are generated with explosive emission cathodes, and the annular beam is topologically more similar to a planar sheet, rather than pencil, beam. The same issues of efficiency vs. perveance therefore do not apply, and we can see that the perveances for annular-beam klystrons are much higher.

Producing high power with an annular-beam klystron while simultaneously controlling the processes that contribute to pulse shortening will require one to address the limitations of these devices in five areas. First, one must provide the requisite current at a manageable voltage, preferably at or below a megavolt. Theoretically, the triaxial configuration has twice the limiting current of the coaxial device, although, as we discussed earlier, the practical current limit for the triaxial geometry is much larger still, given the requirement for a certain beam standoff distance from the wall of the drift tube. Explosive emission diodes can produce the requisite current, albeit at the expense of their well-known problems with gap closure. A thermionic gun operating at 400 Ω (300 kV, 750 A), with a cathode loading of 8 to 10 A/cm², has been considered for further experiments with a triaxial klystron,[58] although operation at this higher impedance negates the advantage these devices have in current-carrying capacity in order to ensure more stable repetitive operation.

Second, having generated the beam, one must properly align the beam and the magnetic field within the drift tube, and the magnetic field must be quite uniform along the drift tube as well. These issues were discussed in Friedman et al.[51] Poor alignment can cause beam electrons to be lost to the walls, which in turn causes outgassing, compromising vacuum quality in the device. Similarly, the beam collector must be isolated to prevent contaminating the drift region with impurities released there; one must also address the issue of x-ray generation in the collector.

Third, one must address the issues of cavity design. In the second phase of annular-beam klystron development, capacitive loading in the bunching cavity — which erodes pulses from the leading edge — and potential energy recovery in the extraction region were addressed at gigawatt levels in coaxial devices. In addition, practical matters of cavity wall materials and radiusing of sharp gap edges were found to increase pulse lengths by preventing erosion from the trailing edge of the pulse. In the set of experiments performed with a triaxial klystron, multigap standing wave cavities were employed to distribute the power and reduce the electric fields. This raises the issue of whether a cavity of this type will generate the unique bunching mechanism described by Equations 9.16 through 9.18. If not, this might well raise a fourth issue: achieving adequate gain in these amplifiers. The lower the gain, the higher the required input power to achieve a given output power, and gains of 30 dB in these devices (as opposed to the gains of 50 to 60 dB in pencil-beam klystrons) would require more input power than commercial microwave sources could supply. The issue was raised somewhat in passing in the experimental paper on the triaxial klystron: it may be necessary to use more than one intermediate bunching cavity to raise the gain if one were to desire tens of gigawatts in a large device maximized for power production.

The fifth issue is one that has been appreciated for some time, but one that has not been fully explored in annular-beam klystrons: improving the vacuum levels from typical values of about 10^{-5} torr. This engineering matter involves the use of ceramic, as opposed to plastic, insulators; different vacuum-pumping technology; and greater attention to system cleanliness, perhaps even involving *in situ* discharge cleaning. The improvements to be gained by lowering the vacuum levels remain to be fully explored in these devices.

One issue that has not been addressed in these devices is tunability. The three-cavity structure — input cavity, bunching cavity, and output structure — limits the opportunities for stagger tuning, as one would in conventional or high power, high-impedance klystrons. Mechanical tuning of the cavities, or the addition of cavities if higher gain or stagger tuning is desired, is a possibility if the intrinsic bandwidth is found to be too narrow for some applications.

9.6.3 Reltrons

Reltrons to date have been compact workhorse tubes capable of producing hundreds of megawatts of output for a very broad range of frequencies, made possible by changing out key parts to achieve large frequency jumps, while using mechanical tuning for finer frequency adjustment within different frequency bands. In this power range, either no magnetic field is required or, at the higher power levels requiring higher beam currents, permanent magnets have been adequate. At lower repetition rates, of the order of 10 Hz, velvet cathodes have been employed. If higher power is required, up to

1 GW or beyond, higher current and voltage in the injector may create the need for a guiding magnetic field. Certainly higher repetition rates will demand metallic explosive emission cathodes (outgassing from the velvet limits repetition rates) or perhaps thermionic emitting cathodes.

9.7   Summary

High power klystrons and reltrons are among the most mature of all HPM sources. The extension of the conventional klystron into the relativistic domain resulted in the production of a 200-MW S-band klystron. More significantly, though, an X-band 75-MW klystron was designed to use periodic permanent magnet focusing of the electron beam, a technological advance that eliminates the power consumption of pulsed or superconducting electromagnets. The next steps in development for these tubes, either a sheet-beam klystron to simplify device design and construction costs or a multibeam klystron to achieve gigawatt-level operation, are stalled, the former because of a decision in 2004 to base the future International Linear Collider on a different technology path, and the latter because of lack of sponsorship at this time for this approach to high power klystrons.

Low-impedance, annular-beam klystrons have progressed through the development of annular-beam devices capable of something more than 10 GW of output in the L-band to a new stage of development based on a proposed triaxial geometry that promises higher-current operation with higher output power, as well as better device scaling to higher frequencies. In addition, this geometry may be amenable to the use of periodic permanent magnet beam focusing. The first experiments have been performed with this new configuration. Device performance is rather well understood below several gigawatts of output power, although pulse shortening is an increasingly troublesome issue as power increases above that, and the realm above 10 GW remains largely uncharted territory.

Problems

1. For an anode–cathode accelerating voltage $V_0 = 1$ MV, drift tube radius $r_0 = 5$ cm, and beam radius $r_b = 4$ cm for a thin annular beam, plot the beam current I_b vs. the relativistic factor for an electron, γ_b.

2. The curve of I_b vs. γ_b has a maximum, which is the maximum current that can be propagated through the drift tube for a given accelerating voltage V_0 and beam radius r_b, assuming a thin annular beam. Derive an expression for the maximum current, which is known as the

space-charge-limited current, I_{SCL}, as a function of the drift tube radius, r_0, r_b, and the factor $\gamma_0 = 1 + eV_0/mc^2$.

3. Compute the space-charge-limited current I_{SCL} (see Problem 2) for $V_0 = 1$ MV, $r_0 = 5$ cm, and

   a. $r_b = 3$ cm

   b. $r_b = 4$ cm

4. For each value of $I_b < I_{SCL}$, Equation 9.2 yields two values of γ_b. Which value of γ_b — the larger or the smaller — is the physically significant solution to Equation 9.2 (i.e., the real one)? Explain your choice.

5. For $V_0 = 0.8$ MV, $r_0 = 5$ cm, and $r_b = 4$cm, compute γ_b and β_b for a beam electron.

6. For $I_b > 0$, $\gamma_b < \gamma_0$. In fact, γ_b decreases as r_b decreases. Where is the "lost" kinetic energy when the kinetic energy of an electron is $(\gamma_b - 1)mc^2 < eV_0$?

7. What magnetic field is required to confine a pencil beam of current $I_b = 400$ A and radius $r_b = 2$ cm that was generated by a 500-kV electron gun? For convenience, assume $\gamma_b = \gamma_0$.

8. Which mode has the lowest cutoff frequency in a cylindrical drift tube (which, of course, acts as a waveguide)? Which mode has the second-lowest cutoff frequency in a cylindrical waveguide?

9. In a cylindrical waveguide, what is the maximum radius at which waves of the following frequencies are cut off?

   a. 1 GHz

   b. 3 GHz

   c. 5 GHz

   d. 10 GHz

10. In the absence of a beam, purely electromagnetic waves propagate in a cylindrical waveguide according to the dispersion relation

$$\omega^2 = \omega_{co}^2 + k_z^2 c^2$$

where $\omega = 2\pi f$ and $k_z = 2\pi/\lambda$, with λ the wavelength along the waveguide axis. Let $f = 1.3$ GHz. Suppose that over a distance of 10 cm, you want a 1.3-GHz electromagnetic wave to be attenuated by at least 30 dB. Assuming that the waveguide walls are perfectly conducting (i.e., ignoring resistive losses in the wall), what must the waveguide radius r_0 be to ensure at least that level of attenuation?

11. Assuming that the gap capacitance C of a coaxial cavity of characteristic impedance Z_0 is small (i.e., $Z_0 \omega C \ll 1$), estimate the three smallest values of the allowable lengths of the cavity at the following frequencies:

    a. 1 GHz

    b. 3 GHz

    c. 5 GHz

    d. 10 GHz

12. The modulation factor M for a planar beam transiting a thin gap is given by Equation 9.9. Compute M for a gap of width 5 cm and the following parameter values, assuming $\gamma_b = \gamma_0$:

    a. $V_0 = 200$ kV, f = 1.3 GHz

    b. $V_0 = 500$ kV, f = 1.3 GHz

    c. $V_0 = 500$ kV, f = 3 GHz

    d. $V_0 = 500$ kV, f = 10 GHz

13. Derive the expression for the modulation factor M in Equation 9.9 for a beam of large cross section transiting a gap of width g between two parallel plates with a time-varying voltage $V = V_1 e^{-i\omega t} = -Ege^{-i\omega t}$ applied across the gap. Assume the field does not vary with distance across the gap and that the change in the electron velocity due to the modulation is much smaller than the velocity of the electrons entering the gap, v_b.

14. Developers of the low-impedance relativistic klystron found that large gaps could be used to mitigate beam-loading effects on the cavities that were eroding the leading edge of the microwave pulse. The gaps were so large that the sign of the electric field reversed in the gap. Derive the modulation factor M for a gap across which the electric field varies as

$$E(z) = E, \; 0 \le z < g/2$$

$$= -E, \; g/2 \le z \le g$$

15. Estimate the bunching length for a 500-kV, 300-A pencil beam (i.e., a beam of circular cross section and radius r_b) used in a 1.3-GHz klystron with circular cross section. For this back-of-the-envelope calculation, use the simplified one-dimensional result in Equation 9.11 and ignore any space-charge depression of the beam energy, so that Equation 9.1 holds. Assume the charge density is a constant across the beam. To determine the beam radius, you will need to determine the drift tube radius. For this, choose the cutoff frequency for the TE_{11} mode of the drift tube to be 1 GHz, and maintain a 5-mm standoff between the outside of the solid electron beam and the wall.

16. The perveance of the beam in Problem 15 is $K = 0.85 \times 10^{-6}$ A/V$^{3/2}$ = 0.85 μperv. Recompute the bunching length as directed in Problem 15 except for a 300-kV beam of the same perveance, 0.85 μperv.

17. Let us examine the low-impedance relativistic klystron of Friedman et al.[23] (a) For a thin annular electron beam of current I_b = 5.6 kA and average radius r_b = 1.75 cm, accelerated in a 500-kV diode, in a drift tube of radius r_0 = 2.35 cm, compute the beam relativistic factor γ_b within the drift tube and the normalized axial velocity $\beta_b = v_b/c$. (b) For that beam, compute the bunching length using Equation 9.15 for a klystron operating at f = 1.328 GHz.

18. Compute the threshold voltage for the unique low-impedance klystron bunching mechanism of Equation 9.18 for the following parameters (see Table 9.6):

    a. 1 MV, 35 kA, r_b = 6.6 cm, r_0 = 7 cm

    b. 500 kV, 5 kA, r_b = 2.35 cm, r_0 = 2.3 cm

19. Derive the expression for the quality factor Q given in Equation 9.22 for the cavity in Figure 9.10 in terms of the lumped circuit elements shown there. Find an expression for the frequency at which the cavity impedance of Equation 9.20 drops to half of its peak value as a function of ω_0 and Q.

20. Assume we have a cavity with the following parameters: $f_0 = \omega_0/2\pi$ = 11.424 GHz, characteristic impedance Z_c = R/Q = 27 Ω, and quality factors Q_b = 230, Q_e = 300, and $Q_c = \infty$ (i.e., we are assuming the cavity walls are perfectly conducting, so that there are no resistive losses in the walls). (a) What is the cavity impedance at f = 11.424 GHz? (b) What is the cavity impedance at f = 11.4 GHz? (c) At what frequencies is the real-valued magnitude of the complex impedance, $|Z|$, equal to half of the value at 11.424 GHz? (d) Assuming the rise time of a microwave signal injected into the cavity is approximately equal to the RC time constant of the cavity (Equation 9.22), estimate the rise time for the injection of an input signal into the cavity.

21. In many klystrons, the designer maximizes power and gain by detuning the penultimate cavity so that its resonant frequency lies above the signal frequency, as indicated in Figure 10.7. This tuning makes the cavity impedance appear to be inductive to the bunches on the beam. This issue is discussed, for example, by Houck et al.[39] Consider the cavity voltage–current relationship of Equation 9.19, with the impedance given in Equation 9.20. Assuming, as in the reference, that the electrons in the leading edge of a bunch are moving faster than those at the trailing edge, show how one adjusts the phase relationship between the voltage and current, through the impedance, in order to slow down the electrons at the front of the bunch and accelerate those at the back.

22. Plot the space-charge-limiting currents for a thin annular beam as a function of wall radius, assuming a 2-mm standoff between wall and beam, for voltages of 1, 1.5, and 2 MV.

23. At a fixed efficiency of 30% for converting beam power to microwave power, operating at a current that is 80% of the space-charge-limiting value, plot the microwave output power at 2 MV as a function of wall radius (assuming a 2-mm standoff between the thin annular beam and the wall). Take account of space-charge depression of the beam energy.

24. To get higher beam power at fixed voltage, one must raise the current. Given that the beam must have some separation from the wall, it seems that one must increase the radius of the drift tube. From the standpoint of basic klystron operating principles, what problem does this raise?

25. To continue to raise beam currents while maintaining cutoff between cavities, the triaxial klystron, with a center conductor in the drift tube, has been proposed. We want to know the cutoff frequencies for the triaxial waveguide configuration. To avoid the headaches of cylindrical geometry and Bessel functions, consider the case where the cylindrical geometry is almost planar, when the inner conductor is close to the outer. (a) Derive the normal modes and dispersion relation for the three classes of waves in a planar waveguide with infinite parallel-plate conductors separated by a distance d: transverse electromagnetic (TEM), transverse electric (TE), and transverse magnetic (TM). (b) Comment qualitatively on the difficulties posed by the TEM waves. If you can, offer a comment about the probability that these waves will couple to a beam and drive axial density perturbations.

26. In the reltron, the modulation cavity fill time delays microwave onset. We define this time as the period from the start of the electron beam to the time when the microwave signal reaches its full-power flattop. In Figure 9.29, estimate the cavity Q and the reduction in energy efficiency that this effect causes.

27. Derive the expression for the space-charge-limiting current in the triaxial klystron given in Equations 9.35 and 9.36.

References

In the following, many publications from the Stanford Linear Accelerator Center, which are available online through the SPIRES database, have SLAC publication numbers, even if they are also printed in the proceedings of the conference at which they were presented.

1. Varian, R. and Varian, S., A high frequency oscillator and amplifier, *J. Appl. Phys.*, 10, 321, 1939.

2. Staprans, A., McCune, E., and Reutz, J., High power linear-beam tubes, *Proc. IEEE*, 61, 299, 1973.
3. Konrad, G.T., High Power RF Klystrons for Linear Accelerators, paper presented at the Linear Accelerator Conference, Darmstadt, Germany, 1984 (SLAC-PUB-3324, 1984).
4. Lee, T.G., Lebacqs, J.V., and Konrad, G.T., A 50-MW Klystron for the Stanford Linear Accelerator Center, paper presented at the International Electron Devices Meeting, Washington, DC, 1983 (SLAC-PUB-3214, 1983).
5. Caryotakis, G., "High-power" microwave tubes: in the laboratory and on-line, *IEEE Trans. Plasma Sci.*, 22, 683, 1994.
6. Sprehn, D., Caryotakis, G., and Phelps, R.M., 150-MW S-Band Klystron Program at the Stanford Linear Accelerator Center, paper presented at the 3rd International Conference on RF Pulsed Power Sources for Linear Colliders, Hayama, Japan, 1996 (SLAC-PUB-7232, 1996).
7. Choroba, S., Hameister, J., and Jarylkapov, S., Performance of an S-Band Klystron at an Output Power of 200 MW, DESY Internal Report M 98-11 (1998), paper presented at the 19th International Linear Accelerator Conference (LINAC 98), Chicago, IL.
8. Sprehn, D. et al., X-band klystron development at the Stanford Linear Accelerator Center, *Proc. SPIE*, 4031, 137, 2000 (SLAC-PUB-8346).
9. See, for example, Lau, Y.Y., Some design considerations on using modulated intense annular electron beams for particle acceleration, *J. Appl. Phys.*, 62, 351, 1987.
10. Allen, M.A. et al., Relativistic Klystron Research for High Gradient Accelerators, paper presented at European Particle Accelerator Conference, Rome, 1988 (SLAC-PUB-4650, 1988; Lawrence Livermore National Laboratory, UCRL-98843).
11. Hamilton, D.R., Knipp, J.R., and Kuper, J.B.H., *Klystrons and Microwave Triodes*, McGraw-Hill, New York, 1948, pp. 81–82.
12. See Shintake, T., Nose-Cone Removed Pillbox Cavity for High-Power Klystron Amplifiers, KEK Preprint 90-53, Tsukuba, Japan, 1990, and references to earlier treatments there.
13. Hamilton, D.R., Knipp, J.R., and Kuper, J.B.H., *Klystrons and Microwave Triodes*, McGraw-Hill, New York, 1948, pp. 79–94.
14. Halbach, K. and Holsinger, R.F., SUPERFISH: a computer program for evaluation of RF cavities with cylindrical symmetry, *Part. Accel.*, 7, 213, 1976.
15. Gilmour, A.S., Jr., *Microwave Tubes*, Artech House, Dedham, MA, 1986, p. 229.
16. Gilmour, A.S., Jr., *Microwave Tubes*, Artech House, Dedham, MA, 1986, p. 216.
17. Branch, G.M., Jr., Electron beam coupling in interaction gaps of cylindrical symmetry, *IRE Trans. Electron Devices*, ED-8, 193, 1961.
18. Branch, G.M. and Mihran, T.G., Plasma frequency reduction factors in electron beams, *IRE Trans. Electron Devices*, ED-2, 3, 1955.
19. Branch, G.M., Jr. et al., Space-charge wavelengths in electron beams, *IEEE Trans. Electron Devices*, ED-14, 350, 1967.
20. Gilmour, A.S., Jr., *Microwave Tubes*, Artech House, Dedham, MA, 1986, p. 228.
21. Gilmour, A.S., Jr., *Microwave Tubes*, Artech House, Dedham, MA, 1986, p. 220.
22. Briggs, R.J., Space-charge waves on a relativistic, unneutralized electron beam and collective ion acceleration, *Phys. Fluids*, 19, 1257, 1976.

23. Friedman, M. et al., Externally modulated intense relativistic electron beams, *J. Appl. Phys.*, 64, 3353, 1988.
24. Lau, Y.Y. et al., Relativistic klystron amplifiers driven by modulated intense relativistic electron beams, *IEEE Trans. Plasma Sci.*, 18, 553, 1990.
25. Hamilton, D.R., Knipp, J.R., and Kuper, J.B.H., *Klystrons and Microwave Triodes*, McGraw-Hill, New York, 1948, pp. 77–80.
26. Lavine, T.L. et al., Transient Analysis of Multicavity Klystrons, paper presented at the Particle Accelerator Conference, Chicago, 1989 (SLAC-PUB-4719 [Rev], 1989).
27. Miller, R.B. et al., Super-reltron theory and experiments, *IEEE Trans. Plasma Sci.*, 20, 332, 1992.
28. Marder, B.M. et al., The split-cavity oscillator: a high-power e-beam modulator and microwave source, *IEEE Trans. Plasma Sci.*, 20, 312, 1992.
29. International Study Group, International Study Group Progress Report on Linear Collider Development (SLAC-Report-559, 2000).
30. Sprehn, D., Phillips, R.M., and Caryotakis, G., The Design and Performance of 150-MW S-Band Klystrons, paper presented at the IEEE International Electron Development Meeting, San Francisco, 1994 (SLAC-PUB-6677, 1994).
31. Sprehn, D. et al., PPM Focused X-Band Klystron Development at the Stanford Linear Accelerator Center, paper presented at the 3rd International Conference on RF Pulsed Power Sources for Linear Colliders, Hayama, Japan, 1996 (SLAC-PUB-7231, 1996).
32. Caryotakis, G., Development of X-Band Klystron Technology at SLAC, paper presented at the Particle Accelerator Conference, Vancouver, BC, 1997 (SLAC-PUB-7548, 1997).
33. From the SLAC Web site: http://www.slac.stanford.edu/grp/kly/.
34. Abdallah, C. et al., Beam transport and RF control, in *High-Power Microwave Sources and Technologies*, Barker, R.J. and Schamiloglu, E., Eds., IEEE Press, New York, 2001, p. 254.
35. Sessler, A. and Yu, S., Relativistic klystron two-beam accelerator, *Phys. Rev. Lett.*, 58, 2439, 1987.
36. Allen, M.A. et al., Relativistic Klystron Research for Linear Colliders, paper presented at the DPF Summer Study: Snowmass '88, High-Energy Physics in the 1990s, Snowmass, CO, 1988 (SLAC-PUB-4733, 1988).
37. Allen, M.A. et al., Relativistic Klystrons, paper presented at the Particle Accelerator Conference, Chicago, 1989 (SLAC-PUB-4861; Lawrence Livermore National Laboratory, UCRL-100634; Lawrence Berkeley National Laboratory, LBL-27147; all 1989).
38. Allen, M.A. et al., Recent Progress in Relativistic Klystron Research, paper presented at the 14th International Conference on High-Energy Particle Accelerators, Tsukuba, Japan, 1989 (SLAC-PUB-5070; Lawrence Livermore National Laboratory, UCRL-101688; Lawrence Berkeley National Laboratory, LBL-27760; all 1989).
39. Houck, T. et al., Prototype microwave source for a relativistic klystron two-beam accelerator, *IEEE Trans. Plasma Sci.*, 24, 938, 1996.
40. Haimson, J. and Mecklenburg, B., Design and construction of a chopper driven 11.4 GHz traveling wave RF generator, in *Proceedings of the 1989 IEEE Particle Accelerator Conference*, Chicago, IL, 1989, p. 243.
41. Dolbilov, G.V. et al., Experimental study of 100 MW wide-aperture X-band klystron with RF absorbing drift tubes, in *Proceedings of the European Accelerator Conference (EPAC-96)*, Bristol, UK, Vol. 3, 1996, p. 2143.

42. Dolbilov, G.V. et al., Design of 135 MW X-band relativistic klystron for linear collider, in *Proceedings of the 1997 IEEE Particle Accelerator Conference*, Vancouver, BC, 1997, p. 3126.

43. Friedman, M. et al., Relativistic klystron amplifier, *Proc. SPIE*, 873, 92, 1988.

44. Levine, J.S., High current relativistic klystron research at Physics International, *Proc. SPIE*, 2154, 19, 1994.

45. Fazio, M.V. et al., A 500 MW, 1 µs pulse length, high current relativistic klystron, *IEEE Trans. Plasma Sci.*, 22, 740, 1994.

46. Hendricks, K.J. et al., Gigawatt-class sources, in *High-Power Microwave Sources and Technologies*, Barker, R.J. and Schamiloglu, E., Eds., IEEE Press, Piscataway, NJ, 2002, p. 66.

47. Serlin, V. and Friedman, M., Development and optimization of the relativistic klystron amplifier, *IEEE Trans. Plasma Sci.*, 22, 692, 1994.

48. Friedman, M. et al., Efficient generation of multigigawatt rf power by a klystronlike amplifier, *Rev. Sci. Instrum.*, 61, 171, 1990.

49. Colombant, D.G. et al., RF converter simulation: imposition of the radiation condition, *Proc. SPIE*, 1629, 15, 1992.

50. Serlin, V. et al., Relativistic klystron amplifier. II. High-frequency operation, *Proc. SPIE*, 1407, 8, 1991.

51. Friedman M. et al., Relativistic klystron amplifier. I. High power operation, *Proc. SPIE*, 1407, 2, 1991.

52. Friedman, M. et al., Intense electron beam modulation by inductively loaded wide gaps for relativistic klystron amplifiers, *Phys. Rev. Lett.*, 74, 322, 1995.

53. Friedman, M. et al., Efficient conversion of the energy of intense relativistic electron beams into rf waves, *Phys. Rev. Lett.*, 75, 1214, 1995.

54. Miller, R.B. and Habiger, K.W., A review of recent progress in reltron tube design, in *Proceedings of the 12th International Conference on High-Power Particle Beams (BEAMS'98)*, Haifa, Israel, 1998, p. 740.

55. Miller, R.B. et al., Super-reltron theory and experiments, *IEEE Trans. Plasma Sci.*, 20, 332, 1992.

56. Caryotakis, G. et al., A 2-GW, 1-µs microwave source, in *Proceedings of the 11th International Conference on High Power Particle Beams (BEAMS'96)*, Prague, Czech Republic, 1996, p. 406.

57. Caryotakis, G., A Sheet-Beam Klystron Paper Design, paper presented at the 5th Modulator-Klystron Workshop for Future Linear Colliders, CERN, Geneva, 2001 (SLAC-PUB-8967, 2001).

58. Pasour, J., Smithe, D., and Ludeking, L., An X-band triaxial klystron, in *High Energy Density and High Power RF, 6th Workshop*, Berkeley Springs, WV, 2003, p. 141.

59. Palmer, R., Herrmannsfeldt, W., and Eppley, K., An immersed field cluster klystron, *Part. Accel.*, 30, 197, 1990.

60. See, for example, Wright, E. et al., Test results for a 10-MW, L-band multiple beam klystron for TESLA, in *Proceedings of EPAC 2004*, Lucerne, Switzerland, 2004, p. 1117.

10

Vircators, Gyrotrons and Electron Cyclotron Masers, and Free-Electron Lasers

10.1 Introduction

This chapter is devoted to a somewhat briefer treatment of three classes of narrowband microwave sources that are unrelated in their underlying operating principles: virtual cathode oscillators, gyrotrons and electron cyclotron masers, and free-electron lasers. The reason for collecting these microwave sources in a single chapter is that at this time we regard these three classes of narrowband sources as somehow less significant in the field of high power microwaves (HPM) than the sources of Chapters 7 to 9. Let us consider each separately and state the reasons for our admittedly subjective judgment, which varies with the source, recognizing that the significance of any one of them could change with shifts in the technology or application needs.

Virtual cathode oscillators are enjoying something of a renaissance at this time, and developments with two variants in particular — tunable vircators and coaxial vircators — may yet overcome two problems that have plagued devices of this type: low efficiency and sensitivity to gap closure in the high-current, explosive emission diodes that are so often used with sources of this type. Nevertheless, despite their efficiency problems, these sources are attractive in applications requiring a simple source configuration — vircators many times employ no applied magnetic field and no slow-wave structure — and low device impedance, which permits high power operation at low voltage or optimizes coupling with low-impedance power sources such as explosive generators, which are a compact, energy-rich power source.

Gyrotrons are an extraordinarily mature source for high-average-power application at the several-megawatt level to the problem of heating magnetic fusion plasmas at electron cyclotron resonance frequencies of 100 GHz or more (see Chapter 3). In fact, gyrotrons join klystrons as the only sources that are produced to continuously generate power levels at 1 MW or above. Another variant of this class of *electron cyclotron masers*, the *gyroklystron*, has been investigated as an alternative to klystrons for power levels around 100 MW and frequencies in the X-band and above. Another variant, the *cyclotron autoresonance maser* (CARM), potentially offers high power at high frequencies, since it features a linear upshift in frequency with beam energy, which compares to the quadratic frequency upshift of free-electron lasers without the complexity of wiggler magnets. In each case, however, these sources are immature at the gigawatt level, lacking the breadth and depth of development of other sources in comparable frequency ranges.

Free-electron lasers (FELs) are potentially the most flexible of all HPM sources, although they are also arguably the most complex. They offer almost unlimited frequency range, albeit at the expense of increasingly high beam energies as the frequency goes up. In one remarkable series of experiments, an FEL was operated at repetition rates of over a kilohertz, with power levels of the order of a gigawatt at frequencies over 100 GHz.

We include these sources in the book more for their strengths than their weaknesses. Current developments in virtual cathode oscillators may result in improvements in efficiency that make them more competitive with the sources treated in earlier chapters here. On the other hand, if future applications arise requiring the higher frequencies for which gyrotrons or ECMs and FELs are better suited, their importance in the field will increase.

10.2 Vircators

Virtual cathode oscillators are actually a class of sources that all depend on radiation generation by one or both of two phenomena that accompany the injection of an electron beam into a waveguide or cavity in which the beam current exceeds the local space-charge-limiting current: virtual cathode and electron reflex oscillations. Sources of this type include the *vircator, reflex triode, reditron, coaxial vircator,* and *feedback vircator* (or *virtode,* as it was originally known). These sources are capable of gigawatt-level output in the 1- to 10-GHz range of frequencies; they are relatively simple to build, since in many cases no magnetic field is required; and they generally operate at relatively low impedances, which makes them rather attractive from the standpoints of producing power at relatively low voltages and of coupling well to low-impedance power sources, notably compact, but single-shot, explosive electrical generators. They are also conspicuously tunable because their operation depends only on the charge density of the beam and not on any resonance condition. A single device can generate radiation over one or two octaves. The wide tunability of the vircator has made it a popular source in HPM testing facilities, where the variation of effects with frequency can be surveyed conveniently. Vircators with staggered tuning ranges have provided a testing capability from 0.5 to 10 GHz.[1] For all of these reasons, vircators are probably the most popular of HPM sources, with research or application efforts in all of the nations with HPM programs: the U.S. and Russia; France, Germany, the U.K, and Sweden; China and Japan; and relatively new nations to HPM, such as India, South Korea, and Taiwan. In a sense, for their simplicity and flexibility, vircators are the "starter source" for an HPM program.

Despite those advantages, these sources have historically suffered from relatively low efficiency, and they have been very sensitive to the problem of gap closure, which causes the diode current to increase with time, so that the frequency chirps upward, upsetting the balance between resonances in a way that causes output to drop (see Problem 1). These problems are being addressed particularly in the development of the coaxial and feedback vircators, so that these sources are currently enjoying something of a renaissance, which may cause their fortunes to rise again in the future.

10.2.1   Vircator History

As recounted in the review of Birdsall and Bridges, the Child–Langmuir relation for charged-particle flow in a diode traces to the early 20th century.[2] In the 1960s and 1970s, studies of virtual cathode formation by beams exceeding the space-charge-limiting current showed that the virtual cathode was unstable and quantified the dependence of the oscillation frequency and position of the virtual cathode on the experimental parameters. Sullivan et al. reviewed the development of vircator theory,[3] while Thode reviewed experiments[4] as of 1987.

A number of early experiments used the reflex triode configuration, in which the anode is pulsed positive and the cathode is grounded. In 1977, Mahaffey and coworkers were the first to produce microwaves using virtual cathode oscillation,[5] followed a year later by the publication of results from a group at the Tomsk Polytechnic Institute in Russia.[6] A group at Harry Diamond Laboratories (HDL) produced gigawatt output in the X-band with a reflex triode in 1980.[7]

The standard vircator, with its negative pulsed cathode and grounded anode, became more popular in the 1980s: Burkhart at Lawrence Livermore National Laboratory (LLNL) reported 4 GW at 6.5 GHz in 1987,[8] Bromborsky et al. at HDL recorded average peak power (averaging out spikes) of 9 GW at frequencies below 1 GHz in 1988 on the 6-MV AURORA accelerator,[9] and Platt et al. at the Air Force Weapons Laboratory (AFWL; now the Air Force Research Laboratory) achieved 7.5 GW at 1.17 GHz in 1989.[10] Two modifications of this basic design demonstrated increased efficiency. The *reditron*, proposed by Kwan and Thode at Los Alamos National Laboratory (LANL) from an earlier suggestion by Sullivan, employed a special anode design to eliminate reflexing by a thin annular electron beam to produce 3.3 GW at 2.15 GHz with a sharper frequency spectrum and an efficiency of about 10%,[11] significantly higher than earlier experiments at that time outside the U.S.S.R. The use of a double anode at LLNL, described in 1991, similarly increased device efficiency by synchronizing, rather than eliminating, electron reflexing with the oscillation of the virtual cathode.[12]

In 1987, Benford and coworkers at Physics International (PI; now L-3 Communications Pulse Sciences)[13] and Sullivan[3] introduced the *cavity vircator* and showed that virtual cathode interaction within a microwave cavity could eliminate mode competition and increase efficiency. Sze and coworkers at PI,[14] as well as Fazio et al. at LANL,[15] used a cavity vircator to achieve about a 5-dB amplification of the output of a relativistic magnetron in 1989. In 1990, Price and Sze phase locked a two-vircator array to the output of a relativistic magnetron operating in the S-band,[16] and Sze and others locked two-cavity vircators to one another, producing about 1.6 GW of total output.[17]

The *virtode* or *feedback vircator* was described in a 1993 publication from a group at the Kharkov Physico-Technical Institute.[18] This variation uses feedback from the microwave generation region through a waveguide back to the diode region where the electron beam is generated. Tunable versions of varying design spanned the frequency spectrum from the S- to the X-band at power levels up to 600 MW and efficiencies that ranged from 3 to 17%. More recently, Kitsanov et al. from the Institute of High-Current Electronics (IHCE) in Tomsk built a 1-GW S-band device of this type.[19]

In 1997, the *coaxial vircator*, in which the virtual cathode is generated by propagating a high-current electron beam radially inward, was proposed in computer simulations from Woolverton et al. at Texas Tech University,[20] followed by a later publication describing experiments.[21]

10.2.2   Vircator Design Principles

A virtual cathode is formed when the magnitude of the current I_b of an electron beam injected into a drift tube exceeds the space-charge-limiting current I_{SCL}. When a virtual cathode is formed, two things happen: (1) the location of the virtual cathode oscillates back and forth at roughly the beam plasma frequency ω_b, and (2) some electrons are transmitted through the virtual cathode while some are reflected back toward the diode, where eventually they are reflected by the cathode potential back toward the virtual cathode, an oscillatory effect known as reflexing. In general, the virtual cathode oscillation occurs at a different frequency than the reflex oscillations, although the two can be tuned to coincide, as we shall see.

Expressions for I_{SCL} were given in Chapters 4 and 9 in Equations 4.115 and 9.3a for a thin annular beam, Equation 4.116 for a beam with a uniform distribution across a circular cross section, and Equations 9.35 and 9.36 for a thin annular beam inside a coaxial drift tube. The general form for I_{SCL} is

$$I_{SCL}\left(kA\right) = \frac{8.5}{G}\left(\gamma_0^{2/3} - 1\right)^{3/2} \qquad (10.1)$$

where γ_0 is related to the anode–cathode voltage in the electron beam diode V_0,

$$\gamma_0 = 1 + \frac{eV_0}{mc^2} = 1 + \frac{V_0\left(kV\right)}{0.511} \qquad (10.2)$$

and G depends on the geometry. For example, for a thin annular beam of mean radius r_b within a circular-cross-section drift tube of radius r_0,

$$G = \ell n\left(r_0 / r_b\right) \text{ (annular beam)} \qquad (10.3\text{a})$$

whereas if the beam is uniformly distributed across the whole beam radius r_b,

$$G = 1 + \ell n\left(r_0 / r_b\right) \text{ (solid beam)} \qquad (10.3\text{b})$$

(see Problems 2, 3, and 5). Alternatively, if we have a thin annular beam within a coaxial drift tube of inner radius r_i and outer radius r_0, then

$$G^{-1} = \frac{1}{\ell n\left(r_o / r_b\right)} + \frac{1}{\ell n\left(r_b / r_i\right)} \text{ (annular beam, coaxial drift tube)} \qquad (10.3\text{c})$$

When $I_b > I_{SCL}$, a virtual cathode is formed and oscillates about a position at a distance downstream of the anode roughly equal to the anode–cathode spacing. The potential fluctuations at the virtual cathode produce a beam with an oscillating current density downstream of the virtual cathode, as well as a fluctuating current that is sent back toward the anode. The operation of a virtual cathode oscillator is best seen in computer simulations. Figure 10.1 shows the time dependence of the position and potential of a virtual cathode formed by a solid-cross-section beam injected through a foil anode into a cylindrical drift space.[22] The spikes in position generate a frequency spectrum that peaks at 10.4 GHz but contains many components. Broadband radiation is a basic characteristic of the simple vircator. We see the oscillation of the virtual cathode about a mean distance somewhat greater than the anode–cathode spacing, and a virtual cathode potential oscillation about a mean of roughly V_0.

From the simulation of a different diode, Figure 10.2 shows, at a single instant in time, the real space and phase-space positions of particles in a particle-in-cell simulation of a virtual cathode oscillator with no applied magnetic field.[4] The cathode on the left is a thin cylinder of metal projecting from a metal stalk; note that there is electron emission from the sides as well as from the tip of the thin cylinder at z = 1.50 cm. In the lower phase-space plot, we see that the electrons reach their maximum axial momentum, labeled P_z, at the location of the anode at z = 2.25 cm, and they decelerate as they propagate farther along the axis. We see that some stop, with $P_z = 0$ at the

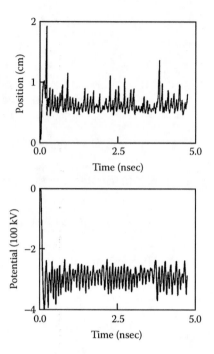

FIGURE 10.1
Time histories of the virtual cathode position and the magnitude of the virtual cathode potential obtained from computer simulation of a 270-kV, 7.3-kA beam of solid cross section and radius 1.25 cm injected through the anode of a diode with a 0.25-cm anode–cathode gap into a drift tube of radius 4.9 cm. (From Lin, T. et al., *J. Appl. Phys.*, 68, 2038, 1990. With permission.)

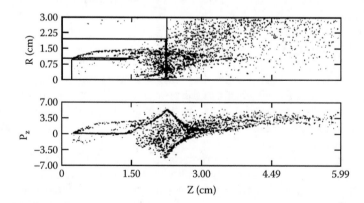

FIGURE 10.2
An r-z plot (top) and phase-space plot (bottom) of an electron beam forming a virtual cathode when injected into a drift space without an axial, guiding magnetic field. (From Thode, L.E., *High Power Microwave Sources*, Granatstein, V.L. and Alexeff, I., Eds., Artech House, Boston, 1987, p. 507. With permission.)

virtual cathode located at about z = 3.00 cm, returning into the gap with negative axial momenta. Those electrons do not reach the cathode because of the energy loss in the foil cathode. Clearly, though, many of the electrons are transmitted through the virtual cathode to pass down the drift tube with a substantial spread in axial momentum.

Virtual cathode oscillations occur at approximately the relativistic plasma frequency

$$f_p = \frac{1}{2\pi}\left(\frac{n_b e^2}{\varepsilon_0 m \gamma_0}\right)^{1/2} = 8.98\times10^3 \left[\frac{n_b\left(cm^{-3}\right)}{\gamma_0}\right]^{1/2} \text{Hz} \qquad (10.4)$$

where n_b is the density of electrons in the beam as it passes through the anode, $-e$ and m are the charge and mass of an electron, and ε_0 is the permittivity of free space. In practical units, we find

$$f_p\left(GHz\right) = 4.10\left[\frac{J\left(kA/cm^2\right)}{\beta\gamma_0}\right]^{1/2} \qquad (10.5)$$

where $\beta = v_b/c = (1 - 1/\gamma_0^2)^{1/2}$, with v_b the electron velocity (see Problem 4). In the nonrelativistic limit, where $\gamma_0 \cong 1$ and $\beta \propto V_0^{1/2}$, Equation 4.1014 shows that $J \propto V_0^{3/2}/d^2$, so that

$$f_{VC} \propto \frac{V_0^{1/2}}{d} \quad \text{(nonrelativistic)} \qquad (10.6)$$

In the strongly relativistic limit, on the other hand, when $\gamma_0 \propto V_0$, $\beta \cong 1$, and, according to Equation 4.107, $J \propto V_0/d^2$, so that

$$f_{VC} \propto \frac{1}{d} \quad \text{(strongly relativistic)} \qquad (10.7)$$

This strongly relativistic regime is reached for higher voltages than we would normally encounter in an HPM system, but the point is that the voltage dependence weakens for voltages significantly above 500 kV. In addition to the oscillation of the virtual cathode itself, trapped electrons reflexing in the potential well formed between the real and virtual cathodes are bunched and consequently radiate at a reflexing frequency

$$f_r = \frac{1}{4T} = \frac{1}{\left(4\int_0^d \dfrac{dz}{v_z}\right)} \tag{10.8}$$

where d is the anode–cathode gap. In practical units, using the nonrelativistic expression for v_z in the anode–cathode gap,

$$f_r\left(GHz\right) \cong 2.5\frac{\beta}{d\left(cm\right)} \quad \text{(nonrelativistic)} \tag{10.9}$$

(see Problem 6). At relativistic voltages, because the electron velocity approaches the final value of βc more quickly, the multiplier approaches the limiting value of 7.5 (at which $v_b = \beta c$ all the way across the gap). The scaling of both f_{VC} and f_r is similar in the nonrelativistic regime, both frequencies falling within the microwave range, so that both phenomena are available to compete with one another. Generally, $f_{vc} > f_r$, and typically $f_{vc} \sim 2f_r$. As we shall discuss, configurations have been explored that either eliminate or exploit this competition.

Weak voltage scaling and the inverse scaling with d, which is the same in either the nonrelativistic or strongly relativistic cases, were confirmed in early experiments.[5] Later experiments by Price et al.[23] verified the inverse scaling with the anode–cathode spacing over a very wide frequency range, as shown in Figure 10.3. The data in the figure were obtained with a vircator that was tunable over about an order of magnitude, albeit at reduced power, as the frequency was extended down to about 400 MHz.

Woo used a two-dimensional treatment to find the beam plasma frequency for a pinched solid beam.[24] His treatment was based on the observation that a relatively low-current beam would not pinch, diverging to the wall and skirting the potential minimum established in the center of the drift tube that would cause reflection of electrons back toward the diode and formation of the virtual cathode. Following the history of the voltage and current during the course of a pulse, he showed that microwave emission began shortly after the time when pinching became significant — determined when the radial and axial current densities in the beam became equal — and that emission terminated around the time that the pinching became so strong that the radial current density became much larger than the axial current density, which is supported by experiments.[25] Using a formalism similar to that of Goldstein et al. for pinched-beam diodes[26] and Creedon for parapotential flow,[27] he derived an expression for the beam plasma frequency that agreed very well with experimental data:

$$f = \frac{c}{2\pi d}\ell n\left[\gamma_0 + \left(\gamma_0^2 - 1\right)^{1/2}\right] = \frac{4.77}{d\left(cm\right)}\ell n\left[\gamma_0 + \left(\gamma_0^2 - 1\right)^{1/2}\right] \text{ GHz} \tag{10.10}$$

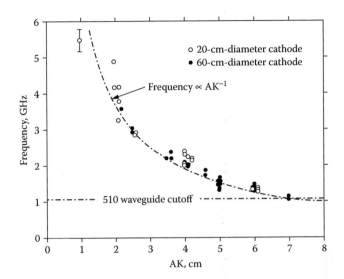

FIGURE 10.3

Tuning range of a vircator vs. the anode–cathode gap spacing, designated AK here (and d in the text). Cathodes of two different diameters were used, and power was extracted into WR-510 waveguide with the cutoff frequency shown. (From Price, D. et al., *IEEE Trans. Plasma Sci.*, 16, 177, 1988. With permission.)

This result is compared to data from a number of experiments in Figure 10.4 (see Problem 7).

In the absence of any downstream resonant structure, the bandwidth of the vircator is large because of several factors: nonuniform voltage, anode–cathode gap closure, and transverse beam energy. Since frequency depends on voltage and gap, the bandwidth is similarly dependent on variations in these two quantities. At nonrelativistic voltages, using Equation 10.6, we can see that

$$\frac{\Delta f_{VC}}{f_{VC}} = \frac{\Delta V_0}{2V_0} - \frac{\Delta d}{d} \tag{10.11}$$

The variation with the anode–cathode gap width d leads to *chirping* of the microwave frequency in diodes when the gap plasma causes closure, so that $\Delta d < 0$. Chirping is the phenomenon whereby the frequency climbs during the course of a pulse, in analogy with the chirp of a bird; the strongest chirping occurs at the highest frequencies, where gap closure is proportionately larger. Chirping results in a lower efficiency and lower gain because (1) the gain–bandwidth product is fixed for a given device with a fixed load and (2) the frequency changes so rapidly that the process generating microwaves does not fully establish itself before the resonant frequency shifts. The introduction of a resonant cavity downstream has been shown to improve the gain and narrow the bandwidth.

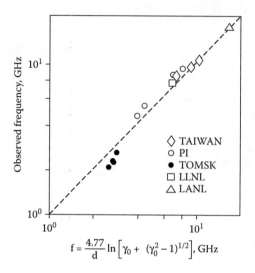

$$f = \frac{4.77}{d} \ln\left[\gamma_0 + (\gamma_0^2 - 1)^{1/2}\right], \text{GHz}$$

FIGURE 10.4
Comparisons of observed vircator frequencies with the two-dimensional analytic model of Woo. (From Woo, W.-Y., *Phys. Fluids*, 30, 239, 1987. With permission.)

Beam temperature, from both axial and angular variations in beam electron velocities, also broadens the bandwidth because the transverse components of the electrons' energy do not interact with the virtual cathode. Beam temperature is produced by both the emission process and scattering in the anode foil. Beam temperature also reduces radiation efficiency. The reduction in output power with beam temperature is illustrated in Figure 10.5, in which increasing the thickness of the anode, through which the beam passes, increases beam temperature by increasing the scattering angle of the electrons.[28]

Because the radio frequency (RF) field components excited by the vircator process are E_z, E_r, and B_θ, the modes excited are TM_{on}. The preferred mode appears to be near cutoff, i.e., the lower-order modes $n \sim 2\, r_a/\lambda$.

Foil-less diodes can also be used for vircator generation as long as the space-charge limit is exceeded. Because foil-less diodes have $\omega_p \approx \omega_c$, the electron cyclotron frequency, due to the presence of an axial magnetic field, foil-less diode vircators can be tuned in frequency simply by changing the magnetic field. The foil-less diode also offers advantages in reduced gap closure and flat impedance characteristics.

10.2.3 Basic Vircator Operational Features

The vircator domain of frequency operation extends from 0.4 to 17 GHz. Lower-frequency devices have larger dimensions. For example, Price et al. have reported a device that operates as low as 0.5 GHz with a chamber inner diameter of 80 cm and a cathode with a 60-cm diameter.[23] A 17-GHz vircator,

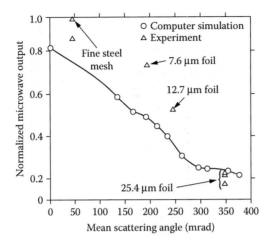

FIGURE 10.5
Decline of microwave power with scattering in the anode. Work done under the auspices of the University of California Lawrence Livermore National Laboratory, Department of Energy. (From Burkhart, S. et al., *J. Appl. Phys.*, 58, 28, 1985. With permission.)

by comparison, used a 6-cm chamber with a 2-cm cathode.[29] Low-frequency operation historically required the use of a velvet-covered cathode. Extrapolation to higher frequencies has been limited by plasma closure caused by the small gaps necessary to produce the very high current densities (see Problem 1). Pulse durations have typically been short (~50 nsec), but the Tomsk group[30] and Coleman and Aurand[31] have reported long-pulse experiments. Burkhart reported 4 GW in the C-band.[8] Platt and coworkers have reported 7.5 GW at 1.17 GHz,[10] and Huttlin and coworkers[32] have reported power in excess of 10 GW below 1 GHz. At the high-frequency end, Davis and coworkers[29] have reported 0.5 GW at 17 GHz.

The two basic vircator geometries for axial and side extraction are shown in Figure 10.6. In the axial vircator, the space-charge cloud couples to radial and longitudinal electric field components, giving TM modes that are extracted along the cylindrical waveguide, usually to a flared horn antenna. Early axial extraction experiments reported a gigawatt at the X-band.

The transverse or side-extracted vircator allows extraction in the TE_{10} mode. The beam is injected through the long wall of a rectangular waveguide forming a virtual cathode. The electric field between the anode and cathode is parallel to the small dimension of the waveguide. Thus, the electric field along the potential well of the virtual cathode system is parallel to the electric field of the TE modes of the waveguide. The rectangular waveguide is overmoded for the frequencies extracted. A remarkable feature of the side-extracted geometry is the purity of the TE_{10} mode; higher-order modes contribute less than 10% of the extracted power. Experiments by the PI group extracted a few hundred megawatts at efficiencies of about half a percent.[23,25] The most ambitious experiments of this type were conducted in the late

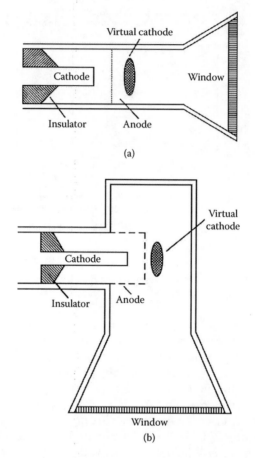

FIGURE 10.6
Basic vircator geometries: (a) axial extraction and (b) side extraction.

1980s at the Army Research Laboratory's now-dismantled Aurora facility. The best results were about 1 kJ in 18 radial arms at an efficiency of 0.6%.[32] The pulse was quite long, typically 100 nsec, with many spikes.

Figure 10.7 shows the basic reflex triode geometry. Note that in the reflex triode, the anode is the center conductor feeding the diode, and it is pulsed positive; by comparison, in vircators, the cathode is the center conductor, and it is pulsed negative. A reflex triode from the Grindsjön Research Center in Sweden, opened to the air, is shown in Figure 10.8.

Thode's survey[4] shows many common aspects of axial vircators. First, the radiation wavelength is typically

$$D < \lambda < 2D \tag{10.12}$$

where D is the waveguide diameter. Second, maximum power occurs with no applied axial magnetic field. If a thin anode is used with a guide field of

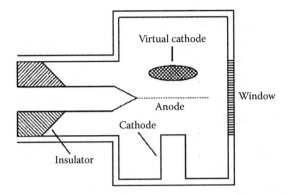

FIGURE 10.7
Basic reflex triode geometry.

FIGURE 10.8
Side view of a reflex triode from the Grindsjön Research Center in Sweden, opened to the air.

~1 kG, little microwave radiation results. This indicates that two-dimensional motion in the potential well is important for microwave generation. However, the Tomsk and Livermore groups have observed high power at high magnetic fields (>5 kG). Poulsen et al.[12] of the Livermore group suggests that increasing the field in a device optimized for $B_z = 0$ simply moves away from a local optimum, reducing power, which can be recovered by reoptimizing geometry. Third, increasing anode thickness reduces microwave power if foil scattering reaches 300 mrad, as we showed in Figure 10.5. Fourth, microwave power depends strongly on the anode–cathode gap. If the gap is too small, the beam pinches and microwaves cease. If the gap is too large, approaching the cathode–wall separation, the beam expands to the outer conductor and is lost. At full-power tuning of the vircator is typically limited by these effects to a range of roughly one octave.

10.2.4 Advanced Vircators

Advanced vircators have been developed in two rather distinct periods. The reditron, the side-extracted vircator, the double-anode vircator, and the cavity vircator were conceived and initially explored in the late 1980s to the early 1990s, while the virtode, or feedback vircator, and coaxial vircator are more recent innovations that are still in the active development stage. Of the three devices in the first generation, only the cavity vircator is still under development, and versions have been sold commercially.

10.2.4.1   Double-Anode Vircators

Researchers at Lawrence Livermore National Laboratory found that efficiency is increased when a resonance occurs in which the virtual cathode and reflex oscillation frequencies are equal, $f_{VC} = f_r$.[12] To achieve this, they employed the double-anode configuration of Figure 10.9 so that the two frequencies can be brought into equality by lengthening the transit time that determines f_r, as in Equation 10.8. Figure 10.10 shows that the resonant condition increases efficiency by about an order of magnitude. They report an output of 100 J from a single source with resonant conditions; however, the resonance condition is quite stringent. The bandwidth of high-efficiency operation was about 50 MHz, so that the frequencies must be equal to within 2%. Almost any diode closure defeated the resonance mechanism; the authors stated that variations of 100 μm in the anode–cathode gap significantly affected the output.

The Livermore group has attacked the problem of diode closure by using a variety of modified cathodes, the most successful being a field-enhanced thermionic emission cathode. They used a cathode with an area of 6 cm² with current densities from 200 to 300 A/cm² to produce microsecond pulse durations with no plasma formation enclosure. Unfortunately, at last reporting the device was not sufficiently reliable to produce sustained high-efficiency, long-duration microwave pulses with full-size cathodes.

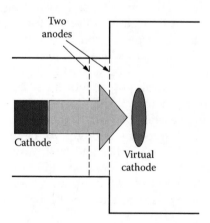

FIGURE 10.9
The double-anode configuration explored at Lawrence Livermore National Laboratory. The added travel time for reflexing electrons through the region between the two anodes is intended to synchronize the virtual cathode and reflex oscillations.

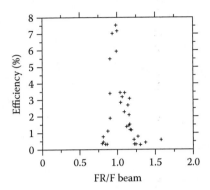

FIGURE 10.10
The resonant efficiency enhancement in the configuration of Figure 10.9 as a function of the ratio of f_r (designated FR in the figure) to f_{VC} (designated F beam in the figure). (From Poulsen, P. et al., *Proc. SPIE*, 1407, 172, 1991. With permission.)

10.2.4.1.1 Reditrons

An entirely different approach to the problem of competition between the virtual cathode and reflexing oscillations was taken at Los Alamos, where the reditron (for reflected electron discrimination microwave generator) was developed with the goal of completely eliminating electron reflexing.[11,29] The basic configuration of the reditron is shown in Figure 10.11.[33] The reditron features a thin annular beam and an otherwise thick anode with a thin annular cutout to allow the beam to pass out of the anode–cathode gap. An axial magnetic field guides the annular electron beam through the cutout in the anode to the downstream drift region. The part of the beam that reflects from the virtual cathode is no longer in force balance and expands radially; thus, it cannot pass through the annular slot and is absorbed in the anode. Reflexing electrons are almost eliminated, and only virtual cathode oscillations can produce radiation. In addition, the *two-stream instability* between outgoing and returning electrons is greatly reduced. When the streaming instability is present, electrons from the diode interact with returning electrons, and their angular divergence and energy spread are increased. Thus, their turning points in the virtual cathode's electric field are smeared out, which reduces the efficiency of the virtual cathode radiation mechanism.

Figure 10.12 shows a reditron configuration and phase-space diagram from a simulation by Kwan and Davis.[34] Compare this figure with Figure 10.2, which has no magnetic field and stronger counterstreaming electron interactions that cause much greater spreading of the electron beam. Simulations such as that in Figure 10.12 predict an efficiency of 10 to 20% and the production of narrow-bandwidth radiation (<3%); the oscillation frequency is ω_p. Since the reditron is an axial vircator, it creates TM modes, and simulations and experiments both show excitation of the lowest modes, typically TM_{01} or TM_{02}.

Elegant experiments by Davis and coworkers at Los Alamos verified the major features of the reditron concept.[33] Efficiency achieved in those exper-

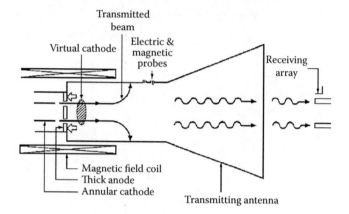

FIGURE 10.11
The reditron experimental arrangement. (From Davis, H. et al., *IEEE Trans. Plasma Sci.*, 16, 192, 1988. With permission.)

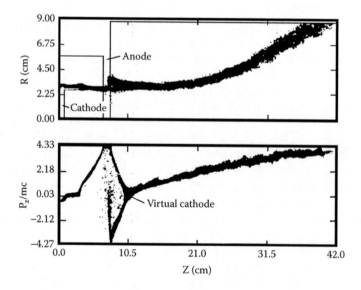

FIGURE 10.12
Simulation of a reditron experiment showing the formation of a virtual cathode, the loss of returning electrons on the thick anode, and the transmitted portion of the beam. (From Kwan, T.J. and Davis, H.A., *IEEE Trans. Plasma Sci.*, 16, 185, 1988. With permission.)

iments was 5.5 to 6%, compared to 1 to 3% when the apparatus was used in the conventional vircator geometry. Calculated diode and transmitted currents were 19 and 12 kA, very close to the measured values of 21 and 12.5 kA in the experiments. The beam current was verified to be the space-charge-limited value, I_{SCL}. Device impedances of 30 Ω to just over 60 Ω were seen in the experiments. Predicted power was 1.1 GW, compared to the 1.6 GW measured. Microwave power quenched well before the end of the electrical pulse. At lower powers, longer durations are observed, pointing to the possibility of breakdown processes. The observed frequency was 2.4 GHz with a bandwidth of 15 MHz, or less than 1%. Concentricity of the anode slot with the cathode was found to be essential for obtaining high power. Surprisingly, the highest power was produced when the beam just grazed the inside edge of the slot, rather than being centered in the slot. The data indicate that rapid diode closure or bipolar flow, due to the emission of ions from the anode, did not occur; the diode operated at constant impedance. Even in its initial experiments, the reditron showed great improvement in the fundamental vircator problems of efficiency, bandwidth, and chirping.

Computer simulations pointed to the possibility of further improvements on the reditron. The Los Alamos group showed that the oscillation of the virtual cathode can produce a highly modulated electron beam, with a modulation fraction as high as 100% occurring at the virtual cathode frequency.[35] Extraction of microwave power from this beam could enhance the output power and overall efficiency of the device. A straightforward way to extract the radiation is by use of an *inverse diode*. A center conductor is placed in the downstream waveguide, and it floats at a potential relative to the wall. The electron beam is absorbed in this conductor, a voltage then develops between the inner and outer conductors, and a TEM wave is transmitted along the coaxial line. Substantial radiation can be extracted downstream if the configuration is optimized. In the Los Alamos simulations, an inverse diode with a 50-Ω impedance was situated downstream of a reditron configuration. With a 1.2-MV diode, the kinetic energy of the leakage electrons impacting the center conductor was about 400 kV, with a highly modulated average current of 9 kA. The simulation gave an average peak power of 1 GW, so that the efficiency of the inverse diode was 28%.

Alternatively, the downstream region could contain a slow-wave structure. The premodulated electron beam from the vircator would induce wave growth at the modulation frequency.[36] This possibility was suggested by the Tomsk group.[37] Again, extraction downstream could increase overall vircator efficiency.

10.2.4.2 Cavity Vircators

A fundamental disadvantage of the vircator is that the downstream region in the basic vircator is large compared to the wavelength. The number of modes available in downstream waveguides or cavities allows many modes to be excited, so that as the vircator changes its characteristics due to variations in voltage or gap (Equation 10.11), there is always a waveguide mode at the appropriate frequency. The basic vircator also does not have one of the fundamental properties of most microwave devices, a feedback mechanism by means of a cavity to allow the field induced in the cavity by the oscillating virtual cathode to feed back on the virtual cathode, forcing it to oscillate at the resonant frequency of the cavity mode. Therefore, enclosing the vircator in a cavity has resulted in a fundamental improvement to the device.

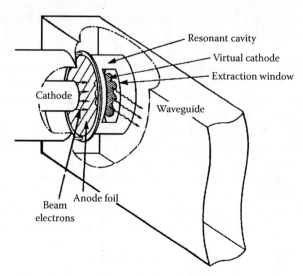

FIGURE 10.13
The resonant cavity vircator. (From Benford, J. et al., *J. Appl. Phys.*, 61, 2098, 1987. With permission.)

Of the many designs for cavities that could be considered, the pillbox cavity with a cylindrical cross section is the only one studied thus far because it is simple, has the same symmetry as the vircator itself, and is easily tuned by moving the end wall. The configuration chosen by the Physics International group (now L-3 Communications Pulse Sciences) in a transversely extracted vircator is shown in Figure 10.13.[13] In the first experiments, expendable aluminum cavities were used. The extraction window size was determined empirically, balancing the requirement for a large window to facilitate extraction and a small window to maximize the cavity Q. The window orientation shown in Figure 10.13 facilitates coupling to the fundamental TE mode of the rectangular extraction waveguide.

In cylindrical resonators, the TM modes are preferred, because E_z couples well to the virtual cathode electric field and B_θ allows convenient coupling through an output slot in the wall. The PI group found that the lowest-frequency TM_{010} mode required too small a cavity, so that the virtual cathode was shorted (the TM_{010} mode is also not tunable by varying the length). The experimental result was that the most tunable modes that were readily excitable were TM_{011}, TM_{020}, and TM_{012}. Perturbation by multiple output ports can influence the choice of mode.[38]

The effect of the cavity on the vircator output is shown in Figure 10.14. Total power in the initial experiments with cavity vircators was 320 MW with the cavity; to compare, the nonresonant cavity gave 80 MW. Another notable feature of the cavity is reduction in bandwidth by a factor of three. Spike structure in the output waveforms observed in earlier experiments, reflecting mode competition, was greatly reduced, and smooth pulse shapes were obtained. The cavity vircator is now a commercial product.

Mile End Library
Queen Mary, University of London

RAG WEEK

If you pay your Library fines on

Tuesday 5th February
or
Friday 8th February

the money will be donated to
the QMUL Hardship Fund
as a celebration of RAG week

Borrowed items 01/02/2013 13:26
XXXXXX1508

Item Title	Due Date
* Handbook of microwave te	31/05/2013
* Handbook of microwave te	31/05/2013
* High power microwaves	31/05/2013

* Indicates items borrowed today
PLEASE NOTE
If you still have overdue books on loan
you may have more fines to pay

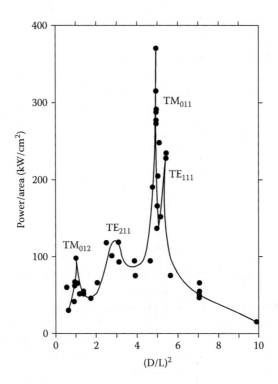

FIGURE 10.14

Radiated microwave power density as a function of the aspect ratio of a resonant cavity. Specific modes are identified. (From Benford, J. et al., *J. Appl. Phys.*, 61, 2098, 1987. With permission.)

The Los Alamos group has explored cavity vircators using a fixed cavity. Tuning is produced by varying the beam current density to bring ω_p in proximity to the passband of a suitable mode of the resonator. The TM_{020} and TM_{012} modes were excited with as much as a 70% variation in injected current.[39] A 1% bandwidth was obtained.

10.2.4.3 Virtodes and Feedback Vircators

In a sense, virtodes and feedback vircators strengthen the feedback mechanism used to increase power and narrow bandwidth in the cavity vircator. Rather than rely on the narrow frequency response of a cavity and the feedback mechanism of the reflexing electrons, the virtode, shown in Figure 10.15, feeds a portion of the microwave signal from the virtual cathode region directly back into the anode–cathode gap.[40] A device like that shown schematically in the figure was operated with a 450- to 500-kV electron beam at a current of up to 14 kA. An applied magnetic field of 0.2 to 0.6 T confined the beam and could be used to tune the output frequency. A maximum power level of 600 MW was achieved, and the efficiency ranged from 3 to 17%. Power without the feedback loop was at least a factor of two

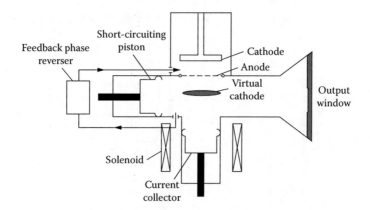

FIGURE 10.15
The virtode. (From Gadestski, P. et al., *Plasma Phys. Rep.*, 19, 273, 1993. With permission.)

smaller. By tuning the phase in the feedback circuit, the power output could be varied by as much as 5 dB. Tunability could also be effected by moving the current collector in and out. To capitalize on this mechanism, the movable current collector was replaced by a slow-wave structure with a beam collector downstream. In this geometry, the feedback loop into the diode gap was disconnected, but backard waves from the slow-wave structure could provide the necessary feedback. Power in this mode of operation reached 500 MW.

The feedback vircator of Figure 10.16 has been explored on both the SINUS-7 and MARINA generators at the Institute of High-Current Electronics in Tomsk.[41] On the latter, a compact inductive-storage machine with explosive fusing, a 1-GW, 2-GHz pulse of about 50-nsec duration was produced. With a voltage of about 1 MV and a current of about 20 kA, the efficiency was about 5%. The output was a quasi-Gaussian beam with the form of a TE_{10} mode. Ion emission in the diode was shown to cause electron beam pinching. By concentrating only on the portion of the beam current that was actually emitted from the cathode face, the authors inferred an effective efficiency of approximately 8 to 10%, closer to actual simulation results.

10.2.4.4 Coaxial Vircators

Coaxial vircators, originally developed at Texas Tech University, are the new entry into the field of advanced vircator development.[42,43] The basic configuration is shown in Figure 10.17. Note the cylindrical geometry and the fact that the electron beam propagates inward radially, taking advantage of the geometric convergence of the beam. The coaxial viractor is expected to have three advantages over traditional vircators. First, note that the anode–cathode region is not physically separated by conductors from the virtual cathode region to the interior of the anode. As a result, it is expected that the open geometry will allow feedback from the microwave generation region to the

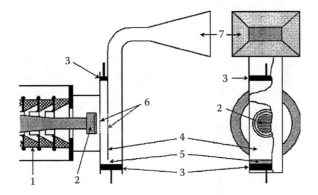

FIGURE 10.16
The feedback vircator used at the Institute of High-Current Electronics in Tomsk. (1) Stack-insulator interface to high-voltage generator, (2) bladed cold cathode, (3) tuning plungers, (4) partition wall, (5) coupling slot, (6) foil or mesh windows, and (7) vacuum horn. (From Kitsanov, S.A. et al., *IEEE Trans. Plasma Sci.*, 30, 1179, 2002. With permission.)

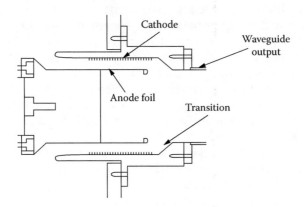

FIGURE 10.17
Side view of a cylindrical coaxial vircator.

beam. Second, the electron motion is radial only, eliminating the power loss associated with the axially transmitted portion of the electron beam in an axial vircator, as one can see in Figure 10.2 and Figure 10.12. Third, the large area of the diode reduces the current density and the beam loading of the cathode and anode, which is expected to increase the lifetime of the device, particularly under repetitive operation.

In the initial experiments, the cathode and anode radii were 13.1 and 9.9 cm; a 3-cm-long velvet cathode was the source of electrons. The diode was pulsed to 500 kV and a peak current of 40 kA for a 30-nsec pulse. The peak electron beam power was about 18 GW, while the peak microwave output was about 400 MW, so that the efficiency was about 2%. Remarkably, however, the microwave power pulse was offset and trailed the electron beam power pulse by about 10 nsec, so that the instantaneous power efficiency at the time of peak microwave power appeared to be about 10%. The 2-GHz output was delivered in the form of a TE_{11} mode that was quite reproducible from shot to shot.

Coaxial vircators are a very active international research topic at this time. Despite the possible advantages of the device, several challenges remain to be solved: determining the optimum configuration; addressing the lack of tunability in current configurations; dealing with difficulties in alignment, particularly in repetitive operation; and addressing the serious mode competition between TE_{11} and TM_{01}.

10.2.4.5   *Phase Locking of Vircators*

Before closing this section, we return to the cavity vircator and consider the phase locking of these devices as an alternative, multisource path to high power. The next logical step beyond the cavity vircator was to inject an external signal into the cavity. Price and coworkers[14] injected the signal from a 100- to 300-MW relativistic magnetron into a vircator that was powered from the same electrical generator. The cavity vircator power varied from 100 to 500 MW. A striking result from this experiment was the extraordinary frequency pulling when the magnetron pulse was injected. When the vircator was radiating alone, the frequency dropped slowly due to gap closure, in good agreement with the two-dimensional model of Woo. The magnetron signal was injected at 2.85 GHz, and the vircator frequency, originally at 2.3 GHz, was pulled up to the magnetron frequency in about 10 nsec, an extraordinary range of frequency pulling. Of course, the injected signal was quite large.

Didenko and coworkers injected a conventional magnetron pulse into a reflex triode and observed about a 100-MHz frequency pulling, in agreement with a driven nonlinear oscillator model.[44] This result was also confirmed by Fazio and coworkers,[15] who reported a 250-MHz frequency pulling when a klystron injected into a vircator of roughly equal power. They also reported gain and linearity of vircator response so that this experiment could be described as a vircator-amplifier experiment with a gain of 4.5 dB. The PI group also reported amplification using a vircator driven by a magnetron.[14] With a drive power of 140 MW, as much as 500 MW was produced in a linear amplification mode.

The demonstration of phase locking of the vircator to an external signal shows that the vircator can be used in arrays, as can other oscillators, such as the magnetron.[14] To understand, recall Adler's inequality, which provides a condition on the phase difference between two oscillators that will permit phase locking. The important parameter is the injection ratio between power input, P_i, and power output, P_o:

$$\rho = \left(\frac{P_i}{P_o}\right)^{1/2}$$

(10.13)

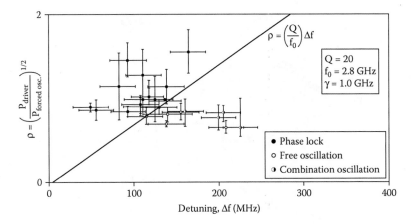

FIGURE 10.18
Adler relation for phase locking of a vircator to an external magnetron. Locked, unlocked, and combined states are shown. (From Price, D. et al., *J. Appl. Phys.*, 65, 5185, 1989. With permission.)

Phase locking occurs when the frequency difference between the oscillators is sufficiently small:

$$\Delta\omega \leq \frac{\omega_0 \rho}{Q} \tag{10.14}$$

where Q is the quality factor of the oscillators (assumed to be equal) and ω_0 is the mean frequency of the two oscillators. Figure 10.18 shows that Equation 10.14 is followed by the magnetron-driven vircator. In addition, there is an intermediate regime called a *combination oscillation* in which the vircator power output is modulated at a frequency corresponding to the frequency difference between the two oscillators. Price and Sze's theoretical analysis of the stability of the locking relation shows that the Adler inequality is a necessary but not sufficient criterion for phase locking.[45] Stable-locked steady-states are achieved in only a fraction of the parameter space. The stable regime occurs for $\rho \lesssim 1$. The rest of the Adler plane is unstable or unlockable. Therefore, overdriving an oscillator produces unstable oscillations and will not be observed as a phase-locked state. Moreover, the overall power of the system will be reduced.

Sze et al. demonstrated phase locking of two cavity vircators to one another.[17] Figure 10.19 shows their apparatus, in which a cathode stalk splits to feed two identical cylindrical cavity vircators, tunable with sliding end walls. Microwave generation starts once the diode currents exceed about 40 kA, about five times the space-charge-limiting current. The coupling bridge between the vircators is of length $7/2 \lambda_g$, where λ_g is the waveguide wavelength at 2.8 GHz. The connecting bridge has a plunger that, when inserted, disconnects the cavities. The cavity frequencies can be tuned to within 20 MHz of one another. With the plunger withdrawn, the cavities mutually

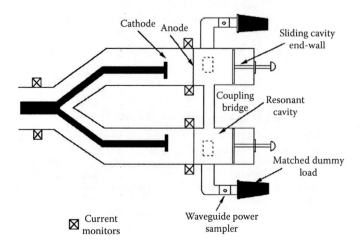

FIGURE 10.19
Dual-vircator phase-locking experiment. (From Sze, H. et al., *J. Appl. Phys.*, 67, 2278, 1990. With permission.)

couple and lock in less than 10 nsec. The relative phase angle varies from shot to shot by ±25°. There is some evidence that mode competition adversely affects the locking process; nevertheless, vircators were demonstrated to lock together in phase for about 25 nsec at a peak power level of 1.6 GW.

10.2.5   Fundamental Limitations and Outlook for Vircators

There will always be a place for vircators, in one form or another, in the field of HPM, because of their simplicity and ruggedness, tunability, versatility, and high power. Their ability to operate at low impedances — note that the Texas Tech coaxial vircator operated at less than 15 Ω — confers some advantage in specialized situations where the power source is an explosively driven flux compression generator with a low source impedance and a large amount of stored energy (see Figure 5.8). The limitation to date, and the question to be addressed by the relatively new cavity, tunable, and coaxial vircators, is the generally low efficiency of this class of devices. Beyond the untapped prospects for the new variants, it may be that some research group will return to some of the ideas explored for recovery of energy from the bunched transmitted beam using slow-wave structures or an inverse-diode arrangement.

10.3   Gyrotrons and Electron Cyclotron Masers

Gyrotrons and the associated class of *electron cyclotron masers* (ECMs), which includes *gyroklystrons* and *cyclotron autoresonance masers* (CARMs), tap the transverse energy of electrons that are gyrating about magnetic field lines. Although relativistic gyrotrons producing 100 megawatts have been explored experimentally, the primary driver for gyrotron development is for their application to magnetic confinement fusion research, where they provide megawatt-level, long-pulse (of the order of seconds to continuous) sources operating in excess of 100 GHz for electron cyclotron resonance heating, current drive, and suppression of instabilities.[46] High power gyroklystrons, approaching 100 MW, have been under development for years as an alternative source of X-band microwave power for radio frequency accelerators for electron-positron colliders. CARMs offer an alternative to free-electron lasers for using higher-energy electron beams to provide relativistic frequency upshifts to reach high power at high frequency. Other variants exist — gyro-BWOs, gyro-TWTs, large-orbit gyrotrons, peniotrons, and magnicons, notably — but none are being explored currently with a goal of producing truly HPM-level output.

10.3.1 History of Gyrotrons and Electron Cyclotron Masers

In the late 1950s, three investigators began a theoretical examination of the generation of microwaves by the electron cyclotron maser interaction:[47,48] Richard Twiss in Australia,[49] Jurgen Schneider in the U.S.,[50] and Andrei Gaponov of the U.S.S.R.[51] In early experiments with devices of this type, there was some debate about the generation mechanism and the relative roles of fast-wave interactions mainly producing rotational electron bunching and slow-wave interactions mainly producing axial electron bunching.[47,48] The predominance of the fast-wave ECM resonance with its rotational bunching in producing microwaves was experimentally verified in the mid-1960s in the U.S.[52] (where the term *electron cyclotron maser* was apparently coined) and the U.S.S.R.[53]

Gyrotrons were the first ECMs to undergo major development, and the requirement for high-average-power, continuous, or long-pulse millimeter-wave sources for electron cyclotron resonance heating, current drive, and instability suppression in magnetically confined plasmas for fusion research continues to drive the development of gyrotrons. As has been noted elsewhere,[54,55] early increases in device power were the result of Soviet-era Russian developments from the early 1970s in magnetron injection guns[56] to electron beams with the necessary transverse energy while minimizing the spread in transverse energies, and in tapered, open-ended waveguide cavities to maximize efficiency by tailoring the electric field distribution in the resonator.[57] Today, the production of megawatt-class gyrotrons at frequencies above 140 GHz has been the result of advances in mode control in extremely high-order-mode cavities, including the use of coaxial inserts; new internal mode converters to produce quasi-optical output coupling; depressed collectors to raise gyrotron efficiencies; and artificial diamond windows with the strength, low losses, and thermal conductivity to tolerate the passage of megawatt microwave output.[58]

The use of relativistic electron beams in gyrotrons followed shortly after the first relativistic beam-driven experiments with BWOs. The record for total output power from a relativistic gyrotron was the result of a collaboration between the Naval Research Laboratory (NRL) and Cornell University; in 1975, as much as 1 GW was reported by summing contributions from several frequencies between 8 and 9 GHz.[59] Gold et al. explored true HPM relativistic gyrotrons for a number of years at NRL as we shall discuss in Section 10.3.3.2. Their work resulted in the production of several hundred megawatts of output power in whispering-gallery devices operating near 35 GHz.

High power gyroklystrons operating in the X-band with power levels approaching 100 MW[60] were meant to provide a possible alternative to klystrons as drivers for a Next Linear Collider. This motivation for the development of these sources may change, however, with the 2004 shelving of the NLC concept in favor of the future International Linear Collider, which is to employ lower-frequency, lower-power RF sources.

Cyclotron autoresonance masers offer the possibility of tapping the large Doppler upshift of frequency afforded by relativistic beams in order to produce high power at high frequencies. Researchers at the Institute of Applied Physics in Nizhny Novgorod conducted the first investigations of these sources in the early 1980s,[61,62] but today they remain relatively immature sources that have not been fully explored.

Two other source types in this class appeared to hold out some promise for high power operation at one time, although they have not been brought to full maturity: quasi-optical gyrotrons and magnicons. Quasi-optical gyrotrons featured a large cavity resonator that was essentially open on the sides and bounded on each end by a mirror to form a Fabry–Perot resonator like that of an optical laser. The goal was to reduce the wall heating that plagued conventional gyrotrons. Originally proposed in the Soviet Union in 1967[63] and analyzed more extensively in the U.S. in the early 1980s,[64] these sources were investigated for a time at the NRL and the Ecole Polytechnique Federale de Lausanne in Switzerland, as well as at Yale University. Magnicons,[65] an enhancement of the original gyrocon,[66] are large-orbit devices originally proposed as drivers for the VEPP-4 accelerator at the Institute of Nuclear Physics in Novosibirsk. The interaction mechanism for these sources promised very high efficiencies, and although they were proposed for operation at tens of megawatts in the gigahertz range of frequencies, they have so far failed to fully realize their promise.

10.3.2 Gyrotron Design Principles

The typical configuration of a gyrotron, the most common type of ECM, is shown in Figure 10.20. Key features are the electron gun or diode, from which the beam is launched, the magnetic field and its axial profile, the interaction region or cavity, the beam collector, and the output extraction region. This diagram shows axial extraction, although as we shall see, side extraction with a mode converter has become far more common. A key

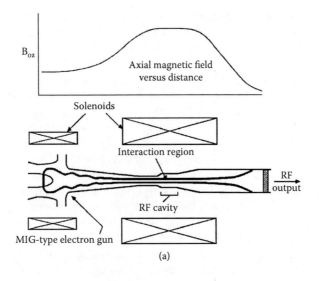

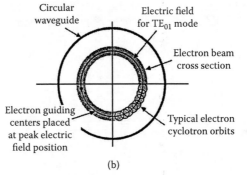

FIGURE 10.20

(a) Configuration of a gyrotron and the axial magnetic field profile. (b) The electron beam cross section showing some sample electron orbits within a gyrotron. (From Baird, J.M., *High-Power Microwave Sources*, Granatstein, V.L. and Alexeff, I., Eds., Artech House, Norwood, MA, 1987, p. 103. With permission.)

requirement of the electron beam in an ECM is that it have a specified value of the perpendicular velocity component. The ratio $\alpha = v_{0\perp}/v_{0z}$ at the entrance to the interaction region is an important parameter governing performance. As the beam moves downstream from the cathode, it is adiabatically compressed in a slowly increasing magnetic field, which raises the current density and increases $v_\perp$ as $B_{0z}^{1/2}$, the square root of the applied axial magnetic field, through the conservation of the electron magnetic moment,

$$\mu = \frac{\left(\dfrac{1}{2}mv_\perp^2\right)}{B_{0z}}$$

(10.15)

Naively, it might be expected that the largest possible value of $v_\perp$ would be desirable; however, in the magnetic compression process, if $v_\perp$ is too large, electrons will be reflected by *magnetic mirroring* if v_z is allowed to vanish (see Problem 8).

In long-pulse or continuous-wave (CW) gyrotrons, the magnetron injection gun (MIG), discussed at length by Baird,[54] is the most common beam source. In relativistic gyrotrons and CARMs, the electron beam is typically generated on an explosive emission cathode with much smaller perpendicular velocity components. Electron rotational energy can then be added after emission by passing the beam through a magnetic wiggler, like that used in an FEL, but short enough to not induce competing oscillations of its own, or through a single pump coil that introduces a nonadiabatic dip in the magnetic field and stimulates perpendicular electron velocity components.

Within the interaction region, the motion of the beam electrons decomposes into three components: a drift along the magnetic field v_z, a slow rotational motion of the beam as a whole about the magnetic axis of the system due to an $\mathbf{E} \times \mathbf{B}_0$ drift involving the radial self-electric field of the beam, and the Larmor rotation of the individual electrons at $v_\perp$ about their guiding centers. The minimum thickness of the beam is the Larmor radius of the electrons,

$$r_L = \frac{v_\perp}{\omega_c} \tag{10.16}$$

where ω_c is the electron cyclotron frequency,

$$\omega_c = \frac{eB}{m\gamma} \tag{10.17}$$

with γ the usual relativistic factor taking into account the total electron velocity (see Problems 9 and 12). Figure 10.20b depicts the rotational electron motion in a beam cross section.

In most cases, the ECM resonant interaction couples a transverse electric (TE) cavity mode to the fast Larmor rotation of the electrons. Signal gain from interactions with the transverse magnetic (TM) modes is also possible, but the coupling into TM modes in relativistic gyrotrons is reduced by β_z^2 relative to that for TE interactions, so that the TM interactions are overwhelmed in weakly relativistic devices. Nevertheless, a relativistic TM mode gyrotron has been built using a special resonator with enhanced Q for TM modes.[67]

The effect of the TE fields on a beamlet of electrons is shown in Figure 10.21.[68] In the figure, the electrons rotate counterclockwise in the x-y plane with a rotational frequency ω_c (we will ignore axial drifts without losing any of the important physics). Initially, as indicated in Figure 10.21a, they are

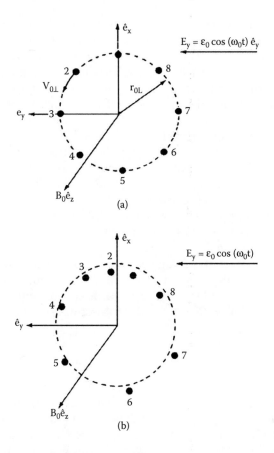

FIGURE 10.21
The bunching mechanism for eight test electrons. (a) The initial state in which the electrons are equally spaced in phase and executing Larmor rotations about a common guiding center. (b) After several cycles, the electrons have begun to bunch about the x-axis. (From Sprangle, P. and Drobot, A.T., *IEEE Trans. Microwave Theory Tech.*, MTT-25, 528, 1977. With permission.)

uniformly arranged in rotational phase on a circle with a radius r_L. The TE field with frequency ω_0 will first decelerate electrons 1, 2, and 8, causing them to lose energy and spiral inward; their rotational frequency, ω_c, will increase, though, because of the decrease in energy and the relativistic factor γ in the denominator of the expression for ω_c (Equation 10.17). Electrons 4, 5, and 6, on the other hand, will gain energy from the field, move outward, and rotate more slowly. Eventually, after a number of field periods, the electrons will bunch in rotational phase about the y axis. This effect, relativistic in origin due to the energy dependence of the rotational cyclotron frequency of the electrons, occurs for electron energies of even tens of keV. If $\omega_0 = \omega_c$, equal numbers of electrons are accelerated and decelerated, and no net energy is exchanged between the electrons and the field. On the other hand, if the oscillation frequency of the field exceeds the rotational velocity, $\omega_0 > \omega_c$, the situation is as depicted in Figure 10.21b, with more electrons decelerated over a field period than are accelerated, and the electrons surrender net energy to the wave, which grows in amplitude. In a more general treatment with v_z taken into account, both azimuthal and axial bunching of the electrons occur. This situation simplifies with certain limits, however, as we shall see shortly. A number of mathematical treatments of gyrotrons and the instability underlying the operations of ECMs can be found in the literature (see, for example, Gaponov et al.[55] for a review of the early Soviet-era theoretical developments, and other references[68–72]).

Another generation mechanism, the Weibel instability, will compete with and, if the phase velocity of the electromagnetic modes in the interaction region is less than the speed of light, dominate the ECM instability.[73] There are devices in which some form of cavity loading is used to slow the waves and employ the Weibel instability as the generating mechanism.[74] Here, however, we will concentrate on fast-wave sources governed by the ECM interaction.

To estimate the operating frequency and illustrate the device types, we work with simplified dispersion diagrams for normal modes of the resonator, typically those of a smooth-walled-waveguide section with dispersion relations of the type given in Equation 4.19, for example, and the fast cyclotron waves on the electron beam

$$\omega = s\omega_c + k_z v_z \,, \, s = 1, 2, 3, \ldots \tag{10.18}$$

where k_z and v_z are the axial wavenumber and axial electron drift velocity, and s in an integer, allowing for interactions at harmonics of the cyclotron frequency (see Problem 11). In fact, the space-charge on the beam and the nonzero $v_\perp$ split the fast cyclotron wave into a positive- and a negative-energy wave (just as space-charge splits the beam line in Cerenkov sources), with the higher-frequency wave of the pair being the negative-energy wave that interacts with the electromagnetic cavity mode to produce microwaves. Thus, their contributions raise the output frequency a bit above the approximate value given by Equation 10.18 (in contrast to Cerenkov sources, where generation occurs below the beam line). The point of intersection between the fast cyclotron wave and the cavity mode determines the type of ECM. Examples are shown in Figure 10.22. We distinguish three main types of devices involving forward-wave interactions in which the group velocity of the electromagnetic mode is positive at resonance. Gyrotrons operate near cutoff, gyro-TWTs have a somewhat larger value of k_z at resonance and a larger forward-directed group velocity at resonance, and CARMs have a large value of k_z and a substantial Doppler upshift in operating frequency (see Problem 14). The greatest gain usually occurs when the fast cyclotron mode intersects the electromagnetic mode tangentially on the dispersion diagram, so that the group velocity approximately matches the electron beam drift velocity. Hence, gyrotrons tend to use beams of lower axial energy in most cases. The greatest efficiency, on the other hand, occurs when the

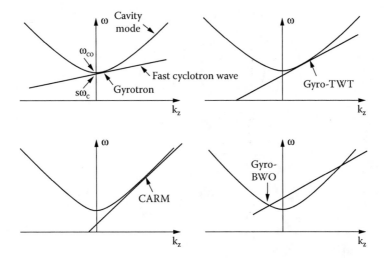

FIGURE 10.22
Dispersion diagrams for the uncoupled cavity mode, for which $\omega^2 = \omega_{co}^2 + k_z^2 c^2$, with ω_{co} the cutoff frequency for the operating mode, and the fast cyclotron wave on the beam, for which $\omega = s\omega_c + k_z v_z$, with ω_c the cyclotron frequency and s the harmonic number.

cyclotron frequency is detuned below the resonance line, so that the fast cyclotron mode line would actually lie somewhat below the line shown.

Historically, gyrotrons typically operated in one of the three types of modes depicted in Figure 10.23:[75] TE_{0p}, TE_{1p}, or a TE_{mp} mode with m >> p. This last type is known as a *whispering-gallery mode*, because its fields tend to be concentrated near the walls of the resonator, much like the acoustic whispering-gallery modes that carry sound near a wall. The TE_{0p} modes, with no azimuthal variation, have the smallest ohmic wall losses (see Figure 4.8). Unfortunately, the higher-order TE_{0p} modes suffer from mode competition, particularly with the TE_{2p} modes,[76–78] which impairs their ability to produce a high power signal. One method for selecting the desired mode under these circumstances is to use a complex cavity with two sections of different radii, essentially to constrain the oscillation to modes with a common frequency in both cavities.[79] The penalty for the use of such cavities is the increased space-charge depression of the beam energy accompanying the large radial separation between the beam and the wall in the larger of the two cavity sections. Another method for controlling the operating mode is to essentially split the cylindrical resonator along its sides to fix the azimuthal mode structure; the diffractive losses of modes with nonzero fields at the slits place them at a competitive disadvantage and effectively select the TE_{1p} modes.[80] The penalties for this method, however, are the somewhat higher ohmic wall losses for these modes, and the fact that their peak electric fields lie near the center of the cavity, so that good beam–field coupling, which requires the beam to be placed near the high-field region, implies substantial space-charge depression of the beam potential. Also, when the radial index p >> 1,

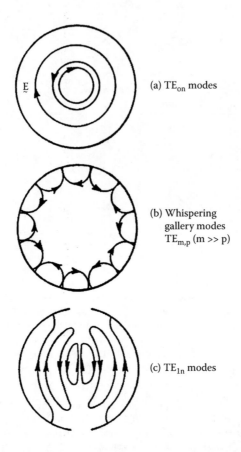

(a) TE_{on} modes

(b) Whispering
gallery modes
$TE_{m,p}$ (m >> p)

(c) TE_{1n} modes

FIGURE 10.23
Three basic types of operating modes of a gyrotron. (From Granatstein, V.L., *High-Power Micro-wave Sources*, Granatstein, V.L. and Alexeff, I., Eds., Artech House, Norwood, MA, 1987, p. 185. With permission.)

the TE_{1p} modes become nearly degenerate in frequency. Finally, the slits fix the mode field pattern and do not allow it to rotate, so that portions of the beam couple poorly to the fields, and the efficiency is reduced.

In recent years, gyrotron designers have turned to high-order whispering-gallery modes for the operation of megawatt-class gyrotrons at frequencies above 100 GHz. For example, a 140-GHz, 1.5-MW gyrotron being developed at MIT for operation in several-microsecond pulses utilizes the $TE_{22,6}$ mode in a cylindrical interaction cavity.[81] On the other hand, a 165-GHz gyrotron being developed at FZK in Karlsruhe, Germany, as a prototype for an expected 170-GHz, 2-MW gyrotron to be required later focuses on a $TE_{31,17}$ mode in a coaxial interaction cavity.[82] In fact, design studies conducted in the early 1990s in Brazil at the Plasma Physics Laboratory at the National Institute for Space Research (Sao Jose dos Campos) even considered a $TE_{42,7}$ mode for a 1-MW, 280-GHz device (see Dumbrajs and Nusinovich[46]). The move to these very high order modes, in cavities that are 20 to 30λ in diameter, is driven primarily by the need to limit the thermal wall loading from cavity losses to 2 to 3 kW/cm^2.[58] Cavities simply must be large to have adequate wall area to handle the power loading, and the requirement for high-frequency operation creates the need to operate in a high-order mode. The whispering-gallery modes have several advantages in this regard.[83–85] Because their fields are concentrated near the walls, the electron beam must also be located near the wall to ensure good coupling. This factor reduces space-charge depression and related velocity spreads in the beam, enhancing efficiency; however, a large-diameter beam is therefore required, and minimizing beam losses to the cavity wall requires careful design. An additional advantage is that in this position, mode competition is reduced because of (1) the poor coupling to the volume modes with fields concentrated toward the interior of the cavity and (2) the relatively larger frequency spacing between other potentially competitive whispering-gallery modes.

The saturation of wave growth in ECMs is the result of two competing phenomena: depletion of the rotational energy and phase trapping.[68] For small values of $v_\perp$, depletion of the rotational kinetic energy is the dominant effect. Wave growth ceases when the perpendicular velocity component is reduced below the threshold for oscillation. When $v_\perp$ is large, on the other hand, the microwave fields build up to large amplitudes, and phase trapping is the dominant saturation mechanism. To explain, return to our discussion of the ECM interaction at the beginning of this section. Remember that some electrons are accelerated and some are decelerated by the fields. At the outset of the generation process, the signal frequency exceeds the cyclotron frequency of rotation, and more electrons are decelerated. As the electrons decelerate and their rotational phases change, the trapped electrons eventually reach a point at which further interaction with the fields causes some electrons to be reaccelerated. At this point, the electrons begin to take energy back from the fields, and the field amplitude passes through a maximum. This mechanism is phase trapping.[68] Figure 10.24, taken from the reference, shows the results of a numerical calculation of the microwave generation efficiency as a function of the initial relativistic factor. The authors of the paper assumed that all electrons had the same axial velocity, and the transformation was made to a moving coordinate frame in which the electrons had no axial drift velocity, only a rotational velocity; thus, $\gamma_{0\perp} = (1 - \beta_{0\perp}^2)^{-1/2}$, with $\beta_{0\perp}$ the initial value of $v_\perp/c$. Shown also are two approximations to the efficiency based on rotational energy depletion and phase trapping. The minimum value for microwave generation to occur is $\gamma_{\perp,crit}$, ω_0 is the signal frequency, and $\Delta\omega$ is the amount by which the signal frequency should lie above ω_c in order for maximum signal growth to occur. The efficiency is maximum for an intermediate value of the rotational energy because of the combination of the two phenomena leading to saturation.

The perpendicular efficiency $\eta_\perp$ measures the microwave conversion efficiency from the electrons' rotational energy perpendicular to the magnetic field. This frequently quoted parameter must be distinguished, however, from the electronic efficiency for conversion of the total kinetic energy of the elec-

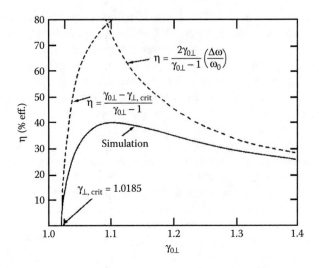

FIGURE 10.24
Calculated efficiency of a gyrotron in a frame of reference moving with the axial beam velocity in which $v_z = 0$. The analytical estimate shown in the dashed line to the left is based on rotational energy depletion as the saturation mechanism, while the estimate to the right assumes that phase trapping is the sole saturation mechanism. (From Sprangle, P. and Drobot, A.T., *IEEE Trans. Microwave Theory Tech.*, MTT-25, 528, 1977. With permission.)

trons to microwaves, η_e. The electronic efficiency is computed from $\eta_\perp$ by making a Lorentz transformation back to the laboratory frame of reference:[68]

$$\eta_e = \frac{\gamma_0\left(\gamma_{0\perp}-1\right)}{\gamma_{0\perp}\left(\gamma_0-1\right)}\eta_\perp \tag{10.19}$$

where $\gamma_0 = (1-\beta_{0\perp}^2-\beta_{0z}^2)^{-1/2}$. For a weakly relativistic beam (i.e., much less than 500 keV per electron), Equation 10.19 can be approximated by

$$\eta_e \approx \frac{\beta_{0\perp}^2}{\beta_{0\perp}^2+\beta_{0z}^2}\eta_\perp = \frac{\alpha^2}{1+\alpha^2}\eta_\perp \tag{10.20}$$

with $\alpha = \beta_{0\perp}/\beta_{0z} = v_{0\perp}/v_{0z}$.

Two primary means of increasing the efficiency of these devices involve tapering either the radius of the cavity or the applied magnetic field in the axial direction.[86] In either case, the basic idea is to begin bunching the beam under less than optimal conditions for maximum energy extraction, and then to pass the bunched beam into a region where it can more effectively surrender its energy to the microwaves. In cavity tapering, the electric field profile in the cavity is tailored in such a way that the highest amplitudes are reached toward the downstream end, by which point the beam has become bunched. In magnetic tapering, the field near the entrance of the cavity is held below a value at which strong resonance can occur, so that bunching is not accompanied by significant energy extraction. Further into the cavity, though, the magnetic field is increased to the resonance value, so that higher-efficiency extraction is accomplished by the bunched beam.

Because of the very high internal power levels in modern gyrotrons, depressed collectors are now used quite frequently to raise efficiency and thus reduce power loss to the cavity and collector walls, as well as to reduce radiation generation.[58] Depressed collectors are placed downstream of the interaction cavity, where the spent electrons are collected and their remaining kinetic energy is converted back to electrical energy to be recycled in the system. Efficiencies of 50% are now regularly seen in megawatt-class gyrotrons using this component (see Problem 10).

10.3.3 Gyrotron Operational Features

In this section, we consider both megawatt-class long-pulse or continuous gyrotrons and the relativistic gyrotrons developed at the NRL.

10.3.3.1 High-Average-Power Gyrotrons

The pace of development for the former is impressive. Consider the development of gyrotrons in the U.S. at CPI in Palo Alto, CA, by comparing the parameters and performance of a 110-GHz gyrotron described in a 1996 paper[87] and a 140-GHz gyrotron described in 2004.[88] A schematic of the former is shown in Figure 10.25. The notable features there are operation in the $TE_{22,6,1}$ mode, with a peak thermal wall loading of about 0.8 kW/cm^2; an internal mode converter using the dimpled-wall or Denisov converter developed at the Institute of Applied Physics in Nizhny Novgorod, which has an efficiency of 95%, as opposed to the earlier Vlasov converter with an efficiency of about 80%; output windows of double-disc sapphire with coolant running between the two discs; and a passive beam collection system. With beam voltage and current of 86 kV and 37.6 A, and an output of 1.013 MW, the efficiency of this gyrotron was 31%. To compare, the 2004 gyrotron operated in the $TE_{28,7,1}$ mode with a peak thermal wall loading of 0.3 kW/cm^2 measured in tests at only 500 kW of output; a dimpled-wall converter capable of coupling out 99% of the tube output; output windows of edge-cooled CVD diamond; and a single-stage depressed collector to recover beam energy. At 80 kV and 40 A in the electron gun, and an output power expected to be 1 MW, the efficiency is actually a bit smaller, at about 31%; however, the depressed collector raises the overall system efficiency to almost 42%.

At FZK Karlsruhe, work on a 1.5-MW, 165-GHz tube is aimed at ultimately making it possible to produce a 2-MW, 170-GHz tube that was expected for ITER, the international tokamak.[82] The 165-GHz tube operates in the $TE_{31,17}$ mode, with a 90-kV, 50-A electron gun and a nominal output of 1.5 MW, for

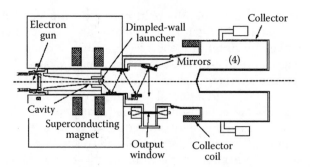

FIGURE 10.25

Schematic diagram of a 110-GHz gyrotron from CPI described in 1996. (From Felch, K. et al., *IEEE Trans. Plasma Sci.*, 24, 558, 1996. With permission.)

an efficiency of 33%; a depressed collector raises the system efficiency still higher, to 48%. This system uses a coaxial cavity.

10.3.3.2 Relativistic Gyrotrons

A series of papers dating to 1985 described high-peak-power experiments at 35 GHz with a relativistic gyrotron at the Naval Research Laboratory.[89–92] The most recent configuration of the experiment is shown in Figure 10.26. An electron beam is generated on an annular, explosive emission cathode on the VEBA pulse-line accelerator. Waveforms for the current and voltage from the accelerator are shown in Figure 10.27; the beam is emitted from the cathode at 1.2 MV and about 20 kA, but the apertured anode passes only about 2.5 kA. In this role, the anode also acts as an emittance filter. As it leaves the diode gap, the beam has very little $v_\perp$, so it is passed through a pump magnet coil, which imparts the perpendicular rotational mode motion to the electrons, as can be seen in the figure. Downstream, the gyrotron cavity is cylindrically symmetric, with a diameter of 3.2 cm and a calculated Q of 180 for the $TE_{6,2}$ mode, in the absence of the beam. Beyond, there is a 5° output taper transition to a drift tube 120 cm in length and 14 cm in diameter, followed by a 1-m output horn flaring to an output window 32 cm in diameter. Operating in this geometry, with a cavity field of 3.2 T, power levels of up to 250 MW at efficiencies as high as 14% have been achieved. The 40-nsec duration of the microwave pulse means that microwaves are produced for roughly the duration of the flat part of the VEBA voltage and current pulse.

The NRL experiments began in the mid-1980s. In the first set of experiments, a 350-kV, 800-A beam was emitted across a magnetic field from an explosive emission cathode to supply some initial transverse energy, and then the beam was adiabatically compressed by the field. At 8% efficiency 20 MW was generated, but the experiment lacked the necessary flexibility to improve that value markedly.[89] In a second version,[90,91] a higher-power beam, 900 kV and 1.6 kA, was generated in a foil-less diode and passed

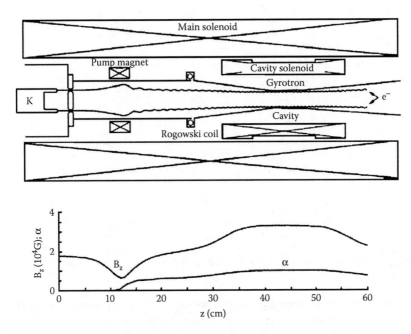

FIGURE 10.26
Configuration of the relativistic gyrotron experiments conducted on the VEBA pulse-line accelerator at NRL. (From Black, W.M. et al., *Phys. Fluids B*, 2, 193, 1990. With permission.)

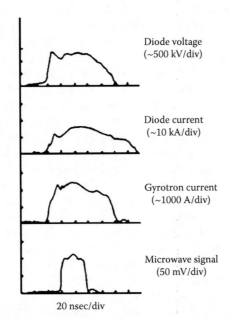

FIGURE 10.27
Accelerator and beam voltage and current waveforms from the experiment of Figure 10.26. (From Black, W.M. et al., *Phys. Fluids B*, 2, 193, 1990. With permission.)

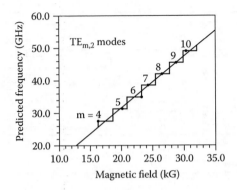

FIGURE 10.28
Step tuning of the NRL relativistic gyrotron through the TE_{m2} whispering-gallery modes by
varying the cavity magnetic field. Points are experimental data, the straight line is a best fit to
the data, and the "stair step" shows ranges of the magnetic field over which each mode is
calculated to start for the beam current used. (From Gold, S.H. et al., *Phys. Fluids*, 30, 2226, 1987.
With permission.)

through a magnetic wiggler to pump up $v_\perp$, which was then further
increased by magnetic compression. This version produced 100 MW at 8%
efficiency in a TE_{62} mode. Further, the gyrotron was shown to be step tunable
through the whispering-gallery modes TE_{n2}, with n = 4 to 10, by varying
the cavity magnetic field, as shown in Figure 10.28. Open-shutter photographs of the gas breakdown pattern in a low-pressure cell verified the field
patterns associated with the expected modes. In this same experiment, 35
MW was produced in the TE_{13} cavity mode by cutting a slot on either side
of the cavity to eliminate competition from the whispering-gallery modes.
A shortcoming of this experiment was the rather low beam quality. Following those experiments, in experiments configured as shown in Figure 10.26,
the group produced 250 MW at 14% efficiency.[92] An additional feature of
those experiments was a set of computer simulations (using a single-cavity
mode) designed to test the response of the gyrotron to the temporal voltage
fluctuations of the electron beam (assuming that the current varied as the
voltage $V^{3/2}$). Calculations of the microwave power envelopes for different
magnetic fields and voltage waveforms with temporal variations that modeled the fluctuations seen in real experiments (i.e., without assuming a
constant voltage during the pulse) showed varying sensitivities of the output
to the voltage glitches of real waveforms, with different microwave power
maxima for different portions of the voltage waveform — depending on the
magnetic field — and complete interruptions of output power in response
to certain voltage variations.

As a final point with regard to the issue of step tuning magnetrons, it is
interesting to compare the rather clean behavior of the NRL gyrotron operating in relatively low order whispering-gallery modes to the situation for
much higher order modes in a higher-frequency gyrotron at MIT.[93] The data
for the latter are shown in Figure 10.29, plotted somewhat differently than

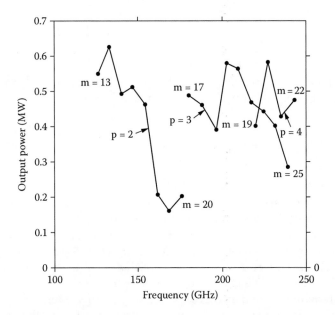

FIGURE 10.29

Peak output power measured at the frequencies shown by step tuning a gyrotron at MIT through the $TE_{m,p,1}$ modes. Operation in the different modes was achieved by varying the cavity magnetic field. (From Kreischer, K.E. and Temkin, R.J., *Phys. Rev. Lett.*, 59, 547, 1987. With permission.)

in Figure 10.28. Nevertheless, we see the complexity of step tuning TE_{mp} modes (remembering that m is the index on azimuthal field variations and p is the index on radial variations) by varying the magnetic field when m ranges as high as 25 for p = 2 to 4, with overlaps in frequency (albeit at different values of the magnetic field).

10.3.4 CARMs and Gyroklystrons

While gyrotrons are the most well known and studied ECMs, they are by no means the only variant of interest. Here we consider briefly two variants of interest from an HPM perspective: the CARM and the gyroklystron.

10.3.4.1 CARMs

Returning to Figure 10.22, we note that at sufficiently high electron energies, which implies large v_z, and resonant phase velocities, the ECM interaction leads to a CARM. The important conceptual feature of CARMs is autoresonance. To understand this concept, we consider the behavior of the cyclotron wave dispersion as the wave loses energy. Differentiating Equation 10.18 for s = 1, we determine the time rate of change of the resonant frequency as energy is drawn from the electrons into the microwave fields:

$$\frac{d\omega}{dt} = k_z \frac{dv_z}{dt} + \frac{d\omega_c}{dt} = k_z \frac{dv_z}{dt} - \omega_c \left(\frac{1}{\gamma} \frac{d\gamma}{dt} \right) \tag{10.21}$$

The derivatives of both $v_z = p_z/m\gamma$ and $\gamma = [1 + \mathbf{p} \bullet \mathbf{p}/(mc)^2]^{1/2}$ are determined from the equation of motion, from which we find

$$\frac{d\omega}{dt} = \frac{e\omega}{\gamma mc^2} v_\perp \bullet E_\perp \left[1 - \left(\frac{k_z c}{\omega} \right)^2 \right] \tag{10.22}$$

where $\mathbf{E}_\perp$ is the electric field component of a TE wave. Thus, $d\omega/dt$ vanishes as the phase velocity, ω/k_z, goes to the speed of light. In this event, as the electrons lose energy to the microwaves, the decrease in the term $k_z v_z$ is compensated by an increase in ω_c, with its inverse dependence on γ.[71,94,95] Alternatively, this effect is viewed as a cancellation of the effects of axial and azimuthal bunching (via the changes in v_z and γ, respectively). Thus, the electrons remain in resonance even as they surrender energy to the microwave fields. By contrast, note that in a gyrotron, $k_z \approx 0$, and the energy loss to azimuthal bunching is uncompensated. A calculation similar to that for FELs shows that in CARMs, the basic electron oscillation frequency, in this case ω_c, is strongly Doppler upshifted by an amount proportional to γ. Note, however, that since the cyclotron frequency is also proportional to γ^{-1}, the upshift in the CARM only scales as γ.

The CARM shown in Figure 10.30, an oscillator from the Institute of Applied Physics, was one of the first high power devices of this type.[62] To shorten the output wavelength from the 4.3 mm of early experiments using the lowest-order TE_{11} mode of the resonator, the CARM shown in the figure employed the TE_{41} mode to achieve a 2-mm output. The electron beam was generated with a pulse-line machine at a voltage of 500 to 600 kV and a current of about 1 kA. The system was immersed in a pulsed 2-T magnetic

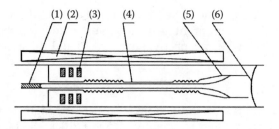

FIGURE 10.30
CARM from the Institute of Applied Physics: (1) cathode, (2) pulsed axial field coil, (3) slotted copper rings, (4) resonator with rippled-wall Bragg reflectors, (5) beam collector, and (6) vacuum window.

field extending from the cathode of the foil-less diode through the interaction cavity. To provide the perpendicular velocity for the beam electrons, a magnetic field wiggler was created by placing three slotted copper rings downstream of the diode with a spacing equal to the Larmor pitch of the beam electrons. Values of $\beta_\perp$ between 0.2 and 0.4 were produced in this manner with a spread of 0.05. The resonator had Bragg mirrors on either end of the interaction region[96]; the diffractive and ohmic Qs for the TE_{41} mode in this cavity were 3000 and 1000, respectively. At $\beta_\perp = 0.2$, the computed starting current for oscillation was 300 A, while the calculated efficiency was 4%. In experiments, a starting current of about 500 A was measured, and the 10-MW output represented an efficiency of 2%. For beam currents below the start current, there was no appreciable microwave emission.

10.3.4.2 Gyroklystrons

Gyroklystrons are the marriage of the gyrotron concept of generating microwaves using the electron cyclotron maser mechanism and the klystron concept of increasing power and efficiency by bunching a beam in one set of cavities and extracting the output from an electromagnetically isolated cavity with the high efficiency that results from the interaction with a prebunched beam. Of course, the difference from klystrons themselves is the microwave-generating mechanism and the fact that the bunching is in the rotational phase of the electrons about their guiding centers.

Significant effort has been devoted to the exploration of gyroklystrons for millimeter-wave radars at the 100-kW level; however, the investigation of high power gyroklystrons at powers approaching 100 MW has been supported by the need for a high power and high-efficiency driver for high-energy electron-positron linear colliders. This work has largely been concentrated at the University of Maryland. The highest power produced by this group — 75 to 85 MW at about 8.6 GHz with an efficiency near 32% and a gain near 30 dB in an amplifier — was described in a 1998 paper by Lawson et al.[60] The device is shown schematically in Figure 10.31. It is a coaxial device, with the center conductor supported by the two tungsten pins, intended to cut off propagation of microwaves at the TE_{01} working mode between cavities. Additional absorber, lossy ceramics made of carbon-impregnated aluminosilicate (CIAS) and 80% BeO/20% SiC, is used to further prevent the propagation of the parasitic, lowest-order TE_{11} mode between cavities; nominal attenuation in the drift tube for this mode in the X-band was 4.7 dB/cm (see Problem 13). The parameters of this device are summarized in Table 10.1. The electron beam is produced in a magnetron injection gun (MIG) that was designed using the EGUN simulation code. Simulations showed that the efficiency of the device declined slowly and monotonically with axial velocity spread, although with the spread shown in the table, the reduction was only about 1%. The output cavity was formed by changing the radius on both the inner and outer walls, and it is bounded on the right by a lip with radii matching those of the upstream drift tube.

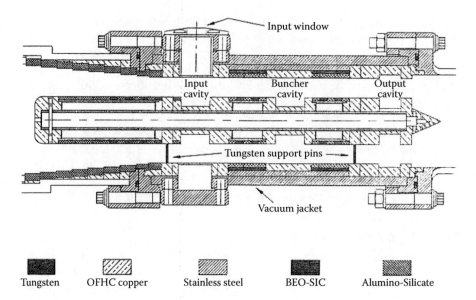

Tungsten   OFHC copper   Stainless steel   BEO-SIC   Alumino-Silicate

FIGURE 10.31
Schematic of the three-cavity microwave circuit for the 80-MW gyroklystron amplifier from the University of Maryland. (From Lawson, W. et al., *Phys. Rev. Lett.*, 81, 3030, 1998. With permission.)

The Q of the input cavity was determined by the combination of the diffractive Q of the input aperture and the ohmic Q of the cavity, while the Q of the buncher cavity was determined by ohmic losses in the nearby absorbing ceramics. The Q of the output cavity was essentially a diffractive Q that could be adjusted by changing the length of the lip at the right of the cavity. Figure 10.32 shows the voltage and current waveforms and microwave output envelope. We note the relative constancy of these parameters in comparison, for example, to the much higher voltage, current, and power traces of a pulse-line-driven experiment such as that shown in Figure 10.26 and Figure 10.27. The gyroklystron gun was driven by a modulator, and mismatches there were the source of the voltage drop in Figure 10.32. Also, the noise on the pulses, particularly notable in the output envelope, was attributed to ground-loop noise.

Subsequent work on this gyroklystron has concentrated on the production of a higher-frequency version operating at 17.1 GHz.[97] The rationale for increasing the frequency was the scaling for the number of sources required to drive a linear electron-positron collider of a given energy: if power and pulse length can be held constant as the frequency f increases, the number of sources decreases as $1/f^2$. The follow-on four-cavity design featured a TE_{011} input cavity driven at half the output frequency, followed by three cavities operating in the TE_{021} with frequency doubling to 17.1 GHz. The device operated at the second harmonic of the cyclotron frequency, allowing a lower magnetic field to be used. An output power of

TABLE 10.1

System Parameters at the Nominal Operating Point for the University of
Maryland Gyroklystron Shown in Figure 10.31

Magnetic Field Parameters		Output Cavity Parameters	
Cathode axial field (T)	0.059	Inner radius (cm)	1.01
Input cavity field (T)	0.569	Outer radius (cm)	3.59
Buncher cavity field (T)	0.538	Length (cm)	1.70
Output cavity field (T)	0.499	Quality factor	135 ± 10
		Cold resonant frequency (GHz)	8.565
Input Cavity Parameters		Lip length (cm)	0.90
Inner radius (cm)	1.10	*Beam Parameters*	
Outer radius (cm)	3.33		
Length (cm)	2.29	Beam voltage (kV)	470
Quality factor	73 ± 10	Beam current (A)	505
Cold resonant frequency (GHz)	8.566	Average velocity ratio (α)	1.05
		Axial velocity spread (%)	4.4
Drift Tube Parameters		Average beam radius (cm)	2.38
		Beam thickness (cm)	1.01
Inner radius (cm)	1.83		
Outer radius (cm)	3.33	*Amplifier Results*	
Length, input to buncher (cm)	5.18		
Length, buncher to output (cm)	5.82	Drive frequency (GHz)	8.60
		Output power (MW)	75
Buncher Cavity Parameters		Pulse length (μsec)	1.7
		Efficiency (%)	31.5
Inner radius (cm)	1.10	Gain (dB)	29.7
Outer radius (cm)	3.33		
Length (cm)	2.29		
Quality factor	75 ± 10		
Cold resonant frequency (GHz)	8.563		

100 MW was planned, although problems with the electron gun limited the
output in early experiments.

Lawson considered the general issue of frequency scaling for devices of
this type in a 2002 paper that looked at 100-MW class K_u-band (17-GHz), K_a-
band (34-GHz), and W-band (91-GHz) devices.[98] He concluded that for a
number of reasons — cathode loading, beam axial spread caused by increasing magnetic compression of the beam, $f^{5/2}$ scaling of cavity wall power
losses in circular modes, manufacturing tolerances, and mode competition
in the W-band — that the gyroklystron as it has been designed is probably
viable up to the K_a-band, around 34 GHz.

Another issue with general application for devices of this type is beam
uniformity in MIGs with thermionic cathodes. Variations in the temperature
and in the local work function for emission contribute to beam nonuniformi-
ties that can drive beam tunnel instabilities where none are expected. Consequently, the group at the University of Maryland is exploring the design
of a space-charge-limited, rather than temperature-limited, electron gun.[99]

Additional prototype experiments at the NRL with a different gyroklystron
illustrated another strength of this type of device in achieving phase locking

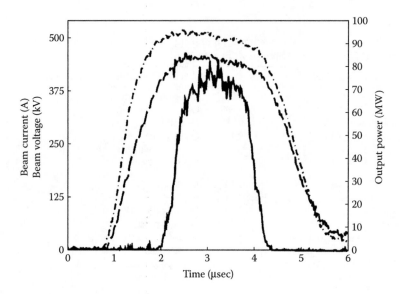

FIGURE 10.32

Voltage (dashed line), current (dash–dot line), and microwave output envelope (solid line) waveforms from the gyroklystron experiment of Figure 10.31. (From Lawson, W. et al., *Phys. Rev. Lett.*, 81, 3030, 1998. With permission.)

of multiple sources.[100] In a single-cavity device, the fractional bandwidth over which phase locking can occur, $\Delta f/f$, is related to the external Q of the cavity, the drive power of the input, P_i, and the output power, P_o, and is given by the Adler inequality (Equation 10.14). Theory shows, though, that injection of the drive signal into a prebunching cavity in a gyroklystron allows a wider phase-locking bandwidth than Adler's inequality would. This prediction was tested in a prototype experiment with the results shown in Figure 10.33. Note that for injection into the third of three cavities, the phase-locking points fall very near the Adler curve. For injection into the input cavity (results plotted in the lower graph), the power required to achieve a given bandwidth is lower than the value indicated by Adler's relation.

10.3.5   Outlook for Electron Cyclotron Masers

Research and development in this class of sources continues to be dominated by 1- to 2-MW-class gyrotrons in the 100- to 200-GHz range for use in magnetic confinement fusion experiments. The developments achieved by an international group of organizations from Russia, Germany, France, the U.S., and Japan have been remarkable over the past decade. Notably, the accomplishments of this effort have included the use and mode control of very high order azimuthal modes in cavities that are tens of free-space wavelengths in diameter; the development and routine inclusion of integrated quasi-optical output couplers that allow the separation of this task

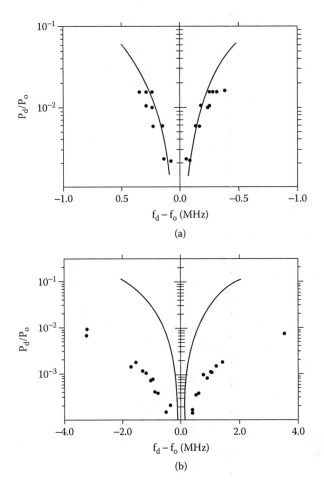

FIGURE 10.33
Phase-locking bandwidth for (a) direct injection to a cavity with an external Q of 1100 and (b) injection into a three-cavity gyroklystron with a final cavity external Q of 375. The solid line is the prediction from Adler's inequality. (From McCurdy, A.H. et al., *Phys. Rev. Lett.*, 57, 2379, 1986. With permission.)

from beam collection; and the continued development of chemical vapor deposition (CVD) artificial diamond windows to get the strength, low loss tangent, and thermal conductivity required for the high power fluxes they must transmit. This last element is particularly important, because the failure of a window breaks the vacuum of the assembly, which not only compromises its cleanliness, but also destroys the cathodes in the electron guns.

Gyroklystrons provide an alternative to klystrons for X-band amplifiers with power in the 100-MW regime and pulse lengths of the order of a microsecond. At higher frequencies, in the K_u- and K_a-bands (up to 35 GHz, roughly), they quite possibly provide the superior alternative to klystrons. As with klystrons, the future development of these sources, at least at this power level, might be called into question by the demise of the Next Linear Collider concept in favor of the International Linear Collider, with lower power and frequency requirements. Radar and other applications, however, might continue to drive the development of this source at lower power levels.

CARMs offer an interesting possibility for the production of high-frequency sources in the HPM regime. Nevertheless, at this point, they remain immature, with relatively little development, and it could be that the requirement they pose for high beam quality could limit the prospects for changing this situation.

10.4 Free-Electron Lasers

Free-electron lasers (FELs) offer an almost unlimited frequency range, and in this regard they are the highest-frequency producer of gigawatt-level microwave power, with an output of 2 GW at 140 GHz while firing repetitively and achieving some tens of percent efficiencies. This remarkable performance comes at a cost in size and volume: FELs require very high beam quality, which militates against high-current, and toward high-voltage, operation. In addition, they are relatively complex sources. More compact devices operating at lower frequencies in the high-current Raman regime with applied axial magnetic fields have been demonstrated as well. All of these sources operate as amplifiers, and as such, they provide tunable alternatives to relativistic klystrons and gyroklystrons for applications with frequency requirements above the X-band. A number of reviews of the field are available, for example, Freund and Antonsen.[101]

10.4.1 History

Although the possibility of using the Doppler upshift of radiation to achieve very high frequencies was realized as early as the 1930s by Kapitsa and Dirac,[102] followed by the recognition that Doppler-upshifted undulator radiation could be used to generate coherent millimeter-wave output,[103] the serious development of the FEL occurred on two fronts, beginning with R.M. Phillips in the late 1950s. Working at the General Electric Microwave Laboratory in Palo Alto, he experimented with the ubitron,[104] a device using a permanent magnet undulator. This work produced a number of megawatt-class devices, both amplifiers and oscillators, at frequencies between 2.5 and 15.7 GHz.

On the second front, researchers at Stanford using higher-energy, lower-current beams from a linear accelerator developed the device they called a free-electron laser. Madey proposed his concept in 1971,[105] and an amplifier

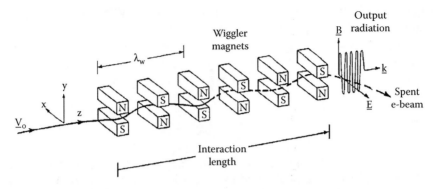

FIGURE 10.34
The basic FEL configuration. Beam electrons undulate in the periodic magnetic field of the wiggler. The electrons radiate and bunch in the ponderomotive potential well formed in the combination of pump and radiation fields.

of 10.6-μm light was announced in 1976.[106] A year later saw the achievement of laser oscillation at 3.4 μm.[107]

In the 1970s, work continued in the microwave and millimeter-wave regimes, with the production of some tens of megawatts of power by groups at the NRL, the Khar'kov Physico-Technical Institute, the Lebedev Institute and the Institute of Applied Physics, and Columbia University. In the 1980s, work continued on relatively high-current FELs at the NRL and MIT, notably, while lower-current, higher-energy beams were employed at the Lawrence Livermore National Laboratory (LLNL) to produce 1 GW at 35 GHz, followed later by 2 GW at 140 GHz, which is the highest brightness experiment conducted, by the measure of the product of power and the square of the frequency, Pf^2. These latter experiments, which we will describe in detail later, also involved the generation of 1-GW pulses at a repetition rate up to several kilohertz.

The LLNL experiments have been discontinued, but work continues in Russia with the Budker Institute of Nuclear Physics in Novosibirsk producing hundreds of megawatts in millimeter-wave radiation in work also involving the Institute of Applied Physics. Also, at the Institute of Applied Physics, the SINUS-6 generator has been employed in research on a new type of FEL concept that has produced power levels approaching 10 MW.

10.4.2 Free-Electron Laser Design Principles

An FEL is configured as shown in Figure 10.34. An electron beam with a z-directed velocity $v_z = v_0$ at the input is injected into a wiggler* magnetic field with a period λ_w. On the axis, the y component of the wiggler magnetic field

---

* There is some debate about the uses of the terms *undulator* and *wiggler* for the periodic magnet inducing radiation in an FEL. We simply use the term *wiggler* everywhere without apology.

deflects the electrons in the x direction. This oscillatory motion has a wave-number k_w and a wiggle frequency Ω_w given by

$$k_w = \frac{2\pi}{\lambda_w} \tag{10.23a}$$

$$\Omega_w = k_w v_z \tag{10.23b}$$

and the transverse acceleration of the moving charges results in forward-directed radiation. The beating of the radiation with wavenumber k_w and frequency ω with the "pump" field of the wiggler creates a potential well (known as the *ponderomotive* potential well[108]) moving at a phase velocity $\omega/(k_z + k_w)$, within which the electrons form bunches. It is this bunching that enforces the coherence of the emitted radiation.

The relationship between the wavelength of the wiggler and the wavelength and frequency of the output radiation is determined approximately from the location of the resonant interaction between the dispersion curves for the Doppler-shifted electron oscillation, which is the wiggler-shifted slow space-charge wave on the beam, and an electromagnetic normal mode. As an initial cut at this estimate, we assume first that the beam density is low and use a wiggler-shifted beam line to approximate the space-charge wave:

$$\omega = \left(k_z + k_w\right)v_z \tag{10.24}$$

For convenience, we approximate the dispersion curve of the electromagnetic wave by the free-space dispersion relation for a wave traveling along the z axis, thus ignoring the effect of the cutoff frequency for $\omega \gg \omega_{co}$:

$$\omega = k_z c \tag{10.25}$$

In Equation 10.24, note that the wiggler shift is shown by the addition of k_w to k_z, and that the space-charge term splitting the fast and slow space-charge waves has been omitted, consistent with our assumption of low beam density. The simultaneous solution of Equations 10.24 and 10.25 provides the frequency and wavenumber at which resonance occurs:

$$\omega_{res} = \left(1+\beta_z\right)\frac{\Omega_w}{1-\beta_z^2} \tag{10.26}$$

with $\beta_z = v_z/c$. Using the definition of the relativistic factor γ, we write

$$1 - \beta_z^2 = \frac{1}{\gamma^2}\left(1 + \beta_x^2 \gamma^2\right) \tag{10.27}$$

Here, given that the wiggler induces electron motion in the x-z plane, $\beta_x = v_x/c$. This transverse wiggle velocity is calculated from the equation of motion, using the fact that for a time-independent wiggler field, $d/dt = v_z(d/dz)$, and assuming a planar wiggler with $B_w \sin(k_w z)$ the y component of the field:

$$v_z \frac{d\left(\gamma\beta_x\right)}{dz} = \frac{e}{mc} v_z B_w \sin\left(k_w z\right) \tag{10.28}$$

This equation is integrated straightforwardly, and the result has a z dependence of $\cos(k_w z)$. Rather than reinsert this term into Equation 10.27, we use the spatial average, with $\langle\cos^2(k_w z)\rangle = 1/2$:

$$\left\langle \gamma^2 \beta_x^2 \right\rangle = \frac{1}{2}\left(\frac{eB_w}{mck_w}\right)^2 \equiv \frac{1}{2} a_w^2 \tag{10.29}$$

Finally, using the fact that $k_z = 2\pi/\lambda$ in free space and $k_w = 2\pi/\lambda_w$, Equations 10.26, 10.27, and 10.29 are combined to yield the characteristic wavelength of the FEL output:

$$\lambda = \frac{\lambda_w}{\beta_z\left(1 + \beta_z\right)\gamma^2}\left(1 + \frac{1}{2} a_w^2\right) \approx \frac{\lambda_w}{2\gamma^2}\left(1 + \frac{1}{2} a_w^2\right) \tag{10.30}$$

In the latter expression, we have made the common replacement as β_z approaches unity for large electron energies, but we note that at the lower energies common for microwave experiments this approximation could introduce some error. Our derivation was for planar wigglers, but the same relation holds for helical wigglers, except without the factor of $1/2$ prior to a_w^2, provided B_w represents the peak value of the field. Equation 10.30 is an important relationship, given the output wavelength as a function of γ, λ_w and b_w. Note that the strong energy dependence of the output frequency, as γ^2, makes device operation very sensitive to spreads in the energy of the electrons in the driving beam. Electrons at energies other than the resonant energy do not contribute to microwave generation at the predominant output frequency. As a consequence, FELs demand a beam of relatively low emittance for efficient operation; this requirement becomes increasingly stringent at shorter wavelengths.

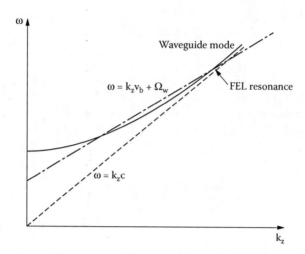

FIGURE 10.35

The uncoupled dispersion curves for a waveguide mode given by $\omega^2 = k_z^2 c^2 + \omega_{co}^2$ (which is asymptotic to the light line, $\omega = k_z c$) and the wiggler-shifted beam line of Equation 10.24.

Our derivation of Equation 10.30 ignored three important features of many HPM FELs: (1) the fact that the electromagnetic normal modes involved in the interaction are actually waveguide or cavity normal modes (see Problem 15), (2) the effect of beam space-charge, and (3) the application of a constant axial magnetic field in many devices. With regard to the first topic, we note that the beam in an FEL can interact with either TE or TM modes; in fact, at high currents, the transverse displacement of the beam couples the TE and TM modes. In general, the wiggler-shifted beam line will intersect a hyperbolic waveguide mode dispersion curve in two places, as illustrated in Figure 10.35. The existence of these two resonances has been experimentally verified.[109] The experimental scaling of the upper resonant frequency with beam energy and the wiggler field is shown in Figure 10.36.[110] In Figure 10.36a, the calculated resonance frequencies for the TE_{11} and TM_{01} modes are compared with output frequencies measured in experiments at the NRL. In Figure 10.36b, the power measured in separate frequency bands using a millimeter-wave grazing spectrometer with a resolution of about 1 GHz is shown; note that as the wiggler field increases in magnitude, the peak power shifts to lower frequencies, consistent with Equation 10.30.

The issue of selecting between the higher- and lower-frequency modes in an FEL is treated elsewhere.[111] The lower-frequency resonance occurs at a point of negative group velocity on the waveguide mode if $\Omega_w > \omega_{co}$. This is potentially problematic, because this interaction with a negative-group-velocity mode can support an absolute instability, which can grow in place to large amplitudes.[112] Possible solutions include modifying the wiggler frequency to push it below the cutoff frequency and keeping the wiggler below a critical length to prevent the oscillation from exceeding its start-current threshold[113] (see Problem 16). In many instances, the lower-frequency

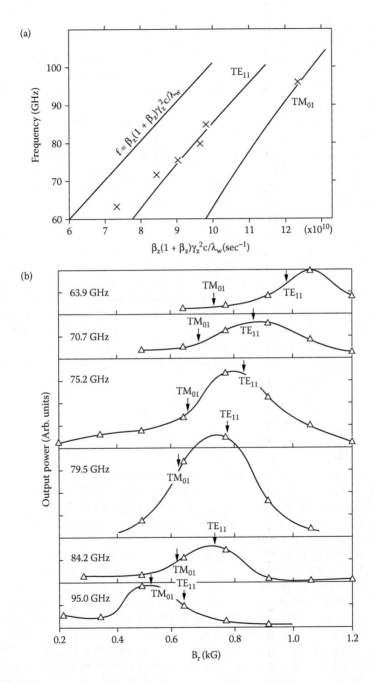

FIGURE 10.36

(a) Solid curves: calculated FEL resonances with the light line and the TE_{11} and TM_{01} waveguide modes. Xs: experimentally determined operating points in the reference. (b) FEL output power (in arbitrary units) in six detector bands as a function of the wiggler field, B_r. (From Gold, S.H. et al., *Phys. Fluids*, 26, 2683, 1983. With permission.)

absolute instability will grow more slowly than the FEL interaction, preventing it from becoming a problem in short-pulse devices.[114]

Space-charge effects are particularly important at microwave frequencies, because many of the experiments in this frequency range use intense electron beams. Depending on the strength of the space-charge and wiggler effects, three regimes of FEL operation are identified: the *Compton, Raman,* and *strong-pump* or *high-gain Compton* regimes.[115] In the Compton regime, the electron beam current and space-charge are very low, coherent single-particle effects dominate, and device gain is low. Compton operation is most appropriate for experiments performed at high energies for wavelengths much shorter than those in the microwave spectrum. The Raman regime is applicable to intense beam experiments where the high space-charge plays a dominant role in the interaction. In this case, the frequencies of the wiggler-shifted fast and slow space-charge waves are split about the frequency given in Equation 10.24, and the interaction occurs at the lower frequency. The strong pump regime is applicable to cases where the effect of a very strong wiggler field overpowers space-charge effects in an otherwise intense electron beam.

The application of an axial magnetic field B_z aids in the transport of the intense electron beam through an FEL. It also has a significant effect on the electron trajectories and, consequently, on the output. A calculation of the single-particle trajectories shows them to be of two types, called group I and group II.[116] For both types, the wiggle velocity perpendicular to the z axis has the magnitude

$$v_w = \frac{\Omega_w}{\gamma(\Omega_0 - \Omega_w)} v_z \tag{10.31}$$

where $\Omega_0 = eB_z/m\gamma$. The difference between the two orbit types lies in the dependence of v_z on B_z, as shown in Figure 10.37.[117] The calculation from which the plot was taken was for a helical wiggler, assuming the form of an ideal wiggler using only the on-axis fields, and an actual wiggler, taking account of the off-axis variation of the fields. Group I orbits occur at low values of the magnetic field, for which $\Omega_0 < \Omega_w$ and have high axial velocities that decline with B_z until the orbits become unstable, when $\Omega_0 = \Omega_w/(1 - v_w^2/v_z^2)$. The ideal group II orbits exist for all positive values of B_z, with an axial velocity that increases with the axial field and is large only at large axial fields. The resonance $\Omega_0 = \Omega_w$ does not occur for either group of trajectories, but it can be approached, in which case the transverse wiggle velocity becomes large, according to Equation 10.31, as does the FEL gain.[118,119] An additional feature of interest is that for smaller values of B_z, electrons with group II orbits have a negative mass characteristic, in that the axial velocity actually *increases* as the energy decreases: $\partial \beta_z/\partial \gamma < 0$. As the axial field increases, however, the sign of the derivative eventually changes. At the lower field values, the interaction between a beam of group II electrons and the electromagnetic field is more complicated, with both space-charge and

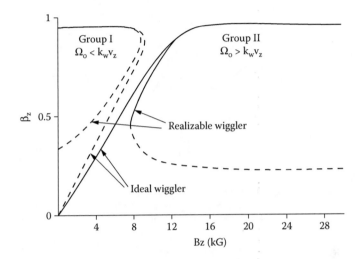

FIGURE 10.37
An example of the variation of β_z with the axial magnetic field for group I and group II orbits.
Calculations were made for both ideal wiggler fields and realizable wigglers, taking account
of the off-axis variation of the fields. (From Jackson, R.H. et al., *IEEE J. Quantum Electron.*, QE-
19, 346, 1983. With permission.)

modified cyclotron wave interactions producing wave growth.[118] Experimental results indicate that even higher power can be achieved by reversing the
axial field, in the sense that the cyclotron motion of the electrons in the axial
field has an opposite rotational sense than the motion in the wiggler field
(removing the possibility of a wiggler cyclotron resonance, since $\Omega_0 < 0$).[120]

The growth of electromagnetic waves in an FEL occurs at the expense of
the kinetic energy of the beam electrons. The beating of the signal against
the wiggler field pump forms a ponderomotive potential with phase velocity
$\omega/(k_z + k_w)$. As the electrons lose energy and the ponderomotive potential
wells deepen, the electrons' axial velocity drops to the phase velocity of the
ponderomotive potential, and some fraction of the electrons are trapped in
the potential wells. The process is depicted in the schematic phase-space plot
of Figure 10.38.[121] Trapping is seen in the swirling of electrons near the
bottom of the ponderomotive potential well. The axial velocity of the trapped
electrons oscillates about the ponderomotive or space-charge wave phase
velocity. This periodic variation, known as *synchrotron oscillations*, is governed by a nonlinear pendulum equation.[108] Figure 10.39 illustrates the effect
of these trapped electrons on the saturated signal: note the variation in the
signal amplitude after it reaches its maximum.[122] This is due to the periodic
exchange of energy between the wave and the trapped electrons, with the
faster of either the wave or the trapped electron accepting energy from the
other at any point in the synchrotron oscillation period.

The output power and efficiency of an FEL can be further increased after
the onset of trapping by continuing to extract energy from the trapped
electrons. This is done by decreasing the phase velocity of the ponderomotive

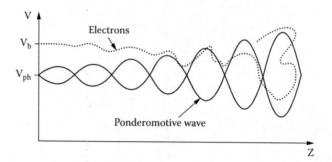

FIGURE 10.38

A schematic phase-space plot of electrons in an FEL illustrating the energy loss and eventual trapping in the growing ponderomotive potential well of the pump and radiation fields. (From Pasour, J.A., *High-Power Microwave Sources*, Granatstein, V.L. and Alexeff, I., Eds., Artech House, Norwood, MA, 1987. With permission.)

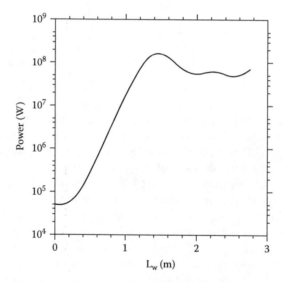

FIGURE 10.39

The growth and saturation of the microwave signal as a function of distance from the input end of an FEL amplifier. Note the small-signal or linear regime of exponential signal growth between 0.4 and 1.4 m and the saturation at peak amplitude. (From Orzechowski, T.J. et al., *Nucl. Instrum. Methods*, A250, 144, 1986. With permission.)

potential; the trapped electrons are slowed as well, surrendering further energy to the electromagnetic wave. Alternatively, in terms of Equation 10.30, as γ decreases for the electrons, signal growth at a fixed wavelength can be maintained if either the wiggler period or the wiggler field magnitude is decreased accordingly. In either case, the process of varying the wiggler parameters is known as *wiggler tapering*. In FELs with an applied axial magnetic field, wiggler tapering becomes more complex, with the added option of tapering the axial field as well to increase the power. The reader seeking more information on this topic is directed to Freund and Ganguly.[123]

Since wiggler tapering increases the power and efficiency of an FEL by extracting energy from the trapped portion of the electron population, it is important to ensure that these electrons remain trapped during the extraction process. One cause of detrapping is the use of an overly rapid taper in the wiggler parameters. A second cause is the *sideband instability*.[124-126] This is a nonlinear interaction between the FEL signal at frequency ω_0 and the synchrotron oscillations of the trapped electrons at ω_{syn}, resulting in the formation of sidebands at $\omega_0 \pm \omega_{syn}$ that grow at the expense of the main output signal. The frequency spacing between the sideband and the signal is roughly[127,128]

$$\Delta\omega = \frac{k_{syn}c}{1 - v_z v_g} \tag{10.32}$$

where v_g is the group velocity of the signal at ω_0 and

$$k_{syn} = \left(\frac{e^2 E_s B_w}{\gamma_z^2 m^2 c^3} \right)^{1/2} \tag{10.33}$$

with E_s the peak electric field of the microwaves and γ_z the resonant value of γ found in Equation 10.30, given λ, λ_w, and a_w. An example of a signal with a sideband in a long constant-parameter wiggler is shown in Figure 10.40.[129] There, the shorter wavelength signal is due to the main FEL resonance, while, of the two possible sidebands, that with the longer wavelength is the one observed. The dashed curve shown in the figure is the computed spatial growth rate of the FEL interaction as a function of wavelength; note that strong sideband formation does not require the main FEL interaction to have growth at the sideband wavelength. The peak power in the main signal and the sideband is plotted as a function of the wiggler field in Figure 10.41. There note the threshold value of the field for FEL output, the analog of the start current in the case of a fixed beam current, and the growth of the sideband power with the wiggler strength.

10.4.3 Operational Features of Free-Electron Lasers

A number of interesting microwave FEL experiments have been performed, but we limit ourselves to the discussion of three: the extraordinarily high brightness experiments, measured by the product Pf^2, conducted on the Experimental Test Accelerator (ETA) at Livermore; a microwave proof of principle of the optical klystron configuration at MIT; and the experiments at the Budker Institute of Nuclear Physics in Novosibirsk.

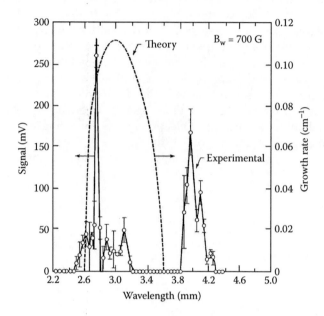

FIGURE 10.40
Spectral output of an FEL with strong sideband growth. The tall peak at shorter wavelength is
the main signal, while the peak on its long-wavelength side is the sideband. The dashed curve
is the calculated spatial growth rate. (From Yee, F.G. et al., *Nucl. Instrum. Methods*, A259, 104,
1987. With permission.)

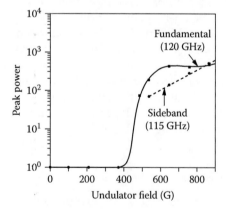

FIGURE 10.41
Power at the main signal frequency and in the sideband as a function of wiggler field in the
same experiments as Figure 10.40. (From Yee, F.G. et al., *Nucl. Instrum. Methods*, A259, 104, 1987.
With permission.)

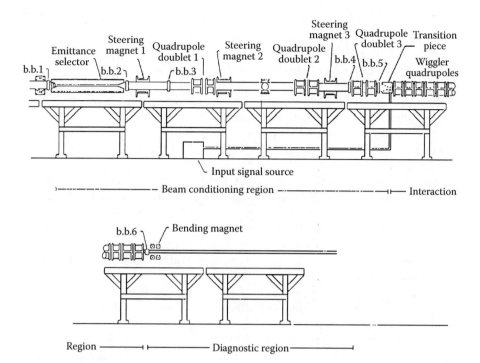

FIGURE 10.42
Layout of the FEL amplifier experiments at LLNL, from the emittance selector beam input to the downstream diagnostic chamber. (From Throop, A.L. et al., Experimental Characteristic of a High-Gain Free-Electron Laser Amplifier Operating at 8-mm and 2-mm Wavelengths, UCRL-95670, Lawrence Livermore National Laboratory, Livermore, CA, 1987. With permission.)

The FEL experiments at LLNL, involving a collaboration with researchers from the Lawrence Berkeley Laboratory (LBL, now the Lawrence Berkeley National Laboratory) as well, produced a remarkable series of results culminating in the production of 2 GW at 140 GHz. The layout of the FEL itself is shown in Figure 10.42,[130] which illustrates the complexity of the FEL. The accelerators creating the electron beam were multicell, linear induction accelerators known as the Experimental Test Accelerator (ETA) and the follow-on ETA II upgrade; both were similarly large and complex. ETA was capable of producing a 4.5-MeV, 10-kA, 30-nsec electron beam[131]; however, at the wavelengths involved, the LLNL FEL required a high-quality, low-emittance electron beam. In order to improve the beam quality by reducing its emittance, the ETA beam was passed into the emittance selector shown at the left of the figure. This was essentially a long, smaller-radius waveguide section that scraped off electrons at large radii and large angular divergence. The maximum normalized emittance acceptance of the selector was 4×10^{-3} π rad-m, and a transmitted 1.6-kA, 10- to 15-nsec beam had a brightness of 2×10^8 A/(m-rad)². Downstream of the selector, steering and quadrupole magnets matched the beam to the planar wiggler. Built of pulsed (1-msec) air-core electromagnets with a period $\lambda_w = 9.8$ cm, the original wiggler had a total length of 3 m (later increased to 4 m) and peak amplitudes of 0.37 T ($a_w = 2.4$) for experiments at 35 GHz and 0.17 T ($a_w = 1.1$) for experiments at 140 GHz. Independent variation of the wiggler field for each of the two periods allowed for tapering of the wiggler strength at constant λ_w. At the highest power levels, optimization of the tapering profile was performed empirically. There was no axial magnetic field; horizontal focusing of the electron beam was provided by the quadrupole doublets. The microwave interaction region was a rectangular stainless steel waveguide with a cross section of 3×10 cm^2.

The keys to maximizing the output power of the Livermore FEL were, first, to minimize the emittance and beam velocity spread so that substantial numbers of electrons would be trapped in the ponderomotive pontential wells and, second, to taper the parameters of the wiggler to continue to draw upon the energy of the trapped electrons. As the beam energy and γ decrease, a corresponding decrease in a_w, as seen in Equation 10.30, can maintain a constant-wavelength resonance condition. Obviously, smaller initial spreads in the electron energies improve the usefulness of this technique by reducing thermal spreading of the trapped bunches. Tapering of the wiggler amplitude at constant λ_w was responsible for raising the output power of the LLNL/LBL FEL from 180 MW without tapering to 1 GW, as shown in Figure 10.43.[132] In terms of the current that entered the wiggler after passage through the emittance filter (Figure 10.42), 850 A at 3.5 MeV, the efficiency was raised from 6% without tapering to 35%. Note that the linear gain over the first 120 cm was about the same in either case: 0.27 dB/cm.[122] The tapering profile for the wiggler is shown in Figure 10.44 (see Problem 17). Significantly, the wiggler tapering does not begin until near the point of saturation of the no-taper signal, because this is the point at which the fraction of electrons trapped in the ponderomotive potential reaches its maximum. Numerical simulations indicated that about 75% of the beam electrons were trapped, and that 45% of their energy was removed.

A further improvement of ETA II, to ETA III, was used for experiments at 140 GHz.[133] The accelerator produced a 6-MeV, 2-kA beam, and peak output power of 2 GW was achieved in single pulses. Bursts of up to 50 pulses were also produced at 2 kHz. Microwave output pulses from such a 45-pulse burst are shown in Figure 10.45.[134] Although higher power levels were predicted in computer modeling of the system, the primary focus of effort was the use of the output for plasma heating in the Microwave Tokamak Experiment (MTX), so the device was not optimized.

Another means of increasing the gain within an FEL draws on a concept proposed originally by Vinokurov and Skrinskiy of the Institute of Nuclear Physics in Novosibirsk.[135] In order to enhance the low-signal gain of Compton FELs driven by high-energy electron beams, the *optical klystron* employs two wigglers separated by a dispersive section. The electron energy modulation imposed in the first wiggler section initiates beam bunching, which proceeds in the dispersive section. The output section then runs at a higher gain and

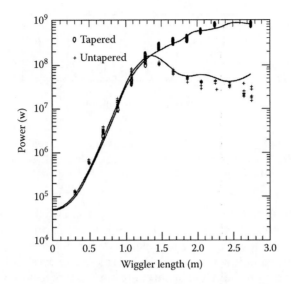

FIGURE 10.43
Microwave power in the LLNL/LBL FEL as a function of distance along the wiggler, both with and without (lower curve to the right) wiggler tapering. (From Orzechowski, T.J. et al., *Phys. Rev. Lett.*, 57, 2172, 1986. With permission.)

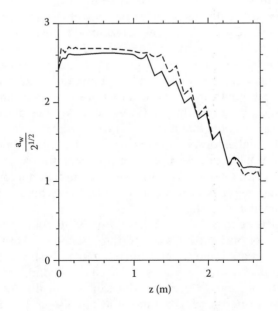

FIGURE 10.44
Axial profile of the wiggler field strength in the LLNL/LBL FEL, optimized both empirically (dashed line) and theoretically (solid line). (From Orzechowski, T.J. et al., *Phys. Rev. Lett.*, 57, 2172, 1986. With permission.)

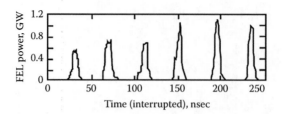

FIGURE 10.45
Pulses from a 45-shot, 2-kHz burst fired on the 140-GHz Livermore FEL powered by the ETA III accelerator. (From Lasnier, C.J. et al., Burst mode FEL with the ETA-III induction linac, in *Proceedings of the 1993 Particle Accelerator Conference*, 1993, Washington, D.C., p. 1554. With permission.)

efficiency with a prebunched beam. Although the initial experiments with this concept were conducted at high energies and short wavelengths, the concept can be adapted as well for operation at microwave frequencies, as was demonstrated in low power, proof-of-principle experiments at MIT.[136,137] A collimated electron beam with axial energy spreads of less than 0.3% passed into an interaction region that was split into two halves by a thin tungsten mesh. The mesh allowed the beam to pass almost unaffected from one half to the other, while the microwaves from the first section were attenuated by 20 dB. In the first section, an injected microwave signal was amplified, and the electron beam began to bunch. The bunched beam passed into the second section, where it began to amplify the small microwave signal that made it through the mesh. The downstream output power as a function of the effective wiggler length (Figure 10.46) was determined by using a movable kicker magnet to eject the beam from the wiggler region after traversing different distances. It must be remembered that power produced upstream of the mesh was measured downstream after it had been attenuated 20 dB by the mesh. The gain per unit length increased dramatically downstream of the mesh, from 0.069 to 0.32 dB/cm for the 188-G wiggler field, thus demonstrating the concept; however, we note that if the mesh were removed, the overall output power would be higher, despite the lower gain due to the removal of the 20-dB attenuation of the mesh. As a proof of principle, the experiment showed the applicability of the optical klystron configuration at microwave frequencies in the Raman regime of operation. Further, the technique could be used to suppress parasitic oscillations in an FEL.[138]

The FEL experiments in Novosibirsk[139] involving collaborators from the Budker Institute of Nuclear Physics and the Institute of Applied Physics in Nizhny Novgorod are interesting in three regards. First, this team has produced 200-MW, 200-J output pulses at 75 GHz. Second, they use a sheet beam — albeit with the small dimensions of 0.3 cm thickness and 12 cm width — which is often proposed, but not as often seen. Third, their system is quite rich in energy, so that they propose to produce multigigawatt output using a very large sheet beam 140 cm in width, which the authors claim has been demonstrated with this system. The weakness of the experiments

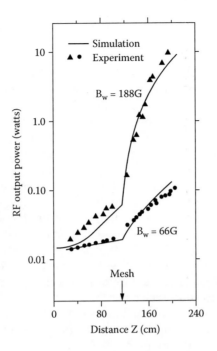

FIGURE 10.46
RF output power as a function of interaction length in the MIT experiment with a prototype optical klystron. (From Leibovitch, C. et al., *IEEE J. Quantum Electron.*, 24, 1825, 1988. With permission.)

described in their 1998 publication is that while the voltage is quite constant during a pulse, the current rises linearly for almost the full duration of the pulse. Thus, the microwave output envelope is quite irregular.

10.4.4 Outlook for Free-Electron Lasers

FELs are a very important source for the production of millimeter-wave HPM output, but the application demanding this output may not exist yet. While plasma heating was the goal for the extraordinary experiments in Livermore, the focus of the effort in support of plasma heating remains the development of long-pulse or continuous gyrotrons. Simply put, the complexity of HPM makes a high-peak-power system with rapid pulsing an uneconomical way to achieve high average power. Nevertheless, FELs retain some untapped potential. For example, the electronic tunability afforded by the ability to vary the wiggler field could offer rapid frequency agility. Further, if the particle accelerator community ever returns to a desire for a high-frequency RF linac to reduce accelerator length, FELs would quite possibly be a superior source for W-band (94-GHz) concepts.

10.5 Summary

The sources in this chapter are all somewhat marginalized in the realm of HPM today, although the reasons for each depend to some degree on factors beyond the control of the device designers. Vircators are genuine workhorse sources for microwave effect testing and simulation because of their broad tunability; however, their low efficiency makes them far less attractive from the standpoint of going into the field, because of the volume and mass associated with the power and energy storage required by a low-efficiency source. In applications driven by explosive flux compression generators, though, given the energy-rich nature of this prime power source, the issue of poor efficiency may be moot. Rather, the lower characteristic impedance of these devices may make them a good electrical match to these single-shot generators. In addition, vircators are enjoying something of a renaissance today, with two new source types — the coaxial vircator and the tunable vircator — that have excited new interest and that permit new approaches to the problem of vircator efficiency.

Gyrotron development is being conducted in a very competitive international environment today, and the vitality of that environment has led to rapid advances in technology in the realms of the use and mode control of very high order azimuthal modes in cavities that are tens of free-space wavelengths in diameter; integrated quasi-optical output couplers that allow the separation of this task from beam collection; and chemical vapor deposition (CVD) artificial diamond windows to get the strength, low loss tangent, and thermal conductivity required for the high power fluxes they must transmit. Gyroklystrons have become competitors to klystrons for applications requiring a 100-MW-class amplifier in the X-band. As the frequency increases above the X-band, gyroklystrons may have a growing competitive advantage over klystrons.

Free-electron lasers offer unmatched peak power performance in the millimeter-wave region, but the number of applications requiring gigawatt millimeter-wave output from a large, complex system is extremely limited. If the particle accelerator community returned to high-frequency concepts, the FEL might be competitive for that application. At this point, though, there are no applications driving the technology in the HPM realm.

We summarize the strengths and weaknesses of the sources in this chapter in Table 10.2.

TABLE 10.2

Comparison of the Strengths and Weaknesses of the Major Source Types in This Chapter

	Strengths	Weaknesses	Notable Accomplishments
Vircators	Relatively simple generation mechanism; broadly tunable configuration; can operate without a magnetic field; low-impedance, high power operation	Low efficiency; gap closure tends to make frequency chirp during pulse	Regularly tunable over an octave, with an order of magnitude demonstrated Multigigawatt output from 1–10 GHz
Gyrotrons	High-frequency (~100 GHz) operation at MW (CW) levels with high efficiency	Immature at hundreds of MW; higher frequencies suffer significant atmospheric attenuation except near 35 and 94 GHz	1 MW at 100 to 170 GHz for long-pulse to continuous operation; 80 MW in an X-band gyroklystron
Free-electron lasers	True HPM amplifier; virtually unlimited frequency range; high power operation; high efficiency	Complicated; multimegavolt operation to achieve high frequencies; demanding beam quality requirements; higher frequencies suffer significant atmospheric attenuation except near 35 and 94 GHz	2 GW in single pulses at 140 GHz, 1-GW pulses at a repetition rate of 2 kHz for 50-pulse bursts

Problems

1. The simplest expression for microwave power produced from an electron beam of efficiency η is:

$$P = \eta VI = \eta VJA$$

The plasma in the diode gap will move across it at velocity v_p:

$$d(t) = d_0 - v_p t$$

where d_0 is the initial gap. Now, assume the temperature of the plasma is produced by ohmic heating of the plasma by the current passing through it:

$$T_p \sim \rho J^2$$

where ρ is the resistivity. When the plasma expands across some fraction of the gap, resonance is destroyed and microwaves cease. Therefore, microwave pulse length scales like the cathode plasma transit time across the gap

$$\tau_\mu \sim \frac{d}{v_p} \sim d\sqrt{\frac{m_p}{T_p}}$$

where T_p is the plasma temperature and m_p is the hottest light ion species, which governs the location of the conducting surface (usually hydrogen). Assume it's a Child's Law diode (Eq. 4.104). Using these relations, derive a scaling relation between microwave power and pulse length:

$$P_\mu \sim \frac{1}{\tau_\mu^{5/3}}$$

where n = 5/3 for a Child's Law diode.

How does pulse energy scale? From this, is there anything to be gained in energy by making longer pulses?

2. For voltages between 100 kV and 3 MV, plot I_{SCL} as a function of voltage for both solid and annular beams, with $r_0/r_b = 0.5$ and 0.8. Add the plots for an annular beam with $r_b/r_i = 0.3$ in both cases.

3. Assume a constant 5-mm standoff of the beam from the wall. Plot I_{SCL} for 1.5-MV solid and annular beams as a function of wall radius, from 1 to 5 cm.

4. A solid-cross-section beam of 30 kA and 1 MV, with a radius of 3 cm, is injected into the interaction region of a virtual cathode oscillator. What is your estimate of the virtual cathode oscillation frequency?

5. Inject a 20-kA, 500-kV beam of radius 2 cm with a solid cross section into a circular waveguide. As you increase the waveguide radius, at

what waveguide radius will a virtual cathode form? At what radius is the beam current equal to $2I_{SCL}$?

6. Derive Equation 10.9.

7. Using Equation 10.10, assuming an anode–cathode gap of 2 cm, plot the expected vircator frequency as a function of diode voltage.

8. The magnetic moment for an electron in the magnetic field of a gyrotron is given in Equation 10.15:

$$\mu = \frac{\left(\frac{1}{2}mv_{\perp}^{2}\right)}{B_{0z}}$$

This quantity is an adiabatic invariant within a gyrotron if the magnetic field varies slowly enough, which we will assume to be true. The statement was made in the text that in designing a gyrotron, one must be careful not to choose too large a value of $v_{\perp}$, because magnetic mirroring could result. Assume that the magnetic field increases along the axis of a gyrotron from a value B_{0z1} at $z = z_1$ to a larger value B_{0z2} at $z = z_2$. Assume also that the electron has a total kinetic energy eV_0, where V_0 is the accelerating voltage in the electron gun. Let $v_{\perp1}$ be the perpendicular velocity component at $z = z_1$. Derive an expression for the maximum value of $\alpha = v_{\perp}/v_z$ at $z = z_1$ in order to prevent v_z at $z = z_2$ from vanishing. Assume that the kinetic energy of an electron is conserved.

9. Plot the electron cyclotron frequency of the beam electrons as a function of accelerating voltage, neglecting space-charge depression, for a magnetic field of 1 T.

10. A gyrotron is built to a requirement that it generate 1 MW of output. If the power efficiency is increased from 33 to 50% by using a depressed collector, what is the reduction in power loss to the cavity and beam collector?

11. What magnetic field is required to achieve resonance in a gyrotron between a 300-kV electron beam and the $TE_{6,2}$ mode of a circular waveguide 3 cm in radius?

12. How much energy is stored in a magnetic field of 0.5 T that fills a cylindrical volume 5 cm in radius and 1 m long?

13. In klystron-like devices, such as a gyroklystron, when it becomes impractical to reduce the radius of the drift space between two cavities in order to cut off propagation in the lowest-order TE_{11} mode, the drift tube is sometimes lined with an absorber. If the attenuation of the absorber to this spurious mode is 5 dB/cm, and

if the spacing between cavities is 5 cm, what is the ratio of the spurious signal at the input and output of the absorbing section?

14. For a CARM using a 2-MeV electron beam, find the magnetic field required to achieve resonance at 50 GHz with a TE_{11} mode of a 3-cm-radius waveguide.

15. Redo the estimate of the frequency of an FEL, replacing the light line in Equation 10.25 with the dispersion curve for a waveguide mode. Ignore the effect of the wiggler dynamics, focusing only on the resonance between the Doppler-shifted wiggler wave and the normal mode of the waveguide. Find an expression for the FEL frequency taking account of the cutoff frequency of the waveguide mode. Find an expression also for the lower-frequency resonance with the waveguide mode.

16. A 2-MeV electron beam is injected into a circular-cross-section drift tube 4 cm in radius. What is the minimum wiggler period allowed in order to avoid backward wave interactions with the TE_{11} mode? These interactions can present a problem because the beam–backward wave interaction supports an absolute instability that can be difficult to control.

17. Consider the plot of a_w for the LLNL FEL operating at 35 GHz. Knowing the output frequency, the wiggle wavelength of $\lambda_w = 9.8$ cm, and the wiggler amplitude upstream of 0.37 T ($a_w = 2.4$), estimate the input beam energy and the energy of the spent beam after microwave generation using the data in Figure 10.44.

References

1. Benford, J., High power microwave simulator development, *Microwave J.*, 30, 97, 1987.
2. Birdsall, C.K. and Bridges, W.B., *Electron Dynamics of Diode Regions*, Academic Press, New York, 1966.
3. Sullivan, D.J., Walsh, J.E., and Coutsias, E.A., Virtual cathode oscillator (vircator) theory, in *High Power Microwave Sources*, Granatstein, V.L. and Alexeff, I., Eds., Artech House, Boston, 1987, p. 441.
4. Thode, L.E., Virtual cathode microwave device research: experiment and simulation, in *High Power Microwave Sources*, Granatstein, V.L. and Alexeff, I., Eds., Artech House, Boston, 1987, p. 507.
5. Mahaffey, R.A. et al., High-power microwaves from a non-isochronic reflexing system, *Phys. Rev. Lett.*, 39, 843, 1977.
6. Didenko, A.N. et al., The generation of high-power microwave radiation in a triode system from a heavy-current beam of microsecond duration, *Sov. Tech. Phys. Lett.*, 4, 3, 1978.

7. Brandt, H. et al., Gigawatt Microwave Emission from a Relativistic Reflex Triode, Harry Diamond Laboratory Report HDL-TR-1917, 1980.

8. Burkhart, S., Multigigawatt microwave generation by use of a virtual cathode oscillator driven by a 1–2 MV electron beam, *J. Appl. Phys.*, 62, 75, 1987.

9. Bromborsky, A. et al., On the path to a terawatt: high power microwave experiments at Aurora, *Proc. SPIE*, 873, 51, 1988.

10. Platt, R. et al., Low-frequency, multigigawatt microwave pulses generated by a virtual cathode oscillator, *Appl. Phys. Lett.*, 54, 1215, 1989.

11. Kwan, T.J.T. et al., Theoretical and experimental investigation of reditrons, *Proc. SPIE*, 873, 62, 1988.

12. Poulsen, P., Pincosy, P.A., and Morrison, J.J., Progress toward steady-state, high-efficiency vircators, *Proc. SPIE*, 1407, 172, 1991.

13. Benford, J. et al., Interaction of a vircator microwave generator with an enclosing resonant cavity, *J. Appl. Phys.*, 61, 2098, 1987.

14. Price, D., Sze, H., and Fittinghoff, D., Phase and frequency locking of a cavity vircator driven by a relativistic magnetron, *J. Appl. Phys.*, 65, 5185, 1989.

15. Fazio, M.V. et al., Virtual cathode microwave amplifier experiment, *J. Appl. Phys.*, 66, 2675, 1989.

16. Price, D., Sze, H., and Fittinghoff, D., Phase and frequency locking of a cavity vircator driven by a relativistic magnetron, *J. Appl. Phys.*, 65, 5185, 1989.

17. Sze, H., Price, D., and Harteneck, B., Phase locking of two strongly-coupled vircators, *J. Appl. Phys.*, 67, 2278, 1989.

18. Gadetskii, N.P. et al., The virtode: a generator using supercritical REB current with controlled feedback, *Plasma Phys. Rep.*, 19, 273, 1993 (*Fiz. Plazmy*, 19, 530, 1993).

19. Kitsanov, S.A. et al., S-band vircator with electron beam promodulation based on compact pulse driver with inductive energy storage, *IEEE Trans. Plasma Sci.*, 30, 1179, 2002.

20. Woolverton, K., Kristiansen, M., and Hatfield, L.L., Computer simulations of a coaxial vircator, *Proc. SPIE*, 3158, 145, 1997.

21. Jiang, W. et al., High-power microwave generation by a coaxial vircator, *Digest of Technol Papers: IEEE International Pulsed Power Conference*, Monterey, CA, 1999, p. 194.

22. Lin, T. et al., Computer simulations of virtual cathode oscillations, *J. Appl. Phys.*, 68, 2038, 1990.

23. Price, D. et al., Operational features and microwave characteristics of Vircator II, *IEEE Trans. Plasma Sci.*, 16, 177, 1988.

24. Woo, W.-Y., Two-dimensional features of virtual cathode and microwave emission, *Phys. Fluids*, 30, 239, 1987.

25. Sze, H. et al., Dynamics of a virtual cathode oscillator driven by a pinched diode, *Phys. Fluids*, 29, 3873, 1986.

26. Goldstein, S.A. et al., Focused-flow model of relativistic diodes, *Phys. Rev. Lett.*, 33, 1471, 1974.

27. Creedon, J.M., Relativistic Brillouin flow in the high v/γ diode, *J. Appl. Phys.*, 46, 2946, 1975.

28. Burkhart, S., Scarpetti, R., and Lundberg, R., Virtual cathode reflex triode for high power microwave generation, *J. Appl. Phys.*, 58, 28, 1985.

29. Davis, H.A. et al., High-power microwave generation from a virtual cathode device, *Phys. Rev. Lett.*, 55, 2293, 1985.

30. Zherlitsyn, A.G. et al., Generation of intense microsecond-length microwave pulses in a virtual-cathode triode, *Sov. Tech. Phys. Lett.*, 11, 450, 1985.

31. Coleman, P.D. and Aurand, V.F., Long pulse virtual cathode oscillator experiments, in *Proceedings of the 1988 IEEE International Conference on Plasma Science*, Seattle, WA, 1988, p. 99.

32. Huttlin, G. et al., Reflex-diode HPM source on Aurora, *IEEE Trans. Plasma Sci.*, 18, 618, 1990, and references therein.

33. Davis, H. et al., Experimental confirmation of the reditron concept, *IEEE Trans. Plasma Sci.*, 16, 192, 1988; Davis, H., Enhanced-efficiency, narrow-band gigawatt microwave output of the reditron oscillator, *IEEE Trans. Plasma Sci.*, 18, 611, 1990.

34. Kwan, T.J. and Davis, H.A., Numerical simulations of the reditron, *IEEE Trans. Plasma Sci.*, 16, 185, 1988.

35. Kwan, T. et al., Beam bunch production and microwave generation in reditrons, *Proc. SPIE*, 1061, 100, 1989.

36. Madonna, R. and Scheno, P., Frequency stabilization of a vircator by use of a slow-wave structure, in *Proceedings of the 1990 IEEE International Conference on Plasma Science*, Oakland, CA, 1990, p. 133.

37. Zherlitsin, A., Melnikov, G., and Fomenko, G., Experimental investigation of the intense electron beam modulation by virtual cathode, in *Proceedings of BEAMS'88*, Karlsruhe, Germany, 1988, p. 1413.

38. Sze, H., Price, D., and Harteneck, B., Phase locking of two strongly coupled vircators, *J. Appl. Phys.*, 67, 2278, 1990.

39. Fazio, M., Hoeberling, R., and Kenross-Wright, J., Narrow-band microwave generation from an oscillating virtual cathode in a resonant cavity, *J. Appl. Phys.*, 65, 1321, 1989.

40. Gadestski, P. et al., The virtode: a generator using supercritical REB current with controlled feedback, *Plasma Phys. Rep.*, 19, 273, 1993.

41. Kitsanov, S.A. et al., S-band vircator with electron beam premodulation based on compact pulse driver with inductive energy storage, *IEEE Trans. Plasma Sci.*, 30, 1179, 2002.

42. Jiang, W. et al., High-power microwave generation by a coaxial virtual cathode oscillator, *IEEE Trans. Plasma Sci.*, 27, 1538, 1999.

43. Jiang, W., Dickens, J., and Kristiansen, M., Efficiency enhancement of a coaxial virtual cathode oscillator, *IEEE Trans. Plasma Sci.*, 27, 1543, 1999.

44. Didenko, A. et al., Investigation of wave electromagnetic generation mechanism in the virtual cathode system, in *Proceedings of BEAMS'88*, Karlsruhe, Germany, 1988, p. 1402.

45. Price, D. and Sze, H., Phase-stability analysis of the magnetron-driven vircator experiment, *IEEE Trans. Plasma Sci.*, 18, 580, 1990.

46. See Dumbrajs, O. and Nusinovich, G.S., Coaxial gyrotrons: past, present, and future (a review), *IEEE Trans. Plasma Sci.*, 32, 934, 2004, and the references therein.

47. Flyagin, V.A. et al., The gyrotron, *IEEE Trans. Microwave Theory Tech.*, MTT-25, 514, 1977.

48. Hirshfield, J.L. and Granatstein, V.L., The electron cyclotron maser: an historical survey, *IEEE Trans. Microwave Theory Tech.*, MTT-25, 522, 1977.

49. Twiss, R.Q., Radiation transfer and the possibility of negative absorption in radio astronomy, *Aust. J. Phys.*, 11, 564, 1958; Twiss, R.Q. and Roberts, J.A., Electromagnetic radiation from electrons rotating in an ionized medium under the action of a uniform magnetic field, *Aust. J. Phys.*, 11, 424, 1958.

50. Schneider, J., Stimulated emission of radiation by relativistic electrons in a magnetic field, *Phys. Rev. Lett.*, 2, 504, 1959.

51. Gaponov, A.V., Addendum, *Izv. VUZ Radiofiz.*, 2, 837, 1959; an addendum to Gaponov, A.V., Interaction between electron fluxes and electromagnetic waves in waveguides, *Izv. VUZ Radiofiz.*, 2, 450, 1959.

52. Hirshfeld, J.L. and Wachtel, J.M., Electron cyclotron maser, *Phys. Rev. Lett.*, 12, 533, 1964.

53. Gaponov, A.V., Petelin, M.I., and Yulpatov, V.K., The induced radiation of excited classical oscillators and its use in high-frequency electronics, *Izv. VUZ Radiofiz.*, 10, 1414, 1967 (*Radiophys. Quantum Electron.*, 10, 794, 1967).

54. Baird, J.M., Gyrotron theory, in *High-Power Microwave Sources*, Granatstein, V.L. and Alexeff, I., Eds., Artech House, Norwood, MA, 1987, p. 103.

55. Gaponov, A.V. et al., Some perspectives on the use of powerful gyrotrons for the electron-cyclotron plasma heating in large tokamaks, *Int. J. Infrared Millimeter Waves*, 1, 351, 1980.

56. See Gaponov, A.V. et al., Experimental investigation of centimeter-band gyrotrons, *Izv. VUZ Radiofiz.*, 18, 280, 1975 (*Radiophys. Quantum Electron.*, 18, 204, 1975).

57. Kisel, D.V. et al., An experimental study of a gyrotron, operating in the second harmonic of the cyclotron frequency, with optimized distribution of the high-frequency field, *Radio Eng. Electron Phys.*, 19, 95, 1974.

58. Litvak, A.G. et al., Gyrotrons for Fusion. Status and Prospects, paper presented at the 18th IAEA Fusion Energy Conference, Sorrento, Italy, 2000 (published online, available from http://www-naweb.iaea.org/napc/physics/ps/conf.htm).

59. Granatstein, V.L. et al., Gigawatt microwave emission from an intense relativistic electron beam, *Plasma Phys.*, 17, 23, 1975.

60. Lawson, W. et al., High-power operation of a three-cavity X-band coaxial gyroklystron, *Phys. Rev. Lett.*, 81, 3030, 1998.

61. Botvinnik, I.E. et al., Free electron masers with Bragg resonators, *Pis'ma Zh. Eksp. Teor. Fiz.*, 35, 418, 1982 (*JETP Lett.*, 35, 516, 1982).

62. Botvinnik, I.E. et al., Cyclotron-autoresonance maser with a wavelength of 2.4 mm, *Pis'ma Zh. Tekh. Fiz.*, 8, 1386, 1982 (*Sov. Tech. Phys. Lett.*, 8, 596, 1982).

63. Rapoport, G.N., Nemak, A.K., and Zhurakhovskiy, V.A., Interaction between helical electron beams and strong electromagnetic cavity-fields at cyclotron-frequency harmonics, *Radioteck. Elektron.*, 12, 633, 1967 (*Radio Eng. Electron. Phys.*, 12, 587, 1967).

64. Sprangle, P., Vomvoridis, J.L., and Manheimer, W.M., Theory of the quasioptical electron cyclotron maser, *Phys. Rev. A*, 223, 3127, 1981.

65. Nezhevenko, O.A., Gyrocons and magnicons: microwave generators with circular deflection of the electron beam, *IEEE Trans. Plasma Sci.*, 22, 756, 1994.

66. Budker, G.I. et al., The gyrocon: an efficient relativistic high-power VHF generator, *Part. Accel.*, 10, 41, 1979.

67. Bratman, V.L. et al., Millimeter-wave HF relativistic electron oscillators, *IEEE Trans. Plasma Sci.*, PS-15, 2, 1987.

68. Sprangle, P. and Drobot, A.T., The linear and self-consistent nonlinear theory of the electron cyclotron maser instability, *IEEE Trans. Microwave Theory Tech.*, MTT-25, 528, 1977.

69. For a simplified analytical treatment, see Lentini, P.J., *Analytic Theory of the Gyrotron*, PFC/RR-89-6, Plasma Fusion Center, MIT, Cambridge, MA, 1989.

70. Nusinovich, G.S. and Erm, R.E., Efficiency of the CRM-monotron with a Gaussian axial structure of the high-frequency field, *Elektronnaya Technika, Elektronika SVCh*, 8, 55, 1972 (in Russian). Recounted in Nusinovich, G.S., *Introduction to the Physics of Gyrotrons*, Johns Hopkins University Press, Baltimore, MD, 2004, p. 74.

71. Bratman, V.L. et al., Relativistic gyrotrons and cyclotron autoresonance masers, *Int. J. Electron.*, 51, 541, 1981.

72. Danly, B.G. and Temkin, R.J., Generalized nonlinear harmonic gyrotron theory, *Phys. Fluids*, 29, 561, 1986.

73. Chu, K.R. and Hirshfeld, J.L., Comparative study of the axial and azimuthal bunching mechanisms in electromagnetic cyclotron instabilities, *Phys. Fluids*, 21, 461, 1978.

74. As an example, see Kho, T.H. and Lin, A.T., Slow-wave electron cyclotron maser, *Phys. Rev. A*, 38, 2883, 1988.

75. Granatstein, V.L., Gyrotron experimental studies, in *High-Power Microwave Sources*, Granatstein, V.L. and Alexeff, I., Eds., Artech House, Norwood, MA, 1987, p. 185.

76. Temkin, R.J. et al., A 100 kW, 140 GHz pulsed gyrotron, *Int. J. Infrared Millimeter Waves*, 3, 427, 1982.

77. Carmel, Y. et al., Mode competition, suppression, and efficiency enhancement in overmoded gyrotron oscillators, *Int. J. Infrared Millimeter Waves*, 3, 645, 1982.

78. Arfin, B. et al., A high power gyrotron operating in the TE_{041} mode, *IEEE Trans. Electron. Dev.*, ED-29, 1911, 1982.

79. See, for example, Carmel, Y. et al., Realization of a stable and highly efficient gyrotron for controlled fusion research, *Phys. Rev. Lett.*, 50, 1121, 1983.

80. See, for example, Luchinin, A.G. and Nusinovich, G.S., An analytical theory for comparing the efficiency of gyrotrons with various electrodynamic systems, *Int. J. Electron.*, 57, 827, 1984.

81. Anderson, J.P. et al., Studies of the 1.5-MW 140-GHz gyrotron experiment, *IEEE Trans. Plasma Sci.*, 32, 877, 2004.

82. Pioszcyk, B. et al., 165-GHz coaxial cavity gyrotron, *IEEE Trans. Plasma Sci.*, 32, 853, 2004.

83. Kreischer, K.E. et al., The design of megawatt gyrotrons, *IEEE Trans. Plasma Sci.*, PS-13, 364, 1985.

84. Danly, B.G. et al., Whispering-gallery-mode gyrotron operation with a quasi-optical antenna, *IEEE Trans. Plasma Sci.*, PS-13, 383, 1985.

85. Flyagin, V.A. and Nusinovich, G.S., Gyrotron oscillators, *Proc. IEEE*, 76, 644, 1988.

86. Chu, K.R., Read, M.E., and Ganguly, A.K., Methods of efficiency enhancement and scaling for the gyrotron oscillator, *IEEE Trans. Microwave Theory Tech.*, MTT-28, 318, 1980.

87. Felch, K. et al., Long-pulse and CW tests of a 110-GHz gyrotron with an internal, quasi-optical converter, *IEEE Trans. Plasma Sci.*, 24, 558, 1996.

88. Blank, M. et al., Demonstration of a high-power long-pulse 140-GHz gyrotron oscillator, *IEEE Trans. Plasma Sci.*, 32, 867, 2004.

89. Gold, S.H. et al., High-voltage K_a-band gyrotron experiment, *IEEE Trans. Plasma Sci.*, PS-13, 374, 1985.

90. Gold, S.H. et al., High peak power K_a-band gyrotron oscillator experiment, *Phys. Fluids*, 30, 2226, 1987.

91. Gold, S.H., High peak power K_a-band gyrotron oscillator experiments with slotted and unslotted cavities, *IEEE Trans. Plasma Sci.*, 16, 142, 1988.

92. Black, W.M. et al., Megavolt multikiloamp K_a band gyrotron oscillator experiment, *Phys. Fluids B*, 2, 193, 1990.

93. Kreischer, K.E. and Temkin, R.J., Single-mode operation of a high-power, step-tunable gyrotron, *Phys. Rev. Lett.*, 59, 547, 1987.

94. Davidovskii, V.Ya., Possibility of resonance acceleration of charged particles by electromagnetic waves in a constant magnetic field, *Zh. Eksp. Teor. Fiz.*, 43, 886, 1962 (*Sov. Phys. JETP*, 16, 629, 1963).

95. Kolomenskii, A.A. and Lebedev, A.N., Self-resonant particle motion in a plane electromagnetic wave, *Dokl. Akad Nauk SSSR*, 149, 1259, 1962 (*Sov. Phys. Dokl.*, 7, 745, 1963).

96. Bratman, V.L. et al., FELs with Bragg reflection resonators: cyclotron autoresonance masers versus ubitrons, *IEEE J. Quantum Electron.*, QE-19, 282, 1983.

97. Yovchev, I.G. et al., Present status of a 17.1-GHz four-cavity frequency-doubling coaxial gyroklystron design, *IEEE Trans. Plasma Sci.*, 28, 523, 2000.

98. Lawson, W., On the frequency scaling of coaxial gyroklystrons, *IEEE Trans. Plasma Sci.*, 30, 876, 2002.

99. Lawson, W., Ragnunathan, H., and Esteban, M., Space-charge-limited magnetron injection guns for high-power gyrotrons, *IEEE Trans. Plasma Sci.*, 32, 1236, 2004.

100. McCurdy, A.H. et al., Improved oscillator phase locking by use of a modulated electron beam in a gyrotron, *Phys. Rev. Lett.*, 57, 2379, 1986.

101. Freund, H. and Antonsen, T.M., *Principles of Free-Electron Lasers*, 2nd ed., Chapman & Hall, London, 1996.

102. Kapitsa, P.L. and Dirac, P.A.M., *Proc. Cambr. Phil. Soc.*, 29, 247, 1933.

103. Motz, H., Applications of the radiation from fast electrons, *J. Appl. Phys.*, 22, 257, 1951.

104. Phillips, R.M., The ubitron, a high-power traveling-wave tube based on a periodic beam interaction in an unloaded waveguide, *IRE Trans. Electron. Dev.*, ED-7, 231, 1960; Phillips, R.M., History of the ubitron, *Nucl. Instrum. Methods*, A272, 1, 1988.

105. Madey, J.M.J., Stimulated emission of bremsstrahlung in a periodic magnetic field, *J. Appl. Phys.*, 42, 1906, 1971.

106. Elias, L.R. et al., Observation of stimulated emission of radiation by relativistic electrons in a spatially periodic transverse magnetic field, *Phys. Rev. Lett.*, 36, 717, 1976.

107. Deacon, D.A.G. et al., First operation of a free-electron laser, *Phys. Rev. Lett.*, 38, 892, 1977.

108. Kroll, N.M., Morton, P., and Rosenbluth, M.N., Free-electron lasers with variable parameter wigglers, *IEEE J. Quantum Electron.*, QE-17, 1436, 1981.

109. See, for example, Fajans, J. et al., Microwave studies of a tunable free-electron laser in combined axial and wiggler magnetic fields, *Phys. Fluids*, 26, 2683, 1983.

110. Gold, S.H. et al., Study of gain, bandwidth, and tunability of a millimeter-wave free-electron laser operating in the collective regime, *Phys. Fluids*, 26, 2683, 1983.

111. Latham, P.E. and Levush, B., The interaction of high- and low-frequency waves in a free electron laser, *IEEE Trans. Plasma Sci.*, 18, 472, 1990.

112. Instabilities can be convective and absolute, a subject beyond our scope here, but the classic treatment is Briggs, R.J., *Electron Stream Interaction with Plasma*, MIT Press, Cambridge, MA, 1964, chap. 2.

113. Liewer, P.C., Lin, A.T., and Dawson, J.M., Theory of an absolute instability of a finite-length free-electron laser, *Phys. Rev. A*, 23, 1251, 1981.

114. Kwan, T.J.T., Application of particle-in-cell simulation in free-electron lasers, *IEEE J. Quantum Electron.*, QE-17, 1394, 1981.

115. Kroll, N.M. and McMullin, W.A., Stimulated emission from relativistic electron passing through a spatially periodic transverse magnetic field, *Phys. Rev. A*, 17, 300, 1978; Sprangle, P. and Smith, R.A., Theory of free-electron lasers, *Phys. Rev. A*, 21, 293, 1980.

116. Friedland, L., Electron beam dynamics in combined guide and pump magnetic fields for free electron laser applications, *Phys. Fluids*, 23, 2376, 1980; Freund, H.P. and Drobot, A.T., Relativistic electron trajectories in free electron lasers with an axial guide field, *Phys. Fluids*, 25, 736, 1982; Freund, H.P. and Ganguly, A.K., Electron orbits in free-electron lasers with helical wiggler and an axial guide magnetic field, *IEEE J. Quantum Electron.*, QE-21, 1073, 1985.

117. Jackson, R.H. et al., Design and operation of a collective millimeter-wave free-electron laser, *IEEE J. Quantum Electron.*, QE-19, 346, 1983.

118. Freund, H.P. et al., Collective effects on the operation of free-electron lasers with an axial guide field, *Phys. Rev. A*, 26, 2004, 1982; Freund, H.P. and Sprangle, P., Unstable electrostatic beam modes in free-electron-laser systems, *Phys. Rev. A*, 28, 1835, 1983.

119. Parker, R.K. et al., Axial magnetic-field effects in a collective-interaction free-electron laser at millimeter wavelengths, *Phys. Rev. Lett.*, 48, 238, 1982.

120. Conde, M.E., Bekefi, G., and Wurtele, J.S., A 33 GHz free electron laser with reversed axial guide magnetic field, in *Abstracts for the 1991 IEEE International Conference on Plasma Science*, Williamsburg, VA, 1991, p. 176.

121. Pasour, J.A., Free-electron lasers, in *High-Power Microwave Sources*, Granatstein, V.L. and Alexeff, I., Eds., Artech House, Norwood, MA, 1987.

122. Orzechowski, T.J. et al., High gain and high extraction efficiency from a free electron laser amplifier operating in the millimeter wave regime, *Nucl. Instrum. Methods*, A250, 144, 1986.

123. Freund, H.P. and Ganguly, A.K., Nonlinear analysis of efficiency enhancement in free-electron laser amplifiers, *Phys. Rev. A*, 33, 1060, 1986.

124. Kroll, N.M. and Rosenbluth, M.N., *Physics of Quantum Elecronics*, Vol. 7, Jacobs, S.F. et al., Eds., Addison-Wesley, Reading, MA, 1980, p. 147.

125. Colson, W.B. and Freedman, R.A., Synchrotron instability for long pulses in free electron laser oscillator, *Opt. Commun.*, 46, 37, 1983.

126. Colson, W.B., The trapped-particle instability in free electron laser oscillators and amplifiers, *Nucl. Instrum. Methods*, A250, 168, 1986.

127. Masud, J. et al., Sideband control in a millimeter-wave free-electron laser, *Phys. Rev. Lett.*, 58, 763, 1987.

128. Yu, S.S. et al., Waveguide suppression of the free electron laser sideband instability, *Nucl. Instrum. Methods*, 259, 219, 1987.

129. Yee, F.G. et al., Power and sideband studies of a Raman FEL, *Nucl. Instrum. Methods*, A259, 104, 1987.

130. Throop, A.L. et al., Experimental Characteristic of a High-Gain Free-Electron Laser Amplifier Operating at 8-mm and 2-mm Wavelengths, UCRL-95670, Lawrence Livermore National Laboratory, Livermore, CA, 1987.
131. Hester, R.E. et al., The Experimental Test Accelerator (ETA), *IEEE Trans. Nucl. Sci.*, NS-26, 3, 1979.
132. Orzechowski, T.J. et al., High-efficiency extraction of microwave radiation from a tapered-wiggler free-electron laser, *Phys. Rev. Lett.*, 57, 2172, 1986.
133. Allen, S.L. et al., Generation of high power 140 GHz microwaves with an FEL for the MTX experiment, in *Proceedings of the 1993 Particle Accelerator Conference*, Washington, D.C., 1993, p. 1551.
134. Lasnier, C.J. et al., Burst mode FEL with the ETA-III induction linac, in *Proceedings of the 1993 Particle Accelerator Conference*, Washington, D.C., 1993, p. 1554.
135. Vinokurov, N.A. and Skrinskiy, A.N., Institute of Nuclear Physics Preprint INP 11-59, Novosibirsk, 1977.
136. Leibovitch, C., Xu, K., and Bekefi, G., Effects of electron prebunching on the radiation growth rate of a collective (Raman) free-electron laser amplifier, *IEEE J. Quantum Electron.*, 24, 1825, 1988.
137. Wurtele, J.S. et al., Prebunching in a collective Raman free-electron laser amplifier, *Phys. Fluids B*, 2, 402, 1990.
138. Fajans, J. and Wurtele, J.S., Suppression of feedback oscillations in a free-electron laser, *IEEE J. Quantum Electron.*, 24, 1805, 1990.
139. Agafonov, M.A. et al., Generation of hundred joules pulses at 4-mm wavelength by FEM with sheet electron beam, *IEEE Trans. Plasma Sci.*, 26, 531, 1998.

Appendix

High Power Microwave Formulary

For plasma physics and much else, see the NRL Plasma Formulary, http://www.ppd.nrl.navy.mil/nrlformulary/. Much microwave information appears in *Microwave Engineers' Handbook*, Volumes 1 and 2, Artech House, Norwood, MA.

A.1 Electromagnetism

- *Definitions of Q:*

    1. $Q = 2\pi \dfrac{\text{average stored energy}}{\text{energy lost per cycle}} = \omega_0 \dfrac{\text{average stored energy}}{\text{power loss rate}}$

    ω_0 is the center frequency of a cavity of resonant line width $\Delta\omega$,

    2. $$Q = \frac{\omega_0}{2\Delta\omega}$$

    3. $$Q_p = \frac{\text{cavity volume}}{(\text{surface area}) \times \delta} \times (\text{geometrical factor})$$

    The geometrical factor is typically of order unity.

    4. The number of cycles in a pulse, the bandwidth, and the Q of the resonant system that produces a pulse are related by

    $$N = 1/\text{percentage bandwidth} = Q$$

5. The radiated energy loss time constant from the cavity T_r, as the ratio of the stored energy to the power lost from the cavity, is

$$Q_r = \omega_0 T_r$$

6. If the cavity has highly reflecting ends with reflection coefficients R_1 and R_2, then

$$T_r \approx \frac{L}{v_g\left(1 - R_1 R_2\right)}$$

- *Skin depth*:

$$\delta = \frac{1}{\left(\pi\sigma\mu w_f\right)^{1/2}} = \frac{1}{\left(\pi\sigma\mu_0 f\right)^{1/2}}$$

In copper, $\sigma = 5.80 \times 10^7$ $(\Omega\text{-m})^{-1}$; skin depth at 1 GHz is 2.1 μm.

TABLE A.1

Conductivities of Some Materials

Material	Conductivity $(\Omega\text{-m})^{-1}$
Aluminum	3.82×10^7
Brass	2.6×10^7
Copper	5.8×10^7
Gold	4.1×10^7
Graphite	7×10^4
Iron	1.03×10^7
Lead	4.6×10^6
Sea water	3–5
Silver	6.2×10^7
Stainless steel	1.1×10^6

A.2 Waveguides and Cavities

TABLE A.2

Expressions for the TM and TE Field Quantities for a
Rectangular Waveguide in Terms of the Axial Field
Component, Derived from Equations 4.1 to 4.4

Transverse Magnetic, $TM_{n,p}$	Transverse Electric, $TE_{n,p}$
$E_z = D\sin\left(\dfrac{n\pi}{a}x\right)\sin\left(\dfrac{p\pi}{b}y\right)$	$B_z = A\cos\left(\dfrac{n\pi}{a}x\right)\cos\left(\dfrac{p\pi}{b}y\right)$
$B_z \equiv 0$	$E_z \equiv 0$
$E_x = i\dfrac{k_z}{k_\perp^2}\dfrac{\partial E_z}{\partial x}$	$E_x = i\dfrac{\omega}{k_\perp^2}\dfrac{\partial B_z}{\partial y}$
$E_y = -i\dfrac{k_z}{k_\perp^2}\dfrac{\partial E_z}{\partial y}$	$E_y = -i\dfrac{\omega}{k_\perp^2}\dfrac{\partial B_z}{\partial x}$
$B_x = -i\dfrac{\omega}{\omega_{co}^2}\dfrac{\partial E_z}{\partial y}$	$B_x = i\dfrac{k_z}{k_\perp^2}\dfrac{\partial B_z}{\partial x}$
$B_y = i\dfrac{\omega}{\omega_{co}^2}\dfrac{\partial E_z}{\partial x}$	$B_y = i\dfrac{k_z}{k_\perp^2}\dfrac{\partial B_z}{\partial y}$

$$\omega_{co} = k_\perp c = \left[\left(\frac{n\pi c}{a}\right)^2 + \left(\frac{p\pi c}{b}\right)^2\right]^{1/2}$$

TABLE A.3

Cutoff Frequencies, Measured in
GHz, for the Lowest-Order
Modes of WR284 Waveguide

n	p	$f_{co} = \omega_{co}/2\pi$ (GHz) a = 7.214 cm, b = 3.404 cm
1	0	2.079
0	1	4.407
1	1	4.873
1	2	9.055
2	0	4.159
2	1	6.059

TABLE A.4

Expressions for the TM and TE Field Quantities for Circular
Waveguide in Terms of the Axial Field Components, Derived from
Equations 4.1 to 4.4

Transverse Magnetic, $TM_{n,p}$ ($B_z = 0$)	Transverse Electric, $TE_{n,p}$ ($E_z = 0$)
$E_z = DJ_p(k_\perp r)\sin(p\theta)$	$B_z = AJ_p(k_\perp r)\sin(p\theta)$
$E_r = i\dfrac{k_z}{k_\perp^2}\dfrac{\partial E_z}{\partial r}$	$E_r = i\dfrac{\omega}{k_\perp^2}\dfrac{1}{r}\dfrac{\partial B_z}{\partial \theta}$
$E_\theta = i\dfrac{k_z}{k_\perp^2}\dfrac{1}{r}\dfrac{\partial E_z}{\partial \theta}$	$E_\theta = -i\dfrac{\omega}{k_\perp^2}\dfrac{\partial B_z}{\partial r}$
$B_r = -i\dfrac{\omega}{\omega_{co}^2}\dfrac{1}{r}\dfrac{\partial E_z}{\partial \theta}$	$B_r = i\dfrac{k_z}{k_\perp^2}\dfrac{\partial B_z}{\partial r}$
$B_\theta = i\dfrac{\omega}{\omega_{co}^2}\dfrac{\partial E_z}{\partial r}$	$B_\theta = i\dfrac{k_z}{k_\perp^2}\dfrac{1}{r}\dfrac{\partial B_z}{\partial \theta}$
$k_\perp = \dfrac{\omega_{co}}{c} = \dfrac{\mu_{pn}}{r_0}$	$k_\perp = \dfrac{\omega_{co}}{c} = \dfrac{\nu_{pn}}{r_0}$
$J_p(\mu_{pn}) = 0$	$J_p'(\nu_{pn}) = 0$

TABLE A.5

Relationship between the Power in a Given Mode, P, and the
Maximum Value of the Electric Field at the Wall, $E_{wall,max}$

Mode	
Rectangular	
TM	$P_{TM} = \dfrac{ab}{8Z_0}\left[\left(\dfrac{n}{a}\right)^2+\left(\dfrac{p}{b}\right)^2\right]\dfrac{1}{\left[1-\left(\dfrac{f_{co}}{f}\right)^2\right]^{1/2}}\cdot\min\left[\left(\dfrac{a}{n}\right)^2,\left(\dfrac{b}{p}\right)^2\right]E_{wall,max}^2$
TE n, p > 0	$P_{TE} = \dfrac{ab}{8Z_0}\left[\left(\dfrac{n}{a}\right)^2+\left(\dfrac{p}{b}\right)^2\right]\left[1-\left(\dfrac{f_{co}}{f}\right)^2\right]^{1/2}\min\left[\left(\dfrac{a}{n}\right)^2,\left(\dfrac{b}{p}\right)^2\right]E_{wall,max}^2$
TE n or p = 0	$P_{TE} = \dfrac{ab}{4Z_0}\left[\left(\dfrac{n}{a}\right)^2+\left(\dfrac{p}{b}\right)^2\right]\left[1-\left(\dfrac{f_{co}}{f}\right)^2\right]^{1/2}\min\left[\left(\dfrac{a}{n}\right)^2,\left(\dfrac{b}{p}\right)^2\right]E_{wall,max}^2$

Continued.

TABLE A.5 *(Continued)*

Relationship between the Power in a Given Mode, P, and the Maximum Value of the Electric Field at the Wall, $E_{wall,max}$

Mode	
Circular	
TM	$$P_{TM} = \frac{\pi r_0^2}{4Z_0} \frac{E_{wall,max}^2}{\left[1-\left(\frac{f_{co}}{f}\right)^2\right]^{1/2}}$$
TE $p > 0$	$$P_{TE} = \frac{\pi r_0^2}{2Z_0}\left[1-\left(\frac{f_{co}}{f}\right)^2\right]^{1/2}\left[1+\left(\frac{v_{pn}^2}{p}\right)^2\right]E_{wall,max}^2$$

Note: Min(x, y) is the smaller of x and y, and $Z_0 = 377\ \Omega$.

TABLE A.6

Standard Rectangular Waveguides

Band (Alternate)	Recommended Frequency Range (GHz)	TE_{10} Cutoff Frequency (GHz)	WR Number	Inside Dimensions, Inches (cm)
L	1.12–1.70	0.908	650	6.50 × 3.25 (16.51 × 8.26)
R	1.70–2.60	1.372	430	4.30 × 2.15 (10.92 × 5.46)
S	2.60–3.95	2.078	284	2.84 × 1.34 (7.21 × 3.40)
H (G)	3.95–5.85	3.152	187	1.87 × 0.87 (4.76 × 2.22)
C (J)	5.85–8.20	4.301	137	1.37 × 0.62 (3.49 × 1.58)
W (H)	7.05–10.0	5.259	112	1.12 × 0.50 (2.85 × 1.26)
X	8.20–12.4	6.557	90	0.90 × 0.40 (2.29 × 1.02)
Ku (P)	12.4–18.0	9.486	62	0.62 × 0.31 (1.58 × 0.79)
K	18.0–26.5	14.047	42	0.42 × 0.17 (1.07 × 0.43)
Ka (R)	26.5–40.0	21.081	28	0.28 × 0.14 (0.71 × 0.36)
Q	33.0–50.5	26.342	22	0.22 × 0.11 (0.57 × 0.28)
U	40.0–60.0	31.357	19	0.19 × 0.09 (0.48 × 0.24)

Continued.

TABLE A.6 *(Continued)*

Standard Rectangular Waveguides

Band (Alternate)	Recommended Frequency Range (GHz)	TE$_{10}$ Cutoff Frequency (GHz)	WR Number	Inside Dimensions, Inches (cm)
V	50.0–75.0	39.863	15	0.15 × 0.07 (0.38 × 0.19)
E	60.0–90.0	48.350	12	0.12 × 0.06 (0.31 × 0.02)
W	75.0–110.0	59.010	10	0.10 × 0.05 (0.25 × 0.13)
F	90.0–140.0	73.840	8	0.08 × 0.04 (0.20 × 0.10)
D	110.0–170.0	90.854	6	0.07 × 0.03 (0.17 × 0.08)
G	140.0–220.0	115.750	5	0.05 × 0.03 (0.13 × 0.06)

A.3 Pulsed Power and Beams

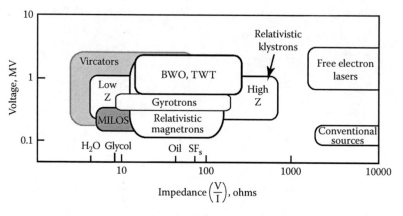

FIGURE A.1

Voltage and impedance of parameter space for HPM and pulsed conventional sources and optimum impedances for widely used pulse line insulators (water, glycol, oil, SF$_6$).

- Below are optimum impedances of pulse-forming lines, maximizing energy density:
  - Oil line, 42 Ω
  - Ethylene glycol line, 9.7 Ω
  - Water line, 6.7 Ω

- On the other hand, liquid energy density is largest for water (0.9 relative to mylar), intermediate for glycol (0.45), and lowest for oil (0.06).

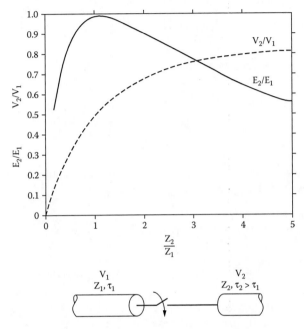

FIGURE A.2
Energy transfer and voltage ratios between transmission line (1) and load (2) as a function of impedance ratio.

Diodes and Beams

- Child–Langmuir current density, impedance:

$$J_{SCL}\left(\frac{kA}{cm^2}\right) = 2.33\frac{\left[V_0\left(MV\right)\right]^{3/2}}{\left[d\left(cm\right)\right]^2}$$

$$I_{SCL}\left(kA\right) = J_{SCL}A = 7.35\left[V_0\left(MV\right)\right]^{3/2}\left[\frac{r_c\left(cm\right)}{d\left(cm\right)}\right]^2$$

$$Z_{SCL}\left(\Omega\right) = \frac{136}{\sqrt{V_0\left(MV\right)}}\left[\frac{d}{r_c}\right]^2$$

- Relativistic Child–Langmuir current density:

$$J\left(\frac{kA}{cm^2}\right) = \frac{2.71}{\left[d(cm)\right]^2}\left[\left(1+\frac{V_0(MV)}{0.511}\right)^{1/2} - 0.847\right]^2$$

- Diode pinch current:

$$I_{pinch}(kA) = 8.5\frac{r_c}{d}\left[1+\left(\frac{V_0(MV)}{0.511}\right)\right]$$

- Parapotential current, impedance; magnetic insulation current I_z must exceed the parapotential current:

$$I_z > I_p(A) = 8500\gamma G\ell n\left(\gamma+\sqrt{\gamma^2-1}\right)$$

where γ is a relativistic factor and G is a geometric factor.

$G = [\ell n(b/a)]^{-1}$, coaxial cylinders of inner and outer radii b and a and $G = W/2\pi d$, parallel plates of width W, gap d.

$Z_p = V/I_p \approx 52G^{-1}$ Ω for voltages ~1 MV.

A.4 Microwave Sources

TABLE A.7

Classification of HPM Sources

	Slow Wave	Fast Wave
O-Type	Backward wave oscillator	Free-electron laser (ubitron)
	Traveling wave tube	Optical klystron
	Surface wave oscillator	Gyrotron
	Relativistic diffraction generator	Gyro-BWO
	Orotron	Gyro-TWT
	Flimatron	Cyclotron autoresonance maser
	Multiwave Cerenkov generator	Gyroklystron
	Dielectric Cerenkov maser	
	Plasma Cerenkov maser	
	Relativistic klystron	
M-Type	Relativistic magnetron	Rippled-field magnetron
	Cross-field amplifier	
	Magnetically insulated line oscillator	
Space-Charge	Vircator	
	Reflex triode	

TABLE A.8

General Features of HPM Sources

Source	Operating Frequency	Efficiency	Complexity	Level of Development
Magnetron	1–9 GHz, BW ~ 1%	10–30%	Low	High
O-type Cerenkov devices	3–60 GHz, somewhat tunable	10–50%	Medium to high	High
Vircator	0.5–35 GHz, wide BW (≤10%), tunable	~1% simple ~10% advanced	Low	Medium
Electron cyclotron masers	5–300 GHz for gyrotrons, higher for CARMs	30% typical	Medium	High for gyrotrons Low for CARMs
FEL	8 GHz and up, tunable	35%	High	Medium to high
Relativistic klystron	1–11 GHz	25–50%	Medium to high	Medium

- *Phase locking* condition, known as *Adler's relation*:

$$\Delta\omega \leq \frac{\omega_0 \rho}{Q}$$

ω_0 is the mean frequency of the two oscillators, which have the same Q. The injection ratio is

$$\rho = \left(\frac{P_i}{P_0} \right)^{1/2} = \frac{E_i}{E_0}$$

Subscripts i and 0 denote the injecting (or master) and receiving (or slave) oscillators. Phase locking occurs when the frequency difference between the two oscillators is sufficiently small. The timescale for locking is given by

$$\tau \sim \frac{Q}{2\rho\omega_0}$$

For $\rho \sim 1$, the locking range $\Delta\omega$ is maximized and the timescale for locking is minimized.

TABLE A.9

Ultrawideband Band Regions

	Bands	Percent Bandwidth, 100 $\Delta f/f$	Bandwidth Ratio, $1 + \Delta f/f$	Example
Narrowband		<1%	1.01	Orion (Figures 5.31 and 7.24)
Ultrawideband	Moderate band (mesoband)	1–100%	1.01–3	MATRIX (Figure 6.11)
	Ultramoderate band (subhyperband)	100–163%	3–10	H-Series (Figure 6.16)
	Hyperband	163–200%	>10	Jolt (Figure 6.17)

A.5 Propagation and Antennas

- *Power density*: Critical fluence for air breakdown for pulses longer than a microsecond:

$$S \left[\frac{MW}{cm^2} \right] = 1.5\rho^2 (atm)$$

where S is microwave power density and ρ is the pressure in units of an atmosphere. This corresponds to a critical field of 24 kV/cm at atmospheric pressure. The field due to microwave power density is

$$E(V/cm) = 19.4 \, [S(W/cm^2)]^{1/2}$$

$$S(W/cm^2)] = E^2(V/cm)/377$$

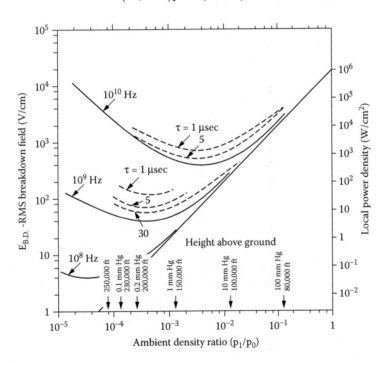

FIGURE A.3
Dependence of breakdown on pressure, frequency, and pulse duration.

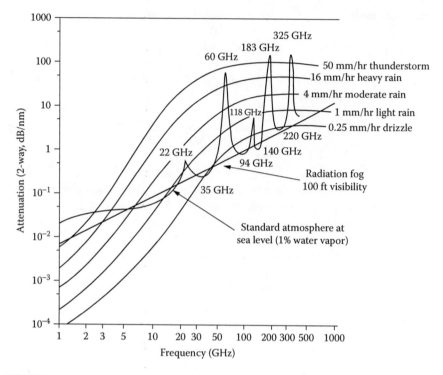

FIGURE A.4
Absorption in the air by water and oxygen molecules gives passbands at 35, 94, 140, and 220 GHz.

- *Directive gain:* The directive gain of an antenna is its ability to concentrate radiated power in a solid angle, Ω_A, producing a power density S at range R due to power P:

$$G_D = \frac{4\pi}{\Omega_A} = \frac{4\pi}{\Delta\theta\Delta\Phi} = \frac{4\pi R^2}{P}S$$

Beam widths in the two transverse dimensions are $\Delta\theta$ and $\Delta\phi$. The *gain*, G, is G_D times the antenna efficiency and can be expressed in terms of wavelength and area, $A = \pi D^2/4$:

$$G = G_D = \frac{4\pi A \varepsilon}{\lambda^2}$$

GP is the *effective radiated power* (ERP). The common way to describe gain is in decibels, G_{dB}. This value corresponds to an actual numerical value of

$$G = 10^{G_{dB}/10}$$

so a gain of 100 is called 20 dB.

The far field of a narrowband antenna begins at

$$L_{ff} = \frac{2D^2}{\lambda}$$

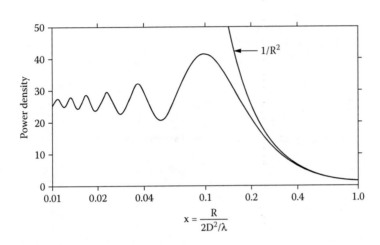

FIGURE A.5
Horn antenna intensity normalized to unity at $2D^2/\lambda$. Its general features are the same for most antenna types.

TABLE A.10

Principal Antenna Types

Antenna	Gain	Comments
Pyramidal and TEM horn	$2\pi ab/\lambda^2$	For standard gain horn; sides a, b; lengths $l_e = b^2/2\lambda$, $l_h = a^2/3\lambda$
Conical horn	$5\,D^2/\lambda^2$	$l = D^2/3\lambda$; D = diameter
Parabolic dish	$5.18\,D^2/\lambda^2$	
Vlasov	$6.36\,[D^2/\lambda^2]/\cos\theta$	Angle of slant-cut θ, $30° < \theta < 60°$
Biconic	$120\cot\alpha/4$	α is the bicone opening angle
Helical	$[148\,N\,S/\lambda]\,D^2/\lambda^2$	N = number of turns; S = spacing between turns
Array of horns	$9.4\,AB/\lambda^2$	A, B are sides of the array

The diffraction-limited beam width is

$$\theta = 2.44\,\lambda/D_t$$

This is the diffraction-limited divergence, giving the first null point of the Bessel function for a circular aperture and including 84% of the beam power. The size of this spot at range R is

$$D_s = \theta R = 2.44 \, \lambda \, R/D_t$$

A.6 Applications

Electronics effects power coupled to internal circuitry, P, from an incident power density S is characterized by a coupling cross section s, with units of area:

$$P = S \, \sigma$$

For frontdoor paths, σ is usually the effective area of an aperture, such as an antenna or slot.

The *Wunsch–Bell relation* for the power to induce failure is

$$P \sim \frac{1}{\sqrt{t}}$$

where t is the pulse duration, applicable between >100 nsec and <10 μsec.

TABLE A.11

Categories and Consequences of Electronic Effects

Failure Mode	Power Required	Wave Shape	Recovery Process	Recovery Time
Interference/ disturbance	Low	Repetitive pulse or continuous	Self-recovery	Seconds
Digital upset	Medium	Short pulse, single or repetitive	Operator intervention	Minutes
Damage	High	UWB or narrowband	Maintenance	Days

Power Beaming

The power–link parameter is

$$Z = \frac{D_t D_r}{\lambda R}$$

where D_t and D_r are the diameters of the transmitting and receiving antennas and R is the distance between the transmitter and receiver. For efficient transfer, the spot size must be $\sim D_r$ and $Z \geq 1$. An approximate analytic expression, good for $Z \sim 1$, is

$$\frac{P_r}{P_t} = 1 - e^{-Z^2}$$

Plasma Heating

Plasma heating resonant frequencies follow:

- *Ion cyclotron resonance heating* (ICRH)

$$f = f_{ci} = \frac{eB}{2\pi m_i}$$

- *Lower hybrid heating* (LHH)

$$f = f_{LH} \cong \frac{f_{pi}}{\left(1 - \dfrac{f_{pe}^2}{f_{ce}^2}\right)^{1/2}}$$

- *Electron cyclotron resonance heating* (ECRH)

$$f = f_{ce} = \frac{eB}{2\pi m_e}, \quad \text{or} \quad f = 2 f_{ce}$$

where m_i and m_e are the ion and electron masses, B is the magnitude of the local magnetic field, $f_{pe} = (n_e e^2 / \varepsilon_0 m_e)^{1/2} / 2$, with n_e the volume density of electrons and ε_0 the permittivity of vacuum, and $f_{pi} = (n_i Z_i^2 e^2 / \varepsilon_0 m_i)^{1/2} / 2$, with $Z = 1$ the charge state of the ions in the fusion plasma. At the densities and field values necessary for fusion, $f_{ci} \approx 100$ MHz, $f_{LH} \approx 5$ GHz, and $f_{ce} \approx 140$ to 250 GHz.

Index